Introduction to Magnetism and Magnetic Materials

Third Edition

Introduction to Magnetism and Magnetic Materials

Third Edition

David Jiles
Iowa State University

CRC Press
Taylor & Francis Group
Boca Raton London New York

CRC Press is an imprint of the
Taylor & Francis Group, an **informa** business

CRC Press
Taylor & Francis Group
6000 Broken Sound Parkway NW, Suite 300
Boca Raton, FL 33487-2742

Printed on acid-free paper
Version Date: 20150720

International Standard Book Number-13: 978-1-4822-3887-7 (Paperback)

Visit the Taylor & Francis Web site at
http://www.taylorandfrancis.com

and the CRC Press Web site at
http://www.crcpress.com

Few subjects in science are more difficult to understand than magnetism.

—Encylopaedia Britannica, 1989

This book is dedicated to Helen, Sarah, Elizabeth, Andrew and Richard

Contents

SECTION I Electromagnetism: Magnetic Phenomena on the Macroscopic Scale

SECTION II Magnetism in Materials: Magnetic Phenomena on the Microscopic Scale

SECTION III Magnetics: Technological Applications

Preface

A new edition of *Introduction to Magnetism and Magnetic Materials* is long overdue, the project having been held up first by the need for a second edition of *Introduction to the Electronic Materials*, and then by the new book *Introduction to the Principles of Materials Evaluation*. This third edition of *Introduction to Magnetism and Magnetic Materials* has been completely revised. The basic science of magnetism evolves slowly, so that most of the content of Chapters 1 through 11 is as relevant today as it was at the time of the second edition. Nevertheless, I have taken the opportunity to add a few more exercises to these chapters with once again complete worked solutions to all exercises at the end of the book. It is my belief that the best way to learn this subject is to read through the chapters, try the exercises that test your knowledge on the subject area of the chapter, and then check your solution against the model answers in the back of the book. That is why I continue to provide complete worked solutions rather than just a numerical answer, because in the latter case if you don't get the answer right, it can be difficult to see where you went wrong.

The book also serves a secondary role as a reference text with a large number of references to key works in magnetism and magnetic materials. In Chapters 12 through 15, which deal with applications of magnetism and magnetic materials, there has been enormous progress in the time since the previous edition—particularly in magnetic recording. In preparing the new edition that is where most of the time and effort had to be devoted. So in these later chapters, you will find new results and many new references—often references to web pages, where a lot of the new results now appear. However, the classic works still provide information on the important milestones in the development of our subject. So I have interspersed references to some of the classic works with newer references in these chapters to provide a more complete treatment of the subject matter.

I would like to acknowledge the assistance of many friends, colleagues and students in preparing this third edition. In particular I would like to thank: Lawrence Crowther, Ravi Hadimani, Orfeas Kypris, Ikenna Nlebedim, Zhen Zhang, Helena Khazdozian, Yan (Michelle) Ni, Neelam Prabhu Gaunkar and Priyam Rastogi.

Finally, thanks to my wife, Helen; our daughters, Sarah and Elizabeth; and our sons, Andrew and Richard, for their patience.

David Jiles
Iowa State University

Acknowledgements

I am grateful to the authors and publishers for permission to reproduce the following figures which appear in this book.

1.3
P. Ruth (1969) *Introduction to Field and Particle,* Butterworths, London.
1.9
G. V. Brown and L. Flax (1964) *Journal of Applied Physics,* **35**, 1764.
2.5
J. A. Osborne (1945) *Physical Review,* **67**, 351.
3.1
D. O. Smith (1956) *Review of Scientific Instruments,* **27**, 261.
3.2, 3.5, 3.6
T. R. McGuire and P. J. Flanders (1969) in *Magnetism and Metallurgy* (eds A. E. Berkowitz and E. Kneller), Academic Press, New York.
3.4
S. Chikazumi (1964) *Physics of Magnetism,* Wiley, New York.
3.9
B. I. Bleaney and B. Bleaney (1976) *Electricity and Magnetism,* Oxford University Press, Oxford.
4.1
F. W. Sears (1951) *Electricity and Magnetism,* Addison–Wesley, Reading, Mass.
4.4
J. Fidler, J. Bernardi and P. Skalicky (1987) *High Performance Permanent Magnet Materials* (eds S. G. Sankar, J. F. Herbst and N. C. Koon), Materials Research Society.
4.6
C. Kittel (1986) *Introduction to Solid State Physics,* 6th edn, Wiley, New York. R. J. Elliott and A. F. Gibson (1974), *An Introduction to Solid State Physics and its Applications,* MacMillan, London.
4.2.1, 5.2
A. E. E. McKenzie (1971) *A Second Course of Electricity,* 2nd edn, Cambridge University Press.
5.7
E. W. Lee (1955) *Rep. Prog. Phys.* **18**, 184. A. E. Clark and H. T. Savage (1983) *Journal of Magnetism and Magnetic Materials,* **31**, 849.
6.4
H. J. Williams, R. J. Bozorth and W. Shockley (1940) *Physical Review,* **75**, 155.
6.7
C. Kittel (1986) *Introduction to Solid State Physics,* 6th edn. Wiley, New York.

6.8

R. M. Bozorth (1951) *Ferromagnetism,* Van Nostrand, New York.

6.9

C. Kittel and J . K. Galt (1956) *Solid State Physics,* **3**, 437.

7.1

C. Kittel (1949) *Reviews of Modern Physics,* **21**, 541.

7.4

R. W. Deblois and C. D. Graham (1958) *Journal of Applied Physics,* **29**, 931.

7.6

S. Chikazumi (1964) *Physics of Magnetism,* Wiley, New York.

8.2

M. Kersten (1938) *Probleme der Technische Magnetisierungs Kurve,* Springer-Verlag, Berlin.

8.5

K. Hoselitz (1951) *Ferromagnetic Properties of Metals and Alloys,* Oxford University Press, Oxford.

8.8, 8.9

J. Degaugue, B. Astie, J. L. Porteseil and R. Vergne (1982) *Journal of Magnetism and Magnetic Material,* **26**, 261.

8.10

A. Globus, P. Duplex and M. Guyot (1971) *IEEE Transactions on Magnetics,* 7, 617. A. Globus and P. Duplex (1966) *IEEE Transactions on Magnetics,* **2**, 441.

8.3.1

S. Chikazumi (1964) *Physics of Magnetism,* Wiley, New York.

9.2, 9.3

C. Kittel (1986) *Introduction to Solid State Physics,* 6th edn. Wiley, New York.

9.5

P. Weiss and R. Forrer (1929) *Annalen der Physik,* **12**, 279.

9.6

D. H. Martin (1967) *Magnetism in Solids,* Illife Books, London.

9.7, 9.9, 9.10

G. E. Bacon (1975) *Neutron Diffraction,* 3rd edn, Clarendon Press, Oxford.

9.8

J. Crangle (1977) *The Magnetic Properties of Solids,* Edward Arnold, London.

9.11

G. L. Squires (1954) *Proceedings of the Physical Society of London,* **A67**, 248.

9.12

D. Cribier, B. Jacrot and G. Parette (1962) *Journal of the Physical Society of Japan,* **17-BIII**, 67.

9.13, 9.14, 9.15

B. D. Cullity (1972) *Introduction to Magnetic Materials,* Addison-Wesley, Reading, Mass.

9.16

C. Kittel (1986) *Introduction to Solid State Physics,* 6th edn, Wiley, New York.

9.19

J. A. Hofman, A. Pashkin, K. J. Tauer and R. J. Weiss (1956) *Journal of Physics and Chemistry of Solids,* **1**, 45.

9.20

D. H. Martin (1967) *Magnetism in Solids,* Illife Books, London.

9.21

S. B. Palmer and C. lsci (1978) *Journal of Physics F. Metal Physics,* **8**, 247. D. C. Jiles and S. B. Palmer (1980) *Journal of Physics F. Metal Physics,* **10**, 2857.

9.22

R. D. Greenough and C. Isci (1978) *Journal of Magnetism and Magnetic Materials,* **8**, 43.

10.6, 10.7

H. Semat (1972) *Introduction to Atomic and Nuclear Physics,* 5th edn, Holt, Rinehart and Winston, New York.

11.1

D. H. Martin (1967) *Magnetism in Solids,* Illife Books, London.

11.5

J. Crangle (1977) *The Magnetic Properties of Solids,* Edward Arnold, London.

11.6

W. E. Henry (1952) *Physical Review,* **88**, 559.

11.10

B. D. Cullity (1972) *Introduction to Magnetic Materials,* Addison–Wesley, Reading, Mass.

11.11, 12.1

R. M. Bozorth (1951) *Ferromagnetism,* Van Nostrand, New York.

11.12

A. Fert, P. Grunberg and A. Barthelmy (1995) J. M. M. M. **140**, 1.

12.1, 12.4, 12.15, 12.24

G. Y. Chin and J. H. Wernick (1980) in *Ferromagnetic Materials,* Vol. 2, (ed. E. P. Wohlfarth), North Holland, Amsterdam.

12.5

M. F. Litmann (1971) *IEEE Transactions on Magnetics,* **7**, 48.

12.6, 12.7

M. F. Litmann (1967) *Journal of Applied Physics,* **38**, 1104.

12.9

E. Adams (1962) *Journal of Applied Physics,* **33**, 1214.

12.8.10, 12.11, 12.12, 12.13

C. Heck (1972) *Magnetic Materials and their Applications,* Crane, Russak and Company, New York.

12.16, 12.18, 12.20

F. E. Luborsky (1980) in *Ferromagnetic Materials,* Vol. 1 (ed. E. P. Wohlfarth), North Holland, Amsterdam.

12.17, 12.19

Reproduced by permission of Allied Signal Company, Morristown, New Jersey.

12.25, 12.26

J. H. Swisher and E. O. Fuchs (1970) *Journal of the Iron and Steel Institute,* August.

12.29

G. W. Elman (1935) *Electrical Engineering,* **54**, 1292.

13.9

Permission of H. A. Davies.

13.5, 13.6, 13.7, 13.8, 13.9, 13.10

D. J. Craik (1971) *Structure and Properties of Magnetic Materials,* Pion, London.

13.12, 13.13

R. J. Parker and R. J. Studders (1962) *Permanent Magnets and their Applications,* Wiley, New York.

13.14, 13.15, 13.16

R. A. McCurrie (1982) in *Ferromagnetic Materials,* Vol. 3 (ed. E. P. Wohlfarth), North Holland, Amsterdam.

13.18, 13.19

M. McCaig and A. G. Clegg (1987) *Permanent Magnets in Theory and Practice,* 2nd edn. Pentech Press, London.

13.22

Permission of H. A. Davis.

14.2, 14.5

E. Grochowshi and D. A. Thompson (1994) *IEEE Trans. Mag.* **30**, 3797.

14.7, 14.8

R. M. White (1985) *Introduction to Magnetic Recording,* IEEE Press, New York.

15.6

R. S. Tebble, I. C. Skidmore and W. D. Corner (1950) *Proceedings of the Physical Society,* **A63**, 739.

15.7

G. A. Matzkanin, R. E. Beissner and C. M. Teller, Southwest Research Institute, Report No. NTIAC-79-2.

15.2

R. A. Langman (1981) *NDT International*, **14**, 255.

15.18

R. E. Beissner, G. A. Matzkanin and C. M. Teller, Southwest Research Institute, Report No. NTIAC-80-1.

15.21

W. E. Lord and J. H. Hwang (1975) *Journal of Testing and Evaluation,* **3**, 21.

Glossary of Symbols

A	Magnetic vector potential
A	Area
	Helmholz energy
	Exchange stiffness coefficient ($= E_{ex}/az$)
a	Distance
	Lattice spacing
	Radius of coil or solenoid
α	Mean field constant
$\alpha_1, \alpha_2, \alpha_3$	Direction cosines of magnetic vector with respect to the applied field
$\boldsymbol{B}$	Magnetic induction
$\boldsymbol{B}_g$	Magnetic induction in air gap
$\boldsymbol{B}_j(x)$	Brillouin function of x
$\boldsymbol{B}_R$	Remanent magnetic induction
$\boldsymbol{B}_s$	Saturation magnetic induction
$\beta_1, \beta_2, \beta_3$	Direction cosines of direction of measurement with respect to the applied field
C	Capacitance
	Curie constant
c	Velocity of light
χ	Susceptibility
$\boldsymbol{D}$	Electric displacement
D	Diameter of solenoid
d	Diameter
δ	Domain wall thickness
$\boldsymbol{E}$	Electric field strength
E	Energy
e	Electronic charge spontaneous strain within a domain
E_a	Anisotropy energy
E_{ex}	Exchange energy per atom
$E_{ex,vol}$	Exchange energy per unit volume ($= NE_{ex}$)
$E_{ex,nn}$	Exchange energy per nearest neighbor spin ($= E_{ex}/Z$)
E_f	Fermi energy
E_H	Magnetic field energy (Zeeman energy)
E_{Hall}	Hall field
E_{loss}	Energy loss
E_p	Potential energy
E_σ	Stress energy
ϵ	Permittivity
ϵ_0	Permittivity of free space
ϵ_{pin}	Domain-wall pinning energy
η	Magnetomotive force
F	Field factor
$\boldsymbol{F}$	Force

(Continued)

f	Frequency
	Current factor
G	Gibbs free energy
g	Spectroscopic splitting factor
	Lande splitting factor
γ	Gyromagnetic ratio
	Domain-wall energy per unit area
$\boldsymbol{H}$	Magnetic field strength
h	Planck's constant
$\boldsymbol{H}_c$	Coercivity
	Critical field
$\boldsymbol{H}_{cr}$	Remanent coercivity
$\boldsymbol{H}_d$	Demagnetizing field
$\boldsymbol{H}_e$	Weiss mean field
$\boldsymbol{H}_{eff}$	Effective magnetic field
$\boldsymbol{H}_g$	Magnetic field in air gap
$\boldsymbol{I}$	Magnetic polarization
	Intensity of magnetization
I	Current
$\boldsymbol{J}$	Current density
	Total atomic angular momentum quantum number
J	Exchange constant
J	Total electronic angular momentum quantum number
$\mathcal{J}$	Coupling between nearest-neighbor magnetic moments
J_{atom}	Exchange integral for an electron on an atom with electrons on several nearest neighbors
J_{ex}	Exchange integral; exchange interaction between two electrons
K	Anisotropy constant
	Kundt's constant
k	Pinning or loss coefficient in hysteresis equation
k_B	Boltzmann's constant
K_{u1}	First anisotropy constant for uniaxial system
K_{u2}	Second anisotropy constant for uniaxial system
K_1	First anisotropy constant for cubic system
K_2	Second anisotropy constant for cubic system
$\boldsymbol{L}$	Inductance
	Length
	Length of solenoid
	Atomic orbital angular momentum
	Electronic orbit length
L	Length
	Orbital angular momentum quantum number
$\mathcal{L}(x)$	Langevin function of x
λ	Wavelength
	Magnetostriction
	Filling factor for solenoid

(*Continued*)

λ_t	Transverse magnetostriction
λ_s	Saturation magnetostriction
λ_0	Spontaneous bulk magnetostriction
$\boldsymbol{M}$	Magnetization
$\boldsymbol{m}$	Magnetic moment
m	Mass
	Momentum
$\boldsymbol{M}_{an}$	Anhysteretic magnetization
m_e	Electronic mass
m_l	Orbital magnetic quantum number
m_o	Orbital magnetic moment of electron
$\boldsymbol{M}_R$	Remanent magnetization
$\boldsymbol{M}_0$	Saturation magnetization (spontaneous magnetization at 0 K)
$\boldsymbol{M}_s$	Spontaneous magnetization within a domain
$\boldsymbol{m}_s$	Spin magnetic moment of electron
m_s	Spin magnetic quantum number
$\boldsymbol{m}_{tot}$	Total magnetic moment of atom
μ	Permeability
μ,ν	Rayleigh coefficients
μ_B	Bohr magneton
μ_0	Permeability of free space
N	Number of turns of solenoid
	Number of atoms per unit volume
n	Number of turns per unit length on solenoid
	Principal quantum number
N_d	Demagnetizing factor
N_0	Avogadro's number
n_s	Number density of paired electrons
υ	Frequency
ω	Angular frequency
P	Pressure
P	Magnetic pole strength
	Angular momentum operator
P	Density
	Resistivity
$\boldsymbol{P}_0$	Orbital angular momentum of electron
$\boldsymbol{P}$S	Spin angular momentum of electron
$\boldsymbol{P}_{tot}$	Total angular momentum of electron
Φ	Magnetic flux
ϕ	Angle
	Spin wave function
Ψ	Total wave function
ψ	Electron wave function
Q	Electric charge
R	Resistance

(Continued)

r	Radius vector
	Radius
	Electronic orbit radius
R_m	Magnetic reluctance
$\boldsymbol{S}$	Atomic spin angular momentum
S	Entropy
s	Electronic spin angular momentum quantum number
σ	Conductivity
	Stress
T	Temperature
t	Time
	Thickness
T_c	Curie temperature
	Critical temperature
t_0	Orbital period of electron
θ	Angle
τ	Torque
	Orbital period
τ_{max}	Maximum torque
U	Internal energy
$\mathbf{u}$	Unit vector
V	Potential difference
	Volume
	Verdet's constant
υ	Velocity
W	Power
W_a	Atomic weight
W_H	Hysteresis loss
x	Distance along x-axis
y	Distance along y-axis
Z	Impedance
	Atomic number
z	Distance along z-axis
	Number of nearest-neighbor atoms

SI Units, Symbols, and Dimensions

Quantity	Unit		MKSA	
	Symbol	Name	Base Units	Dimensions
Length	m	meter	m	L
Mass	kg	kilogram	kg	M
Time	s	second	s	T
Frequency	Hz	hertz	s^{-1}	T^{-1}
Force	N	newton	kg m s^{-2}	MLT^{-2}
Pressure	Pa	pascal	kg m^{-1} s^{-2}	$ML^{-1} T^{-2}$
Energy	J	joule	kg m^2 s^{-2}	$ML^2 T^{-2}$
Power	W	watt	kg m^2 s^{-3}	$ML^2 T^{-3}$
Electric charge	C	coulomb	A s	CT
Electric current	A	ampere	A	C
Electric potential	V	volt	kg m^2 A^{-1} s^{-3}	$ML^2 C^{-1} T^{-3}$
Resistance	Ω	ohm	kg m^2 A^{-2} s^{-3}	$ML^2 C^{-2} T^{-3}$
Resistivity	Ωm	ohm meter	kg m^3 A^{-2} s^{-3}	$ML^3 C^{-2} T^{-3}$
Conductance	S	siemens	A^2 s^3 kg^{-1} m^{-2}	$M^{-1} L^{-2} C^2 T^3$
Conductivity	S m^{-1}	siemens $meter^{-1}$	A^2 s^3 kg^{-1} m^{-3}	$M^{-1} L^{-3} C^2 T^3$
Capacitance	F	farad	A^2 s^4 kg^{-1} m^{-2}	$M^{-1} L^{-2} C^2 T^4$
Electric displacement (flux density)	C m^{-2}	coulomb $meter^{-2}$	A s m^{-2}	$CL^{-2} T$
Electric field	V m^{-1}	volt $meter^{-1}$	kg m A^{-1} s^{-3}	$MLC^{-1} T^{-3}$
Electric polarization	C m^{-2}	coulomb $meter^{-2}$	A s m^{-2}	$CL^{-2} T$
Permittivity	F m^{-1}	farad $meter^{-1}$	A^2 s^4 kg^{-1} m^{-3}	$M^{-1} L^{-3} C^2 T^4$
Inductance	H	henry	kg m^2 A^{-2} s^{-2}	$ML^2 C^{-2} T^{-2}$
Magnetic flux	Wb	weber	kg m^2 A^{-1} s^{-2}	$ML^2 C^{-1} T^{-2}$
Magnetic induction (flux density)	T	Tesla	kg A^{-1} s^{-2}	$MC^{-1} T^{-2}$
Magnetic field	A m^{-1}	ampere $meter^{-1}$	A m^{-1}	CL^{-1}
Magnetization	A m^{-1}	ampere $meter^{-1}$	A m^{-1}	CL^{-1}
Permeability	H m^{-1}	henry $meter^{-1}$	kg m A^{-2} s^{-2}	$MLC^{-2} T^{-2}$

Values of Selected Physical Constants

Avogadro's Number	$N_0 = 6.022 \times 10^{26}$ atoms kg mole^{-1}
Boltzmann's constant	$k_B = 1.381 \times 10^{-23}$ J K^{-1}
Gas constant	$R = 8.314$ J mole^{-1} K^{-1}
Planck's constant	$h = 6.626 \times 10^{-34}$ J s
	$h/2\pi = 1.054 \times 10^{-34}$ J s
Velocity of light in empty space	$c = 2.998 \times 10^{8}$ m s^{-1}
Permittivity of empty space	$\epsilon_0 = 8.854 \times 10^{-12}$ F m^{-1}
Permeability of empty space	$\mu_0 = 1.257 \times 10^{-6}$ H m^{-1}
Atomic mass unit	a.m.u. $= 1.661 \times 10^{-27}$ kg
Properties of electrons	
Electronic charge	$e = -1.602 \times 10^{-19}$ C
Electronic rest mass	$m_e = 9.109 \times 10^{-31}$ kg
Charge to mass ratio	$e/m_e = 1.759 \times 10^{11}$ C kg^{-1}
Electron volt	eV $= 1.602 \times 10^{-19}$ J
Properties of protons	
Proton charge	$e_p = 1.602 \times 10^{-19}$ C
Rest mass	$m_p = 1.673 \times 10^{-27}$ kg
Gyromagnetic ratio of proton	$\gamma_p = 2.675 \times 10^{8}$ Hz T^{-1}
Magnetic constants	
Bohr magneton	$\mu_B = 9.274 \times 10^{-24}$ A m^2 (= J T^{-1})
	$= 1.165 \times 10^{-29}$ J m A^{-1}
Nuclear magneton	$\mu_N = 5.051 \times 10^{-27}$ A m^2 (= J T^{-1})
Magnetic flux quantum	$\Phi_0 = 2.067 \times 10^{-15}$ Wb (= V s)

Introduction

As you study the intricate subject of magnetism in this book, you will find that the journey begins at a familiar level, with electric currents passing through wires, compass needles rotating in magnetic fields, and bar magnets attracting or repelling each other. As the journey progresses, though, in order to understand our observations, we must soon peel back the surface and begin to delve into the materials, to use ever-increasing magnification levels and look in greater and greater detail to explain what is happening. This process takes us from bulk magnets (10^{23}–10^{26} atoms) down to the domain scale (10^{12}–10^{18} atoms) and then down to the scale of a domain wall (10^{3}–10^{2} atoms). In critical phenomena, one is often concerned with the behavior of even smaller numbers (10 atoms or less) in a localized array. Then comes the question of how the magnetic moment of a single atom arises. We must go inside the atom to find the answer by looking at the behavior of a single electron orbiting a nucleus. The next question is, *Why are the magnetic moments of neighboring atoms aligned?* In order to answer this, we must go even further and consider the quantum mechanical exchange interaction between two electrons on neighboring atoms. This then marks the limit of our journey into the fundamentals of our subject. Subsequently, we must ask how this knowledge can be used to our benefit. In Chapters 12 through 15, we look at the most significant applications of magnetism. It is no surprise that these applications deal with ferromagnetism. Ferromagnetism is easily the most important technological branch of magnetism; most scientific studies, even of other forms of magnetism, are ultimately designed to help further our understanding of ferromagnetism, so that we may both fabricate new magnetic materials with improved properties and make better use of existing material.

Finally, I have adopted an unusual format for the book, in which each major heading within a chapter is introduced by a question which the subsequent discussion attempts to answer. Many have said they found this feature useful in focusing attention on the subject matter at hand since it is then clear what the objective of each section is. I have decided therefore to retain this format from my original notes, realizing that though it is unusual for a textbook it would prove helpful to the reader.

Section I

Electromagnetism

Magnetic Phenomena on the Macroscopic Scale

1 Magnetic Fields

In this chapter, we will clarify ideas about what is meant by a *magnetic field* and then show that it is always the result of electrical charge in motion. This will be followed by a discussion of the concept of *magnetic induction*, also known as *magnetic flux density*, and its relation to the magnetic field. We will look at the various unit conventions currently in use in magnetism and finally discuss methods for calculating magnetic fields.

1.1 MAGNETIC FIELD

What do we mean by magnetic field*?*

One of the most fundamental ideas in magnetism is the concept of the magnetic field. When a field is generated in a volume of space, it means that there is a force produced, which can be detected by the acceleration of an electric charge moving in the field, by the force on a current-carrying conductor, by the torque on a magnetic dipole such as a bar magnet, or even by a reorientation of spins on electrons within certain types of atoms. The torque on a compass needle, which is an example of a magnetic dipole, is probably the most widely recognized property of a magnetic field.

1.1.1 Generation of a Magnetic Field

How are magnetic fields produced?

A magnetic field is produced whenever there is an electrical charge in motion. This can be due to an electrical current flowing in a conductor, for example, as was first discovered by Oersted in 1819 [1]. A magnetic field is also produced by a permanent magnet. In this case, there is no conventional electric current, but there are the orbital motions and spins of electrons (the so-called Ampèrian currents) within the magnetic material that lead to a magnetization within the material and a magnetic field outside. The magnetic field exerts a force on both current-carrying conductors and permanent magnets.

In order to explain why a moving charge causes a magnetic field, we need to ask ourselves what forces exist between two electrical charges when they are at rest and when they are in motion, relative to whoever or whatever is measuring the forces between them. In the former case, when the charges are *at rest*, the only force between the charges is the electrostatic Coulomb force. However, when the charges are in motion, there is an additional force on the charges, which we commonly call the *magnetic force* or the *magnetic field*. The existence of this magnetic field can be shown by application of the relativistic Lorentz transformation to the Coulomb force between the charges, which is explained in detail in Appendix A. Therefore, the magnetic field emerges as an extension or correction to the electrostatic Coulomb force.

1.1.2 Biot-Savart Law

Is there any way we can calculate the magnetic field strength generated by an electric current?

The Biot-Savart law, which enables us to calculate the magnetic field $\boldsymbol{H}$ generated by an electrical current, is one of the fundamental laws of electromagnetism. It is a statement of experimental observation rather than a theoretical prediction. In its usual form, the law gives the field contribution generated by a current flowing in an elementary length of conductor

$$\delta \boldsymbol{H} = \frac{1}{4\pi r^2} i\,\delta \boldsymbol{l} \times \mathbf{u} \tag{1.1}$$

where:

i is the current flowing in an elemental length δl of a conductor
r is the radial distance
$\mathbf{u}$ is a unit vector along the radial direction
$\delta \boldsymbol{H}$ is the contribution to the magnetic field at r due to the current element $i\delta l$

The magnetic field decreases with the square of the distance from the current element that produces it.

This form is known as the *Biot-Savart law*, although it was also discovered independently in a different form by Ampère in the same year. For steady currents, it is equivalent to Ampère's circuital law. It is not really capable of direct proof, but is justified by experimental measurements. Notice in particular that it is an inverse square law.

1.1.3 Magnetic Field Due to a Circular Coil

What is the field strength produced by a single-turn circular coil?

The Biot-Savart law can be used to determine the magnetic field $\boldsymbol{H}$ at the center of a circular coil of one turn with a radius of a meters, carrying a current of i amperes. We divide the coil into elements of arc length δl, each of which contributes $\delta \boldsymbol{H}$ to the field at the center of the coil. Since by the Biot-Savart law,

$$\boldsymbol{H} = \sum \frac{1}{4\pi r^2} i\delta l \sin\theta \tag{1.2}$$

and $\sum \delta l = 2\pi a$, while dl is perpendicular to $\mathbf{u}$, so $\theta = 90°$, and hence $\sin\theta = 1$, and $r = a$

$$\boldsymbol{H} = \frac{i}{2a} \quad \mathrm{A\,m^{-1}} \tag{1.3}$$

1.1.4 Definition of Magnetic Field Strength, *H*

What is the unit of magnetic field strength?

There are a number of ways in which the ampere per meter, the unit of magnetic field strength ***H***, can be defined. In accordance with the ideas developed here, we wish to emphasize the connection between the magnetic field ***H*** and the electric current. We shall therefore define the unit of magnetic field strength, the ampere per meter, in terms of the generating current. The simplest definition is as follows:

The ampere per meter: A magnetic field strength of 1 A m^{-1} is produced at the center of a single circular coil of conductor of diameter 1 m when it carries a current of 1 A.

For the time being, we will take the viewpoint that the magnetic field ***H*** is solely determined by the size and distribution of currents producing it and is independent of the material medium. This will allow us to draw a distinction between magnetic field and magnetic induction. However, we shall see in Section 2.4.3 that this assumption needs to be modified under certain circumstances, particularly when demagnetizing fields are encountered in magnetic materials.

1.1.5 Magnetic Field Generated by a Long Current-Carrying Conductor

What is the field strength generated at a fixed distance from a long current-carrying conductor?

We shall determine the magnetic field ***H*** at some point *P* that is at a distance of *a* meters from an infinitely long conductor carrying a current of *i* amperes and calculate the field at a distance of 0.1 m from the conductor when it carries a current of 0.1 A (Figure 1.1).

Using the Biot-Savart law, the contribution $\delta \boldsymbol{H}$ to the field at the point *P*, as shown in Figure 1.2, due to current element $i\delta l$ at an angle α is given by

$$\delta \boldsymbol{H} = \frac{1}{4\pi r^2} i\delta l \sin(90-\alpha) \times \mathbf{u} \tag{1.4}$$

We can write $\delta l = r\delta\alpha/\cos\alpha = a\delta\alpha/\cos^2\alpha$.

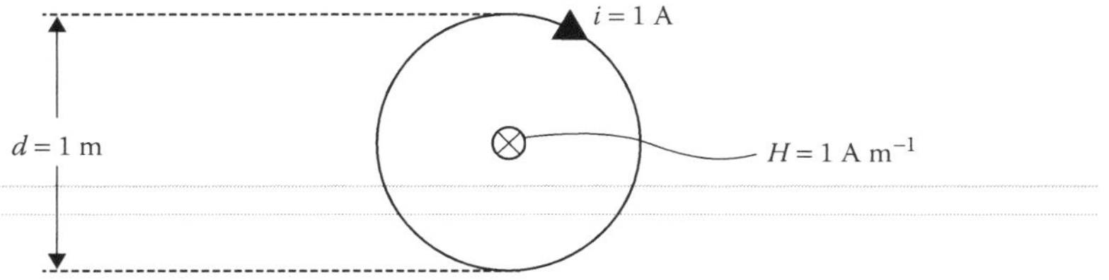

FIGURE 1.1 The magnetic field ***H*** at the center of a single circular coil of diameter 1 m carrying a current of 1 A.

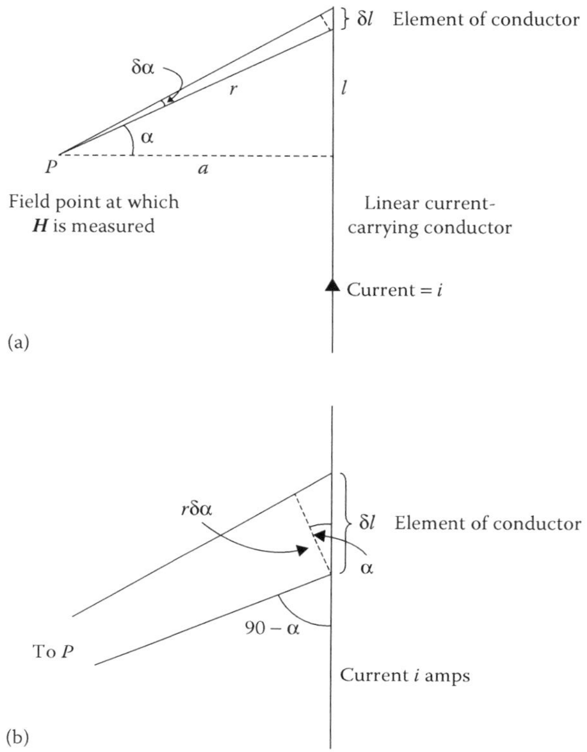

FIGURE 1.2 Magnetic field $\boldsymbol{H}$ due to a long straight conductor carrying an electric current i.

$$\delta \boldsymbol{H} = \frac{i \cos\alpha\, \delta\alpha}{4\pi a} \tag{1.5}$$

Now integrating the expression from $\alpha = -\pi/2$ to $\alpha = \pi/2$ to obtain the total $\boldsymbol{H}$

$$\boldsymbol{H} = \int_{-\pi/2}^{\pi/2} \frac{i}{4\pi a} \cos\alpha \, d\alpha \tag{1.6}$$

$$\boldsymbol{H} = \frac{i}{2\pi a} \ \mathrm{A\,m^{-1}} \tag{1.7}$$

Therefore, if $a = 0.1$ m and $i = 0.1$ A, the field is $1/2\pi$ A m^{-1}, or $\boldsymbol{H} = 0.159$ A m^{-1}.

The direction of this magnetic field is such that it circulates the conductor obeying the right-hand rule. That is, if we look along the conductor in the direction of the conventional current, the magnetic field circulates in a clockwise direction.

1.1.6 Field Patterns around Current-Carrying Conductors

What do these fields look like?

The magnetic field patterns, detected by the magnetic powder, around a bar magnet (magnetic dipole), a straight conductor, a single circular loop, and a solenoid are shown in Figure 1.3a through e. The field circulates around a single current-carrying

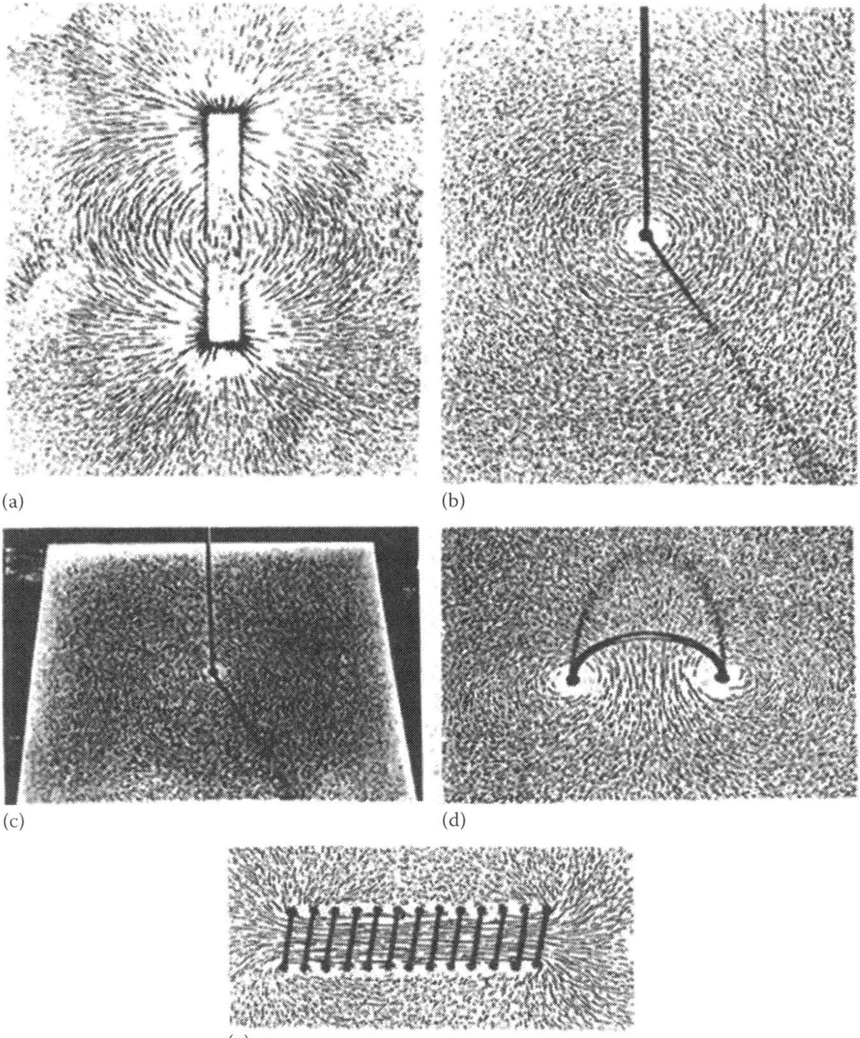

(a) (b) (c) (d) (e)

FIGURE 1.3 Magnetic field patterns in various situations obtained using iron filings: (a) a bar magnet, (b) a straight conductor carrying an electric current, (c) a perspective view of (b), (d) a single circular loop of conductor carrying a current, and (e) a solenoid with an air core.

conductor in a direction given by the right-hand corkscrew rule. The fields around a single current loop and a solenoid are similar to those around a bar magnet.

In a bar magnet, the field emerges from one end of the magnet, conventionally known as the *north pole* and passes through the air making a return path to the other end of the bar magnet, known conventionally as the *south pole*. We can think of the *north pole* of a magnet as a source of magnetic field $\boldsymbol{H}$ while the *south pole* behaves as a field sink. Whether such poles have any real existence is debatable. At present, the convention is to assume that such poles are fictitious, although the concept of the magnetic pole is very useful to those working with magnetic materials. The matter is discussed again in Section 2.1.2.

Notice that the magnetic field produced by a bar magnet is not identical to that of a solenoid. In particular, the magnetic field lines within the bar magnet run in the opposite direction to the field lines within the solenoid. We shall look at this again in Sections 2.3.1 and 2.3.2. It can be explained because the bar magnet has a magnetization $\boldsymbol{M}$ while the solenoid does not, and this magnetization leads to the generation of a magnetic dipole, which acts as a source and sink for the magnetic field.

1.1.7 Ampère's Circuital Law

How can we calculate the strength of a magnetic field generated by an electrical current?

Ampère deduced that a magnetic field is produced by an electrical charge in motion when he read of Oersted's discovery of the effect of an electric current on a compass needle. This was a rather remarkable conclusion considering that until then magnetic fields were known to be generated only by permanent magnets and the Earth and in neither case was the presence of electrical charge in motion obvious.

According to Ampère, the magnetic field generated by an electrical circuit depended on the shape of the circuit (i.e., the conduction path) and the current carried. By assuming that each circuit is made up of an infinite number of current elements, each contributing to the field, and by summing or integrating these contributions at a point to determine the field, Ampère arrived at the following result [2]:

$$Ni = \int_{\text{closed path}} \boldsymbol{H} \cdot d\mathbf{l} \tag{1.8}$$

where N is the number of current-carrying conductors, each carrying a current i amperes. This is the source of the magnetic field $\boldsymbol{H}$, and $\mathbf{l}$ is simply a line vector. The total current Ni equals the line integral of $\boldsymbol{H}$ around a closed path containing the current. We should note that this equation is only true for steady currents.

Ampère's law (Equation 1.8) and the Biot-Savart law can be shown to be equivalent. Consider the field due to a steady current flowing in a long current-carrying conductor. By the Biot-Savart rule, the field at a radial distance r from the conductor is

$$H = \frac{i}{2\pi r} \tag{1.9}$$

while from Ampère's circuital theorem

$$\int \boldsymbol{H} \cdot d\mathbf{l} = i \tag{1.10}$$

and integrating along a closed path around the conductor at a radial distance r leads to

$$\int \boldsymbol{H} \cdot d\mathbf{l} = 2\pi r \boldsymbol{H} = i \tag{1.11}$$

$$\boldsymbol{H} = \frac{i}{2\pi r} \tag{1.12}$$

Furthermore, Ampère's law, which we have really used to define $\boldsymbol{H}$ above, can be shown to be equivalent to one of Maxwell's equations of electromagnetism, specifically $\nabla \times \boldsymbol{H} = \boldsymbol{J}_f$, where $\boldsymbol{J}_f$ is the current density of conventional electrical currents.

1.1.8 Orders of Magnitudes of Magnetic Fields in Various Situations

How large or how small are magnetic fields in different situations?

The magnitudes of the magnetic field strengths in a variety of situations are shown in Table 1.1.

TABLE 1.1
Magnetic Field Strengths (A m^{-1}) in a Variety of Situations, Showing a Range of 19 Orders of Magnitude

10^{14}	Surface of neutron stars
10^{8}	Implosive magnets (microsecond duration)
$2–5 \times 10^{7}$	Pulsed electromagnets (microsecond duration)
$1–3 \times 10^{7}$	High field electromagnets
$1–1.5 \times 10^{7}$	Superconducting magnets
$1–2 \times 10^{6}$	Laboratory electromagnets
1×10^{6}	Strongest permanent magnets
10^{2}	Earth's magnetic field on the surface
10	Stray fields from electrical machinery
1	Urban magnetic noise level
5×10^{-2}	Contours for geomagnetic anomaly maps
10^{-4}	Magnetocardiograms
10^{-5}	Fetal heartbeat
10^{-6}	Magnetic field from human brain
10^{-8}	Limits of detection for superconducting quantum interference devices

1.2 MAGNETIC INDUCTION

How does a medium respond to a magnetic field?

When a magnetic field $\boldsymbol{H}$ has been generated in a medium by a current, in accordance with Ampère's law, the response of the medium is its magnetic induction $\boldsymbol{B}$, also sometimes called the *flux density*. All media will respond with some induction and, as we shall see, the relation between magnetic induction and magnetic field is a property called the *permeability of the medium*. For our purposes, we shall also consider free space to be a medium since a magnetic induction is produced by the presence of a magnetic field in free space.

1.2.1 Magnetic Flux

If a loop of conductor of cross-sectional area A *carries a current* i, *what is the energy associated with it?*

When current passes around a loop of conductor or a solenoid, there is a certain amount of energy associated with this current loop. The magnetic energy is given by

$$E = \frac{1}{2} i\Phi = \frac{1}{2} i^2 L \tag{1.13}$$

where:

- i is the current passing around the loop
- Φ is the amount of magnetic flux generated by the current
- L is the electrical inductance, which has a value $L = \mu_o \mu_r N^2 A/l$ for an air-cored solenoid with N turns, cross-sectional area A, and length l

This result links the energy with a magnetic quantity, flux, and as we shall see later, this will be useful in characterizing the *strength* of a magnetic dipole, such as a single current loop, a solenoid, or a bar magnet, in terms of the quantity Φ, the total magnetic flux that it generates, which can then be related directly to the energy stored in the magnetic dipole.

Whenever a magnetic field is present, there will be a magnetic flux Φ. This magnetic flux is measured in units of webers and its rate of change with time can be measured since it generates a voltage in a closed circuit of conductor through which the flux passes. Small magnetic particles such as iron filings align themselves along the direction of the magnetic flux as shown in Figure 1.3. We can consider the magnetic flux to be caused by the presence of a magnetic field in a medium. We shall see in Chapter 2 that the amount of flux generated by a given field strength depends on the properties of the medium and varies from one medium to another.

The weber: The weber is the amount of magnetic flux that when reduced uniformly to zero in 1 s produces an electromotive force of 1 V in a one-turn coil of conductor through which the flux passes.

1.2.2 Definition of Magnetic Induction

What is the unit of magnetic induction?

The magnetic induction $\boldsymbol{B}$ in webers per square meter is also known as the magnetic flux density and consequently a magnetic induction of 1 Wb m^{-2} is identical to a magnetic induction of 1 T. The magnetic induction is most usefully described in terms of the force on a moving electric charge or an electric current. If the induction is constant, then we can define the tesla as follows.

The tesla: A magnetic induction $\boldsymbol{B}$ of 1 T generates a force of 1 N m^{-1} on a conductor carrying a current of 1 A perpendicular to the direction of the induction.

This definition can be shown to be equivalent to the older definition of the tesla as the couple exerted in newtons per meter on a small current loop when its axis is normal to the field, divided by the product of the loop current and surface area. We shall see in Chapter 2 that there are two contributions to the magnetic induction, one from the magnetic field $\boldsymbol{H}$ and the other from the magnetization $\boldsymbol{M}$ of the medium.

There is often some confusion between the concept of the magnetic field $\boldsymbol{H}$ and the magnetic induction **B**, and since a clear idea of the important difference between these two is essential to the development of the subject presented here, a discussion of the difference is called for. In many media, $\boldsymbol{B}$ is a linear function of $\boldsymbol{H}$.

In particular, in free space, we can write

$$\boldsymbol{B} = \mu_0 \boldsymbol{H} \tag{1.14}$$

where μ_0 is the permeability of free space, which is a universal constant. In the unit convention adopted in this book, $\boldsymbol{H}$ is measured in amperes per meter and $\boldsymbol{B}$ is measured in tesla (= V s m^{-2}), the units of μ_0 are therefore (volt second)/(ampere per meter), also known as *henries per meter*, and its value is $\mu_0 = 4\pi \times 10^{-7}$ H m^{-1}. If the value of $\boldsymbol{B}$ in free space is known, then $\boldsymbol{H}$ in free space is immediately known from this relationship.

However, in another media, particularly ferromagnets and ferrimagnets, $\boldsymbol{B}$ is no longer a linear function of $\boldsymbol{H}$, nor is it even a single-valued function of $\boldsymbol{H}$. In these materials, the distinction becomes readily apparent and important. A simple measurement of the $\boldsymbol{BH}$ loop of a ferromagnet should be all that is necessary to convince anyone of this. Finally, $\boldsymbol{H}$ and $\boldsymbol{B}$ are still related by the permeability of the medium μ through the following equation:

$$\boldsymbol{B} = \mu \boldsymbol{H} \tag{1.15}$$

but now of course μ is not necessarily a constant. We shall see shortly that in paramagnets and diamagnets μ is constant over a considerable range of values of $\boldsymbol{H}$. However, in ferromagnets, μ varies rapidly with $\boldsymbol{H}$.

All of this means that a field $\boldsymbol{H}$ in amperes per meter gives rise to a magnetic induction or flux density $\boldsymbol{B}$ in tesla in a medium with permeability μ measured in henries per meter.

1.2.3 Force per Unit Length on a Current-Carrying Conductor in a Magnetic Field: Ampère's Force Law

How does the presence of a magnetic induction affect the passage of an electric current?

The unit of magnetic induction has been defined in terms of the force exerted on a current-carrying conductor. This will now be generalized to obtain the force $\boldsymbol{F}$ on a current-carrying conductor in a magnetic induction $\boldsymbol{B}$. The force per meter on a conductor carrying a current i in the direction of the unit vector $\mathbf{l}$ caused by a magnetic induction $\boldsymbol{B}$ is given by an equation known as *Ampère's force law*

$$\boldsymbol{F} = i\mathbf{l} \times \boldsymbol{B} \tag{1.16}$$

and hence in free space,

$$\boldsymbol{F} = \mu_0 i\mathbf{l} \times \boldsymbol{H} \tag{1.17}$$

Therefore, if two long wires are arranged parallel at a distance of a meters apart and carrying currents of i_1 and i_2 amperes, the force per meter exerted by one wire on the other is as follows:

$$\boldsymbol{F} = \frac{\mu_0}{2\pi a} i_1 i_2 \tag{1.18}$$

1.2.4 Lines of Magnetic Induction

How can we visualize the magnetic induction?

The lines of magnetic induction are a geometrical abstraction, which help us to visualize the direction and strength of a magnetic field. The direction of the induction can be examined using a small compass needle (magnetic dipole) or a fine magnetic powder such as iron filings. These show that the magnetic induction around a single linear current-carrying conductor are coaxial with the conductor and follow the right-hand, or corkscrew, rule. In a solenoid, the lines are uniform within the solenoid but form a closed return path outside the solenoid. The lines of induction around a bar magnet are very similar to those around a solenoid since both act as magnetic dipoles.

The lines of magnetic induction always form a closed path since we have no direct evidence that isolated magnetic poles exist that would act as sources and sinks of magnetic flux. This means that through any closed surface, the amount of flux entering is equal to the amount of flux leaving. That is the divergence of $\boldsymbol{B}$ is always zero. This is Gauss's law.

$$\int_{\text{closed surface}} \boldsymbol{B} \cdot d\boldsymbol{A} = 0 \tag{1.19}$$

We sometimes say that $\boldsymbol{B}$ is *solenoidal*, which is the same as saying that the lines of $\boldsymbol{B}$ form closed paths—Equation 1.19 is equivalent to another of Maxwell's equations of electromagnetism, specifically $\nabla \cdot \boldsymbol{B} = 0$.

1.2.5 Electromagnetic Induction

Can the magnetic field generate an electrical current or voltage in return?

When the magnetic flux linking an electric circuit changes, an electromotive force is induced and this phenomenon is called *electromagnetic induction*. Faraday and Lenz were two of the early investigators of this effect, and from their work, we have the two laws of induction.

Faraday's law: The voltage induced in an electrical circuit is proportional to the rate of change of magnetic flux linking the circuit.
Lenz's law: The induced voltage is in a direction that opposes the flux change producing it.

The phenomenon of electromagnetic induction can be used to determine the magnetic flux Φ. The unit of magnetic flux is the weber, which has been chosen so that the rate of change of flux linking a circuit is equal to the induced electromotive force in volts.

$$V = -N\frac{d\Phi}{dt} \tag{1.20}$$

where:
Φ is the magnetic flux passing through a coil of N turns
$d\Phi/dt$ the rate of change of flux

Since the magnetic induction is the flux density,

$$\boldsymbol{B} = \frac{\Phi}{A} \tag{1.21}$$

we can rewrite the law of electromagnetic induction as

$$V = -NA\frac{d\boldsymbol{B}}{dt} \tag{1.22}$$

This is an important result since it tells us that an electrical voltage can be generated by a time-dependent magnetic induction. As an example, we can consider the voltage induced in a 50-turn coil of area 1 cm^2 when the magnetic induction linking it changes uniformly from 3 T to zero in 0.01 s.

$$\begin{aligned} V &= -NA\frac{d\boldsymbol{B}}{dt} \\ &= -\frac{(50)\left(1\times10^{-4}\right)(3)}{0.01} \\ &= 1.5\ \text{V} \end{aligned} \tag{1.23}$$

1.2.6 Magnetic Dipole

What is the most elementary entity in magnetism?

As shown above in Ampère's theorem, a current in an electrical circuit generates a field. A circular loop of a conductor carrying an electric current is the simplest circuit that can generate a magnetic field. The most elementary units of the equivalent current and pole models are the current loop and the dipole, as shown in Figure 1.4.

In each case, there is a magnetic moment $\boldsymbol{m}$ associated with the elementary unit of magnetism. In the current loop, the magnetic moment is equal to the product of the current i and the area of the loop A. In the magnetic dipole, the magnetic moment is the product of the pole strength p and the separation between the poles l.

The torque on a magnetic dipole of moment $\boldsymbol{m}$ in magnetic induction $\boldsymbol{B}$ is then simply

$$\tau = \boldsymbol{m} \times \boldsymbol{B} \tag{1.24}$$

This is a variant on Ampère's force law and can be derived directly using Equation 1.6. In free space,

$$\tau = \mu_0 \boldsymbol{m} \times \boldsymbol{H} \tag{1.25}$$

This means that the magnetic induction $\boldsymbol{B}$ tries to align the dipole so that the moment $\boldsymbol{m}$ lies parallel to the induction. Alternatively, we can consider that $\boldsymbol{B}$ tries to align the current loop such that the field produced by the current loop is parallel to it.

If no frictional forces are operating, the work done by the turning force will be conserved. This gives rise to the following expression for the energy of the dipole moment $\boldsymbol{m}$ in the presence of magnetic induction $\boldsymbol{B}$:

$$E = -\boldsymbol{m} \cdot \boldsymbol{B} \tag{1.26}$$

and hence in free space

$$E = -\mu_0 \boldsymbol{m} \cdot \boldsymbol{H} \tag{1.27}$$

The current loop is known as a *magnetic dipole* for historical reasons, since at large distances from the loop, the field produced by such a loop is identical in form to the field produced by calculation from two hypothetical magnetic poles of strength p separated by a distance l, the dipole moment of such an arrangement being

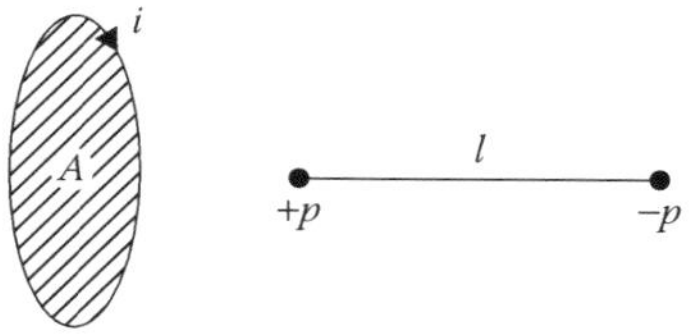

FIGURE 1.4 Configurations of the most basic mathematical entities: the current loop and the linear dipole.

$$m = pl \tag{1.28}$$

We will see in Chapter 2 how important the concept of magnetic dipole moment is in the case of magnetic materials. For in that case, the electrical *current* is caused by the motion of electrons within the solid, particularly the spins of unpaired electrons, which generate a magnetic moment even in the absence of a conventional current.

1.2.7 Unit Systems in Magnetism

What unit systems are currently used to measure the various magnetic quantities?

There are currently three systems of units in widespread use in magnetism and several other systems of units, which are variants of these. The three unit systems are the Gaussian or CGS system and two MKS unit systems, the Sommerfeld and the Kennelly conventions (Table 1.2). Each of these unit systems has certain advantages and disadvantages. The CGS and SI systems of magnetic units have different philosophies. The CGS system took an approach based on magnetostatics and the concept of the *magnetic pole*, while the SI system takes an electrodynamic approach to magnetism based on electric currents. The SI system of units was adopted at the 11th General Congress on Weights and Measures (1960). The Sommerfeld convention was subsequently the one accepted for magnetic measurements by the International Union for Pure and Applied Physics (IUPAP), and therefore, this system has slowly been adopted by the magnetism community. This is the system of units used in this book.

TABLE 1.2
Principal Unit Systems Currently Used in Magnetism

Quantity		SI (Sommerfeld)	SI (Kennelly)	EMU (Gaussian)
Field	H	A m^{-1}	A m^{-1}	oersteds
Induction	B	tesla	tesla	gauss
Magnetization	M	A m^{-1}	–	emu/cc
Intensity of magnetization	I	–	tesla	–
Flux	Φ	weber	weber	maxwell
Moment	m	A m^2	weber meter	emu
Pole strength	p	A m	weber	emu cm^{-1}
Field equation		$B = \mu_0(H + M)$	$B = \mu_0 H + I$	$B = H + 4\pi M$
Energy of moment (in free space)		$E = -\mu_0 m \cdot H$	$E = -m \cdot H$	$E = -m \cdot H$
Torque on moment (in free space)		$\tau = \mu_0 m \times H$	$\tau = m \times H$	$\tau = m \times H$

Note: The intensity of magnetization I used in the Kennelly system of units is merely an alternative measure of the magnetization M, in which tesla is used instead of A m^{-1}. Under all circumstances, therefore, $I = \mu_0 M$.

Conversion factors

$$1\ \text{Oe} = \left(\frac{1000}{4\pi}\right) \text{A m}^{-1} = 79.58\ \text{A m}^{-1} \tag{1.29}$$

$$1\ \text{gauss} = 10^{-4}\ \text{T} \tag{1.30}$$

$$1\ \text{emu cm}^{-3} = 1000\ \text{A m}^{-1} \tag{1.31}$$

To give some idea of the sizes of these units, the Earth's magnetic field is typically $\boldsymbol{H} = 56$ A m^{-1} (0.7 Oe), $\boldsymbol{B} = 0.7 \times 10^{-4}$ T. The saturation magnetization of iron is $\boldsymbol{M}_0 = 1.7 \times 10^6$ A m^{-1}. Remanence of iron is typically 0.8×10^6 A m^{-1}. The magnetic field generated by a large laboratory electromagnet is $\boldsymbol{H} = 1.6 \times 10^6$ A m^{-1}, $\boldsymbol{B} = 2$ T.

1.2.8 Maxwell's Equations of the Electromagnetic Field

Is there a more general and perhaps interconnected set of equations that can be used to describe electromagnetic fields?

In previous sections, we have alluded to a more general form of the Ampère/Biot-Savart law and Gauss's law. In fact, at the classical macroscopic level, electromagnetic fields can be described by four differential equations formulated by Maxwell. These are as follows:

$$\nabla \times \boldsymbol{H} = \boldsymbol{J} + \frac{\partial \boldsymbol{D}}{\partial t} \quad \text{(Ampère's law)} \tag{1.32}$$

$$\nabla \times \boldsymbol{E} = -\frac{\partial \boldsymbol{B}}{\partial t} \quad \text{(Faraday's law)} \tag{1.33}$$

$$\nabla \cdot \boldsymbol{B} = 0 \quad \text{(Gauss's law for magnetic flux density)} \tag{1.34}$$

$$\nabla \cdot \boldsymbol{D} = \rho \quad \text{(Gauss's law for electric flux density)} \tag{1.35}$$

where:

- $\boldsymbol{E}$ is the electric field
- $\boldsymbol{D}$ is the electric flux density
- $\boldsymbol{J}$ is the current density
- ρ is the charge density

The derivation of these equations is given in Appendix B. At this point in the discussion, we can treat them as simply experimental facts, since each was obtained as a result of observation. Only the first three are of concern to us in magnetism. Notice from these equations that there appears to be some correspondence between the magnetic field strength $\boldsymbol{H}$ and the electric field strength $\boldsymbol{E}$, and between the magnetic flux density $\boldsymbol{B}$ and the electric flux density $\boldsymbol{D}$, and that there is one equation for each of $\boldsymbol{H}$, $\boldsymbol{B}$, $\boldsymbol{E}$, and $\boldsymbol{D}$.

1.2.9 Alternating or Time-Dependent Magnetic Fields

What happens if the currents that cause the magnetic field are not steady?

Maxwell's equations form the basis of the description of the electromagnetic field. They apply whether the fields are steady or time dependent, and therefore, the solution of the above equations provides a general result. Two special cases arise, however. First, if the frequency is low enough (typically $<10^{14}$ Hz), then the displacement current term $\partial D/\partial t$ is small, and so the first Maxwell equation reduces to the steady-state Ampère's law. This means that in most practical cases, $\nabla \times \boldsymbol{H}$ is determined by the free current density $\boldsymbol{J}$. Second, if magnetic materials are present, the properties of the materials can significantly alter the penetration of magnetic fields into the material. The most familiar example of this is the penetration of a time-dependent field into an electrically conducting material. In this case, eddy currents are induced in the material by Maxwell's second equation (Faraday's law of induction) and this results in the alteration of the field amplitude with depth into the material. As the frequency of the exciting field increases, the rate of attenuation increases, resulting in a *penetration depth* or *skin depth*, which decreases with increasing frequency. However, this requires the presence of materials that we have not yet addressed in this book, so further discussion of this effect will be deferred.

Before proceeding further, we should derive the equation of a time-dependent magnetic field beginning from Maxwell's equations.

$$\nabla \times \boldsymbol{H} = \boldsymbol{J} + \frac{\partial \boldsymbol{D}}{\partial t} \tag{1.36}$$

and taking the curl of this

$$\nabla \times \nabla \times \boldsymbol{H} = \nabla \times \boldsymbol{J} + \nabla \times \frac{\partial \boldsymbol{D}}{\partial t} \tag{1.37}$$

now the current density $\boldsymbol{J}$ is related to the electric field $\boldsymbol{E}$ via the conductivity σ, $\boldsymbol{J} = \sigma \boldsymbol{E}$, and the electric flux density $\boldsymbol{D}$ is related to the electric field through the permittivity ε. Therefore, substituting for these and assuming that $\boldsymbol{E}$ is a well-behaved function, so that the $\partial D/\partial t$ and ∇ operators commute, we obtain

$$\nabla \times \nabla \times \boldsymbol{H} = \sigma \nabla \times \boldsymbol{E} + \in \frac{\partial}{\partial t}(\nabla \times \boldsymbol{E}) \tag{1.38}$$

and from Maxwell's equation for $\nabla \times \boldsymbol{E}$, we can substitute $-\partial B/\partial t$. Furthermore, $\nabla \times \nabla \times \boldsymbol{H} = \nabla\nabla \cdot \boldsymbol{H} - \nabla^2 \boldsymbol{H}$ for any vector quantity.

$$\nabla^2 \boldsymbol{H} - \nabla\nabla \cdot \boldsymbol{H} = \sigma \frac{\partial \boldsymbol{B}}{\partial t} + \in \frac{\partial^2 \boldsymbol{B}}{\partial t^2} \tag{1.39}$$

Now considering the special case of fields in free space $\boldsymbol{B} = \mu_0 \boldsymbol{H}$ and $\nabla \cdot \boldsymbol{B} = 0$, implying $\nabla \cdot \boldsymbol{H} = 0$ in free space. Therefore, the $\nabla \nabla \cdot \boldsymbol{H}$ term must be identically zero in free space:

$$\nabla^2 \boldsymbol{H} - \sigma \mu_0 \frac{\partial \boldsymbol{H}}{\partial t} - \in \mu_0 \frac{\partial^2 \boldsymbol{H}}{\partial t^2} = 0 \tag{1.40}$$

This is a wave equation for the magnetic field $\boldsymbol{H}$, which is undamped when $\sigma = 0$, but is damped in any electrically conducting medium.

1.3 MAGNETIC FIELD CALCULATIONS

How are magnetic fields of known strength usually produced?

Magnetic fields are usually produced by solenoids or electromagnets. A solenoid is made by winding a large number of turns of insulated copper wire, or a similar electrical conductor, in a helical fashion on an insulated tube known as a *former*. Solenoids are often cylindrical in shape. An electromagnet is made in a similar way except that the windings are made on a soft ferromagnetic material, such as soft iron. The ferromagnetic core of an electromagnet generates a higher magnetic induction $\boldsymbol{B}$ than a solenoid for the same magnetic field $\boldsymbol{H}$.

In view of the widespread use of solenoid of various forms to produce magnetic fields, we shall take some time to examine the field strengths produced by a number of different configurations.

1.3.1 Field at the Center of a Long Thin Solenoid

What is the simplest way to produce a uniform magnetic field?

The simplest way to produce a uniform magnetic field is in a long, thin solenoid. If the solenoid has N turns wound on a former of length L and carries a current i amperes, the field inside it will be

$$\boldsymbol{H} = \frac{Ni}{L} = ni \tag{1.41}$$

where n is defined as the number of turns per unit length.

The magnetic field lines in and around a solenoid are shown in Figure 1.5. A practical method of making an *infinite* solenoid is to make the solenoid toroidal in shape. This ensures that the field is uniform throughout the length of the solenoid. The magnetic field is then

$$\boldsymbol{H} = \frac{N}{2\pi r} i \tag{1.42}$$

where:

N is the total number of turns
r is the radius of the toroid
i is the current flowing in the windings in amperes

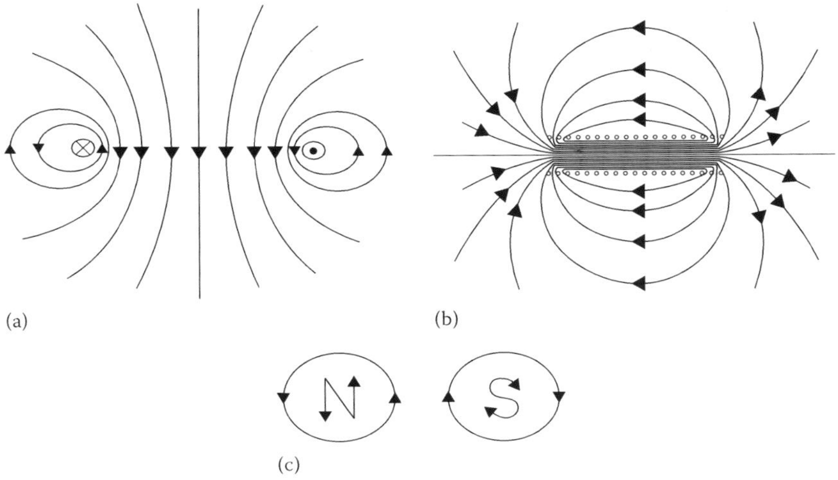

FIGURE 1.5 Magnetic field lines: (a) around a single loop of current carrying a conductor, (b) around a solenoid, and (c) convention for finding which end of a solenoid acts as the north pole (field source) and the south pole (field sink).

1.3.2 Magnetic Field along the Axis of a Circular Coil

What is the field strength along the axis of a single coil of wire carrying a current of i amperes?

The previous calculation of the field at the center of a circular coil can be generalized to obtain the exact expression for the magnetic field on the axis of a circular coil. Using the situation depicted in Figure 1.6, and applying the Biot-Savart rule, the field at the general point P is

$$d\boldsymbol{H} = \frac{1}{4\pi r^2} i\, d\mathbf{l} \times \mathbf{u} \tag{1.43}$$

where $\mathbf{u}$ is a unit vector along the r direction. We can make the substitution

$$r = \frac{a}{\sin\alpha} \tag{1.44}$$

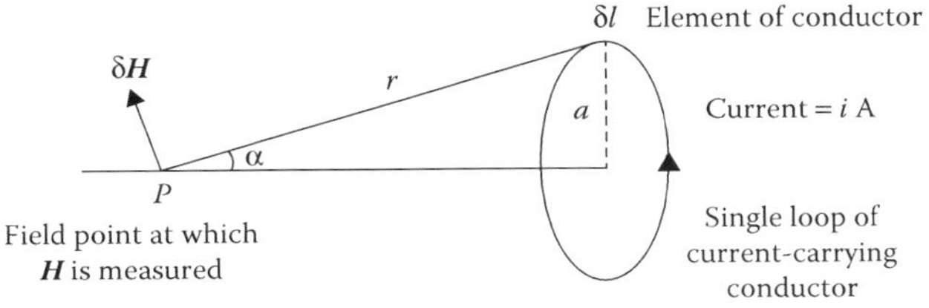

FIGURE 1.6 The magnetic field $\boldsymbol{H}$ due to a single circular coil carrying an electric current i.

which gives

$$dH = \frac{1}{4\pi a^2}\left(\sin^2\alpha\right) i\, d\mathbf{l} \times \mathbf{u} \tag{1.45}$$

The component of the field along the axis, which by symmetry will be the only resultant, is $dH_{\text{axial}} = dH \sin\alpha$

$$dH_{\text{axial}} = \frac{1}{4\pi a^2}\left(\sin^3\alpha\right) i\, d\mathbf{l} \times \mathbf{u} \tag{1.46}$$

Integrating round the coil, $\int d\mathbf{l} = 2\pi a$ and remembering $d\mathbf{l}$ is perpendicular to $\mathbf{u}$

$$H = \frac{i}{2a}\sin^3\alpha \tag{1.47}$$

or, equivalently,

$$H = \frac{ia^2}{2\left(a^2 + x^2\right)^{3/2}} \tag{1.48}$$

where x is the distance along the axis of the coil from its center.

This can be expressed in the form of a series in x and by symmetry all terms of odd order must have zero coefficients so the form of the dipole field becomes

$$H = H_0\left(1 + c_2 x^2 + c_4 x^4 + c_6 x^6 + \cdots\right)$$

where $H_0 = i/2a$ is the field at the center of the coil, and the coefficients have the values $c_2 = -3/2a^2$, $c_4 = 15/8a^4$, and $c_6 = -105/48a^6$.

As an example, we can consider a coil of 100 turns and diameter 0.1 m carrying a current of 0.1 A, and calculate the magnetic field at a distance of 50 cm along the axis of the coil.

When $i = 0.1$ A, $a = 5$ cm, $x = 50$ cm and the coil has 100 turns

$$H = \frac{(100)(0.05)^2(0.1)}{2\left[(0.05)^2 + (0.5)^2\right]^{3/2}} = 0.098\,\text{A m}^{-1} \tag{1.49}$$

Off-axis field of a circular coil

As shown in the above derivations, a simple analytical expression can be obtained for the magnetic field along the axis of a single loop of conductor carrying a current using the Biot-Savart law. However, in the vast majority of cases, there is no closed-form analytic solution for the field generated by a current-carrying conductor. Those that can produce closed-form analytic solutions are only the very simplest types of situation.

To give an example, there is no closed-form analytic solution for the off-axis field of a single circular loop of conductor carrying a steady current, except for the far-field dipole field, which varies with $1/r^3$. This comes at first as somewhat of a surprise given the extreme simplicity of the situation. However, the example does show how very limited the situations are that yield analytical solutions.

In the case of the off-axis field of the single circular loop, the analysis leads to an elliptic integral that has no exact solution. From the Biot-Savart law, the magnetic field contribution at any point $d\boldsymbol{H}$ due to a current element $i\,d\mathbf{l}$ is

$$d\boldsymbol{H} = \frac{i d\mathbf{l} \times \mathbf{u}}{4\pi r^2} \tag{1.50}$$

where r is the distance from the coil.

$$d\boldsymbol{H} = \frac{i\, d\mathrm{l}\sin\theta}{4\pi\left(x^2 + a^2\right)} \tag{1.51}$$

where now a is also a function of θ instead of being a constant. In the case of the off-axis field, the field strength can be calculated from Equation 1.51 by using numerical techniques.

1.3.3 Field Due to Two Coaxial Coils

Which simple coil configurations produce: (i) a constant magnetic field, or (ii) a constant field gradient?

In superposition

Often, when it is necessary to produce a uniform field over a large volume of space, a pair of Helmholtz coils is used. This consists of two flat coaxial coils, each containing N turns, with the current flowing in the same sense in each coil as shown in Figure 1.7. The separation d of the coils in a Helmholtz pair is equal to their common radius a.

The axial component of the magnetic field on the axis of the two coils can be calculated from the Biot-Savart law. The field on the axis of a single coil of N turns and radius a carrying a current i at a distance x from the plane of the coil is

$$\boldsymbol{H} = \left(\frac{Ni}{2a}\right)\left(1 + \frac{x^2}{a^2}\right)^{-1.5} \tag{1.52}$$

If we define one coil at location $x = 0$ and the other at location $x = a$, the field at the center of two such coils wound in superposition is

$$\boldsymbol{H} = \left(\frac{Ni}{2a}\right)\left\{\left[1 + \frac{x^2}{a^2}\right]^{-1.5} + \left[1 + \frac{(a+x)^2}{a^2}\right]^{-1.5}\right\} \tag{1.53}$$

and since for the Helmholtz coils $x - a/2$ at the point on the axis midway between the coils, this gives the axial component of the magnetic field at the midpoint as

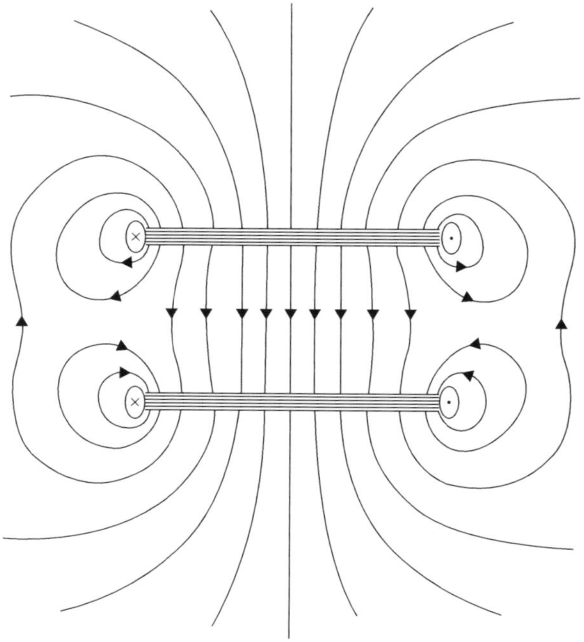

FIGURE 1.7 Two coaxial coils configured as a Helmholtz pair with the separation distance between the coils equal to their common radius.

$$\boldsymbol{H} = \left(\frac{Ni}{2a}\right)\left[(1.25)^{-1.5} + (1.25)^{-1.5}\right] = \frac{0.7155Ni}{a} \qquad (1.54)$$

and by symmetry the radial component on the axis must be zero.

In fact, if a series expansion is made for the axial component of $\boldsymbol{H}$ in terms of the distance x along the axis from the center of the coils, as was given for the single coil in Section 1.3.2, it is found that the term in x^2 disappears when the coil separation d equals the coil radius a, so that the fourth-order correction term becomes the most significant. The series expansion for the field in terms of x is then

$$\boldsymbol{H} = \boldsymbol{H}_0\left(1 + c_4x^4 + c_6x^6 + \cdots\right)$$

This results in a small value of $d\boldsymbol{H}/dx$ at the center of the coils, and consequently a very uniform field along the axis as x is varied close to zero, which is shown in Figure 1.8 for three different values of coil radius a. In addition, the axial component of the magnetic field close to the center of a pair of Helmholtz coils is only very weakly dependent on the radial distance z from the axis. This means that the magnetic field strength $\boldsymbol{H}$ is maintained fairly constant over a large volume of space between the Helmholtz coils.

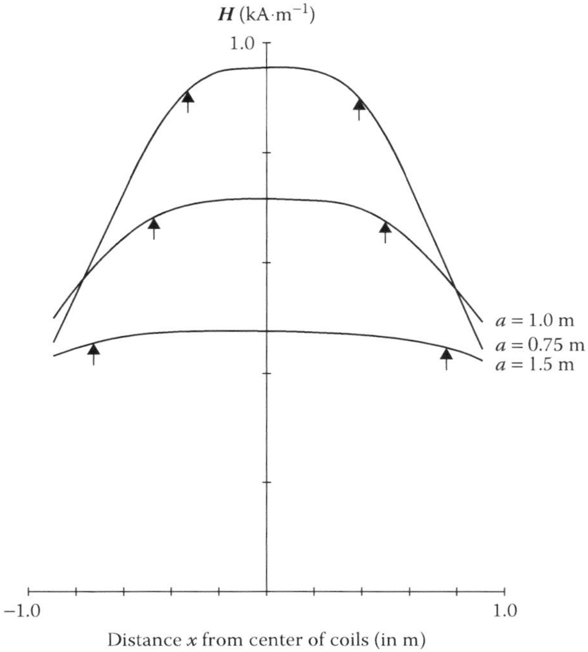

FIGURE 1.8 Axial component of the magnetic field $\boldsymbol{H}$ as a function of position along the axis of a pair of Helmholtz coils for various coil radii. The calculation is for $N = 100$ turns, with the coil carrying a current $i = 10$ A, with coil radii $a = 0.75$, 1, and 1.5 m. The arrows mark the location of the coils in each case.

The useful region of uniform field between a Helmholtz pair can be increased by making the coil spacing slightly larger than $a/2$, although this leads to a slight reduction in field strength over this region.

In opposition

If the current in one of the two coaxial coils described above is reversed, then the magnetic fields generated by the two coils will be in opposition. This is a specific example of a quadrupole field, so called because the form of the field obtained is similar to that obtained from two magnetic dipoles aligned coaxially and anti-parallel.

Under these conditions, the magnetic field along the axis of the pair of coils is given by

$$\boldsymbol{H} = \left(\frac{Ni}{2a}\right)\left[\left(1+\frac{x^2}{a^2}\right)^{-1.5} - \left(1+\frac{(a-x)^2}{a^2}\right)^{-1.5}\right] \tag{1.55}$$

Such a configuration generates a uniform field gradient, which can be useful for applying a constant force to a sample. See, for example, Section 3.2.2.

1.3.4 Field Due to a Thin Solenoid of Finite Length

What field strength is produced in the more practical case of a solenoid of limited length?

So far the field of an infinite solenoid has been considered. Now solenoids of finite length will be considered. A thin solenoid is one in which the inner and outer diameters of the coil windings are equal. So, for example, a solenoid consisting of one layer of windings would be considered as a thin solenoid.

The field of a long thin solenoid has already been calculated in Section 1.3.1. The field on the axis of a thin solenoid of finite length has an analytical solution. If L is the length of the solenoid, D the diameter, i the current in the windings, and x the distance from the center of the solenoid, then the field at x is given by

$$\boldsymbol{H} = \left(\frac{Ni}{L}\right)\left\{\frac{(L+2x)}{2\left[D^2+(L+2x)^2\right]^{1/2}} + \frac{(L-2x)}{2\left[D^2+(L-2x)^2\right]^{1/2}}\right\} \tag{1.56}$$

At the center of the solenoid, $x = 0$ and hence

$$\boldsymbol{H} = \left(\frac{Ni}{L}\right)\left[\frac{L}{\left(L^2+D^2\right)^{1/2}}\right] \tag{1.57}$$

Finally, for a long solenoid, $L >> D$ and $(L^2 + D^2)^{1/2} = L$, so that the result from Section 1.3.1 is a limiting case:

$$\boldsymbol{H} = \frac{Ni}{L} = ni \tag{1.58}$$

The fields generated by solenoids are of course dipole fields.

The field calculations for thin solenoids, that is, solenoids with $L >> D$, at least along the axis, are relatively straightforward and yield analytical solutions as shown. A useful result to remember is that the field at the end of a solenoid is half the value of the field at the center. The field in the middle 50% of a solenoid is also known to be very uniform.

1.3.5 General Formula for the Field of a Solenoid

How can the field strength generated by a solenoid be determined in more general cases?

Not surprisingly, there is no general analytical formula for the magnetic field from a solenoid at a general point in space. However, there are some methods of calculation available. The most obvious method is by the straightforward procedure of using either Ampère's law or the Biot-Savart law, as in Section 1.1.3. This leads to a solution containing an elliptic integral, which can then be solved numerically.

However, there are also some quicker methods that can be used such as those developed by Brown and Flax [3] and by Hart [4]. In the former method, any desired solenoid of finite dimensions can be treated as the superposition of four semi-infinite solenoids as shown in Figure 1.9. The field at any point is the vector sum of the contributions of the four component solenoids.

The advantage of considering four semi-infinite solenoids with no cylindrical hole is that the field contribution of such a solenoid can be expressed in terms of only two variables: the axial distance beyond the end of the solenoid and the radial distance from the unique axis. Therefore, one table can be provided for the field at

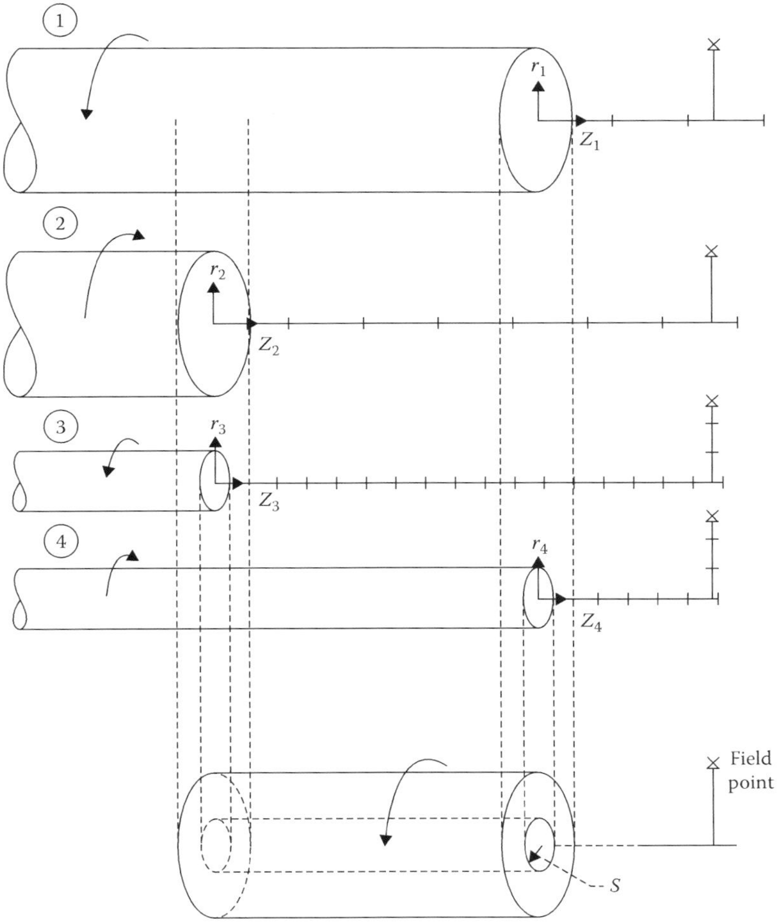

FIGURE 1.9 Method of superposing four semi-infinite solenoids to obtain the mathematical equivalent of one finite solenoid. Curved arrows indicate the direction of current flow, $z_1, \ldots, z_4$ represent the distances along the axis, and $r_1, \ldots, r_4$ indicate the radii of the solenoids. (Data from Brown, G.V., and Flax, L., *J. Appl. Phys.*, 35, 1764, 1964.)

these two reduced coordinates for all semi-infinite solenoids. The field can then be calculated by vector summation of the four contributions.

This method allows rapid and simple calculation of the field of any finite solenoid. Further details of the application of this method can be obtained from the original paper [3].

1.3.6 Field Calculations Using Numerical Methods

How can magnetic field strengths be calculated in more complicated situations, for example, over an entire volume?

Although it has been shown above how the Biot-Savart law can be applied to determine the magnetic field $\boldsymbol{H}$ in various simple situations, it has also been demonstrated that there is not always an analytic solution (e.g., the off-axis field of a single circular current loop). In more general cases, therefore, it is necessary to resort to numerical techniques in order to obtain a solution.

In most cases, the problem amounts to solving Maxwell's equations over a finite region of space, known as the *spatial domain*, which may be three dimensional or two dimensional. Maxwell's equations cannot be solved without an appropriate set of boundary conditions, even though the current density $\boldsymbol{J}$ is known over the entire spatial domain.

These numerical methods for calculating the magnetic field are often used to determine the magnetic field in the air gap of an electrical machine. In this case, there are no field sources in the gap and so

$$\begin{aligned} \nabla \times \boldsymbol{H} &= 0 \\ \nabla \cdot \boldsymbol{B} &= 0 \end{aligned} \tag{1.59}$$

In cases where field sources occur within the region of interest, then the source distributions must be known and included in the calculation. The problem then becomes solving Maxwell's equations with the appropriate boundary conditions.

$$\begin{aligned} \nabla \times \boldsymbol{H} &= \boldsymbol{J} \\ \nabla \cdot \boldsymbol{B} &= 0 \end{aligned} \tag{1.60}$$

The only difference between this and the previous case is the presence of field sources, in the form of the current density $\boldsymbol{J}$.

There are a number of general numerical methods that can be used to solve the equation for the magnetic field $\boldsymbol{H}$. Here, we shall consider finite-difference, finite-element, and boundary-element techniques. Much research effort has been devoted to the boundary-element method, although the finite-element method, which has been in use for over 40 years, is well established and a number of software packages are available. The finite-difference method, which has been in use since the 1940s, and actually traces its origins back to Gauss, is the oldest method, and has largely been superseded by the finite-element method.

It should also be realized that there are a number of analytical methods that have been developed for the calculation of magnetic fields. Some examples of these are series solutions, conformal mappings, and variational formulations. The problem is that these methods suffer from a lack of generality, which restricts their use to only the simplest situations. Often, they are restricted to steady-state conditions and many are applicable to two dimensions only.

A review of the progress in electromagnetic field computation by Trowbridge [5] covers the period 1962–1988. This discusses all three main methods and gives a selection of the most important references during the period. The finite-difference method of calculating magnetic fields was the principal numerical tool from the early days of the digital computer in the 1940s until about 1970. The method has been described by Adey and Brebbia [6] and by Chari and Silvester [7]. It is a technique for the solution of differential equations in which each derivative appearing in the differential equation is replaced by its finite-difference approximation at regularly spaced intervals over the volume of space of interest. This means that a continuous region of space must be replaced by a regular grid of discrete points at which the field values are calculated. The process is known as *discretization*.

In order to ease the computation, a regularly spaced orthogonal or polar grid is used for discretization, although in principle, the use of curvilinear grids is possible. However, as a result of the practical restriction of grids, the use of finite differences in the case of complicated geometries has encountered difficulties. Also, in situations with large field gradients in order to obtain sufficient accuracy, the number of nodes needs to be increased over the whole volume in order to maintain a regular grid (not just over the region of high gradient) and this results in a rapid increase in computation time and memory requirements. Despite these difficulties, the finite-difference method has been successfully implemented for field calculations and some excellent examples of its use have been reported [8].

In the finite-element method, the spatial domain is divided into triangle-shaped elements and the field values are computed at the three nodes of each element. The sizes of the elements can be varied over the region of interest (unlike the grid in the finite-difference calculations), so that more elements can be included in regions where the field gradient is large. Furthermore, the elements used to discretize the spatial domain need not be triangular, they can be of any polygonal shape; however, triangles remain the most popular form of element.

An introductory survey of the finite-element technique for the nonspecialist has been given by Owen and Hinton [9], while an excellent and a more detailed review has been given by Silvester and Ferrari [10]. The method first came to attention in the work of Winslow [11] and began to be used on a regular basis for field calculations from about 1968. From this time onward, it gradually supplanted the older finite-difference method.

The advantages of numerical techniques such as the finite-element method over analytical methods of field computation were demonstrated in the case of leakage fields by Hwang and Lord [12]. This was the first successful attempt to use numerical field calculations for the determination of fields in the vicinity of defects in materials. The implementation of both finite-difference and finite-element methods to two-dimensional nonlinear magnetic problems have been compared by Demerdash and Nehl [13]. They concluded that the finite-element technique was superior in that it required less computation time and less memory. However, although this conclusion that finite elements seem to be preferable is on the whole true, the relative performance of various numerical methods is highly dependent on the nature of the specific problem under consideration.

Boundary-element techniques for field calculations are well developed. The general method has been discussed by Brebbia and Walker [14] and its application to the problem of magnetic field calculations in particular by Lean and Wexler [15]. A comparison of integral and differential equation methods has been made by Simkin [16]. A comprehensive survey of these techniques is given by Hoole [17], which provides a guide to the whole subject of numerical methods of field calculation.

1.3.7 Developments in Magnetic Field Computation

In practice, how are magnetic fields calculated in more complicated and realistic situations for which no analytic formulae exist?

Finite-element calculations are the most common techniques employed for numerical field calculations at present. There has been enormous progress in the area of computational magnetics. So much in fact that now entire conferences are devoted solely to the subject. An introduction to this subject is provided by Lowther and Silvester [18], while Hoole [17] gave a comprehensive coverage of a range of numerical techniques that can be used for magnetic field calculations, including finite elements. A similar comprehensive reference work by Binns and Trowbridge [19] also provides a good overview of the subject. Given the enormous progress in this branch of the subject, it is difficult to give viable coverage that would do justice to the area within a book of this type. Therefore, reference to the above-mentioned sources is recommended.

Developments that are specific to this type of modeling have been discussed by Hoole [20] and Brauer [21]. A review by Chari et al. [22] has covered the range of applications that finite-element methods are being used to address today. A review by Trowbridge [23] summarized the main technical developments in computational magnetics. These have included: three-dimensional solutions to problems using finite-element modeling; extension of computational algorithms to dynamic (i.e., time-dependent) problems; two- and three-dimensional solutions using boundary-element methods; automatic mesh generation for complex problems; error analysis; and validation of model calculations using experimental results, which provides a critical check of whether the final results of the model calculations are realistic.

REFERENCES

1. Oersted, H. C. Experiment on the effects of a current on the magnetic needle, *Ann. Philos.*, 16, 1820.
2. Ampère, A. M. *Theorie Mathematique des Phenomenes Electrodynamiques Uniquement Diduite de l'Experience*, 1827. Reprinted by Blanchard, Paris, France, 1958.
3. Brown, G. V., and Flax, L. *J. Appl. Phys.*, 35, 1764, 1964.
4. Hart, P. J. *Universal Tables for Magnetic Fields of Filamentary and Distributed Circular Currents*, Elsevier, New York, 1967.
5. Trowbridge, C. W. *IEEE Trans. Magn.*, 24, 13, 1988.

6. Adey, R. A., and Brebbia, C. A. *Basic Computational Techniques for Engineers*, Pentech Press, London, 1988.
7. Chari, M. V. K., and Silvester, P. P. *Finite Elements in Electrical and Magnetic Field Problems*, Wiley, New York, 1980.
8. Fuchs, E. F., and Erdelyi, E. A. *IEEE Trans. PAS*, 92, 583, 1973.
9. Owen, D. R. J., and Hinton, E. *A Simple Guide to Finite Elements*, Pineridge Press, Whiting, NJ, 1980.
10. Silvester, P. P., and Ferrari, R. L. *Finite Elements for Electrical Engineers*, Cambridge University Press, Cambridge, 1983.
11. Winslow, A. M. *Magnetic Field Calculation in an Irregular Mesh*, Lawrence Radiation Laboratory, Livermore, CA, Report UCRL-7784-T, 1965.
12. Hwang, J. H., and Lord, W. *J. Test Eval.*, 3, 21, 1975.
13. Demerdash, N. A., and Nehl, T. W. *IEEE Trans. Magn.*, 12, 1036, 1976.
14. Brebbia, C. A., and Walker, S. *Boundary Element Techniques in Engineering*, Newnes-Butterworths, London, 1980.
15. Lean, M. H., and Wexler, A. *IEEE Trans. Magn.*, 18, 331, 1982.
16. Simkin, J. *IEEE Trans. Magn.*, 18, 401, 1982.
17. Hoole, S. R. H. *Computer Aided Analysis and Design of Electromagnetic Devices*, Elsevier, New York, 1989.
18. Lowther, D. A., and Silvester, P. P. *Computer Aided Design in Magnetics*, Springer-Verlag, New York, 1986.
19. Binns, K. J., and Trowbridge, C. W. *The Analytical and Numerical Solution of Electric and Magnetic Fields*, Wiley, New York, 1992.
20. Hoole, S. R. H. (ed.). *Finite Elements, Electromagnetics and Design*, Elsevier, New York, 1989.
21. Brauer, J. *Finite Element Analysis*, Marcel Dekker, New York, 1993.
22. Chari, M. V. K., Bedrosian, G., d'Angelo, J., and Konrad, A. *IEEE Trans. Magn.*, 29, 1306, 1993.
23. Trowbridge, C. W. *IEEE Trans. Magn.*, 32, 627, 1996.

FURTHER READING

Bennet, G. A. *Electricity and Modern Physics*, Arnold, Cambridge, London, 1968, Ch. 5.

Cendes, Z. (ed.). *Computational Electromagnetics*, North Holland Publishing, Amsterdam, the Netherlands, 1986.

Chari, M. V. K., and Silvester, P. P. (eds.). *Finite Elements in Electrical and Magnetic Field Problems*, Wiley, New York, 1980.

Grover, F. W. *Inductance Calculations*, Dover, New York, 1946.

Hoole, S. R. H. *Computer Aided Analysis and Design of Electromagnetic Devices*, Elsevier, New York, 1989.

Lorrain, P., and Corson, D. R. *Electromagnetism*, W. H. Freeman, San Francisco, CA, 1978.

McKenzie, A. E. E. *A Second Course of Electricity*, 2nd edn, Cambridge University Press, Cambridge, Chs. 3 and 4, 1961.

Reitz, J. R., and Milford, F. J. *Foundations of Electromagnetic Theory*, 3rd edn, Addison-Wesley, Reading, MA, Ch. 8, 1980.

EXERCISES

1.1 *Definition of ampere.* Define the ampere and show that this definition leads to a value of $4\pi \times 10^{-7}$ H m^{-1} for the permeability μ_0 of free space. What is the physical significance of μ_0?

1.2 *Difference between **H** and **B**.* Explain the difference between magnetic field ***H*** and magnetic induction ***B***. Include the basic laws of electromagnetism that govern these quantities and also include definitions of the two units.

1.3 *Units in magnetism.* Discuss in an essay the problem of units and definitions in magnetism. Why did this problem arise, and why are scientists and engineers reluctant to accept one particular unit system? What are the long-term consequences of failure to resolve this problem? What solutions do you recommend?

1.4 *Magnetic field at the center of a long solenoid.* Prove that the magnetic field ***H*** at the center of a *long* solenoid is $\boldsymbol{H} = ni$, where n is the number of turns per meter and i is the current flowing in the coils in amperes.

1.5 *Force on a current-carrying conductor*

a. Calculate the force per unit length between two parallel current-carrying conductors 1 m apart when each carries a current of 1 A.

b. Find the force exerted on a straight current-carrying conductor of length 3.5 cm carrying a current of 5 A and situated at right angles to a magnetic field of 160 kA m^{-1}.

1.6 *Torque on a current-loop dipole.* Find the torque on a circular coil of area 4 cm^2 containing 100 turns when a current of 1 mA flows through it and the coil is in a field of magnetic induction of 0.2 T.

1.7 *Force between two flat coaxial coils.* Calculate the force between two flat circular coils each containing 100 turns of wire carrying 0.5 A of current in the same direction of circulation, which have mean radii of 0.05 m and are placed coaxially 0.005 m apart.

1.8 *Forces on a dipole.* What type of force will be experienced by a magnetic dipole of fixed magnetic moment $\boldsymbol{m} = 5 \times 10^{-4}$ A m^2 as it is moved along the axis of a single circular coil of radius $a = 0.05$ m carrying a constant current $i = 0.5$ A. Calculate the magnitude of the force on this dipole at different distances x along the axis from the center of the coil if (1) ***m*** is always oriented along the axis, and (2) ***m*** is always oriented perpendicular to the axis. At what distance does the force have a maximum value? What is the value of this maximum force?

1.9 *Field generated by a square coil.* Calculate the ***H*** field (1) at the center of a square coil with length of side a meter carrying a current i ampere, (2) at a distance x from the center of the coil along the perpendicular axis, and (3) at the center of a circle of radius r and carrying a current i. (4) Give numerical values of the field in cases (1), (2), and (3) when $a = 1$ m, $x = 1$ m, and $i = 1$ A.

1.10 *Force on a rod carrying a current.* A uniform rod of mass m per unit length is suspended vertically so that it can turn freely about its upper end. A current i amperes is sent through it along its length. What will be the equation for the magnetic force on the rod if the Earth's field ***B*** at that point is horizontal? Give numerical value for this force if $m = 200$ g m^{-1}, $i = 1$ A, and $\boldsymbol{B} = 0.1 \times 10^{-3}$ T. What will be its angular deflection from the vertical under these conditions?

2 Magnetization and Magnetic Moment

In the previous chapter, we were concerned with magnetic field $\boldsymbol{H}$ and magnetic flux density $\boldsymbol{B}$ in free space. We now consider the effect that a magnetic material has on the magnetic induction $\boldsymbol{B}$. This is represented by the magnetization $\boldsymbol{M}$. Materials can alter the magnetic flux density either by making it larger, as in both paramagnets and ferromagnets, or by making it smaller, as in diamagnets. The relative permeability of the material indicates how the material changes the magnetic flux density compared with the magnetic flux density that would be observed in free space.

2.1 MAGNETIC MOMENT

How do we measure the response of a material to a magnetic field?

When going on to consider magnetic materials, it is first necessary to define quantities that represent the response of these materials to the magnetic field. These quantities are magnetic moment and magnetization. Once that has been done, we can consider another property, the susceptibility, which is closely related to the permeability.

2.1.1 Force on a Magnet

Can we use the torque on a specimen in a field of known strength to define its magnetic properties?

In the previous chapter, we have defined the magnetic moment $\boldsymbol{m}$ of a current loop dipole and shown that the torque τ on the dipole in the presence of a magnetic field $\boldsymbol{H}$ in free space is given by $\tau = \boldsymbol{m} \times \boldsymbol{B}$. Therefore, the magnetic moment can be expressed as the maximum torque on a magnetic dipole τ_{max} divided by $\boldsymbol{B}$.

$$\boldsymbol{m} = \frac{\tau_{max}}{\boldsymbol{B}} \tag{2.1}$$

and hence in free space

$$\boldsymbol{m} = \frac{\tau_{max}}{\mu_0 \boldsymbol{H}} \tag{2.2}$$

where the magnetic moment $\boldsymbol{m}$ in the convention we are using is measured in ampere meter squared, A m^2. This formula applies equally to a current loop or to a bar magnet.

2.1.1.1 Unit of Magnetic Moment

A magnetic moment of 1 A m^2 experiences a maximum torque of 1 N m when oriented perpendicular to a magnetic induction of 1 T.

In the case of a current loop, as in Figure 2.1, $\boldsymbol{m} = iA$, where i is the current flowing and A is the cross-sectional area of the loop. In the case of a bar magnet, as in Figure 2.2, $\boldsymbol{m} = pl$, where p is the pole strength in ampere meters and l is the dipole length in meters. The *pole strength* is a term arising from the CGS treatment of magnetism.

The magnetic moment vector $\boldsymbol{m}$ in a bar magnet tends to align itself with $\boldsymbol{B}$ under the action of the torque as shown in Figure 2.2. For this reason, a bar magnet in the field generated by a second bar magnet experiences a torque, which aligns it parallel to the local direction $\boldsymbol{B}$. It is this force that gives rise to the most widely recognized phenomenon in magnetism that unlike magnetic poles attract each other while

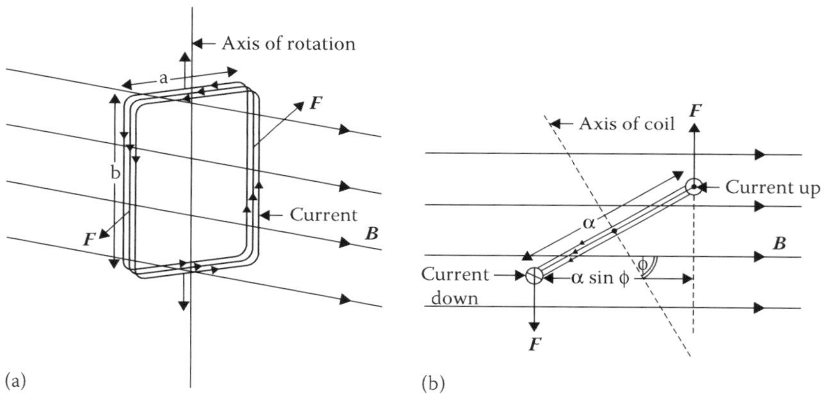

FIGURE 2.1 The torque on a current loop in an external magnetic field: (a) side view and (b) top view. If the loop is free to rotate, the torque turns the loop until its plane is normal to the field direction.

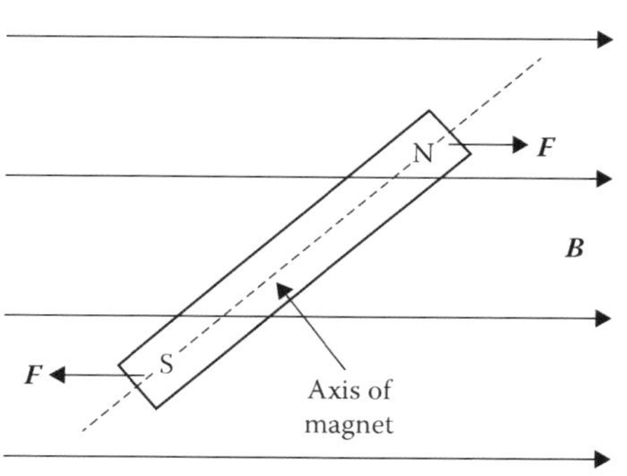

FIGURE 2.2 The torque on a bar magnet in an external field. If the bar is free to rotate, the torque turns the bar until its plane is parallel to the field direction.

like magnetic poles repel each other. However, our difficulty in explaining the exact meaning of magnetic *poles* remains.

As discussed in Section 1.2.6, the basic unit of magnetism is the magnetic dipole. The basic unit of electricity is the electric charge. There have been searches for the magnetic monopole, which has a theoretical value, according to some authors [1], of $p = 3.29 \times 10^{-9}$ A m ($\mu_0 p = 4.136 \times 10^{-15}$ Wb) and, if it exists, would be the magnetic analogue of electric charge. The discovery of the monopole would have important consequences but the search has so far been inconclusive [2,3].

However, if the magnetic monopole is discovered, this will lead to fundamental changes in our understanding of magnetism. In particular, the much maligned concept of the magnetic pole, which is a useful concept for those working in magnetic materials, would become more acceptable. Maxwell's equations of electromagnetism would also need to be altered to allow the divergence of $\boldsymbol{B}$ to be nonzero. A paper on the subject has been written by Carrigan and Trower [4].

2.1.2 Pole Strength of a Dipole p

What measure should be used to classify the strength *of a dipole?*

It has been shown in Chapter 1 that the energy associated with a single loop of conductor carrying a current depends on the product of the current i with the magnetic flux Φ. It seems reasonable therefore to classify the *strength* of a permanent magnet in terms of the magnetic flux that it produces. From this magnetic flux, we can define a pole strength p in units of ampere meters given by

$$p = \frac{\Phi}{\mu_0} \tag{2.3}$$

where μ_0 is again permeability of free space.

2.1.3 Magnetic Dipole Moment m

What is the magnetic moment of a dipole in terms of its pole strength?

In a dipole consisting of two opposite poles each of strength p, separated by a length l, the dipole magnetic moment is

$$\boldsymbol{m} = pl \tag{2.4}$$

where $\boldsymbol{m}$ is in units of ampere meter squared A m^2, and consequently

$$\boldsymbol{m} = \frac{\Phi l}{\mu_0} \tag{2.5}$$

Comparing this with the previous expression (Equation 2.4) for the magnetic moment of a current loop, we obtained $\boldsymbol{m} = iA$, and hence we can begin to equate some of the magnetic properties of the dipole with those of the current loop.

$$\frac{\Phi l}{\mu_0} = iA \tag{2.6}$$

This brings us quickly to a realization of the limitations of describing magnetic materials either solely in terms of current loops or magnetic dipoles. The current loop has no effective length, while the dipole has no effective cross-sectional area. These models must then simply represent mathematical abstractions of the real situation chosen primarily because of the simplicity of their description. The real situation is likely to be that a magnetic moment in a material has some characteristics of both these extremes of mathematical abstraction.

2.1.4 Force on a Dipole Suspended in a Magnetic Field

How can the force on a dipole be determined?

Consider a dipole as shown Figure 2.2 with poles p_n and p_s separated by a distance l. If the applied magnetic field is uniform, then the force on the first pole is

$$\boldsymbol{F}_{\mathrm{N}} = \mu_0 p_n \boldsymbol{H} \tag{2.7}$$

and on the second pole is

$$\boldsymbol{F}_{\mathrm{S}} = \mu_0 p_s \boldsymbol{H} \tag{2.8}$$

If p_n and p_s are equal and of opposite signs, then there is no net translational force, only a net torque τ

$$\tau = (\mu_0 p_s \boldsymbol{H} + \mu_0 p_n \boldsymbol{H})\frac{1}{2}\sin\theta\hat{\tau} \tag{2.9}$$

where $\hat{\tau}$ is a unit vector along the direction $l \times \boldsymbol{H}$. If the *poles* are of same strength, then $p_s = -p_n = p$

$$\tau = \mu_0 p l \times \boldsymbol{H} \tag{2.10}$$

where pl is the magnetic moment $\boldsymbol{m}$, which now gives the equation for the torque as

$$\tau = \mu_0 \boldsymbol{m} \times \boldsymbol{H} \tag{2.11}$$

Therefore, the magnetic moment is seen to be the product of the pole strength p and the length of the dipole l, and is directed along the length of the dipole.

2.1.5 Force on a Current Loop Suspended in a Magnetic Field

How can the force on a current loop be determined?

Probably the simplest calculation is the case of a rectangular current loop of sides a meters and b meters carrying a current i oriented at an angle θ to the field, as shown in Figure 2.1. The force per unit length on each side will be, according to Ampère's force law,

$$\boldsymbol{F} = i\boldsymbol{l} \times \boldsymbol{B} \tag{2.12}$$

$$\boldsymbol{F} = i\boldsymbol{a} \times \boldsymbol{B} \tag{2.13}$$

and the turning force will be twice the product of this with the distance $b/2$ from the axis

$$\tau = ib\boldsymbol{a} \times \boldsymbol{B} \tag{2.14}$$

and since there is an equal torque on both sides these can be summed to give

$$\tau = iba\,\boldsymbol{B}\sin\theta \tag{2.15}$$

where θ is the angle between $\boldsymbol{B}$ and a unit vector normal to the plane of the current loop $\boldsymbol{m}$. Replacing iba with $\boldsymbol{m}$ gives

$$\tau = \boldsymbol{m} \times \boldsymbol{B} \tag{2.16}$$

so that the magnetic moment $\boldsymbol{m}$ of the current loop is the product of the current and its cross-sectional area and is directed normal to the plane of the loop.

2.2 MAGNETIC POLES AND AMPÈRIAN BOUND CURRENTS

What means do we have for describing the origin of the magnetic moment of a material?

In Chapter 1, it has been shown how a magnetic field can be produced by an electric current, that is, charge in motion. However, it is well known that a magnetic field can also be produced by a permanent magnet, and this occurs without the presence of a real electric current. The generation of the magnetic field in this case is closely linked to the magnetic properties of the material. The question then becomes how to describe this.

It is possible to describe the generation of a magnetic field by a permanent magnet in two ways. First, the magnetic material could be replaced by a set of equivalent currents, which generate exactly the same external magnetic field. This is the *Ampèrian* or *bound* current approach, and can easily be understood in terms of the description of field generation by electric currents as provided in Chapter 1. The second approach is to describe the magnetic material in terms of an assembly of magnetic poles, which are hypothetical sources of magnetic field that generate exactly the same external field.

2.2.1 Existence of Poles and Bound Currents

Are magnetic poles or bound currents real?

Neither magnetic poles nor bound currents are real in a physical sense. Instead, both are merely mathematical artifacts, or approximations, which allow calculation of magnetic fields and magnetic moments in a wide range of different situations.

2.2.2 Usefulness of the Pole and Bound Current Models

If neither approach is correct, why bother at all with these concepts?

Ultimately, neither poles nor Ampèrian currents are considered to be quite correct, so it must be asked what function do they perform. This can be answered quite clearly. The mathematical concepts of magnetic poles and Ampèrian currents provide a framework for calculating the magnetization and magnetic fields generated by magnetic materials. These calculations proceed of course on a classical macroscopic basis. The pole/bound current models provide a limiting boundary between the classical description of magnetism in materials and the quantum description, so that the question of which of these is more *correct* cannot be answered on a classical basis. The classical level calculations merely show that, at distances sufficiently far from the magnetic dipole or current loop, the magnetic fields generated by these two mathematical models are equivalent. It is curious to note, therefore, that from time to time either the poles or the bound currents become more fashionable as a description of the magnetic properties of magnetic materials at the limit of the classical theory. It is recommended here that both concepts should be considered and the most appropriate one should be chosen for solving any specific calculation.

2.2.3 Relative Advantages of Poles or Bound Currents

Are poles or bound currents better?

It can be shown that the description of fields due to magnetic poles or Ampèrian currents are mathematically equivalent so long as the calculation is made at field points, which are sufficiently distant from the poles or the current sources. Therefore, in general, neither approach is better or worse than the other, although in different situations, it may be easier to perform a particular calculation using one approach rather than the other. Nevertheless, the end result will be the same, since, as shown in Section 2.2.6, the two formalisms are mathematically equivalent at points far from the field source.

2.2.4 Magnetic Field Due to Poles

How can we calculate the magnetic field from a magnet in terms of an assembly of magnetic poles?

Although it has been stated by the report of the Coulomb's Law Committee [5] that strictly Coulomb's law cannot be considered quite correct for magnetic poles, it can still provide a useful approximation. The rationale for its consideration here is that in the final analysis neither poles nor Ampèrian currents are adequate as a description of magnetism in materials; they only serve to enable us to perform calculations on a classical macroscopic level, without addressing or considering the fundamental origin of magnetic moments in materials.

The contribution $d\boldsymbol{H}(\boldsymbol{r})$ to the magnetic field at a point $\boldsymbol{r}$ from a dipole dp is given by

$$dH(r) = \frac{1}{4\pi} \frac{dp}{r^2} \hat{r} \tag{2.17}$$

where:

$\hat{r}$ is the unit vector along r

p is the pole strength of the dipole

If this expression is integrated over all poles,

$$\begin{aligned} H(r) &= \frac{1}{4\pi} \sum_{\substack{\text{all} \\ \text{poles}}} \frac{dp}{r^2} \hat{r} \quad \text{(discrete)} \\ &= \frac{1}{4\pi} \int_{\substack{\text{all} \\ \text{poles}}} \frac{dp}{r^2} \hat{r} \quad \text{(continuous)} \end{aligned} \tag{2.18}$$

2.2.5 Magnetic Field Due to Equivalent, Bound, or Ampèrian Currents

How can we calculate the magnetic field from the equivalent current distribution?

If the equivalent current approach is used, then we can simply invoke the Biot-Savart law that was used for real currents in Chapter 1. In this case, the contribution to the magnetic field dH due to current i in an element dl is

$$dH(r) = \frac{1}{4\pi} \frac{i dl \times \hat{r}}{r^2} \tag{2.19}$$

This expression can then either be summed or integrated over all current elements to give the total field

$$H(r) = \frac{1}{4\pi} \sum_{\substack{\text{all} \\ \text{currents}}} \frac{i dl \times \hat{r}}{r^2} \quad \text{(discrete)} \tag{2.20}$$

$$H(r) = \frac{1}{4\pi} \int_{\substack{\text{all} \\ \text{currents}}} \frac{i dl \times \hat{r}}{r^2} \quad \text{(continuous)} \tag{2.21}$$

2.2.6 Equivalence of Fields Due to Pole and Current Distribution

How do the fields due to a pole and a current distribution differ?

It is clear from a comparison of the above equations that, provided we are only interested in the field values far from the dipole, these descriptions are completely equivalent. In the limit as $r \to 0$, of course, they cannot be evaluated, and in fact the geometry suggests that at these distances the results are likely to be different. An obvious example of the eventual difference at small distances arises if we consider a simple dipole. On the pole model, this is a linear structure consisting of two poles of strength $+p$ and $-p$, separated by a distance l. On the current model, this consists of a planar structure with circulating current i bounding a surface of area A. Therefore, in the limit as r becomes comparable to the dimensions of the dipole or current loop the two models give different results.

2.2.7 Calculation of Field Due to a Current Loop and a Dipole

What is the simplest direct comparison of the field due to a current loop and a dipole?

Probably the simplest comparison to make is that of the on-axis field of a current loop with the field along the line of a dipole. In fact, the calculation for the on-axis field of the current loop was given in Section 1.3.2, so we will just quote the result here:

$$\boldsymbol{H} = \frac{ia^2}{2\left(a^2 + x^2\right)^{3/2}} \tag{2.22}$$

where:
- i is the current
- a is the radius of the circular loop
- x is the distance from the plane of the current loop

The calculation for a dipole is very simple. Using Equation 2.22 for a dipole field, we need to consider only two poles $+p$ and $-p$, as shown in Figure 2.3. These are separated by a distance l. Now consider a field point distant x from the center of the dipole. The two contributions to the field at x are

$$\boldsymbol{H}_+ = \frac{1}{4\pi}\frac{p}{\left[x-(l/2)\right]^2} \tag{2.23}$$

$$\boldsymbol{H}_- = \frac{1}{4}\frac{-p}{\left[x+(l/2)\right]^2} \tag{2.24}$$

Therefore,

$$\boldsymbol{H} = \frac{1}{4\pi}\left\{\frac{p}{\left[x-(l/2)\right]^2} - \frac{-p}{\left[x+(l/2)\right]^2}\right\} \tag{2.25}$$

$$\boldsymbol{H} = \frac{pl}{2\pi}\left\{\frac{x}{\left[x+(l/2)\right]^2\left[x-(l/2)\right]^2}\right\} \tag{2.26}$$

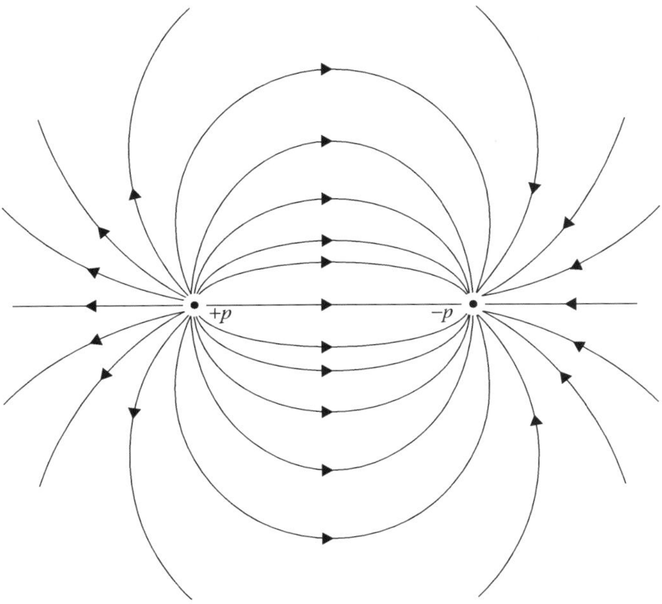

FIGURE 2.3 Field due to a linear dipole, a useful mathematical abstraction.

Note immediately that the expressions for the field due to a dipole or a current loop are different in the near field—that is, when x is small compared with the size of the dipole. Furthermore, while it is relatively easy to obtain an analytic expression for the off-axis field of a dipole, there is no analytic expression for the off-axis field of a current loop. The latter has to be solved numerically. Now, however, consider the field at large distances from the current loop or from the dipole. For the current loop, the far-field expression is

$$\lim_{x\to\infty} \boldsymbol{H} = \frac{ia^2}{2x^3} \tag{2.27}$$

and for the dipole the far-field expression is

$$\lim_{x\to\infty} \boldsymbol{H} = \frac{pl}{2\pi x^3} \tag{2.28}$$

These expressions are identical and furthermore imply that $pl = \pi ia^2$, which of course is the representation of the magnetic moment of the dipole or current loop.

Summarizing the conclusions from this analysis, we note that the pole and equivalent current models are fundamentally different, but at large distances (i.e., on a classical macroscopic scale), they give identical results. Neither is in any physical sense more correct than the other since there is as of yet no evidence of the existence of bound currents or magnetic poles. These two models are, however, extremely useful for making field calculations on the classical scale. But if they are taken too literally, they lead to contradictions (e.g., imposing the Ampèrian current model on a spinning electron simply proves that the description is inadequate as shown later). Therefore, we should treat these models as merely a useful boundary between the classical and quantum description of magnetism.

2.3 MAGNETIZATION

*How are the magnetic properties of the material and the magnetic induction **B** related?*

We can define a new quantity $\boldsymbol{M}$, the magnetization, as the magnetic moment per unit volume of a solid.

$$\boldsymbol{M} = \frac{\boldsymbol{m}}{V} \tag{2.29}$$

From the relationship between magnetic moment $\boldsymbol{m}$ and flux given in Section 2.1.3, a simple relationship between $\boldsymbol{M}$ and $\boldsymbol{B}$ can be found. A bar magnet with flux density Φ at the center, dipole length l, and with a cross-sectional area A has a magnetic moment $\boldsymbol{m}$ given by $\boldsymbol{m} = \Phi l/\mu_0$. The magnetization $\boldsymbol{M}$ is therefore given by $\boldsymbol{M} = \boldsymbol{m}/Al$. Hence

$$\boldsymbol{M} = \frac{\Phi}{\mu_0 A} = \frac{\boldsymbol{B}}{\mu_0} \tag{2.30}$$

In this case, there are no conventional external electric currents present to generate an external magnetic field and so $\boldsymbol{B} = \mu_0\boldsymbol{M}$. We see therefore that the magnetization $\boldsymbol{M}$ and magnetic field $\boldsymbol{H}$ contribute to the magnetic induction in a similar way. If both magnetization and magnetic field are present, then their contributions can be summed.

2.3.1 Relationship between *H*, *M*, and *B*

Can we define a constitutive equation relating these three magnetic quantities: field, induction, and magnetization?

We have seen that the magnetic induction $\boldsymbol{B}$ consists of two contributions: one from the magnetic field $\boldsymbol{H}$ and the other from the magnetization $\boldsymbol{M}$. The magnetic induction or flux density in free space is $\mu_0\boldsymbol{H}$, while in the convention, which we are following, the contribution to the induction from the magnetization of a material is $\mu_0\boldsymbol{M}$. The magnetic induction is then simply the vector sum of these

$$\boldsymbol{B} = \mu_0(\boldsymbol{H} + \boldsymbol{M}) \tag{2.31}$$

where:

$\boldsymbol{B}$ is in tesla

$\boldsymbol{H}$ and $\boldsymbol{M}$ are in amperes per meter

Equation 2.31, which relates these three basic magnetic quantities, is true under all circumstances in SI unit system, but is different in CGS unit system.

The magnetic field $\boldsymbol{H}$ is generated by electrical currents outside the material either from a solenoid, an electromagnet, or from a permanent magnet. The magnetization is generated by the resultant (uncompensated) spin and orbital angular momentum of electrons within the solid. The origin of the net angular momentum of the electrons requires further development of ideas before it can be explained. Discussion is therefore deferred until Chapters 9 through 11.

A related quantity, the magnetic polarization or intensity of magnetization $\boldsymbol{I}$, is used in the Kennelly convention. This is defined by

$$\boldsymbol{I} = \mu_0 \boldsymbol{M} \tag{2.32}$$

Although it is not often employed when the Sommerfeld system is used, it is a useful unit. Measuring the magnetic polarization $\boldsymbol{I}$ of a material in tesla is often more convenient than measuring the magnetization $\boldsymbol{M}$ in amperes per meter. Crangle [6] has remarked that since the Sommerfeld and Kennelly systems are not mutually exclusive, this unit can easily be incorporated into the International Union of Pure and Applied Physics system without contradictions.

In the SI system of units [7,8], $\boldsymbol{M}$ is usually measured in amperes per meter (the Sommerfeld convention, which we are using) but you will sometimes find it measured, as indicated above, in tesla (the Kennelly convention). This means that the torque equation in free space is different in the two conventions by a factor of μ_0, being $\tau = \boldsymbol{m} \times \boldsymbol{H}$ in the Kennelly convention but $\tau = \boldsymbol{m} \times \boldsymbol{B} = \mu_0 \boldsymbol{m} \times \boldsymbol{H}$ in the Sommerfeld convention. Similarly, the magnetic moment in the Sommerfeld convention is measured in amperes meter squared, whereas in the Kennelly convention, it is measured in weber meters. The relative merits of the two conventions have been discussed at length [9–11], and each has its own advantages and disadvantages. In the Sommerfeld convention, the definition of susceptibility is useful, but in the Kennelly system the susceptibility is an awkward unit. However, in the Kennelly system, the unit of magnetic polarization is more convenient than the unit of magnetization in the Sommerfeld convention. Also, the energy of a magnetic moment in a field and the torque on a magnetic moment in a field are simpler in the Kennelly system because μ_0 does not enter the equations.

2.3.2 Saturation Magnetization

Is there a limit to the magnetization that a given material can reach?

If a material has n elementary atomic magnetic dipoles per unit volume each of magnetic moment $\boldsymbol{m}$, then the magnetic moment per unit volume of the material when all these moments are aligned parallel is termed the *saturation magnetization* $\boldsymbol{M}_0$. This is equal to the product of n and $\boldsymbol{m}$.

A distinction can be made between technical saturation $\boldsymbol{M}_s$ and complete saturation $\boldsymbol{M}_0$. In order to fully understand this distinction, a discussion of domain processes must first be presented. At this stage, we shall merely note that technical saturation magnetization is achieved when a material is converted to a single magnetic domain, but at higher fields, the magnetization increases very slowly beyond technical saturation. This slow increase of magnetization at high fields is due to an increase in the spontaneous magnetization within a single domain, as discussed in Section 6.2.5, and is known as *forced magnetization*.

2.3.3 Permeability and Susceptibility

How can we represent the response of a magnetic material to a magnetic field?

We can now make a general statement for the permeability μ and susceptibility χ. The permeability is defined as

$$\mu = \frac{\boldsymbol{B}}{\boldsymbol{H}} \tag{2.33}$$

and the susceptibility is defined as

$$\chi = \frac{\boldsymbol{M}}{\boldsymbol{H}} \tag{2.34}$$

and the differential permeability and susceptibility are defined as

$$\mu' = \frac{d\boldsymbol{B}}{d\boldsymbol{H}} \tag{2.35}$$

$$\chi' = \frac{d\boldsymbol{M}}{d\boldsymbol{H}} \tag{2.36}$$

Since $\boldsymbol{B}$ and $\boldsymbol{M}$ may or may not be linear functions of $\boldsymbol{H}$, depending on the type of material or medium, it should be noted here that the permeability and the susceptibility may or may not be constant.

Sometimes, you will find the term *relative permeability* used, particularly in SI units. The relative permeability of a medium, denoted μ_r, is given by

$$\mu_r = \frac{\mu}{\mu_0} \tag{2.37}$$

where μ_0 is of course the permeability of free space $\mu_0 = 4\pi \times 10^{-7}$ H m^{-1}. The relative permeability of free space is 1. The relative permeability is closely related to the susceptibility and Equation 2.38 is always true in SI (Sommerfeld) units:

$$\mu_r = \chi + 1 \tag{2.38}$$

Other commonly encountered properties are the initial permeability μ_{in} and the initial susceptibility χ_{in}. These are the values of the respective quantities at the origin of the initial magnetization curve:

$$\mu_{in} = \left(\frac{d\boldsymbol{B}}{d\boldsymbol{H}}\right)_{\boldsymbol{B}=0,\boldsymbol{H}=0} = \left(\frac{\boldsymbol{B}}{\boldsymbol{H}}\right)_{\boldsymbol{B}\to 0,\boldsymbol{H}\to 0} \tag{2.39}$$

$$\chi_{in} = \left(\frac{d\boldsymbol{M}}{d\boldsymbol{H}}\right)_{\boldsymbol{M}=0,\boldsymbol{H}=0} = \left(\frac{\boldsymbol{M}}{H}\right)_{\boldsymbol{M}\to 0,\boldsymbol{H}\to 0} \tag{2.40}$$

In general, physicists and materials scientists are more interested in magnetization and susceptibility, while engineers who work mainly with ferromagnets are usually more concerned with magnetic induction and permeability.

Example 2.1: Permeability of and Magnetic Induction in Iron

In a magnetic field of 400 A m^{-1}, the relative permeability of a piece of soft iron is 3000. Calculate the magnetic induction in the iron at this field strength.

$$\boldsymbol{B} = \mu_0(\boldsymbol{H}+\boldsymbol{M}) = \mu_0\mu_r \boldsymbol{H} = 0.48\pi \text{ T} \tag{2.41}$$

Example 2.2: Flux Differences between Air Core and Iron Core

Calculate the magnetic induction $\boldsymbol{B}$ and flux Φ at the center of a toroidal solenoid with mean circumference of 50 cm and a cross-sectional area of 200 mm^2, wound with 800 turns of wire carrying 1.0 A, (1) when the solenoid has an air core and (2) when the solenoid has a soft iron core of relative permeability 1000.

The magnetic field $\boldsymbol{H}$ will be given by

$$H = ni = 1600 \text{ A m}^{-1} \tag{2.42}$$

1. In air $\boldsymbol{B} = \mu_0 H$

$$B = \left(4\pi\times10^{-7}\right)1600 = 2.0\times10^{-3} \text{ T} \tag{2.43}$$

$$\Phi = \boldsymbol{B}A$$

$$\Phi = \left(2.0\times10^{-3}\right)\left(2.0\times10^{-4}\right) = 4.0\times10^{-7} \text{ Wb} \tag{2.44}$$

2. In iron with $\mu_r = 1000$

$$\boldsymbol{B} = \mu_r\mu_0\boldsymbol{H} = (1000)\left(2.0\times10^{-3}\right)\text{T} = 2.0\text{T} \tag{2.45}$$

$$\Phi = \boldsymbol{B}A$$

$$\Phi = (2.0)\left(2.0\times10^{-4}\right) = 4.0\times10^{-4} \text{ Wb} \tag{2.46}$$

2.4 MAGNETIC CIRCUITS AND THE DEMAGNETIZING FIELD

How does a magnetic material alter the field and flux density in its vicinity?

In a given magnetic field $\boldsymbol{H}$, the presence of a magnetic material affects the magnetic induction $\boldsymbol{B}$ due to its permeability μ as indicated in Chapter 1. For determining the magnetic flux in magnetic circuits, a useful concept is the reluctance R, which is the magnetic analog of electrical resistance. Furthermore, if the magnetic material has finite length, the generation of *magnetic poles* near its ends gives rise to a magnetic field opposing the applied field. This opposing field is called the *demagnetization field* and its strength is dependent on the geometry and magnetization $\boldsymbol{M}$ of the material.

2.4.1 Flux Lines around a Bar Magnet

How does the presence of a magnetic material alter the flux lines in its vicinity?

The magnetic flux lines around a bar magnet can be mapped using a small dipole such as a compass needle or iron filings. The flux lines are continuous throughout the material and have the form shown in Figure 2.4. The flux lines have a similar form to the flux lines in and around a current loop dipole such as a single turn of current-carrying conductor or short solenoid.

2.4.2 Field Lines around a Bar Magnet

Are the magnetic field lines identical to the magnetic flux lines?

The field lines around a bar magnet are the same as the flux lines outside the material as shown in Figure 2.4, since $\boldsymbol{B} = \mu_0 \boldsymbol{H}$ in free space. However, inside the material, they are different and in fact $\boldsymbol{B}$ and $\boldsymbol{H}$ point in different directions because of the magnetization of the material $\boldsymbol{M}$. This can be proved using Ampère's circuital law. We can view the magnetization as being the effect of aligning the magnetic dipoles within the material, which creates magnetic *poles* near the ends of a finite specimen.

The magnetization vector $\boldsymbol{M}$ inside a magnetized ferromagnet points from the *south pole* to the *north pole* since this is the convention adopted for the definition of magnetic moment for a magnetic dipole. The magnetic field $\boldsymbol{H}$ always points from the *north pole* to the *south pole*.

2.4.3 Demagnetizing Fields

Does the shape of a specimen have any effect on the $\boldsymbol{H}$ *field?*

In view of the fact that the magnetization $\boldsymbol{M}$ and the magnetic field $\boldsymbol{H}$ point in opposite directions inside a magnetized material of finite dimensions, due to the presence of the magnetic dipole moment, it is possible to define a demagnetizing field $\boldsymbol{H}_{\mathrm{d}}$, which is present whenever magnetic poles are created in a material. This demagnetizing field can be detected during hysteresis measurements on finite-length samples

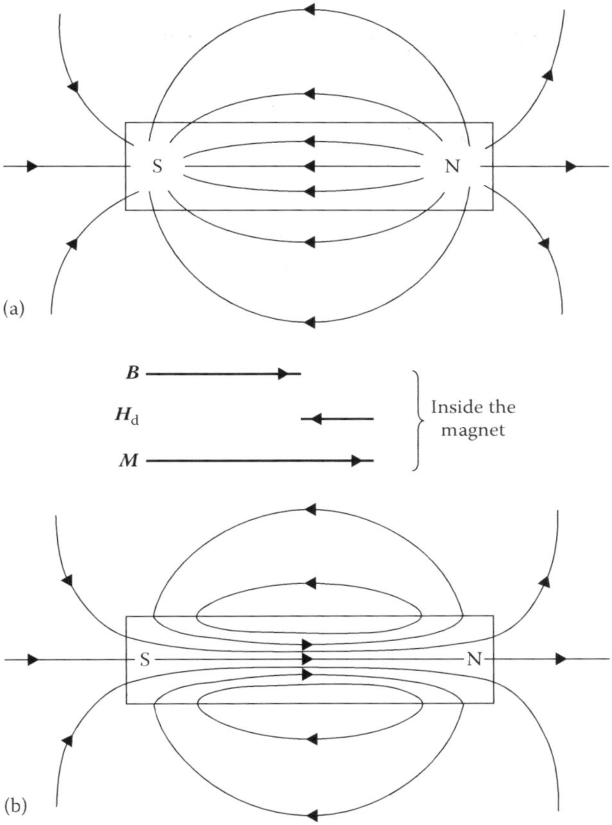

FIGURE 2.4 (a) Magnetic field $\boldsymbol{H}$ both inside and outside a bar magnet, (b) magnetic induction $\boldsymbol{B}$ both inside and outside a bar magnet. Notice in particular that the magnetic field and induction lines are identical outside the material, but inside they are quite different (they even point in opposite directions).

when the applied field is reduced to zero but the measured field is negative due to the remanent magnetization.

The demagnetizing field depends on two factors only. These are the magnetization in the material (i.e., the pole strength) and the shape of the specimen (i.e., the pole separation, which is determined by sample geometry). The demagnetizing field is proportional to the magnetization and is given by the expression

$$\boldsymbol{H}_{\mathrm{d}} = N_{\mathrm{d}}\boldsymbol{M} \tag{2.47}$$

where N_{d} is the demagnetizing factor, which is calculated solely from the sample geometry. It should be remembered that the numerical value of N_{d} depends on the units used for $\boldsymbol{M}$ and $\boldsymbol{H}_{\mathrm{d}}$. In the unit convention used in this book, both $\boldsymbol{M}$ and $\boldsymbol{H}_{\mathrm{d}}$ are measured in amperes per meter, so that N_{d} is simply a dimensionless number.

2.4.4 Demagnetizing Factors

What value does the demagnetizing factor have in different cases?

Exact analytic solutions for N_d can only be obtained in the case of second-order shapes, that is, spheres and ellipsoids [12]. However, tables or charts of approximately calculated demagnetizing factors are available for solids of various shapes, as shown in Figure 2.5 and Table 2.1.

2.4.5 Field Correction Due to Demagnetizing Field

How do we make a correction to the field if we have specimens of finite dimensions?

We have shown above how the demagnetizing field arises in a sample of given shape with magnetization $\boldsymbol{M}$ in zero field. When dealing with samples of finite dimensions in an applied magnetic field $\boldsymbol{H}_{app}$, it is necessary to make some demagnetizing field correction to determine the exact internal field in the solid $\boldsymbol{H}_{in}$.

$$\boldsymbol{H}_{in} = \boldsymbol{H}_{app} - N_d \boldsymbol{M} \tag{2.48}$$

2.4.6 Effect of Demagnetizing Field on Measurements

What effect does the demagnetizing field actually have on the measured properties of a material?

In a magnetic material, there will be a relationship between the magnetization $\boldsymbol{M}$ and the magnetic field $\boldsymbol{H}$, which is a characteristic of the material. This relationship is structure sensitive, but nevertheless, it is a material's property and could be obtained, for example, by measuring the variation of magnetization with field using a continuous closed path of material such as a toroid.

However, whenever a finite-length specimen of the material is used, the variation of magnetization with field is found to be different, and in particular the measured susceptibility χ will decrease. From a given length of specimen, if progressively shorter specimens are cut, it is found that the measured susceptibility decreases as the length of the specimen decreases (always assuming that the cross-sectional area and shape remain constant).

This means that any attempt to determine the intrinsic magnetic properties of a material on a specimen of finite length will be in error, and the error will increase as the length of specimen decreases. The resulting effect is usually described as *shearing* of the magnetizing curve or hysteresis loop, which simply means that, on the M–H plane on which the magnetic properties are usually plotted, the shortening of the specimen causes the magnetization curve or hysteresis loop to appear to be stretched along the H-axis, with a resulting decrease in susceptibility. A good example of this *shearing* of the magnetization curve is given by Cullity [13].

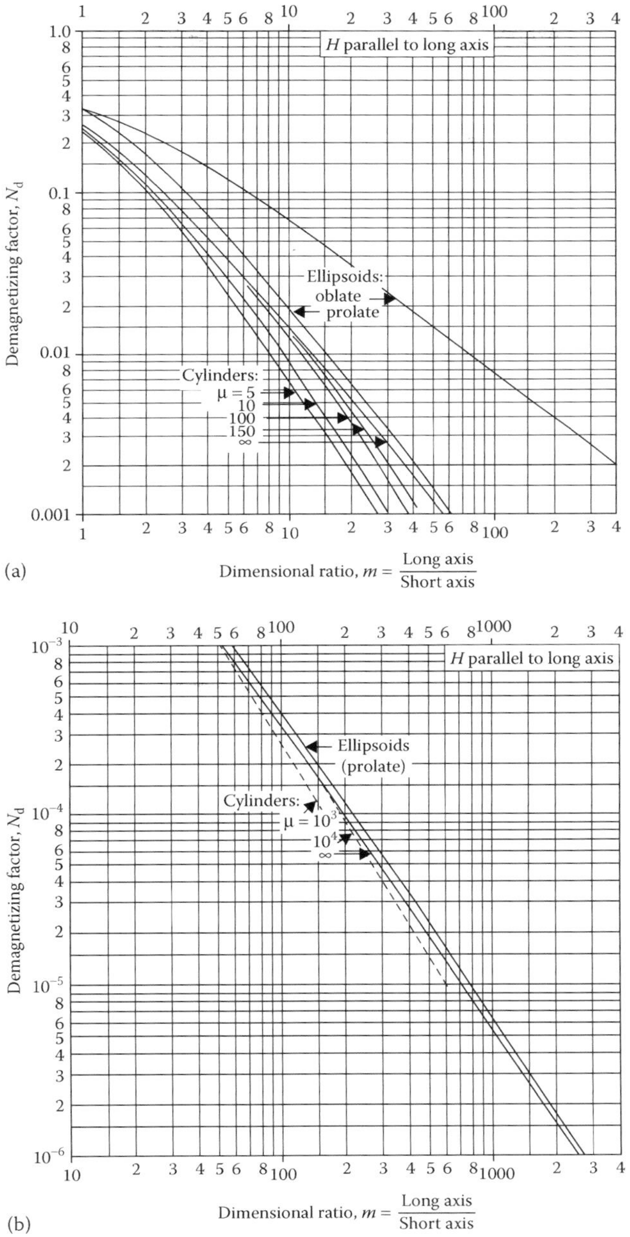

FIGURE 2.5 Calculated demagnetizing factors N_d for (a) ellipsoids and (b) cylinders. Note that the values of the demagnetizing factor are dependent on the permeability as well as on the shape.

TABLE 2.1
Demagnetizing Factors for Various Simple Geometries

Geometry	l/d	N_d
Toroid		0
Long cylinder		0
Cylinder	20	0.00617
Cylinder	10	0.0172
Cylinder	8	0.02
Cylinder	5	0.040
Cylinder	1	0.27
Sphere		0.333

The demagnetizing factor is an attempt to provide a means for calculating the intrinsic magnetic properties of a material from measurements conducted on finite specimens. The approach is necessarily an approximate one, since the idea of making a single uniform correction to the field throughout a specimen is an oversimplification. Nevertheless, in many cases, this approach can give good practical results.

Using Equation 2.48 for the demagnetizing field correction, it is easily shown that in terms of the magnetic induction $\boldsymbol{B}$, we obtain

$$\boldsymbol{H}_{\text{in}} = \frac{\boldsymbol{H}_{\text{app}} - (N_{\text{d}}\boldsymbol{B}/\mu_0)}{1 - N_{\text{d}}} \tag{2.49}$$

2.4.7 Corrections to the Susceptibility and the Permeability

How do we correct χ and μ_r for demagnetizing effects?

Again beginning from the equation that corrects the magnetic field for demagnetizing effects, we can define a true susceptibility $\chi_{\text{true}} = \boldsymbol{M}/\boldsymbol{H}_{\text{in}}$, which is a material's property, and a measured susceptibility $\chi_{\text{meas}} = \boldsymbol{M}/\boldsymbol{H}_{\text{app}}$, which is dependent on geometry. From this, it is seen that

$$\chi_{\text{true}} = \frac{\chi_{\text{meas}}}{1 - N_{\text{d}}\chi_{\text{meas}}} \tag{2.50}$$

Similarly with permeability, if we define the true permeability $\mu_{\text{true}} = \boldsymbol{B}/\boldsymbol{H}_{\text{in}}$ as a material's property and the measured permeability $\mu_{\text{meas}} = \boldsymbol{B}/\boldsymbol{H}_{\text{app}}$, which is geometry dependent, then

$$\mu_{\text{true}} = \frac{\mu_{\text{meas}}(1 - N_{\text{d}})}{1 - [N_{\text{d}}(\mu_{\text{meas}}/\mu_0)]} \tag{2.51}$$

2.4.8 Further Developments in Demagnetizing Field Calculations

What progress is being made in determining the demagnetizing factors for various shapes?

The work by Chen [14,15] has summarized developments in the determination of demagnetizing factors. The first of these papers deals primarily with cylinders, but the extensive references provide a review of the most important developments in demagnetizing field calculations for a wider range of shapes. The second paper, while not so easily available, is nevertheless much more general in its approach to the whole problem of demagnetizing effects, including demagnetizing effects under time-dependent fields. All of the above, however, assume that the magnetic material is homogeneous, but in real materials, this is rarely the case. Therefore, the determination of effective demagnetizing factors in inhomogeneous media provides the next challenge.

2.4.9 Magnetic Circuits and Reluctance

How do we calculate magnetic field or flux in situations where we have an air gap or two materials with different magnetic properties?

Situations in which a magnetic flux path is interrupted by an air gap are of practical importance because they occur in engineering applications of permanent magnets, electric motors and generators, and in testing of materials. The problems encountered here are more complicated than in calculating the flux in a single material; however, the ideas of demagnetizing fields together with some generalizations of the equations relating magnetic flux to magnetic field can be employed to provide solutions.

The magnet engineer is often in the situation of calculating the magnetic flux in magnetic circuits [16,17] with a combination of an iron and air core. Consider, for example, the situation where a ring of iron is wound with N turns of a solenoid, which carries a current i as shown in Figure 2.6a. In this case, the magnetic field will be Ni/L, where L is the average length of the ring and from this, and knowledge of the permeability, the magnetic flux or flux density passing in the ring can be calculated as

$$\boldsymbol{B} = \mu_0 (\boldsymbol{H} + \boldsymbol{M}) \tag{2.52}$$

and from Ampère's circuital law, in this simple situation $\boldsymbol{H} = Ni/L$, and therefore,

$$\boldsymbol{B} = \mu_0 \left(\frac{Ni}{L} + \boldsymbol{M} \right) \tag{2.53}$$

and consequently

$$Ni = \left(\frac{\boldsymbol{B}}{\mu_0} - \boldsymbol{M} \right) L = \frac{\boldsymbol{B}L}{\mu} \tag{2.54}$$

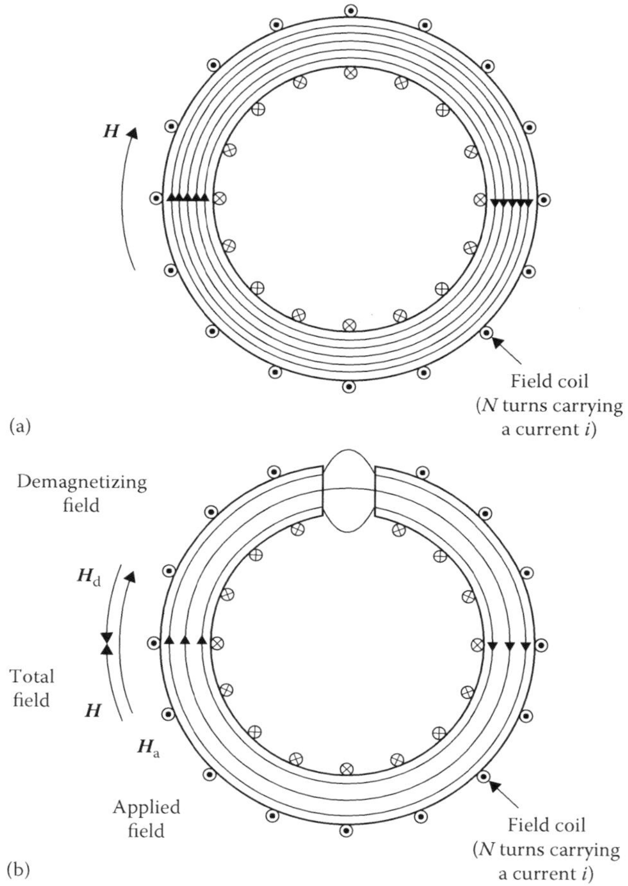

FIGURE 2.6 Magnetic flux path in (a) an iron toroid forming a closed magnetic path and (b) an iron toroid with an air gap. The flux lines are shown as concentric circles within the iron. The air gap increases the magnetic reluctance of the magnetic circuit and reduces the flux density in the iron as well as in the gap.

We can define here the magnetomotive force η, which for a solenoid is Ni, where N is the number of turns and i is the current flowing in the solenoid. From this, we can formulate a general equation relating the magnetic flux Φ, the magnetic reluctance R_m, and the magnetomotive force η as follows:

$$\Phi = \frac{\eta}{R_m} \tag{2.55}$$

Equation 2.55 is a magnetic circuit analog of Ohm's law for electrical circuits. If the iron ring has cross-sectional area A square meter and permeability μ with N turns of solenoid on a length L, we can derive an expression for the reluctance. Starting from the relationship between flux, magnetic induction, and magnetic field,

$$\Phi = \boldsymbol{B}A = \mu\boldsymbol{H}A = \mu\left(\frac{Ni}{L}\right)A = \frac{Ni}{(L/\mu A)} \tag{2.56}$$

From our definitions above, it follows that the term $L/\mu A$ is the magnetic reluctance of the path. This is analogous to electrical resistance in an electrical circuit, which means that in a magnetic circuit the magnetic reluctances in series may be added.

A saw slot can be introduced in the ring as shown in Figure 2.6b to provide an air gap. If the air gap is small, there will be little leakage of the flux at the gap, but a single equation $\boldsymbol{B} = \mu\boldsymbol{H}$ can no longer apply since the permeability of air and the iron ring are very different.

Ignoring demagnetizing effects for the present calculation and starting from Ampère's circuital law

$$Ni = \int_{\text{closed path}} \boldsymbol{H} \cdot d\mathbf{l} \tag{2.57}$$

and for a two-component magnetic circuit consisting of an iron ring and an air gap

$$Ni = \boldsymbol{H}_{\text{i}}L_{\text{i}} + \boldsymbol{H}_{\text{a}}L_{\text{a}} \tag{2.58}$$

where:

L_{i} is the path length in the iron
L_{a} is the path length in the air
$\boldsymbol{H}_{\text{i}}$ and $\boldsymbol{H}_{\text{a}}$ are correspondingly the respective fields in the iron and in the air

For the ring with the air gap, the flux density in the gap equals the flux density in the iron, but the magnetization in the air gap is necessarily zero. Therefore, we can write a similar equation for the magnetic induction:

$$Ni = \left(\frac{\boldsymbol{B}}{\mu_0} - \boldsymbol{M}\right)L_{\text{i}} + \frac{\boldsymbol{B}L_{\text{a}}}{\mu_{\text{a}}} = \boldsymbol{B}\left(\frac{L_{\text{i}}}{\mu_{\text{i}}} + \frac{L_{\text{a}}}{\mu_{\text{a}}}\right) \tag{2.59}$$

It is clear from this that there is a discontinuity in $\boldsymbol{H}$ across the air gap while at the same time there is continuity in $\boldsymbol{B}$. Rewriting the equation in terms of flux Φ leads to

$$Ni = \Phi\left(\frac{L_{\text{i}}}{A_{\text{i}}\mu_{\text{i}}} + \frac{L_{\text{a}}}{A_{\text{a}}\mu_{\text{a}}}\right) \tag{2.60}$$

Therefore, from the above definition of magnetic reluctance as the ratio of flux to magnetomotive force, the reluctance of this magnetic circuit with an air gap is

$$R_{\text{m}} = \frac{L_{\text{i}}}{\mu_{\text{i}}A_{\text{i}}} + \frac{L_{\text{a}}}{\mu_{\text{a}}A_{\text{a}}} \tag{2.61}$$

Then the equation for the flux passing through the toroid with an air gap is

$$\Phi = \frac{Ni}{\left[\left(L_i/\mu_i A_i\right) + \left(L_a/\mu_a A_a\right)\right]} \tag{2.62}$$

The magnetic flux in the ring is reduced when the air gap is introduced because it requires more energy to drive the same flux across the air gap than through an equal volume of the iron due to the much lower permeability of the air.

2.4.10 Magnetic Field Calculations in Magnetic Materials

How can the magnetic field be calculated in more complex situations such as those with both magnetic materials and air gaps of different shapes?

At the end of Chapter 1, we showed how a variety of numerical techniques could be used to calculate the magnetic field in free space. This was a particularly simple situation. In the presence of magnetic materials, the situation becomes more difficult because the magnetization of the material needs to be taken into account. Nevertheless, the same general techniques, such as finite-element analysis [18], can be used successfully.

The effect of hysteresis adds a further complication to the problem since the magnetization $\boldsymbol{M}$ of the material then depends not only on magnetic field $\boldsymbol{H}$ but also on the field history. Even in the cases where hysteresis occurs, methods have been devised for calculation of magnetic fields in devices [19], which take hysteresis into account.

2.5 PENETRATION OF ALTERNATING MAGNETIC FIELDS INTO MATERIALS

Do time-dependent fields penetrate into materials in the same way as static fields?

We have seen that when a magnetic material is magnetized by an external static magnetic field, an internal demagnetizing field is created, which is determined by the level of magnetization and the shape of the piece of material. This demagnetizing field opposes the applied field and can be described in terms of the generation of magnetic poles near the ends of the piece of material.

In the case of an alternating or time-dependent magnetic field, there is another effect that occurs when the field impinges on an electrically conducting material. It has been shown above by the Faraday-Lenz law of electromagnetic induction that a time-dependent magnetic flux will generate an opposing e.m.f. or voltage. In an electrically conducting material, this gives rise to induced currents, known widely as *eddy currents*, which oppose the penetration of the applied field. The attenuation of the magnetic field in the material then depends on the rate of change of the applied field (often expressed via its frequency) and the electrical conductivity of the material and its permeability (since it is the rate of change of magnetic induction $\boldsymbol{B}$ that generates the eddy currents, not the rate of change of $\boldsymbol{H}$). Therefore, the attenuation of the field in the material increases with conductivity and permeability of the material and with rate of change of the applied field.

The simplest case to analyze is a sinusoidal plane electromagnetic field of frequency ω impinging on a semi-infinite plane surface of a material with conductivity σ and relative permeability μ_r. The wave equation for the magnetic field is then

$$\nabla^2\boldsymbol{H} - \varepsilon\mu_0\mu_r\frac{\partial^2\boldsymbol{H}}{\partial t^2} - \sigma\mu_0\mu_r\frac{\partial\boldsymbol{H}}{\partial t} = 0 \qquad (2.63)$$

and if

$$\boldsymbol{H} = \boldsymbol{H}_0\,e^{-i\omega t} \qquad (2.64)$$

then

$$\nabla^2\boldsymbol{H} - \omega^2\varepsilon\mu_0\mu_r\boldsymbol{H} + i\omega\sigma\mu_0\mu_r\boldsymbol{H} = 0 \qquad (2.65)$$

and if we have a planar wave front incident on a planar medium

$$\nabla^2\boldsymbol{H} = \frac{\partial^2\boldsymbol{H}}{\partial z^2} \qquad (2.66)$$

so that

$$\frac{\partial^2\boldsymbol{H}}{\partial z^2} + \omega^2\varepsilon\mu_0\mu_r\boldsymbol{H} + i\omega\sigma\mu_0\mu_r\boldsymbol{H} = 0 \qquad (2.67)$$

and the solution of such an equation is

$$\boldsymbol{H} = \boldsymbol{H}_0\,e^{(\alpha+i\beta)z} \qquad (2.68)$$

giving, for frequency below the optical range,

$$\beta = \sqrt{\frac{\omega\sigma\mu_0\mu_r}{2}} \qquad (2.69)$$

The reciprocal of this term is the depth δ at which the magnetic field decays to $1/e$ of its value at the surface. By convention, this term is called the *skin depth* δ. It is a measure of the rate of decay of the time-dependent magnetic field as it enters an electrically conducting magnetically permeable medium.

$$\delta = \sqrt{\frac{2}{\omega\sigma\mu_0\mu_r}} \qquad (2.70)$$

Note that this expression for the skin depth δ only applies to a plane sinusoidal magnetic field impinging on a flat surface. In practical situations, where these conditions are seldom fulfilled, the skin depth or rate of attenuation of the magnetic field will be

TABLE 2.2
Classical Skin Depth Values for Penetration of Different Frequencies of a Planar Time-Dependent Electromagnetic Field into a Material of Fixed Permeability

Material	μ_r	σ $(\Omega\ \text{m})^{-1}$	$\omega/2\pi$ (Hz)	Δ (mm)
Iron	1,000	10×10^6	10	1.6
			100	0.5
			1000	0.16
Permalloy (Ni-Fe)	1,000,000	5×10^6	10	0.07
			100	0.02
			1000	0.007
Copper	1	60×10^6	10	20
			100	6.5
			1000	2

different. The important point, however, is that the field amplitude is attenuated as it passes into a material and therefore the field is confined to a surface layer, the thickness of which decreases with frequency of the incident magnetic field (Table 2.2), and also with the conductivity and permeability of the material.

2.5.1 Eddy Currents in Conducting Magnetic Materials

How is the field energy dissipated by the generation of eddy currents in a magnetic material?

When a time-dependent magnetic field impinges on a conducting magnetic medium, the generation of eddy currents causes a dissipation of power. The classical eddy current loss W_{ec} can be calculated by solving Maxwell's equations in a homogeneous magnetic material of fixed permeability. From Faraday's law of induction,

$$\nabla \times \boldsymbol{E} = \frac{1}{A}\oint \boldsymbol{E} \cdot d\mathbf{l} = -\frac{d\boldsymbol{B}}{dt} \tag{2.71}$$

and for a magnetic field impinging on a lamina of width w and thickness d, with the field oriented normal to this area,

$$\frac{1}{A}\oint \boldsymbol{E} \cdot d\mathbf{l} = \frac{2(w+d)}{wd}\boldsymbol{E} \tag{2.72}$$

assuming that the $\boldsymbol{E}$ field is uniform throughout. If $w >> d$, then $2(w + d)/wd \cong 2/d$ so that

$$\boldsymbol{E} = -\frac{d}{2}\frac{d\boldsymbol{B}}{dt} \tag{2.73}$$

and remembering that $\boldsymbol{J} = \boldsymbol{E}/\rho$, where $\boldsymbol{J}$ is the current density and ρ is the resistivity

$$\boldsymbol{J} = -\frac{d}{2\rho}\frac{d\boldsymbol{B}}{dt} \tag{2.74}$$

From this, the power loss can be calculated from the integral of the product of current density $\boldsymbol{J}$ and electric field strength $\boldsymbol{E}$ over the volume

$$W_{ec} = \frac{1}{ld}\int_0^d\int_l^l \boldsymbol{J}\cdot\boldsymbol{E}\,dl\,dd = \frac{d^2}{12\rho}\left(\frac{d\mathbf{B}}{dt}\right)^2 \tag{2.75}$$

A similar analysis for a cylindrical conductor, with the applied field oriented along the unique axis, gives

$$W_{ec} = \frac{1}{\pi r_0^2}\int_0^{r_0} \boldsymbol{J}\cdot\boldsymbol{E}\,2\pi r\,dr = \frac{d^2}{32\rho}\left(\frac{d\boldsymbol{B}}{dt}\right)^2 \tag{2.76}$$

yielding a similar result, namely, that the power loss per unit volume depends on $(d\boldsymbol{B}/dt)^2$ and the cross-section d^2 while being inversely dependent on ρ.

In general, the classical eddy current loss can be expressed as

$$W_{ec} = \frac{d^2}{2\beta\rho}\left(\frac{d\boldsymbol{B}}{dt}\right)^2 \tag{2.77}$$

where β is a *shape factor*, which has the value $\beta = 6$ for laminations, $\beta = 16$ for cylinders, and $\beta = 20$ for spheres.

We will see in Chapter 12 that this equation forms the basis for determining one of the principal causes of power dissipation in alternating current electromagnetic energy conversion devices such as transformers and motors.

REFERENCES

1. Dirac, P. A. M. *Proc. Roy. Soc. Lond. A*, 133, 60, 1931.
2. Fleischer, R. L., Hart, H. R., Jacobs, I. S. et al. *J. Appl. Phys.*, 41, 958, 1970.
3. Trower, W. P. *IEEE Trans. Magn.*, 19, 2061, 1983.
4. Carrigan, R. A., and Trower, W. P. *Magnetic Monopoles*, Plenum, New York, 1983.
5. Brown, W. F., Frank, N. H., Kemble, E. C. et al. *Am. J. Phys.*, 18, 1, 1950.
6. Crangle, J. *The Magnetic Properties of Solids*, Edward Arnold, London, 1977, Ch. 1, pp. 13–15.
7. Vigoureux, P. *Units and Standards in Electromagnetism*, Wykeham Press, London, 1971.
8. Hopkins, R. A. *The International (SI) Metric System*, 3rd edn, AMJ Publishers, Tarzana, CA, 1975.
9. Giacolletto, L. J. *IEEE Trans. Magn.*, 10, 1134, 1974.
10. Graham, C. D. *IEEE Trans. Magn.*, 12, 822, 1976.
11. Brown, W. F. *IEEE Trans. Magn.*, 20, 112, 1984.
12. Osborne, J. A. *Phys. Rev.*, 67, 351, 1945.

13. Cullity, B. D. *Introduction to Magnetic Materials*, Addison-Wesley, Reading, MA, 1972, p. 62.
14. Chen, D. X., Brug, J. A., and Goldfarb, R. B. *IEEE Trans. Magn.*, 27, 3601, 1991.
15. Chen, D. X. *Proceedings of the Advances in Measurement Techniques and Instrumentation for Magnetic Properties Determination*, May, Ames, IA, 1994.
16. *Proceedings of the IEEE Workshop on Applied Magnetics*, Washington, DC. IEEE, New York, 1972.
17. *Proceedings of the IEEE Workshop on Applied Magnetics*, Milwaukee, WI. IEEE, New York, 1975.
18. Preis, K., Stogner, H., and Richter, K. R. *IEEE Trans. Magn.*, 17, 3396, 1981.
19. Saito, Y. *IEEE Trans. Magn.*, 18, 546, 1982.

FURTHER READING

Bennet, G. A. *Electricity and Modern Physics*, Arnold, London, 1968, Ch. 6.
Bradley, F. N. *Materials for Magnetic Functions*, Hayden, New York, 1971.
Lorrain, P., and Corson, D. R. *Electromagnetism*, W. H. Freeman, San Francisco, CA, 1978.
McKenzie, A. E. E. *A Second Course of Electricity*, 2nd edn, Cambridge University Press, Cambridge, Chs. 5 and 6, 1961.
Reitz, J. R., and Milford, F. J. *Foundations of Electromagnetic Theory*, 3rd edn, Addison-Wesley, Reading, MA, 1980, Ch. 9.
Rieger, H. *The Magnetic Circuit*, Heyden & Son, London, 1978.

EXERCISES

2.1 *Currents and poles.* Explain the *equivalent current* and *magnetic pole* models for describing magnetization in materials. Use these to derive an expression for the magnetic field produced by a simple linear magnetic dipole, such as a bar magnet, as a function of position from the center of the dipole.

2.2 *Demagnetizing field.* Discuss what is meant by the *demagnetizing field* in a magnetic material and explain why it should exist at all. What factors affect the strength of the demagnetizing field? Two cylinders of the material Terfenol, with constant relative permeability of $\mu_r = 5$ up to a saturation magnetization of $\boldsymbol{M}_s = 0.8 \times 10^6$ A m^{-1}, have length to diameter ratios of 5:1 and 2:1. Calculate the demagnetizing field in each when it is magnetized to saturation along the cylinder axis, and calculate the strength of the magnetic field needed to saturate the material in each of these two cases.

2.3 *Demagnetizing field calculation.* How strong is the magnetic field needed to magnetize an iron sphere to its saturation magnetization ($\boldsymbol{M}_s = 1.69 \times 10^6$ A m^{-1}), assuming that the field needed to overcome the demagnetizing field is much greater than the field needed to saturate the material in toroidal form.

2.4 *Demagnetizing effects at different field strengths.* A specimen of Terfenol has a length to diameter ratio of 8:1. At a field strength of $\boldsymbol{H} = 80$ kA m^{-1}, the magnetic induction is 0.9 T, while at a field strength of $\boldsymbol{H} = 160$ kA m^{-1}, the induction is 1.1 T. Calculate the internal magnetic field $\boldsymbol{H}_{in}$ in each case and compare with the applied field.

If we consider the fractional error in the observation of the uncorrected field, what do you conclude about this error at higher fields? Do demagnetizing effects become more or less of a problem as the applied field increases?

2.5 *Flux density in an iron ring with and without an air gap.* An iron toroid has a mean path length of 0.5 m with a cross-sectional area of 2×10^{-4} m^2. The number of turns on the coil wound on the toroid is $N = 800$ and this carries a current of 1 A. If the relative permeability of the iron is 1500, estimate the flux in the iron ring. An air gap of length 0.0005 m is then cut in the ring. Estimate the flux under these conditions. Calculate the current needed to restore the flux to its original value in the uncut toroid.

2.6 *Force between two permanent magnets.* In the situation depicted in Figure 2.6.1 express the force $\boldsymbol{F}$ between the two identical permanent magnets in terms of the properties of the magnets. The factors you may wish to consider may include some or all of the following: the magnetic field $\boldsymbol{H}$, flux magnetic induction $\boldsymbol{B}$, active surface area of the magnet Φ, the permeability the distance between the ends d, and/or any other parameters that apply. State any assumptions that you make.

2.7 *Magnetic field and magnetic flux density.* A helical coil of 500 turns is wound on a ring of demagnetized magnetic material. The ring of material has a mean diameter 0.1 m and a cross-sectional area of 1×10^{-5} m^2. When

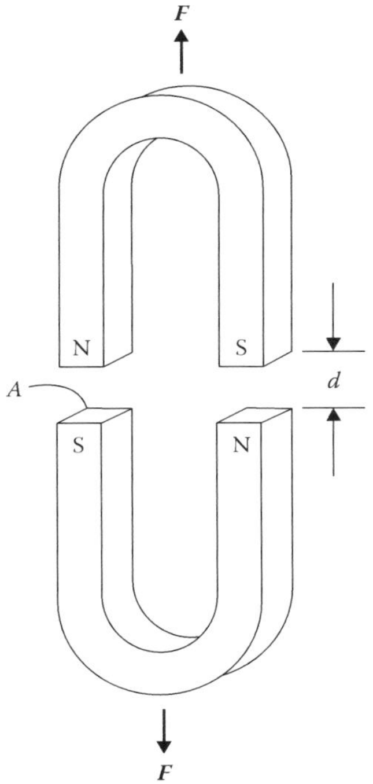

FIGURE 2.6.1 Two identical horseshoe magnets at a distance d attract each other with a force F.

a current of 1 A is passed through the coil, a flux change of 0.2×10^{-4} Wb is detected. Calculate the magnetic field $\boldsymbol{H}$, magnetic flux density $\boldsymbol{B}$, magnetization $\boldsymbol{M}$, and the permeability μ_r of the material under these conditions.

2.8 *Change in **B** caused by the demagnetizing field.* A piece of ferromagnetic material with constant relative permeability of 1000 is placed in a uniform magnetic field. The material is in the shape of a spheroid and is placed with its longest axis parallel to the field. What is the ratio of the flux density $\boldsymbol{B}$ inside the body to the original flux density outside the body when the axial ratio of the spheroid is (1) 1:1 (a sphere), (2) 5:1 (a prolate spheroid), and (3) 100:1 (also a prolate spheroid)?

2.9 *Divergence of **B**.* If the spatial components of the magnetic induction $\boldsymbol{B}$ are as follows:

$$\boldsymbol{B}_{\mathrm{x}} = axi$$

$$\boldsymbol{B}_{\mathrm{y}} = ay\,j$$

$$\boldsymbol{B}_{\mathrm{z}} = 0k$$

where a is a constant. Sketch the shape that the lines of $\boldsymbol{B}$ would take and explain why such an expression for the $\boldsymbol{B}$ field is unphysical.

2.10 *Demagnetizing field and the directions of **H** and **B**.* Explain what is meant by the demagnetizing field in a magnetized material. Under what circumstances is there no demagnetizing field? Show that inside a uniformly magnetized permanent magnet, $\boldsymbol{H}$ and $\boldsymbol{B}$ are always in opposite directions. Calculate the magnitudes and directions of $\boldsymbol{H}$, $\boldsymbol{B}$, and $\boldsymbol{M}$ inside a rod of iron with length to diameter ratio 5:1 that is magnetized to saturation.

3 Magnetic Measurements

This chapter is concerned with the various means of measuring the magnetic field, magnetic induction, or magnetization. There are several methods available for measuring the magnetic induction but these can be classified into three categories of measurement as follows. Those that depend on Faraday's law of electromagnetic induction to generate measurable voltages in coils; those that depend on Ampère's force law in one or other of its forms, and therefore measure force in order to determine the magnetic induction; and those that depend on measuring changes in properties of materials caused by the presence of a magnetic field. The measurements of magnetization are either force measurements such as in the torque magnetometer and torsion balance or difference measurements, such as the vibrating sample magnetometer (VSM), which measures the difference in magnetic induction with and without the sample present.

3.1 INDUCTION METHODS

How can the strength of an external field be measured from the voltage generated in an electrical circuit due to a change in flux linking the circuit?

The induction methods of measuring magnetic flux are all dependent on Faraday's law of electromagnetic induction, which has been discussed in Section 1.2.2. This states that the voltage induced in a circuit is equal to the rate of change of flux linking the circuit

$$V = -N\frac{d\Phi}{dt} \tag{3.1}$$

If A is the cross-sectional area of the coil and N is the number of turns, the magnetic induction is, then $\boldsymbol{B} = \Phi/A$.

$$V = -NA\frac{d\boldsymbol{B}}{dt} \tag{3.2}$$

Note that these coil methods measure the magnetic flux passing through the coil Φ, and from a knowledge of the cross-sectional area A, the magnetic induction $\boldsymbol{B}$ can be found. The induced voltage is increased if $\boldsymbol{B}$ is increased while $\boldsymbol{H}$ is maintained constant by inserting a high-permeability core into the coil. In free space, of course, $\boldsymbol{B} = \mu_0\boldsymbol{H}$ and so

$$V = -\mu_0 NA\frac{d\boldsymbol{H}}{dt} \tag{3.3}$$

3.1.1 Stationary Coil Methods

How can the rate of change of field be found using the electromotive force generated in a stationary coil?

A stationary coil method can only measure the rate of change of the magnetic induction by measuring the induced voltage. Such devices do have applications but if magnetic induction measurements are required, it is necessary to include some form of time integration [1]. Integrating voltmeter/fluxmeter devices are commercially available. These measure the magnetic induction from the relation

$$\boldsymbol{B} = -\frac{1}{NA}\int V\,dt \tag{3.4}$$

These instruments can be highly sensitive and are used extensively in hysteresis graphs for the measurement of the magnetic properties of soft magnetic materials [2]. These instruments work well but in setting them up, it is necessary to pay attention to the problem of drift, which can be adjusted using an offset voltage control. If this is not done, the fluxmeter will continue to integrate a small out-of-balance voltage with time, giving the impression of a linearly varying magnetic induction. This can be a problem for high-sensitivity measurement when the magnetic flux needs to be measured to better than about 10^{-10} Wb (0.01 maxwell).

3.1.2 Moving Coil (Extraction) Method

How is the magnetic induction measured when a search coil is placed in a magnetic field and rapidly removed?

As we know from Faraday's law of electromagnetic induction (Section 1.2.2), the induced voltage in a coil resulting from a change in flux linking the coil is given by

$$V = -NA\frac{d\boldsymbol{B}}{dt} \tag{3.5}$$

Integrating this gives

$$\int V\,dt = -NA\left(\boldsymbol{B}_{\mathrm{f}} - \boldsymbol{B}_{\mathrm{i}}\right) \tag{3.6}$$

where:

$\boldsymbol{B}_{\mathrm{i}}$ is the initial magnetic induction
$\boldsymbol{B}_{\mathrm{f}}$ is the final magnetic induction

Therefore, if a search coil is moved from the location where the field strength needs to be measured (e.g., between the poles of an electromagnet) to a region of *zero* field (e.g., outside the electromagnet), $\int Vdt$ will be proportional to the magnetic induction.

A ballistic (moving-coil) galvanometer can be used to measure the induction since, providing the time of oscillation of the ballistic galvanometer is long compared with the voltage pulse from the search coil, the angular deflection ϕ is proportional to $\int Vdt$. The device needs calibrating in order to determine the coefficient of proportionality.

$$\phi = \text{constant} \times \int V dt = \text{constant} \times NA\left(\boldsymbol{B}_\text{f} - \boldsymbol{B}_\text{i}\right) \tag{3.7}$$

The deflection of the galvanometer is therefore proportional to the change in magnetic induction. After calibration, this method is accurate to better than 1%.

3.1.3 Rotating Coil Method

Can we make use of a rotating coil to generate the necessary induced electromotive force in a static field?

In order to obtain a measurement of the magnetic induction, it is also possible to use various moving-coil instruments. The simplest of these is the rotating coil, which rotates at a fixed angular velocity ω. Under these conditions, the flux linking the coil is

$$\boldsymbol{B}(t) = \boldsymbol{B}\cos\omega t \tag{3.8}$$

and the voltage generated is

$$V = -NA\frac{d\boldsymbol{B}}{dt} = -\mu_0 NA\frac{d\boldsymbol{H}}{dt} = -\mu_0 NA\omega\boldsymbol{H}\sin\omega t \tag{3.9}$$

Therefore, the amplitude of the voltage generated by the rotating coil is proportional to the magnetic induction and therefore the amplitude can be used to measure $\boldsymbol{B}$ or $\boldsymbol{H}$ in free space. The signal can be read directly as an ac voltage or converted into a dc voltage, which is proportional to the amplitude. Typical inductions for this instrument range from 10 to 10^{-7} T. The electrical connections to the rotating coil include slip rings, which are a source of error in dealing with small voltages. The precision is of the order of one part in 10^4.

3.1.4 Vibrating Coil Magnetometer

How can the linear displacement of a coil in a magnetic field be used to measure the strength of the field?

The vibrating coil magnetometer [3,4] is no longer used to any great extent. It has been superseded by the VSM. The device is based on the same principles as the previous technique, but is used primarily as a method of determining the magnetization $\boldsymbol{M}$. The arrangement is shown in Figure 3.1.

The coil vibrates between the sample and a region of free space and thereby acts as a gradiometer by measuring the difference in induction in the two positions. While surrounding the sample the magnetic induction is

$$\boldsymbol{B}_\text{m} = \mu_0(\boldsymbol{H} + \boldsymbol{M}) \tag{3.10}$$

whereas the induction linking the coil when it has moved away from the sample is

$$\boldsymbol{B}_0 = \mu_0\boldsymbol{H} \tag{3.11}$$

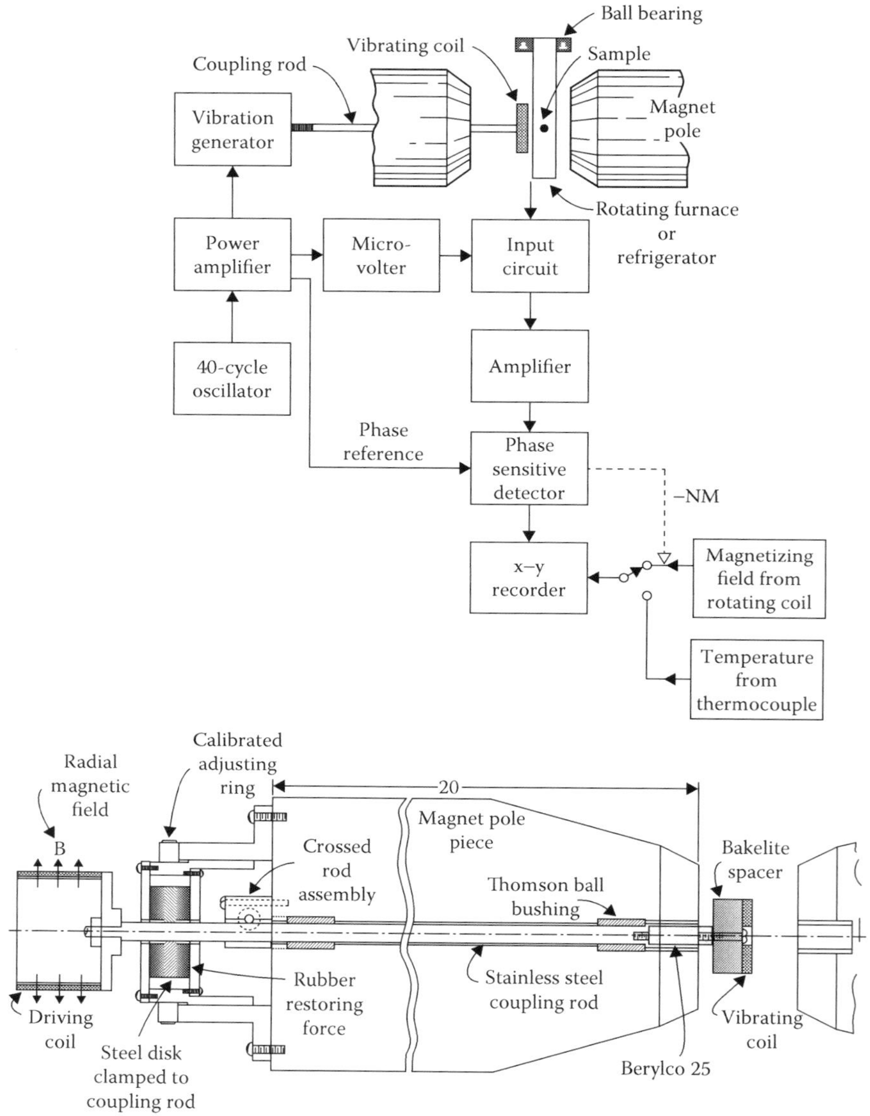

FIGURE 3.1 A schematic of a vibrating coil magnetometer.

The change in induction is then simply

$$\Delta \boldsymbol{B} = \mu_0 \boldsymbol{M} \tag{3.12}$$

The method therefore depends on the flux change caused when the coil is removed from the specimen

$$\int V dt = -NA\mu_0 \boldsymbol{M} \tag{3.13}$$

and consequently the output of the vibrating coil magnetometer is independent of $\boldsymbol{H}$, but is dependent on $\boldsymbol{M}$. This device is subject to noise caused by variation of the magnetic field $\boldsymbol{H}$ with time and is rarely used in situations in which it is possible to vibrate the sample instead of the coil as in the VSM (Section 3.1.5).

3.1.5 Vibrating Sample Magnetometer

If the specimen is moved instead of the coil, how can the induced voltage be used to determine the magnetization of the specimen?

The VSM is identical in principle to the vibrating coil magnetometer except that the sample is moved instead of the coil. The VSM was first described by Foner [5] and has now almost completely superseded the vibrating coil device.

A VSM is really a gradiometer measuring the difference in magnetic induction between a region of space with and without the specimen. It therefore gives a direct measure of the magnetization $\boldsymbol{M}$.

A schematic of a typical VSM is shown in Figure 3.2. The specimen in general has to be rather short to fit between poles of the electromagnet. The method is, therefore, in most cases not well suited to the direct determination of the intrinsic magnetization curve or hysteresis loop because of the demagnetizing effects associated with the short specimen. However, it is well suited for the determination of the saturation magnetization $\boldsymbol{M}_s$.

The detected signal, being an ac signal of fixed frequency, is measured using a lock-in amplifier. A reference signal is provided for the lock-in amplifier as shown in Figure 3.2 using a permanent magnet and a reference pick-up coil. Magnetic moments as small as 5×10^4 A m^2 (5×10^{-5} emu) are measurable with a VSM. Its accuracy is better than 2%. A description of the VSM has been given by Foner, who invented the instrument [6].

3.1.6 Fluxgate Magnetometers

How can the nonlinear magnetization characteristics of ferromagnets be used to determine external field strength?

Fluxgate magnetometers, also known as *saturable-core magnetometers*, were first developed for measurement of the Earth's magnetic field. There are several different forms of these fluxgates all based on the same basic principle of detecting asymmetry in response of the sense coils due to nonlinearity of their magnetic core in the presence of a magnetic field. The measured field is the dc or quasi-dc field along the direction of the sense coils.

The most usual configuration used for fluxgates is the two-core arrangement with the coils wound in opposition (the gradiometer configuration). We consider here one of the simplest types, the Forster gradiometer, which consists of two *identical* cores of high-permeability material, each wrapped with *identical* excitation coil but with the two coils wound in opposition to each other. These are each surrounded by a sense coils as shown in Figure 3.3.

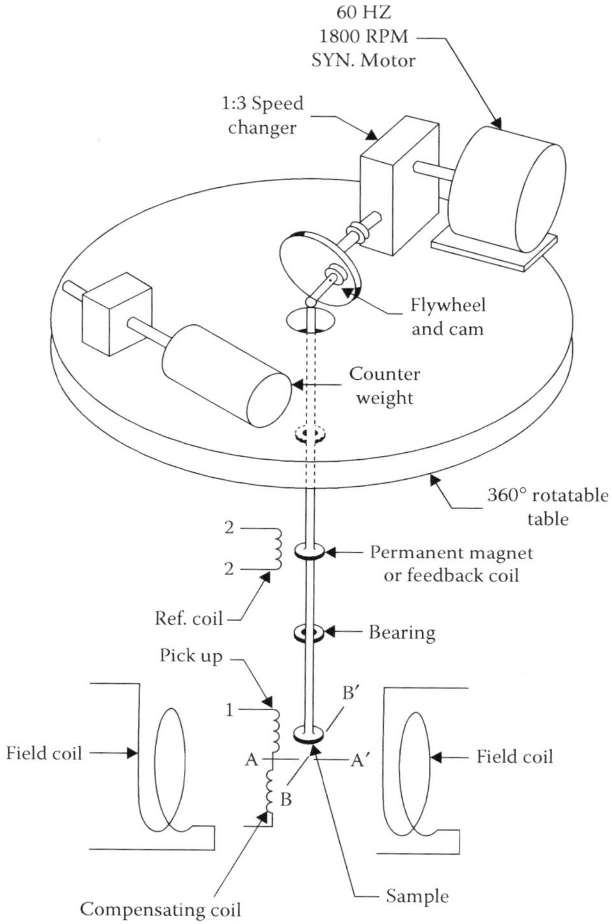

FIGURE 3.2 A schematic of a vibrating sample magnetometer.

The drive coil is excited with a periodic waveform, which saturates the magnetization, sending the core through an alternating cycle of magnetic saturation (i.e., magnetized—unmagnetized—oppositely magnetized—unmagnetized—magnetized). When the cores are not exposed to an applied magnetic field along their length, they are unbiased and therefore their output voltages are identical but in opposite directions. When subtracted, they cancel resulting in a null output.

When the cores are exposed to an external dc magnetic field along the direction of the sense coils, their responses are biased by the dc field. They will no longer traverse the same path on their hysteresis loops. Each core will be more easily saturated by the ac excitation in the direction aligned parallel with that of the dc bias field and less easily saturated when in opposition to it. So, the response of

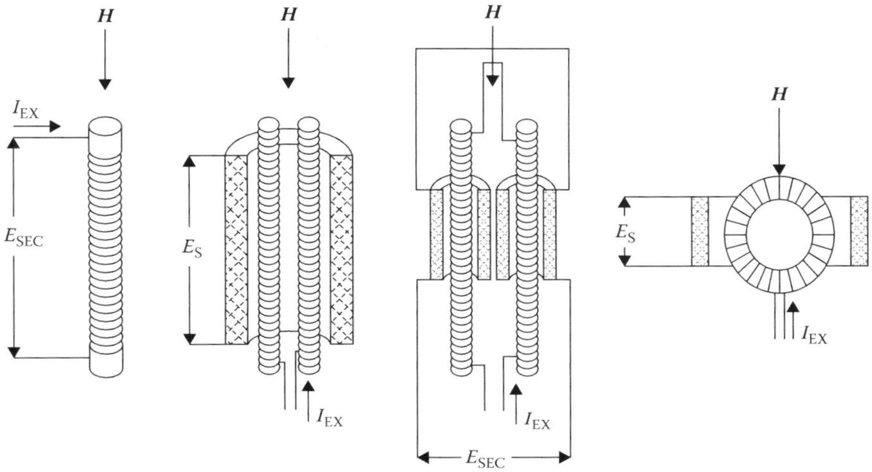

FIGURE 3.3 Diagrams of various forms of fluxgate magnetometer. These include the single-core configuration, the Vacquier gradiometer configuration with a single secondary coil, the Forster gradiometer configuration with separate secondary coils, and the Aschenbrenner and Goubau configurations with a toroidal core.

the sense coils to the ac excitation when added together is different in the presence of an applied magnetic field. This difference can be measured and calibrated to determine the field strength of the applied field along the direction of the cores.

The extent of the asymmetry of response of the two coils depends on the strength of the applied magnetic field and the nonlinearity of the magnetization curves of the cores. The degree of asymmetry of this voltage, as measured by the third harmonic component of the output voltage, can then be calibrated to measure the strength of the external field.

Fluxgates only measure the component of field strength parallel to the coils, since any field component perpendicular to this direction does not affect the symmetry of the sense coil output. The range of field strengths measurable by fluxgate magnetometers is down to 10^{-4} A m^{-1} (10^{-6} Oe). These magnetometers are mainly used for measuring magnetic inductions over the range of 10^{-10}–10^{-7} T, corresponding to magnetic fields over the range of 10^{-4}–10^{-1} A m^{-1} in free space. Reviews of fluxgate magnetometry have been given by Ripka [7,8].

3.2 FORCE METHODS

In this section, we look at four additional techniques: two earlier methods, the torque magnetometer, and the force balance method, which depend on the force on a magnetic dipole in a field and two later methods based on similar principles, the alternating gradient force magnetometer and the atomic force microscope.

3.2.1 Torque Magnetometers

How can the magnetic moment or magnetization be found from the torque exerted by a known external field?

The torque magnetometer is used mainly for anisotropy measurements on short specimens. It is based on the fact that a magnetic dipole $\boldsymbol{m}$ in an external magnetic field $\boldsymbol{H}$ in free space experiences a torque τ

$$\tau = \mu_0 \boldsymbol{m} \times \boldsymbol{H} \tag{3.14}$$

as described in Section 1.2.3. The torque must be measured in a uniform magnetic field. The specimen is aligned so that its magnetization lies in the plane of rotation of the field (if the field is to be rotated) or the sample is rotated in the plane defined by the magnetic field $\boldsymbol{H}$ and the magnetization $\boldsymbol{M}$.

A restoring torque is used to maintain the specimen in position. In some instruments, notably the earlier ones, this restoring torque was provided by twisting a torsion fiber. The angle θ through which the torsion fiber is twisted is dependent on the length of the fiber and its shear modulus, as well as the torque. ϕ is therefore proportional to the turning force on the specimen.

$$\theta = \text{constant} \times \tau = \text{constant} \times \mu_0 \boldsymbol{m} \times \boldsymbol{H} \tag{3.15}$$

If α is the angle between $\boldsymbol{m}$ and $\boldsymbol{H}$

$$\theta = \mu_0 \boldsymbol{m} \boldsymbol{H} \sin\alpha \times \text{constant} \tag{3.16}$$

The instrument is calibrated using a specimen of known anisotropy, and consequently since μ_0, $\boldsymbol{H}$, and the constant are known, the angle α can be measured and the magnetic moment $\boldsymbol{m}$ can be calculated.

A schematic diagram of a torque magnetometer is shown in Figure 3.4. In this instrument, the restoring torque is provided by the torque coil in the presence of a field from the permanent magnet, and the torque is proportional to the current in the torque coil.

3.2.2 Force Balance Methods

How can the magnetization be found from the force exerted on a specimen by a known field?

Two types of force balance methods have been devised for determining the magnetization $\boldsymbol{M}$ or equivalently the susceptibility χ. These are the analytical balance and the torsion balance. Both depend on measuring the linear force on a sample suspended in a magnetic field gradient.

The specimen is suspended using a long string in a magnetic field with a constant field gradient. The force on the specimen of volume V and magnetization $\boldsymbol{M}$ is then

$$\boldsymbol{F}_x = \mu_0 \boldsymbol{m} \frac{d\boldsymbol{H}}{dx} = -\mu_0 V \boldsymbol{M} \frac{d\boldsymbol{H}}{dx} \tag{3.17}$$

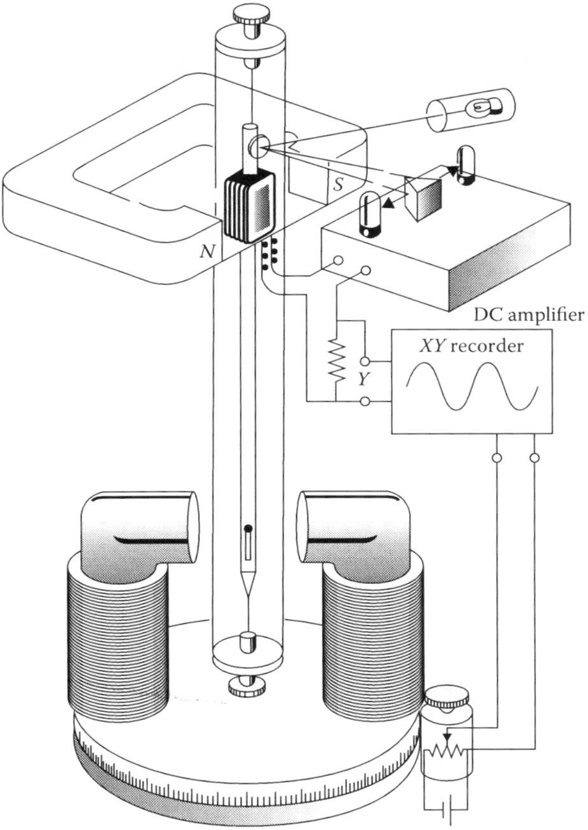

FIGURE 3.4 A schematic of a torque magnetometer.

which is obtained by differentiating the energy equation given in Section 1.2.6. Since we can write the susceptibility as

$$\chi = \frac{\boldsymbol{M}}{\boldsymbol{H}} \tag{3.18}$$

this leads to

$$\boldsymbol{F}_x = -\mu_0 \chi V \boldsymbol{H} \frac{d\boldsymbol{H}}{dx} \tag{3.19}$$

The force on the specimen is therefore proportional to its susceptibility.

This method can usually only be used on a short specimen, so that the whole specimen can be located between the pole pieces of an electromagnet. In addition, since the field must have a constant field gradient, the spatial extent of the specimen perpendicular to the field must also be small.

In the analytical balance method, in order to measure the force, the specimen is suspended by a string from one arm of an analytical balance (hence the name of the method). In zero field, the weight of the specimen is counterbalanced and when the field is switched on, the force on the specimen can be measured directly. A diagram of the experimental arrangement for this method is shown in Figure 3.5.

The torsion balance magnetometer is shown in Figure 3.6 and is a variation on the same linear force method. It can be designed to have a very small restoring torque

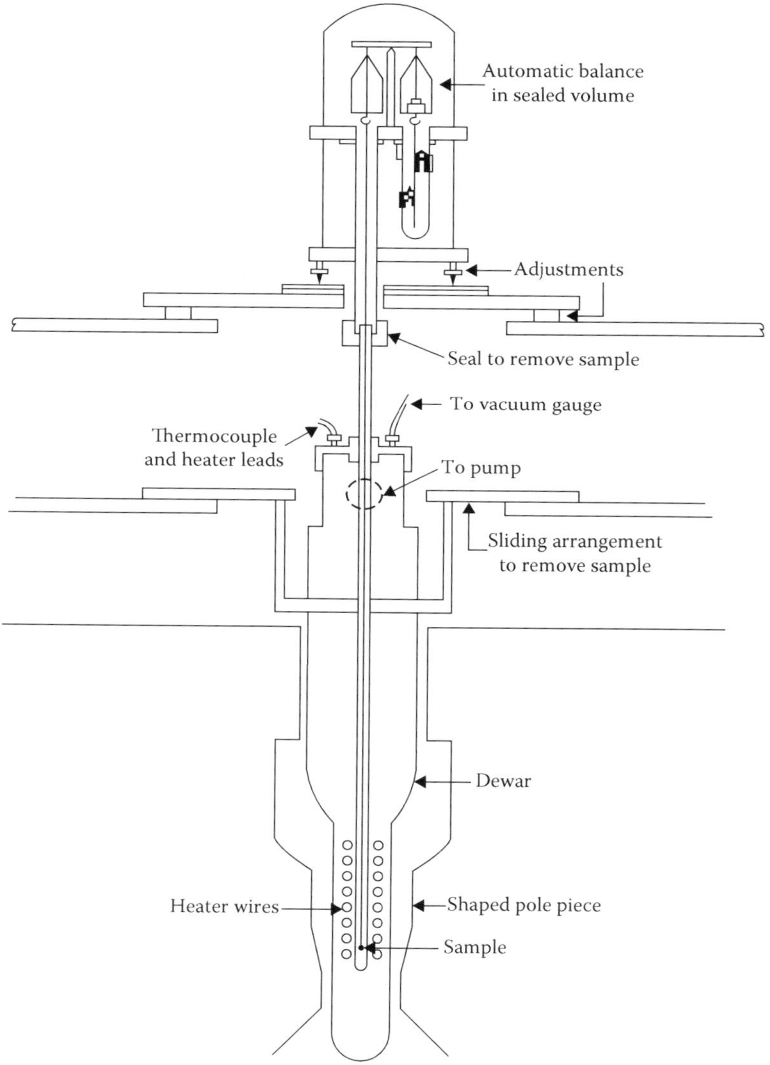

FIGURE 3.5 A schematic of an analytical balance magnetometer.

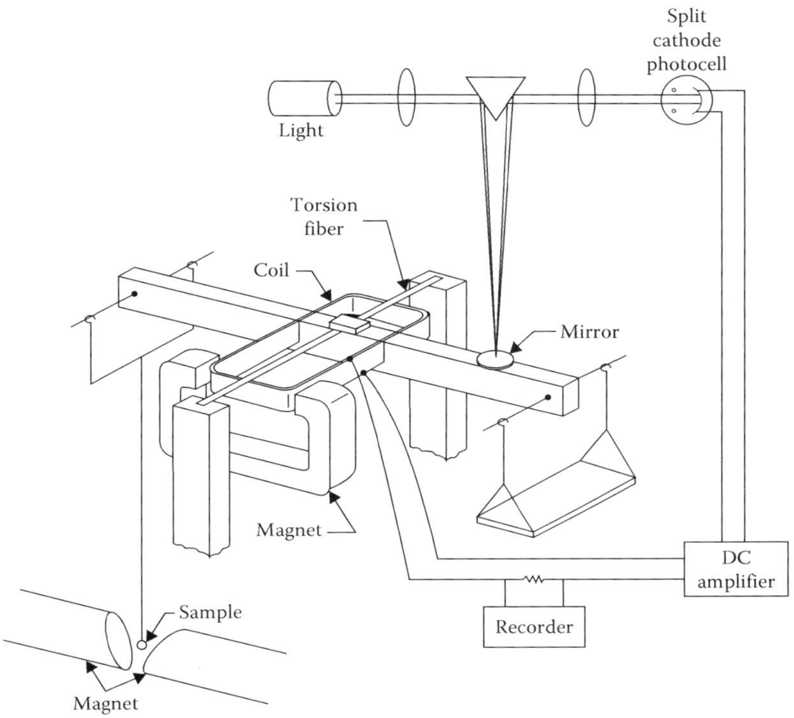

FIGURE 3.6 A schematic of a torsion balance magnetometer.

and therefore a high sensitivity; however, it cannot support such a large mass as the analytical balance and is limited to specimens of a few grams.

3.2.3 Alternating Gradient Force Magnetometer

What happens if the magnetic field gradient changes periodically with time?

Another more recent force technique for measuring magnetization is the alternating gradient force magnetometer (AGFM). The idea of this method is to utilize the fact that the net translational force on a magnetic dipole depends on the gradient of the magnetic field, and then by varying this gradient in a controlled way, for example, sinusoidally, to produce a time-varying force at the same frequency. This signal can then be measured with the advantages of noise and offset rejection that are possible by detecting only signals that occur at the known excitation frequency.

The AGFM, as introduced by Flanders [9], is in many respects a variant of the VSM described above. However, its sensitivity exceeds 10^{-11} A m^2 (10^{-8} emu) for measurement of magnetic moment, and is therefore typically 1000 times more sensitive than a conventional VSM. The output of the AGFM is obtained by measuring the amplitude of the motion of the specimen. This amplitude depends on the force experienced by the specimen, and if the field gradient is known, the magnetic moment can be calculated. The optimization of the coil design for the AGFM is identical to that of the VSM as discussed by Mallinson [10].

3.2.4 Magnetic Force Microscopy

Is it possible to use the force on a small magnetic dipole to scan the variation in magnetic induction across the surface of a material?

The magnetic force microscope is a derivative of the atomic force microscope, which first came to prominence in 1986 [11]. The magnetic variant was first reported by Martin and Wickramasinghe [12], in which a small magnetic dipole is attached to a flexible cantilever beam and the image is formed by scanning the probe in a raster fashion across the surface of the sample and measuring the magnetostatic force at different locations across the surface. The spatial resolution of this method is quite adequate for imaging bits of data stored on magnetic recording media and domain walls in a range of magnetic materials such as NdFeB, CoCr, Fe, and NiFe [13]. The spatial resolution of the current instruments is typically 50 nm. Theoretical analysis of the performance of the magnetic microscope has been reported by Proksch and Dahlberg [14] and by Gomez et al. [15,16].

3.3 METHODS DEPENDING ON CHANGES IN MATERIAL PROPERTIES

How can magnetic field strengths be determined from changes in material properties?

While in the previous sections, the measurement of magnetic field was dependent on the change in flux linking a circuit, or the force produced on a dipole by a magnetic field, in the following sections, the measurement depends on changes in the properties of materials under the action of a magnetic field.

3.3.1 Hall Effect Magnetometers

How does the presence of a magnetic field alter the motion of charge carriers?

The Hall effect magnetometers are perhaps the most versatile and widely used form of magnetometer. The range of fields measurable by these devices is typically 0.4 up to 4×10^6 A m^{-1} (equivalently 5×10^{-3} up to 5×10^4 Oe). Accuracy of measurements is typically 1%.

When a magnetic field is applied to a conducting material carrying an electric current, there is a transverse Lorentz force on the charge carriers given by

$$\boldsymbol{F} = \mu_0 e\boldsymbol{v} \times \boldsymbol{H} \tag{3.20}$$

as discussed in Section 1.2.3. The expression here is for the force on a single charge e with velocity $\boldsymbol{v}$ in a field of strength $\boldsymbol{H}$. Since the force on a charge e can be expressed as

$$\boldsymbol{F} = e\boldsymbol{E} \tag{3.21}$$

where $\boldsymbol{E}$ is the electric field, we can consider that the force is due to an equivalent electric field $\boldsymbol{E}_{\text{Hall}}$, known as the *Hall field*

$$\boldsymbol{E}_{\text{Hall}} = \mu_0 \boldsymbol{v} \times \boldsymbol{H} \tag{3.22}$$

and hence to a Hall electromotive force, V_{Hall} is in the direction perpendicular to the plane containing i and $\boldsymbol{H}$. The Hall electromotive force therefore depends linearly on the magnetic field $\boldsymbol{H}$ if the current is kept constant and this provides a very convenient measure of magnetic field $\boldsymbol{H}$, as the following analysis shows.

If the electric current passes in the x-direction and the magnetic field passes in the y-direction of a slab of semiconductor of dimensions l_x, l_y, and l_z, the Hall electromagnetive force will be along the z-axis as shown in Figure 3.7.

If n is the number of charge carriers per unit volume, then the current density will be

$$\boldsymbol{J} = ne\boldsymbol{v} \tag{3.23}$$

$$\boldsymbol{v} = \frac{\boldsymbol{J}}{ne} \tag{3.24}$$

and so the Hall field is

$$\boldsymbol{E}_{\text{Hall}} = \mu_0 \boldsymbol{J} \times \frac{\boldsymbol{H}}{ne} \tag{3.25}$$

and by replacing l/*ne* by the term R_{H}, which is called the *Hall constant*, we get

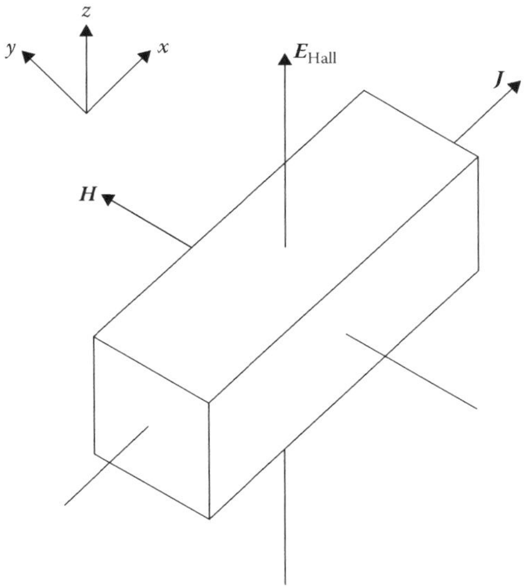

FIGURE 3.7 Generation of a Hall electromagnetive force in a slab of a conducting material. The electric current density $\boldsymbol{J}$ passes along the x-axis, the magnetic field $\boldsymbol{H}$ lies along the y-axis, and the Hall field $\boldsymbol{E}_{\text{Hall}}$ is generated along the z-axis. The sign of the Hall electromagnetive force depends on the sign of the charge carriers.

TABLE 3.1
Values of the Hall Coefficient for Various Materials

Material	R_H (m^3 C^{-1})
Li	-1.7×10^{-10}
In	$+1.59 \times 10^{-10}$
Sb	-1.98×10^{-9}
Bi	-5.4×10^{-7}

Note: The sign of R_H gives the sign of the charge carriers.

$$\boldsymbol{E}_{\text{Hall}} = \mu_0 R_H \boldsymbol{J} \times \boldsymbol{H} \tag{3.26}$$

and since the electric field $\boldsymbol{E}$ in volts per meter can be expressed by the form $\boldsymbol{E} = V/l_z$, where V is the potential difference over a distance l_z

$$V_{\text{Hall}} = \mu_0 R_H l_z \boldsymbol{J} \times \boldsymbol{H} \tag{3.27}$$

The value of the Hall coefficient R_H is typically 10^{-10} m^3 C^{-1}. Table 3.1 presents the values of the Hall coefficient for various materials.

The Hall effect magnetometers can be fabricated with very small active areas, down to 10^{-6} m^2, which can therefore be used to measure the magnetic field with high spatial resolution. Another important factor is that unlike coils, which measure the flux linkage and therefore need to be scaled appropriately for their cross-sectional area in order to determine the magnetic induction, the Hall magnetometers measure the field strength directly.

The only significant difficulties with Hall probes arise from the temperature dependence of the response. Most commercial Hall probes are made of InAs or InSb.

3.3.2 Magnetoresistors

How does the presence of a magnetic field alter the resistance of a material?

Magnetoresistance is the change in electrical resistance of a material when subjected to a magnetic field. Once the variation of resistance with field is known, then resistance measurements can be made in order to determine the field strength. Generally, the resistance increases when a field is applied but is nonlinear. The main advantage of this method is that very small probes can be fabricated to measure the field at a point. Magnetoresistive probes are particularly useful for field measurements at low temperatures.

In all materials, the effect of magnetic field on resistance is greater when the field is perpendicular to the direction of current flow. In ferromagnetic materials, the change in resistance can be, typically $\Delta R/R = 2\%$ at saturation in nickel and 0.3% at saturation in iron.

The earliest application of this effect as a field-measuring device was the use of bismuth in which the magnetoresistance increases by 150% in a field of 9.5×10^5 A m^{-1} (1.2 T). Materials that are rather more sensitive have been found such as the eutectic compound of InSb-NiSb, which undergoes a 300% change in magnetoresistance in a field of 2.3×10^5 A m^{-1} (0.3 T). Unfortunately, the effects of temperature on the resistance of this material are also large and this limits its applications.

Despite these drawbacks, the measurement of resistance is fairly simple and hence the method does have great merit in situations where the temperature can be well controlled (e.g., cryogenic applications) and where the range of flux density is limited to less than 16 T but greater than 2 T. The accuracy of field measurements made with this method is about 1%.

There are a number of thin-film magnetoresistive devices available commercially that are capable of a resolution of 10^{-5} A m^{-1} (10^{-7} Oe). These devices look very much like strain gauges, being flat plates of typically 5 × 5 mm with two terminals attached.

3.3.3 Magnetostrictive Devices

Can the change in length of a ferromagnet in the presence of a field be used to measure that field?

When a specimen of magnetic material is subjected to a magnetic field, changes in the shape of the specimen occur [17]. This phenomenon is known as *magnetostriction* and it is most often demonstrated by measuring the fractional change in length $\Delta l/l$ of a specimen as it is magnetized. The effect is quite small in most materials, but in ferromagnets, it is typically of the order of $\Delta l/l = 10^{-6}$, which is measurable by resistive strain gauges or by optical techniques. Recent materials with much higher magnetostrictions have been discovered [18], with $\Delta l/l$ up to 2500×10^{-6}.

The magnetostriction and magnetoresistance are closely related, both being generated by the spin orbit coupling, so that the electron distribution at each ionic site is rotated. This change in electron distribution alters the scattering undergone by the conduction electrons (magnetoresistance). The rotation of the moments also leads to a change in the ionic spacing (magnetostriction).

If the magnetostriction of a material as a function of field is known, this can be used as a measurement of the magnetic field. High-resolution measurements of magnetostriction down to $dl/l \approx 10^{-10}$ can now be made with magnetostrictive amorphous ferromagnetic materials [19]. The drawback with this is that the magnetostriction is nonlinear, and furthermore exhibits hysteresis.

3.3.4 Magneto-Optic Methods

How can the changes in optical properties of media under the action of an external field be used to determine field strength or magnetization?

The two principal magneto-optic effects are the Faraday effect, which occurs when light is transmitted through a transparent medium in the presence of a magnetic field along the direction of propagation of the light, and the Kerr effect, which occurs when light is reflected from a ferromagnetic medium. Both involve rotation of the angle of polarization of linearly polarized light [20]. Another phenomenon, the Cotton–Mouton effect, is related to the Faraday effect, and occurs when the magnetic field is perpendicular to the direction of propagation of light transmitted in the medium. However, the magnitude of the rotation observed in the Cotton–Mouton effect is much smaller than in the Faraday effect.

The Faraday effect is easily adapted as a technique for measurement of field strength since the rotation of polarization of light passing through a transparent paramagnetic material [such as $MgCe(PO_4)_2$] can give a measure of the local magnetic field. The rotation of the plane of polarization is given by

$$\theta = V \boldsymbol{H} t \tag{3.28}$$

where:
- V is the Verdet constant (V = 0.001 to 0.1 min A^{-1}, or equivalently 0.1–10 min $Oe^{-1}\ cm^{-1}$)
- $\boldsymbol{H}$ is the field strength
- t is the thickness of the specimen or more precisely the path length of light in the material

In ferromagnetic or ferrimagnetic materials, the angle of rotation θ can also be related to the magnetization $\boldsymbol{M}$ by

$$\theta = K \boldsymbol{M} t \tag{3.29}$$

where:
- K is Kundt's constant
- $\boldsymbol{M}$ is the magnetization of the material
- t is the path length

Both the Faraday effect and the Cotton–Mouton effect can be used for domain observation. However, these techniques are limited to thin sections of ferromagnetic materials in order that sufficient light is transmitted.

The Kerr effect is used to observe domain patterns in ferromagnetic materials and is discussed in more detail in Chapter 6. A beam of linearly polarized light is incident on the surface of the specimen. It is important that the beam is not normal to the surface of the ferromagnet or ferrimagnet since there must be a component of the magnetization in the surface or the specimen parallel to the direction of propagation of the light, and it is well known that surface domains are oriented with their magnetizations in the plane of the surface. The magnetization within a domain rotates the plane of polarization of the beam that is reflected from the surface through the angle θ, which is related to the magnetization $\boldsymbol{M}$ by

$$\theta = K_r \boldsymbol{M} \tag{3.30}$$

Typical rotations are 9 min of arc from saturated nickel and 20 min of arc from saturated iron and cobalt. This technique can only be used for the determination of field strength in situations where the magnetization can be directly related to the field.

3.3.5 Thin-Film Magnetometers

Can the presence of magnetic anisotropy be used to measure external field strength?

Magnetic fields can be measured using thin-film magnetometers [21]. Thin films that are in the range of 200–5000 Å thick are usually fabricated for these purposes from a nonmagnetostrictive alloy such as Ni-20%Fe. They have uniaxial anisotropy with the easy axis parallel to the direction of applied field during deposition.

Magnetization measurements are made along the hard axis while the field being measured is applied along the easy axis. Since the applied field effectively alters the anisotropy, this leads to changes in the magnetization characteristics. The film is used as the inductor in the frequency-controlling circuit of an oscillator. The output frequency is then a function of the external field. The field ranges over which the thin-film devices are useful are typically from 10^{-7} to 10^{-3} T [22].

3.3.6 Magnetic Resonance Methods

How can we utilize the magnetic properties of elementary particles to determine field strength?

Resonance magnetometers include all magnetic field measurement techniques based on electron spin resonance, nuclear magnetic resonance, and proton procession. A review of these has been given by Seiden [23]. The sensitivities of these instruments can be of the order of 10^{-14} T (10^{-10} gauss). These methods have the advantage of measuring the total magnetic field in a region of space. That is, they are not dependent on the orientation of the field for measurement, as are most other techniques, which can only measure the component of magnetization along a given direction.

The discrete energy levels of electrons in materials are changed by the presence of a magnetic field. This was first discovered by Zeeman [24,25]. In electron spin resonance (ESR), electrons can be excited from one state to another by high-frequency radiation and hence will exhibit absorption or resonance at those characteristic frequencies.

The resonance frequency v_0 can be used as a measure of the magnetic field strength since it is related to the field by the expression

$$\omega_0 = 2\pi v_0 = \gamma \boldsymbol{B} = \gamma \mu_0 \boldsymbol{H} \tag{3.31}$$

where γ is known as the *gyromagnetic ratio*. The value of γ for a free electron (as in electron paramagnetic spin resonance) is 1.762×10^{11} rad s^{-1} T^{-1} (or 28.043×10^9 Hz T^{-1}) and by this method very weak fields may be measured.

The value of γ for an electron is related to its fundamental properties as we shall discuss in Chapter 10. From a classical perspective, the gyromagnetic ratio is related to the electron charge e and the electron mass m_e by the equation $\gamma = -e/(2m_e)$. In the quantum description of electrons, it is found that $\gamma = -2\boldsymbol{m}/(h/2\pi) = -\mu_B g/(h/2\pi)$, where $\boldsymbol{m}$ is the magnetic moment of the electron, μ_B is a Bohr magneton (9.274×10^{-24} A m^2), g is the Lande splitting factor of the electron ($g = 2$ for a free electron), and h is Planck's constant (6.626×10^{-34} J s).

The nuclear magnetic resonance technique depends on resonance of the magnetic moment of the nucleus in a radio frequency field in a similar way to ESR. The energy levels of the nucleus are quantized and are altered by the presence of a magnetic field as in ESR. Therefore, resonant absorption is observed when the r.f. energy equals the difference in energy level between these quantized states. The resonant frequency is therefore a measure of the field strength.

Once again, as in ESR, the resonant frequency is proportional to the field; however, the gyromagnetic ratio γ_n of the particular nucleus must be used in the following equation:

$$\omega_0 = \gamma_n \mu_0 \boldsymbol{H} \tag{3.32}$$

The material used as the medium in this method need not be very special; in fact, water is often used. In this case, the nuclei are protons for which $\gamma_n = 2.675 \times 10^8$ rad s^{-1} T^{-1} (or 42.579×10^6 Hz T^{-1}). Values of γ for various nuclei and for free electrons are given in Table 3.2.

The experimental setup for these techniques for field measurement involves two mutually orthogonal pairs of coils: the r.f. coils, which emit the resonant radio frequency, and the receiver coils, which indicate when resonance occurs. The resonance methods typically have a precision of 1 part in 10^6. This means that highly accurate field measurements can be made under the right conditions, but because of the high precision there is a need for a highly homogeneous field throughout the specimen being used.

In practice, when these methods are used for field measurement, as distinct from material property measurements, there are two ways of deducing the field. In one method, the field is calculated from the resonance frequency with γ already known; in the other method, the field is determined from an amplitude measurement with ω fixed and γ known. Under small-amplitude field modulations $\Delta\boldsymbol{H}$ of the quasi-dc (*static*) field close to the resonance $\boldsymbol{H}_0$, the voltage response of the pickup coil ΔV is

$$\Delta V = K \Delta \boldsymbol{H} \tag{3.33}$$

where K is a constant of the material, which must be known. In this way, $\Delta\boldsymbol{H}$ can be measured to a high degree of accuracy [26].

A closely related technique based on proton precession is widely used by geophysicists for precise measurements of the local strength of the Earth's field. Proton

TABLE 3.2
Values of the Gyromagnetic Ratio γ for Various Nuclei and for Free Electrons

Particle	γ (rad s^{-1} T^{-1})	$\gamma/2\pi$ (Hz T^{-1})
^{1}H (proton)	2.6753×10^8	42.579×10^6
^{2}D (deuteron)	4.1064×10^7	6.536×10^6
^{7}Li	1.0396×10^8	16.546×10^6
Free electrons	1.762×10^{11}	28.043×10^9

precession magnetometers are used for flux densities in the range of 10^{-9}–10^{-4} T. A container of water is placed in a coil perpendicular to the field to be measured. This coil is then pulsed to align the magnetic moments of the protons. When it is switched off, the nuclear moments of the protons precess about the field being measured. This precession generates an electromagnetive force in the coil at the precession frequency. Measurement of this frequency enables the field strength to be calculated.

3.4 SUPERCONDUCTING QUANTUM INTERFERENCE DEVICES

What is a superconducting quantum interference device (SQUID) and how can it be used to measure a magnetic field?

At present, SQUIDs provide the ultimate in resolution for field measurements. The SQUID consists of a superconducting ring with a small insulating layer known as the *weak link*, as shown in Figure 3.8. The weak link is also known as the *Josephson junction*. The resolution of these devices is below 10^{-14} T (10^{-10} gauss) [27]. The flux passing through the ring is quantized once the ring has gone superconducting but the weak link enables the flux trapped in the ring to change by discrete amounts. Changes in the pickup voltage occur as the flux is incremented in amounts of $\Delta\Phi = 2.067 \times 10^{-15}$ Wb. The device can thereby be used to measure very small changes in flux. In fact, it can be used to count the changes in flux quanta in the ring.

With no weak link flux can not enter the ring, as we know from superconductivity, and so the field passing through the ring remains at the value it was when the ring became superconducting. If the link is very thick, so that no supercurrent can flow, then the flux in the ring will be exactly that which is expected from the applied field. The presence of the weakly superconducting link typically restricts the value of the supercurrent flowing in the ring to less than 10^{-5} A. Therefore, with a weak link some magnetic flux can enter the ring. The supercurrent in the weak link tries to oppose the entry of flux but because it is limited by the weak link, it cannot achieve

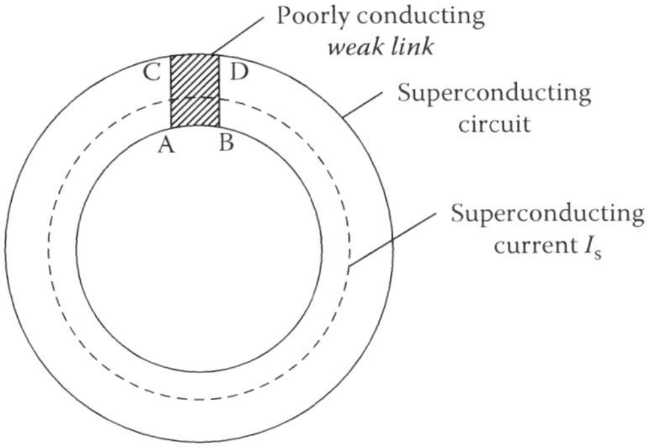

FIGURE 3.8 A Josephson junction device, which consists of a superconductor with a poorly conducting *weak link*, ABCD.

this entirely as the flux is increased. It therefore becomes a periodic function of the flux threading the superconducting ring.

The relation between the flux density in the ring and the flux density due to the applied field is

$$\Phi = \Phi_a + LI_s \tag{3.34}$$

where:

Φ is the flux passing through the ring
Φ_a is the flux due to the applied field
L is the inductance of the ring
I_s is the superconducting current, which produces a flux of $\Phi_a + LI_s$

In the Josephson junction, the superconducting current in the ring is related to the critical current I_c determined by the properties of the weak link

$$I_s = I_c \sin\theta \tag{3.35}$$

where θ is the phase difference of the electron wave functions across the weak link. Therefore,

$$\Phi = \Phi_a + LI_c \sin\theta \tag{3.36}$$

In a completely superconducting ring, the flux is an integral number of flux quanta. Therefore, if Φ_0 is the flux quantum of 2.067×10^{-15} Wb

$$\Phi = N\Phi_0 \tag{3.37}$$

where N is an integer. With the weak link the phase angle θ across the link depends on the flux in the following way:

$$\theta = 2\pi N - 2\pi\left(\frac{\Phi}{\Phi_0}\right) \tag{3.38}$$

Since N is an integer

$$\sin\theta = \sin\left(-2\pi\frac{\Phi}{\Phi_0}\right) = -\sin\left(2\pi\frac{\Phi}{\Phi_0}\right) \tag{3.39}$$

and therefore,

$$\Phi = \Phi_a - LI_c \sin\left(2\pi\frac{\Phi}{\Phi_0}\right) \tag{3.40}$$

and the relation between Φ and Φ_a is given in Figure 3.9. From this graph, we see that the SQUID counts flux quanta of the applied field in units of 2.067×10^{-15} Wb. Each time $\sin(2\pi\Phi/\Phi_0) = 0$, then Φ/Φ_0 and Φ_a/Φ become equal; however, at values in between when the flux is not an integral multiple of 2.067×10^{-15} Wb, they are

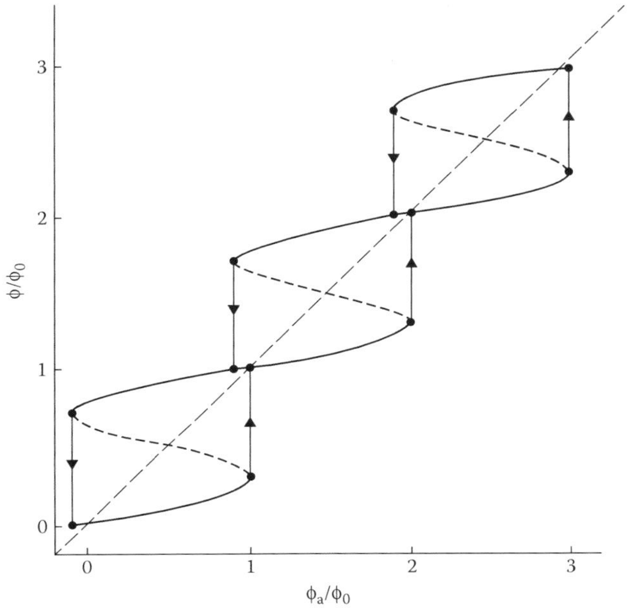

FIGURE 3.9 The relation between ϕ, the flux in the ring, and ϕ_a, the flux due to the applied field, in a SQUID magnetometer. These are normalized by the flux quantum ϕ_0.

not equal. If a loop of wire or a coil is placed around the superconducting ring, then a voltage pulse is induced in the coil at each quantum jump, and this pulse can be used to measure the applied field.

The SQUID is a highly sensitive device and is therefore really best suited to measuring very small changes in magnetic field.

REFERENCES

1. DeMott, E. G. *IEEE Trans. Magn.*, 6, 269, 1970.
2. Gordon, D. I., Brown, R. E., and Haben, J. F. *IEEE Trans. Magn.*, 8, 48, 1972.
3. Smith, D. O. *Rev. Sci. Inst.*, 27, 261, 1956.
4. Fert, C., and Gautier, P. *C. R. Acad. Sci.*, Paris, France, 233, 148, 1951.
5. Foner, S. *Rev. Sci. Inst.*, 30, 548, 1959.
6. Foner, S. *J. Appl. Phys.*, 79, 4740, 1996.
7. Ripka, P. Review of fluxgate sensors, *Sensor Actuat. A*, 33, 129, 1992.
8. Ripka, P. Advances in fluxgate sensors, *Sensor Actuat. A*, 106, 8, 2003.
9. Flanders, P. J. *J. Appl. Phys.*, 63, 3940, 1988.
10. Mallinson, J. C. *IEEE Trans. Magn.*, 27, 4398, 1991.
11. Binnig, G., Quate, C. F., and Gerber, C. *Phys. Rev. Letts.*, 56, 930, 1986.
12. Martin, Y., and Wickramasinghe, K. *Appl. Phys. Letts.*, 50, 1455, 1987.
13. Hartmann, U. et al. *IEEE Trans. Magn.*, 26, 1512, 1990.
14. Proksch, R., and Dahlberg, E. D. *J. Magn. Magn. Mater.*, 104, 2123, 1992.
15. Gomez, R., Adly, A. A., Mayergoyz, I. D., and Burke, E. R. *IEEE Trans. Magn.*, 29, 2494, 1993.

16. Gomez, R., Burke, E. R., and Mayergoyz, I. D. *J. Appl. Phys.*, 79, 6441, 1996.
17. Cullity, B. D. *J. Metals*, 23, 35, 1971.
18. Clark, A. E., and Abbundi, R. *IEEE Trans. Magn.*, 13, 1519, 1977.
19. Wun Fogle, M., Savage, H. T., and Clark, A. E. *Sensor Actuat.*, 12, 323, 1987.
20. Craik, D. J., and Tebble, R. S. *Ferromagnetism and Ferromagnetic Domains*, North Holland, Amsterdam, the Netherlands, 1965.
21. Irons, H. R., and Schwee, L. J. *IEEE Trans. Magn.*, 8, 61, 1972.
22. Bader, C. J., and Fussel, R. L. *Proceedings 1965 Intermag Conference*, New York, 1965.
23. Seiden, P. E. in *Magnetism and Metallurgy*, Vol. 1 (eds. A. E. Berkowitz and E. Kneller), Academic Press, New York, 1969.
24. Zeeman, P. *Phil. Magn.*, 43, 226, 1897.
25. Herzberg, G. *Atomic Spectra and Atomic Structure*, 2nd edn, Dover, New York, 1944, p. 96.
26. Hartman, F. *IEEE Trans. Magn.*, 8, 66, 1972.
27. Webb, W. *IEEE Trans. Magn.*, 8, 51, 1972.

FURTHER READING

Berkowitz, A. E., and Kneller, E. *Magnetism and Metallurgy,* Vol. 1, Academic Press, New York, 1969, Ch. 4.

Crangle, J. *The Magnetic Properties of Solids*, Edward Arnold, London, 1977.

Cullity, B. D. *Introduction to Magnetic Materials*, Addison-Wesley, Reading, MA, 1972.

Janicke, J. Magnetic measurements, in *IEEE Workshop on Applied Magnetics*, IEEE, Milwaukee, WI, 1975.

EXERCISES

3.1 *Search coil method.* Explain how the magnetic induction in a region of space (e.g., between the poles of an electromagnet) can be measured using the search-coil method. Describe the law on which this method is based. What is the quantity of charge induced in a coil of 50 turns in a circuit of total resistance 250 Ω due to a flux change of 3 m Wb?

3.2 *Dipole oscillations.* Define the magnetic moment of a dipole and state the factors that determine the rate of oscillation of a suspended dipole that is free to rotate in a magnetic field. A long solenoid is placed horizontally with its axis pointing north–south. A small dipole at its center makes 20 oscillations per minute when a current is passed through the solenoid, and 6 oscillations per minute when the Earth's field alone acts. If the solenoid has 1500 turns m^{-1}, calculate the possible values of the current passing through it. (You may assume that the horizontal component of the Earth's field is 14 A m^{-1}.)

3.3 *Torque magnetometer.* Explain how to measure the strength of a magnetic field $\boldsymbol{H}$ using a magnetic dipole of known strength. Define the magnetic moment of a coil and a bar magnet and show that $\boldsymbol{m} = \phi l/\mu_0$. A magnetized rod of magnetic moment 0.318 A m^2 is suspended horizontally in a magnetic field of 14 A m^{-1}. What is the torque required to keep it at an angle of (1) 90° and (2) 30° to the field direction?

3.4 *Magnetic resonance.* Describe the phenomenon of electron spin resonance and explain briefly why it occurs. Explain the following:

a. Bohr magneton
b. Gyromagnetic ratio
c. Lande splitting factor

A dilute paramagnetic material has a ground state $S = 1/2$. Calculate the gyromagnetic ratio γ and hence determine the expected resonance frequency in a field of 0.796×10^5 A m^{-1} (equivalent to a free-space magnetic induction of 0.1 T).

3.5 *Induction coil method.* The initial magnetization curve of a specimen of iron of length 20 cm and diameter 0.5 cm is measured using a ballistic galvanometer. The galvanometer has a sensitivity of 0.17×10^{-4} Wb turns mm^{-1}, and the solenoid used to generate the field gives a $\boldsymbol{H}$ field of 400 A m^{-1} for each ampere flowing in the coil.

A 40 turn induction coil is wound on the specimen to measure the flux passing through it. The current reversed in the solenoid i and the deflection d were as in Table 3.5.1.

a. Find the demagnetization factor N_d.
b. Plot the initial magnetizing curves $\boldsymbol{B}$ against $\boldsymbol{H}_{app}$ and $\boldsymbol{B}$ against $\boldsymbol{H}_{in}$.
c. Find the true and apparent permeability at $\boldsymbol{B} = 10$ T.

3.6 *Hall effect.* Explain the Hall effect and show that the direction of the Hall field can be used to determine the sign of the charge carriers in a material. The Hall coefficient of indium is 1.6×10^{-10} m^3 C^{-1}. Determine whether the conduction takes place by holes or electrons in this material and how many electrons or holes per atom take part in electrical conduction. (Assume the atomic mass is 115 and the density 7280 kg m^{-3}.)

3.7 *Hall effect sensor.* Explain what is meant by the Hall effect and how this can be used either to measure magnetic fields or to give information about the charge carrier density and type. A Hall probe is needed in the form of a cube of side 1 mm to measure magnetic fields. If the required sensitivity is 1 V T^{-1} and the charge carrier mobility is 0.5 m^2 V^{-1} s^{-1}, calculate the

TABLE 3.5.1
Deflection of the Galvanometer *d* in mm for Different Values of the Reversal Current *i* in Amperes

i (A)	*d* (mm)
1.5	24.0
3.1	49.2
4.9	77.6
8.5	103.7
11.0	107.5
12.7	109.1

necessary charge carrier density for the Hall probe material to meet this requirement. Why do we not use metals for Hall effect sensors?

3.8 *Susceptibility balance.* Describe how the translational force on a magnetic material is related to the magnetic field and the properties of the material. The mass of a sample of paramagnetic material is being determined using a conventional force balance (i.e., by measuring its weight). However, in this case, there is a magnetic field of strength 2×10^6 A m^{-1} present with a vertical field gradient of 1×10^7 A m^{-2}. If the susceptibility of the material is 2×10^{-4}, calculate the force on the sample and hence the error in the measured mass.

3.9 *Lorentz force and cyclotron resonance.* A charged particle with charge e and mass m is projected with velocity $\boldsymbol{v}$ at right angles to a uniform magnetic induction $\boldsymbol{B}$. What are the resulting orbital magnetic dipole moment and the orbital angular momentum that are acquired by the particle as a result of its subsequent motion? If the particle is an electron with mass $m = 9.1 \times 10^{-31}$ kg, charge $e = 1.6 \times 10^{-19}$ C, velocity $v = 10$ km^{-1}, and the magnetic induction $\boldsymbol{B} = 10$ T, give the numerical values of the resulting orbital magnetic dipole moment and the orbital angular momentum. (Ignore the spin angular momentum of the electron for this calculation.)

3.10 *Flux measurement using coil and integrating voltmeter.* A toroid of magnetic material with a cross-sectional area of 1 cm^2 has a saturation magnetization of $\boldsymbol{M} = 0.875 \times 10^6$ A m^{-1} in a field of 1000 A m^{-1}. The flux density will be measured by a fluxmeter (integrating voltmeter). What should be the number of turns on the flux detecting coil to ensure an output signal of $\Delta V = 0.5$ V as the ring sample is driven from positive saturation to negative saturation in 1 s?

4 Magnetic Materials

The macroscopic behavior of magnetic materials can be classified using a few magnetic parameters. We look at the most significant of these, give some definitions, and show how the most important class of magnetic materials, the ferromagnets, can be classified on this basis. We then survey the main uses of ferromagnets and indicate how the macroscopic properties determine the suitability of a material for a given application.

4.1 CLASSIFICATION OF MAGNETIC MATERIALS

4.1.1 Diamagnets, Paramagnets, and Ferromagnets

How are the different types of magnetic materials classified?

The different types of magnetic materials are traditionally classified according to their bulk susceptibility. The first group are materials for which χ is small and negative $\chi \approx -10^{-5}$. These materials are called *diamagnetic*; their magnetic response opposes the applied magnetic field. Examples of diamagnets are copper, silver, gold bismuth, and beryllium. Superconductors form another special group of diamagnets for which $\chi \approx -1$.

A second group of materials for which χ is small and positive and typically $\chi \approx 10^{-3}$–10^{-5} are the paramagnets. The magnetization of paramagnets is weak but aligned parallel with the direction of the magnetic field. Examples of paramagnets are aluminum, platinum, and manganese.

The most widely recognized magnetic materials are the ferromagnetic solids for which the susceptibility is positive, much greater than 1, and typically can have values $\chi \approx 50$ to 10,000. Examples of these materials are iron, cobalt, nickel, and several rare earth metals and their alloys.

4.1.2 Susceptibilities of Diamagnetic and Paramagnetic Materials

What are typical values of μ and χ in diamagnets and paramagnets?

At constant temperature and for relatively low values of magnetic field $\boldsymbol{H}$, the magnetic susceptibilities of diamagnets and paramagnets are constant. Under these conditions, the materials are called *linear*, that is, $\boldsymbol{M}$ is proportional to $\boldsymbol{H}$. Consequently, it is possible to write

$$\boldsymbol{M} = \chi \boldsymbol{H} \tag{4.1}$$

$$\boldsymbol{B} = \mu_0(1+\chi)\boldsymbol{H} = \mu_0\mu_r\boldsymbol{H} = \mu\boldsymbol{H} \tag{4.2}$$

Clearly then μ_r is slightly greater than one in the paramagnets and slightly less than one in the diamagnets. At the same time χ is slightly more than zero in the paramagnets and slightly less than zero in the diamagnets.

The linear result is a useful one in that it allows us to write a proportional relation between ***B*** and ***H*** in these materials at low fields. However, we shall see later in the classical Langevin theories of diamagnetism and paramagnetism that the linear approximation no longer holds at higher fields and in fact paramagnets exhibit saturation of magnetization at very high fields.

4.1.3 Values of μ_r and χ for Various Materials

What values of permeability and susceptibility do various metals have?

Values of the susceptibilities and permeabilities of various diamagnets, paramagnets, and ferromagnets are shown in Table 4.1.

In ferromagnets, neither χ nor μ_r has a constant value. Both permeability and susceptibility in ferromagnets are strongly affected by the prevailing magnetic field ***H*** and the previous history of the material. For example, the change in the permeability of annealed iron along its initial magnetization curve is shown in Figure 4.1.

4.1.4 Other Types of Magnetic Materials

Are there other types of magnetic materials that the above classification fails to identify?

There are some other types of magnetic materials apart from the three classes of diamagnets, paramagnets, and ferromagnets as given above. These other materials are all very closely related to ferromagnets because they are magnetically ordered, as explained in Chapter 6. They are ferrimagnets, antiferromagnets, helimagnets, and

TABLE 4.1
Susceptibilities and Permeabilities of Various Elements

		χ	μ_r
Diamagnets	Bi	-1.31×10^{-6}	0.9999987
	Be	-1.85×10^{-6}	0.9999982
	Ag	-2.02×10^{-6}	0.9999980
	Au	-2.74×10^{-6}	0.9999973
	Ge	-0.56×10^{-6}	0.9999994
	Cu	-0.77×10^{-6}	0.9999992
Paramagnets	β-Sn	0.19×10^{-6}	1.0000002
	W	6.18×10^{-6}	1.0000062
	Al	1.65×10^{-6}	1.0000016
	Pt	21.04×10^{-6}	1.0000210
	Mn	66.10×10^{-6}	1.0000660

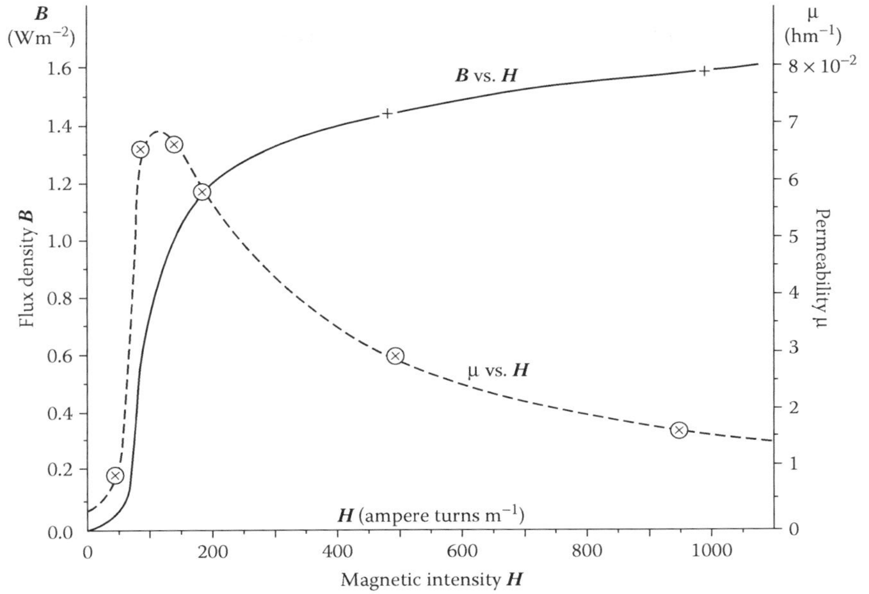

FIGURE 4.1 Initial magnetization curve and the permeability along the same curve for annealed iron.

superparamagnets. All were discovered many years after the three classical groups of magnetic materials discussed above. From bulk magnetic measurements, the ferrimagnets are indistinguishable from ferromagnets, while the antiferromagnets and helimagnets were for many years mistaken for paramagnets. These magnetic types will be discussed in Chapter 9.

4.2 MAGNETIC PROPERTIES OF FERROMAGNETS

Which magnetic materials are the most significant for applications?

By far, the most important class of magnetic materials is the ferromagnets including the ferrimagnets. We can make this statement unreservedly both from the practical and the theoretical viewpoints. The applications that these materials find are very diverse and have been discussed in two significant ferromagnetic materials reference texts by Heck [1] and Wohlfarth [2]. In engineering applications, the ferromagnets are used because of their high permeabilities, which enable high magnetic inductions to be obtained with only modest magnetic fields; their ability to retain magnetization and thereby act as a source of field, and of course, the torque on a magnetic dipole in a field can be used in electric motors. It is perhaps somewhat surprising that the few ferromagnetic elements in the periodic table, such as iron, cobalt, nickel, and several of the lanthanides, are so technologically vital.

At this stage, we are still considering the materials on a macroscopic scale and consequently all that we have really discussed so far is the definition that ferromagnets have in general very large values of relative permeability μ_r and susceptibility χ. These are important quantities, and later we shall go on to consider what the special factors are that cause such high permeabilities. However, we will now pause to consider some of the more characteristic features of ferromagnets, which are noticeable on the everyday, macroscopic scale.

4.2.1 Permeability

What is the most important single property of ferromagnets for applications?

By far, the most important single property of ferromagnets is their high relative permeability. The permeability of a ferromagnet is not constant as a function of magnetic field in the way that the permeability of a paramagnet is. Instead, in order to characterize the properties of a given ferromagnetic material, it is necessary to measure the magnetic induction $\boldsymbol{B}$ as a function of $\boldsymbol{H}$ over a continuous range of $\boldsymbol{H}$ to obtain a hysteresis curve.

We can still make some comments about permeabilities; however, for ferromagnets, the initial relative permeabilities usually lie in the range of $10–10^5$. The highest values occur for special alloys such as Metglas or permalloy and supermalloy, the latter two being nickel-iron alloys. These materials are useful as flux concentrators. Permanent magnet materials do not have such high permeabilities, but their applications depend on their retentivity, which is the next most important property.

4.2.2 Retentivity

What is the most characteristic property of ferromagnets?

It is well known that ferromagnets can be magnetized. That is to say that once exposed to a magnetic field, they retain their magnetization even when the field is removed. Probably this is the most widely recognized property of ferromagnets since we have all spent time magnetizing pieces of iron using a permanent magnet. The retention of magnetization distinguishes ferromagnets from paramagnets, which although they acquire a magnetic moment in an applied field $\boldsymbol{H}$, cannot maintain the magnetization after the field is removed.

4.2.3 Hysteresis

How can we best represent the magnetic properties of ferromagnets?

The most common way to represent the bulk magnetic properties of a ferromagnetic material is by a plot of magnetic induction $\boldsymbol{B}$ for various field strengths $\boldsymbol{H}$. Alternatively, plots of magnetization $\boldsymbol{M}$ against $\boldsymbol{H}$ are used, but these contain the same information since $\boldsymbol{B} = \mu_0(\boldsymbol{H} + \boldsymbol{M})$. Hysteresis in iron was first observed by Warburg [3]. The term *hysteresis*, meaning to lag behind, was introduced by Ewing [4], who was the first to systematically investigate it. A typical hysteresis loop is shown in Figure 4.2.

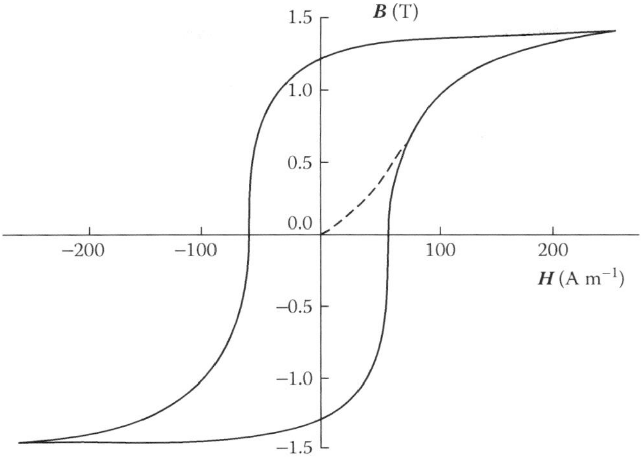

FIGURE 4.2 A typical hysteresis loop of a ferromagnetic material.

The suitability of ferromagnetic materials for applications is determined principally from characteristics shown by their hysteresis loops. Therefore, materials for transformer applications need to have high permeability and low hysteresis losses because of the need for efficient conversion of electrical energy. Materials for electromagnets need to have low remanence and coercivity in order to ensure that the magnetization can easily be reduced to zero as needed. Permanent magnet materials need high remanence and coercivity in order to retain the magnetization as much as possible.

4.2.4 Saturation Magnetization

Is there an upper limit to the magnetization of a ferromagnet?

From the hysteresis plot, it can be seen that the ferromagnet in its initial state is not magnetized. Application of a field $\boldsymbol{H}$ causes the magnetic induction to increase in the field direction. If $\boldsymbol{H}$ is increased indefinitely, the magnetization eventually reaches saturation at a value, which we shall designate as $\boldsymbol{M}_0$. This represents a condition where all the magnetic dipoles within the material are aligned in the direction of the magnetic field $\boldsymbol{H}$. The saturation magnetization is dependent only on the magnitude of the atomic magnetic moments $\boldsymbol{m}$ and the number of atoms per unit volume n.

$$\boldsymbol{M}_0 = n\boldsymbol{m} \quad (4.3)$$

$\boldsymbol{M}_0$ is therefore dependent only on the materials present in a specimen, it is not structure sensitive, nor is it temperature sensitive. Some typical values of saturation magnetization for different materials are shown in Table 4.2. A distinction is drawn here between saturation magnetization $\boldsymbol{M}_0$ and spontaneous magnetization $\boldsymbol{M}_S$. Spontaneous magnetization, which will be encountered later in Section 9.2.3, is the

TABLE 4.2
Saturation Magnetization of Various Ferromagnets

Material	M_0 (10^6 A m^{-1})
Iron	1.71
Cobalt	1.42
Nickel	0.48
78 Permalloy ($Ni_{78}Fe_{22}$)	0.86
Supermalloy ($Ni_{80}Fe_{15}Mo_5$)	0.63
Metglas 2605 ($Fe_{80}B_{20}$)	1.27
Metglas 2615 ($Fe_{80}P_{16}C_3B_1$)	1.36
Permendur ($Co_{50}Fe_{50}$)	1.91

magnetization within a domain. This is temperature sensitive, but at sufficiently low temperatures is very close in value to the saturation magnetization. Often $\boldsymbol{M}_S$ is used to represent saturation magnetization, but this is not quite correct.

4.2.5 Remanence

What happens to the magnetic induction of a ferromagnet when the magnetic field is switched off?

When the field is reduced to zero after magnetizing a magnetic material, the remaining magnetic induction is called the *remanent induction* $\boldsymbol{B}_R$ and the remaining magnetization is called the *remanent magnetization* $\boldsymbol{M}_R$.

$$\boldsymbol{B}_R = \mu_0 \boldsymbol{M}_R \tag{4.4}$$

A convention seems to be emerging in which a distinction is drawn between the remanence and the remanent induction or magnetization. The remanence is used to describe the value of the remaining induction or magnetization when the field has been removed after the magnetic material has been magnetized to saturation. The remanent induction or magnetization is used to describe the remaining induction or magnetization when the field has been removed after magnetizing to an arbitrary level. The remanence therefore becomes the upper limit for all remanent inductions or magnetizations.

4.2.6 Coercivity

How is the magnetic induction reduced to zero?

The magnetic induction can be reduced to zero by applying a reverse magnetic field of strength $\boldsymbol{H}_c$. This field strength is known as the *coercivity*. It is strongly dependent on the condition of the sample, being affected by factors such as heat treatment and deformation.

TABLE 4.3
Relative Permeabilities (Dimensionless) and Coercivities (in A m^{-1}) for Various Traditional Soft and Hard Magnetic Materials

Material	Date Developed	Permeability, μ_r	Coercivity, H_c (A m^{-1})
Metglas	1976	7,000,000	0.3
Supermalloy	1947	100,000	0.4
Sendust	1936	50,000	5
78 Permalloy	1923	10,000	6
Hypersil	1934	2000	8
Iron	pre 1900	100	80
Nickel	pre 1900	100	50
Iron with 5% tungsten	1885	50	8000
KS steel	1923	20	32,000
MK steel	1931	10	40,000
Alnico V	1940	8	70,000
Ferroplatinum	1936	2	400,000
Samarium cobalt	1967	1	800,000
Neodymium iron boron	1984	1	2,000,000

As with the remanence, a distinction is drawn by some authors between the coercive field (or coercive force), which is the magnetic field needed to reduce the magnetic induction to zero from an arbitrary level, and the coercivity, which is the magnetic field needed to reduce the magnetic induction to zero from saturation [5]. In this nomenclature, the coercivity becomes an upper limit for all values of coercive force.

The intrinsic coercivity, denoted $\boldsymbol{H}_{ci}$, is defined as the field strength at which the magnetization $\boldsymbol{M}$ is reduced to zero. In soft magnetic materials, for which coercivity is typically less than 1 kA m^{-1}, $\boldsymbol{H}_c$ and $\boldsymbol{H}_{ci}$ are so close in value that usually no distinction is made between coercivity and intrinsic coercivity. However, in hard magnetic materials, for which coercivity is typically above 10 kA m^{-1}, there is a clear difference in value between them, with $\boldsymbol{H}_{ci}$ always being larger than $\boldsymbol{H}_{c.}$

Coercivity and permeability are inversely related as shown in Table 4.3, where hard magnets have high coercivity and low permeability, while soft magnetic materials have low coercivity and high permeability.

4.2.7 Differential Permeability

How useful is permeability when considering ferromagnets?

We should note in passing that the permeability μ is not a particularly precise parameter for characterization of ferromagnets, since by virtue of the hysteresis loop, almost any value of μ can be obtained, including $\mu = \infty$ at the remanence

$B = B_R$, $H = 0$, and $\mu = 0$, and at the coercivity, $B = 0$ and $H = H_c$. The differential permeability $\mu' = dB/dH$ is a more useful quantity, although it should be remembered that its value also varies with field. The maximum differential permeability μ'_{max}, which usually occurs at the coercive point $H = H_c$ and $B = 0$, and the initial differential permeability μ'_{in}, which is the slope of the initial magnetization curve at the origin, are much more useful since it is possible to relate them to other material properties such as the number and strength of pinning sites [6] and applied stress [7].

4.2.8 Curie Temperature

What happens if a ferromagnet is heated?

All ferromagnets when heated to sufficiently high temperatures become paramagnetic. The transition temperature from ferromagnetic to paramagnetic behavior is called the *Curie temperature*. At this temperature, the permeability of the material drops suddenly and both coercivity and remanence become zero. This property of ferromagnets was known long before the work of Curie. In fact, the existence of a transition temperature was first reported by Gilbert [8], who was the author of the first treatise on magnetism. The Curie temperatures of some widely used magnetic materials are shown in Table 4.4.

The reasons for the sudden transition from ferromagnetism to paramagnetism will be discussed in detail in Chapter 9. At this stage, however, we are interested in this merely as an empirical observation based on the macroscopic magnetic properties that the permeability suddenly decreases at a characteristic temperature.

TABLE 4.4
Curie Temperatures of Various Materials

Material	Curie Temperature (°C)
Iron	770
Nickel	358
Cobalt	1130
Gadolinium	20
Terfenol $Tb_xDy_{1-x}Fe_2$	380–430[a]
$Nd_2Fe_{14}B$	312
Alnico	850
$SmCo_5$	720
Sm_2Co_{17}	810
Hard ferrites	400–700
Barium ferrite	450

[a] Depending on composition.

4.3 DIFFERENT TYPES OF FERROMAGNETIC MATERIALS FOR APPLICATIONS

Where do ferromagnetic materials find their main uses?

We now consider the different applications of ferromagnets such as in permanent magnets, electrical motors, magnetic recording, power generation, and inductors. The objective of this section is to give a concise summary of the types of magnetic materials and their uses before treating the subject in detail in Chapters 12 through 14 and before discussing the underlying mechanisms behind the observed macroscopic magnetic properties of these materials in Chapters 5 through 8. A review of the development and the role of magnetic materials in science and technology have been given by Enz [9].

4.3.1 Hard and Soft Magnetic Materials

Is there a simple method of classification of the various ferromagnetic materials?

We can make a simple classification of ferromagnetic materials on the basis of their coercivity. Coercivity is a structure-sensitive magnetic property, which is to say that it can be altered by subjecting the specimen to different thermal and mechanical treatments, in a way that, for example, saturation magnetization cannot.

It has been noticed in the past that iron and steel specimens that were mechanically hard also had high coercivity, while those that were soft had low coercivity. Therefore, the terms *hard* and *soft* were used to distinguish ferromagnets on the basis of their coercivity. Broadly speaking, *hard* magnetic materials were those with coercivities above 10 kA m^{-1} (125 Oe), while *soft* magnetic materials were those with coercivities below 1 kA m^{-1} (12.5 Oe).

Values of relative permeability and coercive force for various materials are given in Table 4.3.

4.3.2 Electromagnets

Where do soft magnetic materials find uses?

Soft magnetic materials find applications in electromagnets, motors, transformers, and relays. The properties of various soft magnetic materials in use at present have been discussed by Chin and Wernick [10]. The criterion for consideration of materials for electromagnets is that the core material should have high permeability, to enable high magnetic induction to be achieved, while having a low coercivity, so that the induction can easily be reversed.

In electromagnets, soft iron is used almost exclusively. Its coercivity is typically 80 A m^{-1} (1 Oe), as can be seen from Table 4.3, and this when coupled with its high saturation magnetization of 1.7×10^6 A m^{-1} make it the ideal material.

Electromagnets are used in the laboratory for generating high magnetic fields. A typical laboratory electromagnet is capable of generating fields of up to 2.0 T without any special configuration and, with small air gaps, fields of up to 2.5 T can be produced. Sometimes, the pole tips of the electromagnet are made from a cobalt-iron alloy, which has a higher saturation magnetization in order to achieve slightly higher fields in the air gap ($M_0 = 1.95 \times 10^6$ A m^{-1} for an alloy of 35% Co and 65% Fe, compared with $M_0 = 1.7 \times 10^6$ A m^{-1} for iron). The material most often used for these pole tips is an alloy containing 49% Fe, 49% Co, and 2% V. For magnetic inductions above 3 T, electromagnets are not very useful because the iron cannot contribute much additional field. Therefore, for higher field strengths, either water-cooled iron-free magnets, known as *Bitter magnets*, or *superconducting magnets* are used.

4.3.3 Transformers

What are the characteristics of ferromagnetic material needed in power generation?

At first sight, it would appear that the requirements for transformer materials were identical to those for electromagnets. However, this is not quite true. Transformers operate under ac conditions and therefore, although high permeability of the core material is desirable, it is also necessary to reduce the eddy current losses by employing as low a conductivity material as possible.

The material that is used exclusively for transformer cores is grain-oriented silicon-iron. This contains about 3%–4% by weight silicon to reduce conductivity. The material is usually hot-rolled then cold-worked twice followed by an anneal to improve the grain orientation, increasing permeability along the rolling direction.

One of the most important parameters for transformer steels is the total core loss at a frequency of 50 or 60 Hz. Engineers usually measure this in watts per kilogram. Losses decrease with increasing silicon content but the material also becomes more brittle.

Table 4.5 shows the electrical core losses in watts per kilogram at a frequency of 60 Hz for sheets of the various materials of thickness 0.3–0.5 mm. These thicknesses

TABLE 4.5
Core Losses of Selected Soft Magnetic Materials

Material	Core Loss at 1.5 T and 60 Hz (W kg^{-1})
Low-carbon steel	2.8
Nonoriented silicon-iron	0.9
Grain-oriented silicon-iron	0.3
80 Permalloy ($Ni_{80}Fe_{20}$)	0.2[a]
Metglas	0.2–0.3[b]

[a] At saturation magnetization of 0.86 MA m^{-1} (1.1 T).
[b] At 1.4 T since loss increases rapidly above this.

are equal to or less than the skin depth δ in the materials, which at 60 Hz is typically $\delta = 0.3$–0.7 mm.

Amorphous metals have been developed for use in electromagnetic devices [11]. These alloys, such as Metglas, have found applications in some smaller, lower-power devices, but have not been successful in replacing silicon-iron in transformers except in some cases where distribution transformers have been required in locations where fuel costs are high. Many of these *Metglas transformers* have been built and sold; however, that remains a very small fraction of the market for transformers. Large-scale adoption of these materials as transformer cores depends not so much on performance as cost, both for the materials themselves and the fabrication costs in producing the transformers. Also efforts have continued to improve the properties of silicon-iron [12].

4.3.4 Electromagnetic Switches or Relays

What magnetic properties are useful in magnetic switches and control devices?

A relay is a form of electromagnet that can be used as a switch for opening or closing an electrical circuit. The control circuit that opens and closes the switch and the operating circuit are electrically isolated from each other. Consequently, a very weak current in the control circuit can be used to control a much larger current in the operating circuit.

The control circuit of the relay consists of a coil with a magnetizable core and a movable component called the *armature*, which is used to make or break the circuit. The yoke and core materials of relays have much the same requirements as electromagnets, that is, low coercivity, low remanence, and high magnetic induction. This in addition leads to low core loss and high permeability. Relay materials are almost always iron or iron-based alloys such as Fe-Si and Fe-Ni. Unalloyed iron is the most frequently used material for relays. The addition of silicon to iron reduces the coercivity from typically 100 A m^{-1} to a few amperes per meter. The addition of nickel to iron can reduce the coercivity to as low as 1 A m^{-1}.

4.3.5 Magnetic Recording Materials

What are the desirable characteristics of recording media?

Magnetic recording materials have some characteristics in common with permanent magnets in that to be useful they need to have a relatively high remanence and a sufficiently high coercivity to prevent unanticipated demagnetization resulting in the loss of information stored on the magnetic tape or disk. Magnetic recording can either be analog, as in audio recording of signals on magnetic tape, or digital recording, as used in the storage of information of data on magnetic disks and tapes for computer applications. A review of magnetic recording media has been given by Bate [13].

Gamma ferric oxide is used in both equiaxed and acicular forms. Equiaxed gamma ferric oxide particles used for magnetic recording have diameters of 0.05–0.3 μm. Magnetic recording tapes and floppy disks contain small needle-shaped particles of one of these oxides. The particles are embedded in a flexible binding material and at present the needles lie in the plane of the tape or disk.

The needle-shaped particles are aligned by a magnetic field during the fabrication process. The final tapes of γ-Fe_2O_3 have coercivities typically of 20–24 kA m^{-1}, and the acicular particles have lengths ranging from 0.1 to 0.7 μm [13], with length-to-diameter ratios from 3:1 to 10:1. Tapes made from CrO_2 have coercivities of 36–44 kA m^{-1}. The chromium dioxide particles have dimensions ranging from 0.5 × 0.03 μm to 0.2 × 0.02 μm, which are significantly smaller than the typical sizes of gamma iron oxide particles used in recording tapes. In all cases, the ferromagnetic particles used in magnetic recording are too small to contain a domain wall and we therefore have single-domain particles. Research into perpendicular recording media has been conducted [14], in particular much attention has been directed toward CoCr layers for this purpose.

The hysteresis loops that are desirable for magnetic recording materials are generally square loops, with high remanence, moderately high coercivity, and rapid switching from one state to the other, as shown in Figure 4.3, which is the hysteresis loop of a metallic magnetic recording medium. In this case, the coercivity of 56 kA m^{-1} and remanence of 0.9×10^6 A m^{-1} are substantially higher than for γ-Fe_2O_3 particles.

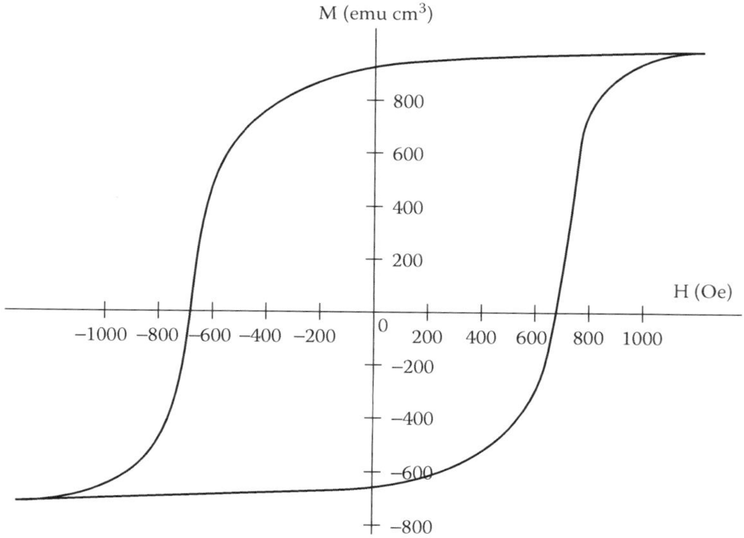

FIGURE 4.3 Hysteresis loop for a typical metallic magnetic recording material.

4.3.6 Permanent Magnets

Where do we use ferromagnetic materials that remain permanently magnetized such as alnico, neodymium-iron-boron, or samarium-cobalt?

Permanent magnets are one of the three most important classes of magnetic materials, the others being electrical steels and magnetic recording media. They find applications in electrical motors and generators, loudspeakers, television tubes, moving-coil meters, magnetic suspension devices, and clamps [15,16].

Clearly, the application determines the choice of the magnetic material based on its hysteresis characteristics. The properties of these materials are usually represented by the *demagnetization curve*, which is the portion of the hysteresis curve in the second quadrant as the magnetization is reduced from saturation. The demagnetization curves for some permanent magnet materials are shown in Figure 4.4. It is important to realize that the final magnetic properties depend as much on the metallurgical treatment and processing of the material as on its chemical composition.

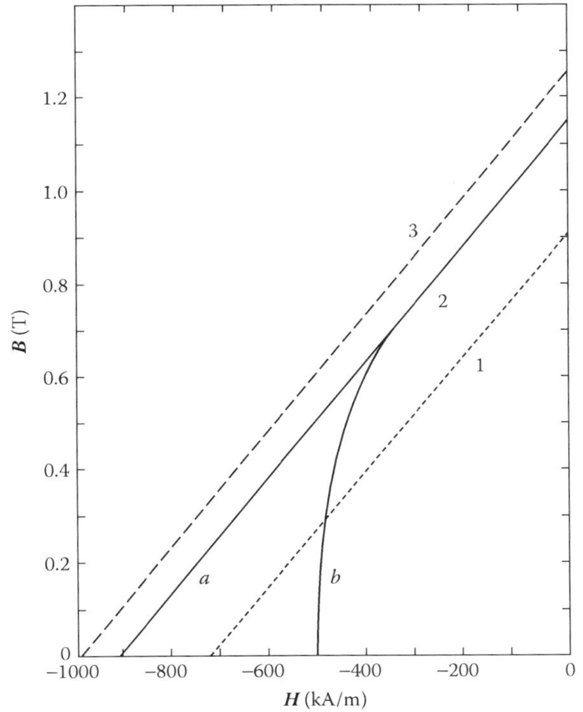

FIGURE 4.4 Second quadrant magnetization curves of specimens of samarium-cobalt (1 and 2) and neodymium-iron-boron (3).

The permanent magnet materials based on neodymium-iron-boron [17] have magnetic properties that are superior for many applications when compared with samarium-cobalt [18]. For example, its coercivity can be as high as 1.12×10^6 A m^{-1} (14 kOe) compared with 0.72×10^6 A m^{-1} (9 kOe) for samarium-cobalt.

In addition to the coercivity, another parameter of prime importance to permanent magnet users is the maximum energy product $\boldsymbol{BH}_{max}$. This is obtained by finding the maximum value of the product $|\boldsymbol{BH}|$ in the second, or demagnetizing, quadrant of the hysteresis loop. It represents the magnetic energy stored in a permanent magnet material. We will discuss its significance to permanent magnet users in detail in Chapter 13. Generally, the maximum energy product by itself does not give sufficient information for permanent magnet users to decide on the suitability of a material for a particular application, but it is a parameter that is widely quoted when comparing various permanent magnet materials.

For many years, the maximum energy product of available materials was in the range of 50×10^3 J m^{-3} (a few megagauss-oersted). The development of samarium-cobalt permanent magnets raised this to about 160×10^3 J m^{-3} (20 megagauss-oersted) and later neodymium-iron-boron magnets were developed with energy products of typically 500×10^3 J m^{-3} (60 megagauss-oersted). Progress over the years has resulted in permanent magnets with higher energy products, and therefore smaller magnets with the same energy, as shown in Figure 4.5.

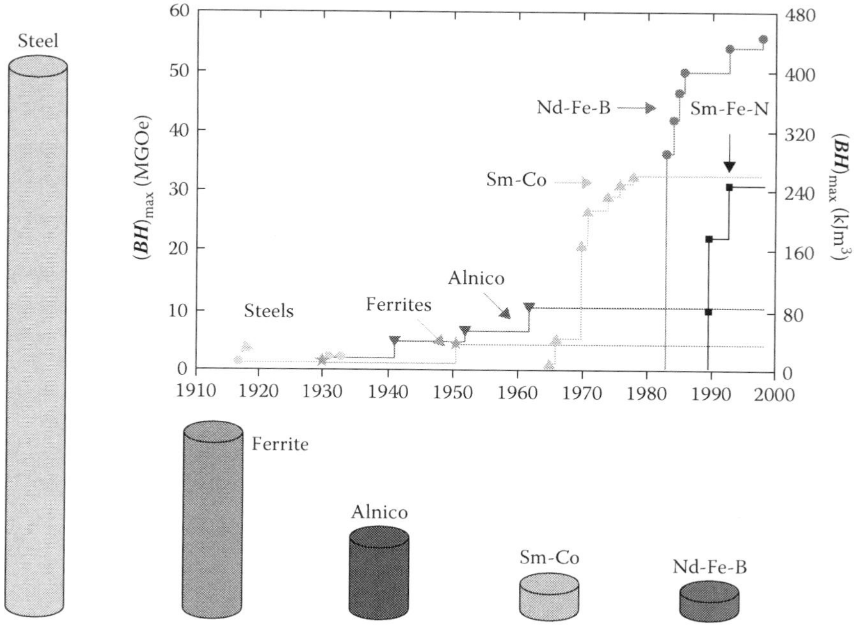

FIGURE 4.5 Increases in maximum energy product, and consequent reduction in size of magnets, of different permanent magnet materials over the years from 1920 to 2000. (Data from Poudyal, N., and Liu, J.P., *J. Phys. D: Appl. Phys.*, 46, 043001, 2013.)

In most applications, the stability of the permanent magnet is an important consideration and therefore the material must be operated sufficiently far from its Curie point since the spontaneous magnetization decreases rapidly with temperature above about 75% of the Curie temperature. This is one of the problems that has arisen with neodymium-iron-boron magnets for higher temperature applications. In this respect, the samarium-cobalt magnets, particularly Sm_2Co_{17}, are superior.

4.3.7 Inductance Cores

What additional properties are needed for high-frequency applications of ferromagnets?

Soft magnetic materials are also used as cores for induction coils. They enhance the flux density inside the coil and thereby improve inductance. When inductors are required to operate at high frequencies then, due to the skin depth, only nonconducting or finely laminated magnetic materials can be used. This usually means soft ferrites are used. These soft ferrites are magnetic materials with high electrical resistivity and high permeability, which for many years were thought to be ferromagnets. This was because their bulk magnetic properties are very similar to ferromagnets. It is now known that these materials are ferrimagnets, which are fundamentally different from ferromagnets and this difference is discussed in Chapter 9. Ferrite-cored inductors are used extensively in frequency-selective circuits, so that the resonant frequency of the circuit ensures that it only responds to the given frequency.

Another application of soft ferrites is in antennae for radio receivers. These have an internal ferrite-cored antenna consisting of a short solenoidal coil, which enhances the electromotive force in the circuit for a given amplitude of field. Typical values of μ_r for these applications are $\mu_r \approx 100$–1000.

Soft ferrites came into commercial production in 1948. They are composed of a compound oxide that consists of iron oxide (Fe_2O_3) together with other oxides such as manganese, nickel, and magnesium. For example, nickel ferrite has the composition $NiO\text{-}Fe_2O_3$. In their final form, they are usually a brown-colored ceramic. Their saturation magnetization is typically $\boldsymbol{M}_0 = 0.2 \times 10^6$ A m^{-1} ($\boldsymbol{B}_s = 0.25$ T), with coercivities of the order of 8 A m^{-1} (0.1 Oe) and maximum permeability $\mu_r = 1500$.

4.3.8 Ceramic Magnets

Which permanent magnet materials should be used where the demagnetizing effects are large?

The hard ferrites, which are also known as *ceramic magnets*, are in widespread use in motors, generators, and other rotating machines, loudspeakers, and various holding or clamping devices. They are also used as the magnetic strip on credit cards. The ceramic magnets are usually made from barium or strontium ferrite. They are very cheap to produce and can be powdered and included in a plastic binder to form the so-called plastic magnets, which can be formed easily into any desirable shape. They have very high coercivity, typically 200 kA m^{-1}, so that they can be usefully used in the form of a short magnet even though the demagnetizing effects are large.

4.4 PARAMAGNETISM AND DIAMAGNETISM

What uses do paramagnets and diamagnets find?

Paramagnets do not find nearly as many applications as ferromagnets and therefore our discussion at this stage will be somewhat limited. The description of paramagnetism is, however, of vital importance in the understanding of magnetism. The reason for this is that paramagnetism is a much simpler phenomenon to describe than ferromagnetism and quite reasonable theories of paramagnetism have been developed on the basis of very simple models and these simple theories give good agreement with experimental observation. In the limiting case, the atomic magnetic moments of paramagnets can be treated as noninteracting (i.e., *dilute paramagnetism*), an approximation that simplifies the modeling greatly.

Diamagnets generally do not find many applications that depend on their magnetic properties either, except for the special case of the superconductors, which are perfect diamagnets with $\chi = -1$.

4.4.1 Paramagnets

How do paramagnets differ fundamentally from ferromagnets?

The study of paramagnetism allows us to investigate the atomic magnetic moments of atoms almost in isolation, since unlike ferromagnetism, paramagnetism is not a cooperative phenomenon. Solid-state physicists are therefore more familiar with the underlying theories of paramagnetism such as the temperature dependence of paramagnetic susceptibility, and its description using the classical expression, the Langevin function (see Section 9.1.5), or its quantum mechanical analog, the Brillouin function (see Section 11.2.2). Materials exhibiting paramagnetism are usually atoms and molecules with an odd number of electrons so that there is an unpaired electron spin, giving rise to a net magnetic moment. These include atoms and ions with partially filled inner shells, such as transition elements. Some elements with even numbers of electrons are paramagnetic.

Examples of paramagnetic materials are platinum; aluminum; oxygen; various salts of the transition metals such as chlorides, sulfates, and carbonates of manganese, chromium, iron, and copper, in which the paramagnetic moments reside on the Cr^{3+}, Mn^{2+}, Fe^{2+}, and Cu^{2+}, respectively; and hydrated salts such as potassium-chromium alum $KCr(SO_4)_2 \cdot 12H_2O$. These salts obey the Curie law, which states that the susceptibility χ is inversely proportional to the temperature T, because the magnetic moments are localized on the metal ions, while the presence of the water molecules in the hydrated salts ensures that the interactions between these electrons on neighboring metal ions are weak.

Salts and oxides of rare earth (lanthanide) elements are strongly paramagnetic. In these solids, the magnetic properties are determined by highly localized 4f electrons. These are closely bound to the nucleus, and are effectively shielded by the outer electrons from the magnetic field at the ionic site caused by the other atoms in the crystal lattice, that is, the crystal field. Rare earth metals are also paramagnetic for the same reasons; however, if the temperature is reduced, many of them exhibit ordered states such as ferromagnetism.

All ferromagnetic metals such as cobalt, iron, and nickel become paramagnetic above their Curie points, as do the antiferromagnetic metals such as chromium and manganese above their transition temperatures of 35°C and −173°C, respectively. Paramagnetic metals that do not exhibit a ferromagnetic state include all the alkali metals (sodium series), and the alkaline earth metals (calcium series) with the exception of beryllium. The remaining 3d, 4d, and 5d transition metals are all paramagnetic with the exception of copper, zinc, silver, cadmium, and mercury, which are diamagnetic.

4.4.2 Temperature Dependence of Paramagnetic Susceptibility

How does the susceptibility vary with environmental factors such as temperature?

In many paramagnets, the susceptibility is inversely proportional to temperature. This dependence, which was discovered empirically, is known as the *Curie law.*

$$\chi = \frac{C}{T} \tag{4.5}$$

where:

T is the temperature in Kelvin

C is a constant known as the *Curie constant*

In other paramagnets, the susceptibility is independent of temperature. Two theories have evolved to deal with these two types of paramagnetism: the localized moment model, which leads to the Curie law, and the conduction band electron model due to Pauli, which leads to temperature-independent and rather weaker susceptibility. The dependence of the susceptibility on temperature of some paramagnetic solids is shown in Figure 4.6.

4.4.3 Field and Temperature Dependence of Magnetization in a Paramagnet

How does a magnetic field affect the magnetization of a paramagnet?

In paramagnets subjected to magnetic fields other than very high fields, the magnetization $\boldsymbol{M}$ is proportional to the field $\boldsymbol{H}$. That is the susceptibility is virtually constant and lies in the range from 10^{-3} to 10^{-5}. In most cases, the individual magnetic moments, or spins, are not coupled or are only weakly coupled so that

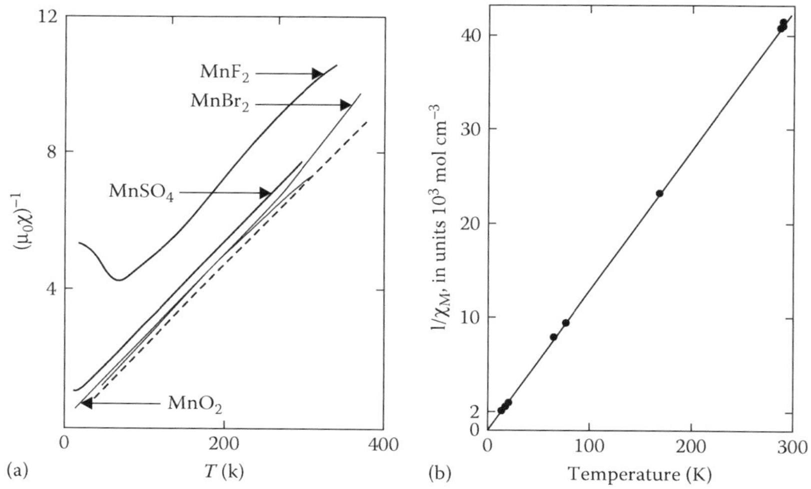

FIGURE 4.6 Temperature dependence of the reciprocal magnetic susceptibility of some magnetic materials: (a) manganese compounds; (b) Gd $(C_2H_5SO_4)_3 \cdot 9H_2O$. (Data from de Hass et al.; Jackson and Kamerlingh Onnes.)

to a good approximation, they can be considered independent. The reason for this is the thermal energy is sufficiently great to cause random alignment of the moments among the available states in zero field. When a field is applied, the atomic moments begin to align with the field, but the net alignment in the field direction remains small for all practical field strengths. At moderate to strong fields, the susceptibility is still constant and saturation only occurs at very high field strengths.

The dependence of magnetization $\boldsymbol{M}$ on magnetic field $\boldsymbol{H}$ and temperature T in a paramagnet can be expressed using Maxwell–Boltzmann statistics. The probability $p(E)$ of an individual magnetic moment having an energy $E = -\mu_0 \boldsymbol{m}\,\boldsymbol{H}$ in the field is accordingly

$$p(E) = p_0 \exp\left(-\frac{E}{k_B T}\right) = p_0 \exp\left(\frac{\mu_0 \boldsymbol{mH}}{k_B T}\right) \tag{4.6}$$

where:

k_B is Boltzmann's constant

T is the temperature in degrees Kelvin

p_0 is a normalizing constant so that the probabilities over all possible states integrate to 1

In the case of a paramagnet in which the magnetic moments are constrained to lie along a single axis, even in the absence of an applied field (a condition known as *uniaxial anisotropy*), the solution is particularly simple. For this two microstate system, the only possible states for any magnetic moment are *spin up* or *spin down*. The moments distribute themselves among the available states, with more in the low energy parallel or *spin up* state and less in the higher energy antiparallel or *spin down* state. As the strength of the magnetic field $\boldsymbol{H}$ is increased, more magnetic moments align parallel to the field and fewer magnetic moments align antiparallel to the field. The resulting magnetization $\boldsymbol{M}$ is therefore given by

$$\boldsymbol{M} = N\boldsymbol{m}\left[\frac{\exp(\mu_0 \boldsymbol{mH}/k_B T) - \exp(-\mu_0 \boldsymbol{mH}/k_B T)}{\exp(\mu_0 \boldsymbol{mH}/k_B T) + \exp(-\mu_0 \boldsymbol{mH}/k_B T)}\right] \tag{4.7}$$

where:

N is the number of atoms per unit volume

$\boldsymbol{m}$ is the magnetic moment per atom

The expression in brackets on the right-hand side is equivalent to the hyperbolic tangent function, and therefore, the equation for the resulting magnetization has the form shown in Equation 4.8.

$$\boldsymbol{M} = N\boldsymbol{m} \tan h\left(\frac{\mu_0 \boldsymbol{mH}}{k_B T}\right) \tag{4.8}$$

The variation of the magnetization with temperature and field of such a paramagnet is shown in Figure 4.7.

For isotropic paramagnets in which the magnetic moments can point in any direction in the absence of an applied field, the dependence of magnetization $\boldsymbol{M}$ on magnetic field $\boldsymbol{H}$ and temperature T can also be expressed using Maxwell–Boltzmann

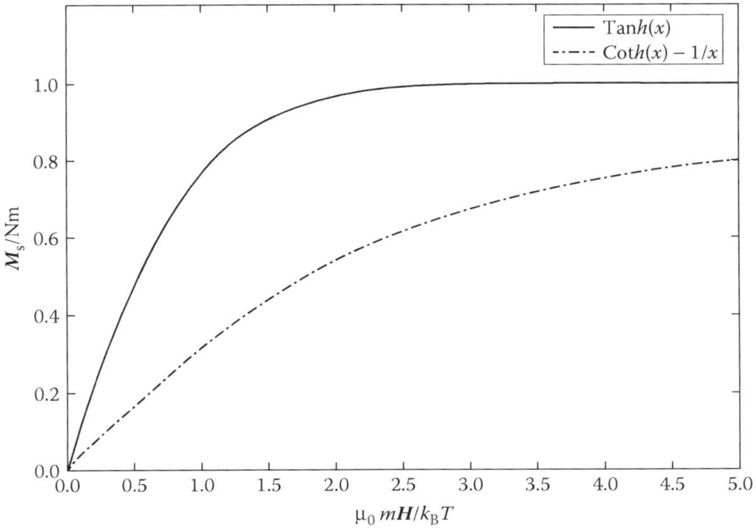

FIGURE 4.7 Variation of magnetization of a typical paramagnet with temperature and magnetic field using the classical Maxwell–Boltzmann statistics.

statistics resulting in a similar function, the Langevin function, which is shown in Equation 4.9.

$$M = Nm\left[\coth\left(\frac{\mu_0 mH}{k_B T}\right) - \left(\frac{k_B T}{\mu_0 mH}\right)\right] \quad (4.9)$$

The derivation of this equation is given in Chapter 9, Section 9.1.5. Equation 4.9 leads to the following approximate expression for the susceptibility, which works well at high temperatures:

$$\chi = \frac{\mu_0 n m^2}{3k_B T} \quad (4.10)$$

This is the Curie law, of the same form as shown in Equation 4.5, but now with some connection between the Curie constant C and the properties of the material because $\mu_0 nm^2/3k_B$ is the Curie constant. This is described in more detail in Chapter 9. A more fundamental expression for the variation of magnetization with temperature and field is obtained using the Brillouin function as described in Chapter 11.

4.4.4 Applications of Paramagnets

Where do paramagnets find uses?

There are very few applications of paramagnetic materials on the basis of their magnetic properties. Their use occurs primarily in the scientific study of magnetism since they help in our understanding of the much more important phenomenon of ferromagnetism by allowing us to study the electronic properties of materials with net atomic magnetic moments in the absence of strong cooperative effects.

There is an increasing use of paramagnetic materials in electron spin resonance for the purpose of measuring magnetic fields in which the magnetic properties of the material are already well characterized (rather than studying the resonance of the material to determine its electronic energy states).

Another application is in the production of very low temperatures. The use of paramagnetic salts to achieve ultra-low temperatures was first suggested by Debye [19] and Giauque [20]. A paramagnetic salt is magnetized isothermally and then cooled to as low a temperature as possible by conventional cryogenic means, for example, using liquid helium at reduced pressure. It is then thermally isolated and adiabatically demagnetized whereupon the temperature drops even further. By this means, temperatures down to the millikelvin range can be achieved.

4.4.5 Diamagnets

How do diamagnets differ fundamentally from paramagnets and ferromagnets?

Elements without permanent atomic electronic magnetic moments are unable to exhibit paramagnetism or ferromagnetism. These atoms have filled electron shells and therefore no net magnetic moment. When subjected to a magnetic field, their induced magnetization opposes the applied field, in the manner described by Lenz's law, and so they have negative susceptibility.

The dependence of the magnetization on applied field in diamagnets, that is, the susceptibility, is according to the classical Langevin theory of diamagnetism (Section 9.1.2) given by

$$\chi = -\frac{\mu_0 Z e^2 n \left(r^2\right)}{6m_e} \tag{4.11}$$

where:

n is the number of atoms per unit volume
Z is the number of electrons per atom
e is the electronic charge
m_e is the electronic mass
(r^2) is the mean square atomic radius, which is typically 10^{-21} m^2

Diamagnetic susceptibility is substantially independent of temperature.

4.4.6 Superconductors

How are superconductors classified among magnetic materials?

Superconductors are diamagnets that find many applications; however, they are a unique class of diamagnet in which the susceptibility is caused by macroscopic currents circulating in the material that oppose the applied field rather than changes in the orbital motion of closely bound electrons. They therefore represent a very special case. Clearly, their susceptibility is temperature dependent since above their critical temperature, they are no longer perfect diamagnets.

REFERENCES

1. Heck, C. *Magnetic Materials and Their Applications*, Crane and Russak, New York, 1974.
2. Wohlfarth, E. P. (ed.). *Ferromagnetic Materials*, Three-volume series, North Holland, Amsterdam, the Netherlands, 1980, 1982.
3. Warbug, E. *Ann. Physik*, 13, 141, 1881.
4. Ewing, J. A. *Magnetic Induction in Iron and Other Metals*, 3rd edn, The Electrician, London, 1900; *Proc. Roy. Soc.*, 33, 21, 1881.
5. Chen, C. W. *Magnetism and Metallurgy of Soft Magnetic Materials*, North Holland, Amsterdam, the Netherlands, 1977.
6. Hilzinger, H. R., and Kronmuller, H. *Physica*, 86–88B, 1365, 1977.
7. Jiles, D. C., Garikepati, P., and Chang, T. T. *IEEE Trans. Magn.*, 24, 2922, 1988.
8. Gilbert, W. *De Magnete* (trans. *On the Magnet*) (1600). Republished by Dover, New York, 1958.
9. Enz, U. Magnetism and magnetic materials: Historical developments and present role in industry and technology, in *Ferromagnetic Materials*, Vol. 3 (ed. E. P. Wohlfarth), North Holland, Amsterdam, the Netherlands, 1982.

10. Chin, G. Y., and Wernick, J. H. Soft magnetic metallic materials, in *Ferromagnetic Materials*, Vol. 2 (ed. E. P. Wohlfarth), North Holland, Amsterdam, the Netherlands, 1980.
11. Luborsky, F. E. Amorphous ferromagnets, in *Ferromagnetic Materials*, Vol. 1. (ed. E. P. Wohlfarth), North Holland, Amsterdam, the Netherlands, 1980.
12. Shiozaki, M. Recent trends in nonoriented and grain oriented electrical steel sheets in Japan, in *Proceedings of the 6th Annual Conference on Properties and Applications of Magnetic Materials*, Illinois Institute of Technology, Chicago, IL, 1987.
13. Bate, G. Recording materials, in *Ferromagnetic Materials*, Vol. 2 (ed. E. P. Wohlfarth), North Holland, Amsterdam, the Netherlands, 1980.
14. Bernards, J. P. C., Schrauwen, C. P. G., Luitjens, S. B., Zieren, V., and de Bie, R. W. *IEEE Trans. Magn.*, 23, 125, 1987.
15. McCaig, M. *Permanent Magnets in Theory and Practice*, Wiley, New York, 1977.
16. Moskowitz, L. R. *Permanent Magnet Design and Application Handbook*, Cahner, Boston, MA, 1976.
17. Croat, J. J., Herbst, J. F., Lee, R. W., and Pinkerton, F. E. *J. Appl. Phys.*, 55, 2078, 1984.
18. Strnat, K., Hoffer, G., Olson, J., Ostertag, W., and Becker, J. J. *J. Appl. Phys.*, 38, 1001, 1967.
19. Debye, P. *Ann. Physik*, 81, 1154, 1926.
20. Giauque, W. F. *Am. Chem. Soc.* 49, 1864, 1927.

FURTHER READING

Bozorth, R. M. *Ferromagnetism*, Van Nostrand, New York, 1951.

Fidler, J., Bernard, J., and Skalicky, P. in *High Performance Permanent Magnet Materials* (eds. S. G. Sankar, J. F. Herbst, and N. C. Koon), Materials Research Society, New York, 1987, p. 181.

Heck, C. *Magnetic Materials and Their Applications*, Crane Russak, New York, 1974.

Hummel, R. *Electronic Properties of Materials*, Springer-Verlag, Berlin, Germany, 1985, Ch. 17.

Jorgensen, F. *The Complete Handbook of Magnetic Recording*, 3rd edn, TAB, Pennsylvania, PA, 1988.

McCaig, M. *Permanent Magnets in Theory and Practice*, Wiley, New York, 1977.

Snelling, E. C. *Soft Ferrites, Properties and Applications*, 2nd edn, Butterworths, London, 1988.

EXERCISES

4.1 *Properties of ferromagnets.* Explain the difference between a ferromagnet and a paramagnet on the basis of macroscopic measurements. What is the saturation magnetization and how does it differ as a material is heated through its Curie point?

What criteria are used to distinguish between hard and soft magnetic materials? In deciding on a material for use in (1) an electromagnet and (2) a transformer, what magnetic properties would you take into consideration?

4.2 *Use of initial magnetization curve to find flux in core.* Explain the meaning of hysteresis loss, coercivity, and remanence. How do the magnitudes of these quantities determine the suitability of a ferromagnet

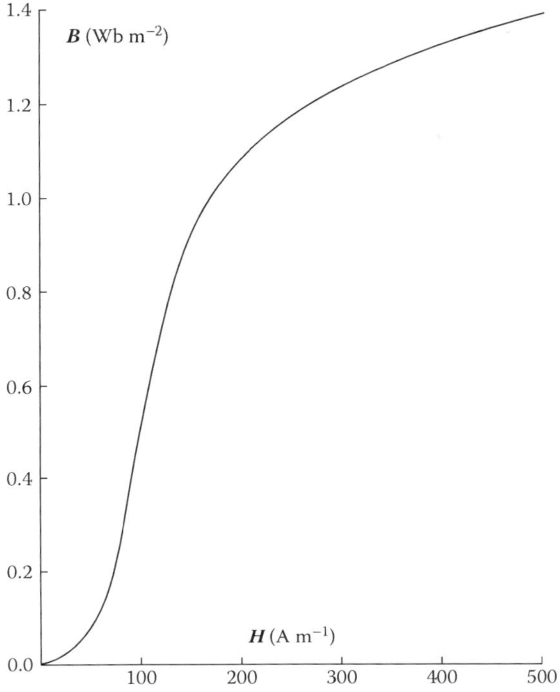

FIGURE 4.2.1 Initial magnetization curve for a low-carbon steel.

for applications? Explain what is meant by the initial magnetization curve. The initial magnetization curve for a specimen of steel is given in Figure 4.2.1, a toroid of this steel has a mean circumference of 40 cm and a cross-sectional area of 4 cm^2. It is wound with a coil of 400 turns. What will be the total flux in the material for currents of 0.1, 0.2, 0.3, 0.4, and 0.5 A?

4.3 *Calculation of atomic magnetic moment.* The saturation magnetization of iron is 1.7×10^6 A m^{-1}. If the density of iron is 7970 kg m^{-3} and Avogadro's number is 6.025×10^{26}/Kg mol, calculate the magnetic moment per iron atom in amperes per square meter and in Bohr magnetons. (1 Bohr magneton $\mu_B = 9.27 \times 10^{-24}$ J T^{-1} or 1.16×10^{-29} J m A^{-1}. Relative atomic mass of iron = 56.)

4.4 *Force due to a field gradient.* What gradient of magnetic field is needed to exert a force of 1 N on a 1 mm diameter sphere of iron that is magnetized to saturation ($\boldsymbol{M}_0 = 1.7 \times 10^6$ A m^{-1})?

4.5 *Elementary dipole moments.* Explain the origin of magnetic moments in solids. In your description comment on whether the *elementary* atomic dipole moments are permanent or induced by a field, and explain the differences between a ferromagnet, a paramagnet, and a diamagnet on the scale of a few atoms.

4.6 *Curie's Law.* State the Langevin equation for magnetization of a paramagnet as a function of magnetic field and temperature, and justify its use. Derive the Curie law for susceptibility at high temperatures from this equation.

4.7 *Differences between diamagnets, paramagnets, and ferromagnets.* Explain the difference between diamagnets, paramagnets, and ferromagnets in terms of their bulk magnetic properties and in terms of their internal magnetic order. Give examples of equations that can be used to describe the susceptibility of a paramagnet and of a diamagnet.

4.8 *Susceptibility of a paramagnetic gas.* Calculate the paramagnetic susceptibility of an *ideal gas* of noninteracting particles each with a magnetic moment of 1 Bohr magneton, at a pressure of 1 atm (10^5 Pa) at 0°C.

(One gram mole of *ideal gas* occupies a volume of 22.4 L at 1 atm and 0°C. Avogadro's number is 6.02×10^{23}/g mol = 6.02×10^{26}/kg mol.)

4.9 *Spontaneous magnetization.* Iron has a magnetic moment of 2 Bohr magnetons/ atom, a relative atomic mass of 56, and a density of 8×10^3 kg m^{-3}. Calculate the expected bulk spontaneous magnetization of iron in amperes per meter at 0 K and describe any assumptions that you have made. How would you expect this bulk magnetization of iron to vary as temperature is increased away from 0 K. How does the number of Bohr magnetons per atom change as the temperature increases from 0 K to 300 K? Why does a piece of iron typically not exhibit high magnetization at room temperature unless it has been magnetized?

4.10 *Relationship between Curie temperature and internal exchange coupling.* In a particular ferromagnetic material, the internal coupling energy causing parallel alignment of the magnetic moments of any individual atom is 0.27 eV. Calculate the Curie temperature of the material. Explain any assumptions that you made in making this calculation (1 eV = 1.6×10^{-19} J and $k_B = 1.38 \times 10^{-23}$ J K^{-1}).

Section II

Magnetism in Materials

Magnetic Phenomena on the Microscopic Scale

5 Magnetic Properties

We will now look at the causes of hysteresis in ferromagnets and how the variation of magnetization with magnetic field can be quantified in restricted cases such as at low field and in the approach to saturation. High-resolution measurements of the variation of $\mathbf{M}$ with $\mathbf{H}$ indicate that there are discontinuities. This is known as the *Barkhausen effect*. We will also consider the change in length of a ferromagnet as it is magnetized, that is, the magnetostriction, and discuss anisotropy in relation to magnetostriction.

We have shown in the previous chapter that most of the important macroscopic magnetic properties of ferromagnets can be represented on a **B-H** plot or hysteresis loop. From this, the magnetization can be calculated at every point on the hysteresis curve using the general formula $\mathbf{B} = \mu_0(\mathbf{H} + \mathbf{M})$. As the magnetization curve is traversed, there are discontinuous, irreversible changes in the magnetization known as the *Barkhausen effect* after its discoverer. In recent years, the Barkhausen effect has become an important tool for investigating the microstructural properties of ferromagnetic materials.

One important bulk property of interest, which is not contained in the hysteresis plot, is the magnetostriction. This is the change in length of a material either as a result of a magnetic order (spontaneous magnetostriction) or under the action of a magnetic field (field-induced magnetostriction). This will also be discussed in this chapter.

5.1 HYSTERESIS AND RELATED PROPERTIES

5.1.1 Information from the Hysteresis Curve

Which are the most important macroscopic magnetic properties of ferromagnets?

From the hysteresis curve such as the one shown in Figure 5.1, we can define a number of magnetic properties of the material, which can be used to characterize the hysteresis loop. A question immediately arises: How many degrees of freedom are there in a hysteresis loop? Or to put the question another way: How many parameters are needed to characterize a hysteresis loop? Clearly, there is no general answer to this, but for the commonly encountered sigmoid-shaped hysteresis loop such as the one in Figure 5.1, we can start to enumerate the important properties and thereby make an estimate.

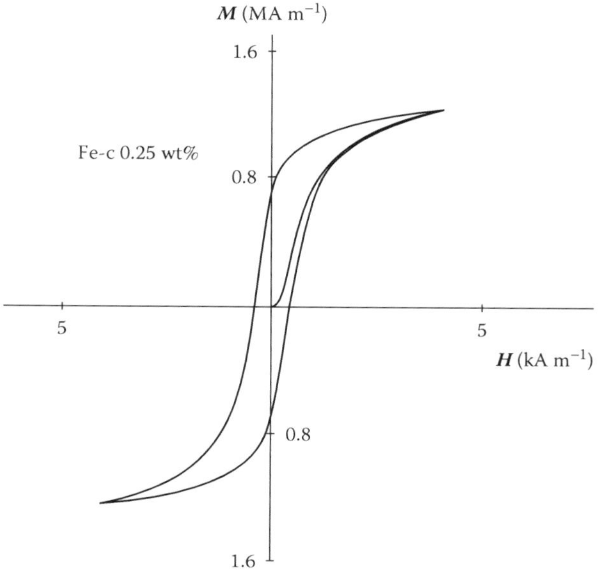

FIGURE 5.1 Typical sigmoid-shaped hysteresis curve of a specimen of iron containing 0.25 wt% carbon.

5.1.2 Parametric Characterization of Hysteresis

Which are the parameters that can be used to define hysteresis?

First of all, the saturation magnetization $\boldsymbol{M}_0$ will give us an upper limit to the magnetization that can be achieved. At temperatures well below the Curie point, the technical saturation $\boldsymbol{M}_s$ can be used instead. The width of the loop across the $\boldsymbol{H}$ axis, which is twice the coercivity $\boldsymbol{H}_c$, is also an independent parameter since this can be altered by heat treatment and hence is not dependent on $\boldsymbol{M}_s$. The height of the curve along the $\boldsymbol{B}$-axis, that is, the remanence is also an independent parameter since it is not wholly dependent on $\boldsymbol{M}_s$ and $\boldsymbol{H}_c$. The orientation of the hysteresis curve on the $\boldsymbol{B}$-$\boldsymbol{H}$ plane, which can be represented by μ'_{max}, the slope of the curve at the coercive point, is also independent of the other parameters.

The hysteresis loss $\boldsymbol{W}_H$ may also be an independent parameter as may the initial permeability μ'_{in}. Finally, the curvature of the sides of the hysteresis loop, which although not immediately obvious as an independent parameter, is clearly not dependent on factors such as coercivity and maximum differential permeability. This parameter emerges more clearly from a consideration of anhysteretic magnetization given below, which requires at least two independent parameters in addition to $\boldsymbol{M}_s$ in order to characterize it.

From the above simplistic considerations, we may expect to be able to characterize the bulk magnetic properties of a material in terms of perhaps five or six independent parameters. In fact, we often find that when the magnetic properties of ferromagnetic

TABLE 5.1
Magnetic Properties[a] of Various High-Permeability Ferromagnetic Materials

Material	μ_{2T}	μ_{max}	B_s (T)	W_H (J m^{-3})	H_c (A m^{-1})
Purified iron	5,000	180,000	2.15	30	4
Iron	200	5,000	2.15	500	80
Carbonyl iron	55	132	2.15	–	–
Cold-rolled steel	180	2,000	2.1	–	144
Iron-4% silicon	500	7,000	1.97	350	40
45 permalloy	2,500	25,000	1.6	120	24
Hipernik	4,500	70,000	1.6	22	4
Monimax	2,000	35,000	1.5	–	8
Sinimax	3,000	35,000	1.1	–	–
78 permalloy	8,000	100,000	1.07	20	4
Mumetal	20,000	100,000	0.65	–	4
Supermalloy	100,000	800,000	0.8	–	0.16
Permendur	800	5,000	2.45	1,200	160
2V permendur	800	4,500	2.4	600	160
Hiperco	650	10,000	2.42	–	80
Ferroxcube	1,000	1,500	2.5	–	8

Note: [a] Magnetic properties are represented in terms of relative permeability at a magnetic induction of 2 T (μ_{2T}), maximum relative permeability (μ_{max}), saturation magnetic induction ($\boldsymbol{B}_s$), hysteresis loss (W_H), and coercivity (H_c).

materials are displayed in a tabular form, the properties are represented in terms of coercivity, remanence, hysteresis loss, initial permeability, maximum permeability, and saturation magnetization or saturation magnetic induction, as in Table 5.1.

5.1.3 Causes of Hysteresis

What are the underlying mechanisms behind hysteresis?

It is well known that if a specimen of iron or steel is subjected to cold working, both the hysteresis loss and the coercivity increase (Figure 5.2). It is also well known that the addition of other nonmagnetic elements to iron, such as carbon in making steel, increases the hysteresis loss and coercivity. These empirical facts were known long before theories of hysteresis were suggested.

From these results, it would appear that *imperfections*, whether in the form of dislocations or impurity elements in the metal, cause an increase in the energy lost during the magnetization process, in the form of a kind of internal friction. It is these *imperfections* that give rise to hysteresis.

Another mechanism that gives rise to hysteresis is caused by magnetocrystalline anisotropy. Ferromagnetic materials with higher anisotropy have greater hysteresis. This is well known by those working with permanent magnets. In an anisotropic

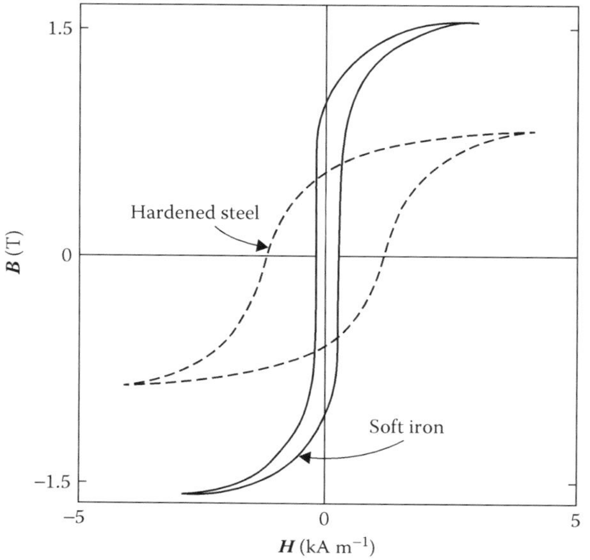

FIGURE 5.2 Dependence of the hysteresis loop of iron or steel on hardness caused by the addition of carbon or other nonmagnetic material or by cold working.

solid, certain crystallographic axes are favored by the magnetic moments, which will prefer to lie along these directions as this leads to a lower energy. The magnetic moments can be dislodged from the direction they are occupying by application of a magnetic field, but when this occurs, they jump to crystallographically equivalent axes that are closer to the field direction, and hence of lower energy. This results in discontinuous and irreversible rotation of the magnetic moments, which leads to a kind of switching action.

In order to discuss this second process properly, we need some additional background knowledge of domain processes, which have yet to be discussed (Chapter 8). Therefore, we will defer the discussion until after the necessary background material has been presented.

5.1.4 Anhysteretic, or Hysteresis-Free, Magnetization

What happens in the case of a material without defects such as dislocations or impurities?

If we accept the hypothesis that it is the imperfections, whatever their nature, that cause hysteresis, then we must also ask ourselves what the magnetization curve would look like if the material were devoid of all imperfections. The answer is that, ignoring anisotropic effects for the moment, it would be hysteresis free. That is, the magnetic induction would be a single-valued function of the magnetic field ***H***. The magnetization curve would therefore be reversible.

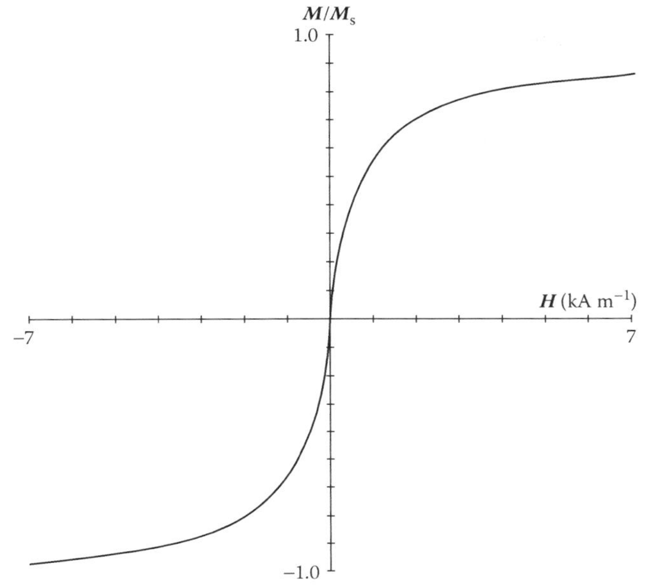

FIGURE 5.3 Anhysteretic magnetization curve. This is antisymmetric with respect to the magnetic field. The differential susceptibility is greatest at the origin and decreases monotonically with increasing field. The curve has no hysteresis and is completely reversible.

We can briefly speculate on the form of such a curve before presenting a simple model for it. Suppose we consider a plot of magnetization against field for such an ideal case. Since the magnetization of a ferromagnet saturates, it is clear that as $\boldsymbol{H}$ increases so $\boldsymbol{M}$ tends toward $\boldsymbol{M}_s$. Furthermore, we would expect that at first the magnetization would change fairly rapidly with $\boldsymbol{H}$. However, as $\boldsymbol{H}$ increases, the rate of change $d\boldsymbol{M}/d\boldsymbol{H}$ would decrease, since this is in the nature of physical systems that saturate. Therefore, we would expect $\boldsymbol{M}$ to be a monotonically increasing function of $\boldsymbol{H}$, while $d\boldsymbol{M}/d\boldsymbol{H}$ would be a monotonically decreasing function of $\boldsymbol{H}$. This would give the S-shaped curve of Figure 5.3.

5.1.5 Fröhlich-Kennelly Relation

Can we find a simple equation for the anhysteretic?

A quantitative relationship between magnetization $\boldsymbol{M}$ and magnetic field $\boldsymbol{H}$ is clearly highly desirable since any such equation provides a means of telling how the magnetization or the magnetic induction of a material will change with field. An empirical relationship between $\boldsymbol{M}$ and $\boldsymbol{H}$ along the anhysteretic magnetization curve was suggested by Fröhlich [1] and later in a different, but equivalent, form by Kennelly [2]. The Fröhlich equation for the anhysteretic magnetization is

$$M = \frac{\alpha H}{1 + \beta H} \tag{5.1}$$

where $\alpha/\beta = M_s$ since $H \to \infty$ the magnetization must tend to M_s.

Independently, Kennelly arrived at an expression for the high-field susceptibility as the magnetization approached saturation. If Kennelly's expression is converted to SI units, it becomes

$$\frac{1}{\mu - \mu_0} = a + bH \tag{5.2}$$

which can easily be shown to be equivalent to the Fröhlich equation, in which $\mu_0 a = 1/\alpha$ and $\mu_0 b = \beta/\alpha = 1/M_s$.

This equation can also be rewritten in the form of a series.

$$M = M_s\left[1 - \left(\frac{cM_s}{H}\right) + \left(\frac{cM_s}{H}\right)^2 \cdots\right] \tag{5.3}$$

This is the form of equation used by Weiss [3] for finding M_s from magnetization curves by extrapolation, using only the terms up to $1/H$. It is also of interest in Section 5.1.8 when we compare it with the law of approach to saturation given much later by Becker and Doring [4].

5.1.6 Measurement of Anhysteretic Magnetization

If the anhysteretic is so important, how do we measure it?

Elimination of all defects within a material is not usually practicable; however, there is a way of reaching the anhysteretic magnetization by other means. This is done by cycling the magnetization by applying an alternating field of gradually decreasing amplitude superimposed on the dc field of interest. As the ac field is cycled, the hysteresis is gradually removed and the magnetization converges on the anhysteretic value for the prevailing dc field strength. This procedure can be thought of as *shaking* the magnetization, so that it overcomes the internal frictional forces, and anisotropic or switching hysteresis effects, and reaches its true equilibrium value. The same effect can also be brought about by stress cycling, although this is generally more difficult to achieve. It can also be brought about by *thermally demagnetizing.*

In Chapter 14, the use of field cycling is well known as a method of reaching the anhysteretic magnetization [5]. The anhysteretic susceptibility at the origin is typically an order of magnitude greater than the normal dc susceptibility at the origin. The anhysteretic magnetization also varies linearly with field at low values, which is one of the prime considerations for magnetic recording.

5.1.7 Low-Field Behavior: Rayleigh Law

Is there a simple equation for the initial magnetization curve?

We now go on to consider the initial magnetization curve, that is, the variation of magnetization with field obtained when a dc field is first applied to a demagnetized ferromagnet. It was noticed by Rayleigh [6] that in the low-field region of the initial magnetization curve, the permeability could be represented by an equation of the form

$$\mu(\boldsymbol{H}) = \mu(0) + v\boldsymbol{H} \tag{5.4}$$

which leads to the following parabolic dependence of $\boldsymbol{B}$ on $\boldsymbol{H}$ along the initial magnetization curve:

$$\boldsymbol{B}(\boldsymbol{H}) = \mu(0)\boldsymbol{H} + v\boldsymbol{H}^2 \tag{5.5}$$

According to Rayleigh, the term $\mu(0)\boldsymbol{H}$ represented the reversible change in magnetic induction while the term $v\boldsymbol{H}^2$ represented the irreversible change in magnetic induction. Furthermore, Rayleigh indicated that low-amplitude hysteresis loops could be represented by parabolic curves that have a reversible differential permeability at the loop tips, which is equal to $\mu(0)$, as shown in Figure 5.4. It follows from this assumption and the Rayleigh law that in the low-field region, the small-amplitude hysteresis loops can be described by an equation of the following form:

$$\boldsymbol{B} = \left[\mu(0) + v\boldsymbol{H}_{\mathrm{m}}\right]\boldsymbol{H} \pm \left(\frac{v}{2}\right)\left(\boldsymbol{H}_{\mathrm{m}}^2 - \boldsymbol{H}^2\right) \tag{5.6}$$

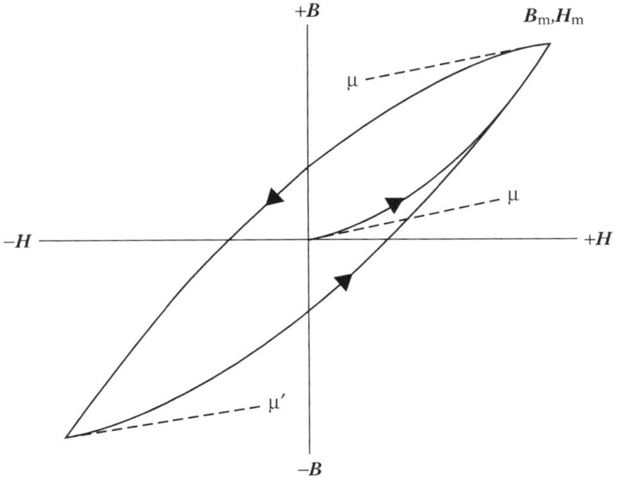

FIGURE 5.4 Hysteresis loops of low-field amplitude in the Rayleigh region.

where $\boldsymbol{H}_{\mathrm{m}}$ is the maximum field at the loop tip. Low-amplitude hysteresis loops for which this parabolic relation applies are known as *Rayleigh loops*.

This leads to two further results of interest: expressions for the hysteresis loss W_{H} and the remanence B_{R}. In SI units these are

$$W_{\mathrm{H}} = \int \boldsymbol{H} \cdot d\mathbf{B} = \left(\frac{4}{3}\right) \nu \boldsymbol{H}_{\mathrm{m}}^{3} \tag{5.7}$$

and

$$B_{\mathrm{R}} = \left(\frac{\nu}{2}\right) \boldsymbol{H}_{\mathrm{m}}^{2} \tag{5.8}$$

It must be remembered of course that these relations only hold true in the low-field region. As $\boldsymbol{H}_{\mathrm{m}}$ is increased, the parabolic relation breaks down. In order to model the hysteresis behavior over a wider range of $\boldsymbol{H}$, it is necessary to gain further insight into the microscopic mechanisms occurring within the material.

5.1.8 High-Field Behavior: Law of Approach to Saturation

Is there an equation for the magnetization at high fields?

In the high-field region, the magnetization approaches saturation. The first attempt to give an equation describing the behavior in this region was Lamont's law [7],

$$\chi' = \frac{d\boldsymbol{M}}{d\boldsymbol{H}} = (\boldsymbol{M}_{\mathrm{s}} - \boldsymbol{M}) \times \text{constant} \tag{5.9}$$

which states simply that at high fields, the differential susceptibility is proportional to the displacement from saturation. This was later shown to be equivalent to the Fröhlich-Kennelly relation. This is of interest in relation to the development of the hysteresis model in Section 8.2.

Later work indicated that the high-field behavior can be modeled by the law of approach to saturation as given by Becker and Doring [4] and Bozorth [7]. This is expressed in the form of a series.

$$\boldsymbol{M} = \boldsymbol{M}_{\mathrm{s}}\left(1 - \frac{a}{\boldsymbol{H}} - \frac{b}{\boldsymbol{H}^{2}} - \cdots\right) + k\boldsymbol{H} \tag{5.10}$$

where the final term $k\boldsymbol{H}$ represents the forced magnetization, that is, the field-induced increase in spontaneous magnetization, which is a very small contribution except at high fields. Typically, $\boldsymbol{H}$ has to be of the order of 10^5 or 10^6 Oe before this last term becomes significant.

It is interesting to note that this law, which was only derived at high magnetizations, is also very similar to the series form of the Fröhlich-Kennelly relation.

The reason for this is that at high fields, the initial magnetization curve, the upper and lower branches of the hysteresis loop, and the anhysteretic magnetization approach each other asymptotically.

5.2 BARKHAUSEN EFFECT AND RELATED PHENOMENA

5.2.1 Barkhausen Effect

Does the magnetization change smoothly with the magnetic field?

The Barkhausen effect is the phenomenon of discontinuous changes in the flux density $\boldsymbol{B}$ within a ferromagnet as the magnetic field $\boldsymbol{H}$ is changed continuously. This was first observed in 1919 [8] when a secondary coil was wound on a piece of iron and connected to an amplifier and loudspeaker. As the $\boldsymbol{H}$ field was increased smoothly, a series of clicks were heard over the loudspeaker, which was due to small voltage pulses induced in the secondary coil. These voltages were caused through the law of electromagnetic induction by small changes in flux density through the coil arising from discontinuous changes in magnetization $\boldsymbol{M}$ and hence in the induction $\boldsymbol{B}$.

If the initial magnetization curve, which looks to be a smooth variation of $\boldsymbol{B}$ with $\boldsymbol{H}$ under normal circumstances, is greatly magnified, then the discontinuous changes in $\boldsymbol{B}$, which constitute the Barkhausen effect, can be observed directly as in Figure 5.5. At first, these discontinuities in induction were attributed to the sudden discontinuous rotation of the direction of magnetization within a domain, a mechanism known as *domain rotation*, but it is now known that discontinuous domain boundary motion is the most significant factor contributing to Barkhausen emissions [9]. Nevertheless, both of these mechanisms do occur and contribute to the Barkhausen effect.

The Barkhausen emissions are greatly affected by changes in the microstructure of the material and also by stress. Therefore, Barkhausen measurements have found an important role in materials evaluation as discussed by Matzkanin, Beissner, and Teller [10].

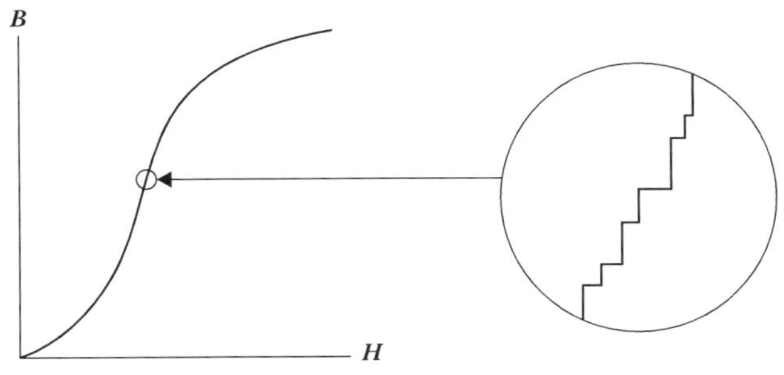

FIGURE 5.5 Barkhausen discontinuities along the initial magnetization curve observed by amplifying the magnetization.

5.2.2 Theory of Barkhausen Effect

How can we describe the discontinuous changes in magnetization with field?

The quantitative description of the Barkhausen effect, at least in terms of defining equations, has proved to be extremely difficult principally because of its random nature. Nevertheless, progress has been made in recent years, most notably by Bertotti and coworkers [11], who have now developed a comprehensive model of the effect based on stochastic processes. In this description, the Barkhausen activity depends on the *internal potential* experienced by the magnetic domain walls as they move through the material. This *internal potential* has a random nature, but the amplitude and periodicity of the fluctuations can be characterized on average, and these two parameters control the number and intensity of the Barkhausen emissions when subject to a changing field. From the solution of the model equations, it is found that for a given rate of change of magnetization with time the Barkhausen activity in a given time interval is correlated with the activity in the previous time interval, but varies by an amount, which is random in nature. Depending on the magnitude of this random component, the fluctuations in the Barkhausen activity may be greater or lesser, with correspondingly less or more correlation with the activity in the previous time interval. The results of these model calculations agree well with experimental observations, which therefore seem to confirm the basic ideas behind the theory. A recent extension of the model [12] has shown that the Barkhausen effect can be described in terms of two components: a deterministic component, which is reproducible and depends on the differential susceptibility, and a stochastic or random component.

5.2.3 Magnetoacoustic Emission

What other effects occur as a result of sudden discontinuous domain-wall motion?

Magnetoacoustic emission, also known sometimes as the *acoustic Barkhausen effect*, is closely related to the magnetic Barkhausen effect described above. Magnetoacoustic emission consists of bursts of low-level acoustic energy generated by sudden discontinuous changes in magnetization involving localized strains or magnetostriction (Section 5.3). These can be detected by a broadband ultrasonic transducer. They are caused by microscopic magnetostrictive pulses as the domain walls move, and in nonmagnetostrictive materials such as Fe-80% Ni, they are almost nonexistent. Therefore, the effect depends on both sudden discontinuous domain processes and magnetostriction.

Magnetoacoustic emission was first observed by Lord [13]. Subsequently, investigations were carried out by Ono and Shibata [14] and others [15]. Because the effect depends on magnetostriction, it cannot be generated by 180° domain-wall motion or rotation, as these involve no change in magnetostriction. These 180° domain boundaries exist between neighboring domains, in which the magnetization vectors point in exactly opposite directions. The relative number density of 180° and non-180° domain walls is affected by the application of uniaxial stress. Therefore, the method has been suggested as a means of detecting stress in ferromagnetic materials and this has been shown to be viable by Buttle et al. [16].

5.3 MAGNETOSTRICTION

Do the dimensions of a specimen change when it is magnetized?

The magnetization of a ferromagnetic material is in nearly all cases accompanied by changes in dimensions. The resulting strain is called the *magnetostriction* (λ). From a phenomenological viewpoint, there are really two main types of magnetostriction to consider: (1) spontaneous magnetostriction, arising from the ordering of magnetic moments into domains at the Curie temperature, and (2) field-induced magnetostriction. These are manifestations of the same effect but can be usefully treated as distinct.

In both cases, the magnetostriction is simply defined as λ, the fractional change in length.

$$\lambda = \frac{dl}{l} \tag{5.11}$$

The existence of magnetostriction was first discovered by Joule [17,18]. Spontaneous magnetostriction within domains arises from the creation of domains as the temperature of the ferromagnet passes through the Curie (or ordering) temperature. Field-induced magnetostriction arises when domains that have spontaneous magnetostriction are reoriented under the action of a magnetic field. Magnetostriction can be measured using resistive strain gauges or optical techniques.

We will first consider the magnetostriction of a hypothetical isotropic solid, since this leads to the simplest mathematical results.

5.3.1 Spontaneous Magnetostriction in Isotropic Materials

Does the length of a specimen change when it becomes ferromagnetic?

When a ferromagnetic material is cooled through its Curie point, the previously disordered magnetic moments, which had completely random alignment above the Curie point, become ordered over volumes containing large numbers (typically, 10^{12}–10^{18}) of atoms. These volumes in which all moments lie parallel are called *domains* and can be observed under a microscope. The direction of spontaneous magnetization $\boldsymbol{M}_s$ varies from domain to domain throughout the material to ensure that the bulk magnetization is zero.

The transition to ferromagnetism is said to cause the onset of *long-range order* of the atomic moments. By this, we mean of course long range compared with atomic dimensions, since the range is still microscopic, and three or four orders of magnitude smaller than the range of magnetic order imposed when the material is magnetically saturated.

Let us therefore consider spherical volumes of unstrained solid within the solid above the Curie temperature and hence in the disordered phase as shown in Figure 5.6a. When the material becomes ferromagnetic at the Curie point, spontaneous

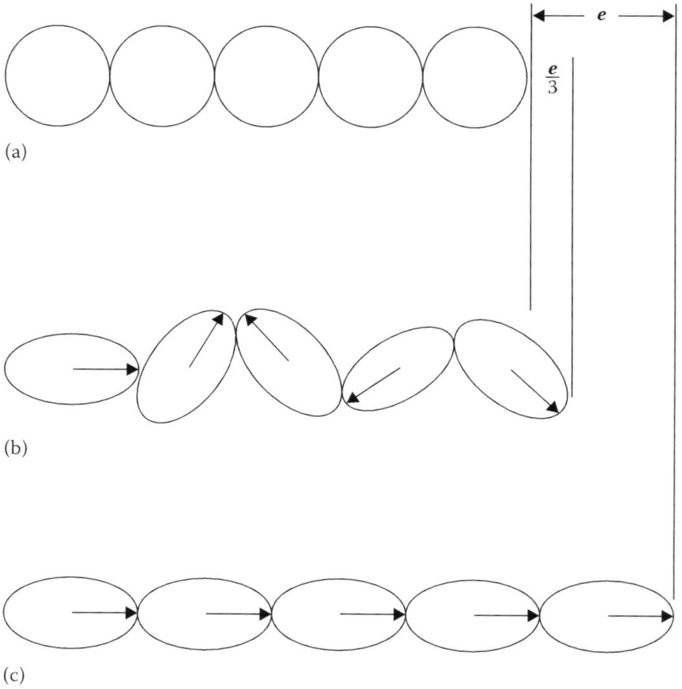

FIGURE 5.6 A schematic diagram illustrating the magnetostriction in: (a) the disordered (paramagnetic) regime; (b) the ferromagnetic regime demagnetized; and (c) the ferromagnetic regime, magnetized to saturation.

magnetization appears within the domains and with it an associated spontaneous strain e or magnetostriction λ_0, along a particular direction, as shown in Figure 5.6b.

For the present isotropic case, the amplitudes of these spontaneous magnetostrictions are independent of crystallographic direction. Within each *isotropic* domain, the strain varies with angle θ from the direction of spontaneous magnetization according to the following relation:

$$e(\theta) = e\cos^2\theta \tag{5.12}$$

The average deformation throughout the solid due to the onset of spontaneous magnetostriction can then be obtained by integration, assuming that the domains are oriented at random, so that any particular direction is equally likely.

$$\begin{aligned}\lambda_0 &= \int_0^{\pi/2} e\cos^2\theta\sin\theta\, d\theta \\ &= \frac{e}{3}\end{aligned} \tag{5.13}$$

This then is the spontaneous magnetostriction caused by the ordering of the magnetic moments at the onset of ferromagnetism. We should note that since we have assumed an isotropic material, the domains are arranged with equal probability in any direction and therefore the strain is equivalent in all directions. Therefore, in this case, although the sample undergoes changes in dimensions, its shape remains the same.

5.3.2 Saturation Magnetostriction

What is the maximum change in length when a ferromagnet is magnetized?

Next, we will consider saturation magnetostriction, which is the fractional change in length between a demagnetized ferromagnetic specimen and the same specimen in a magnetic field sufficiently strong to saturate the magnetization along the field direction. In this case, there will be a change of shape since the applied field generates a preferred direction.

Using the very simple model above, we cause the transition from the ordered but demagnetized state to the ordered saturated state by application of a magnetic field. In the saturated state, of course, the magnetic moments in the domains are all aligned parallel to the field and hence the strains are parallel as shown in Figure 5.6c.

$$\begin{aligned}\lambda_s &= e - \lambda_0 \\ &= \frac{2}{3}e\end{aligned} \tag{5.14}$$

This gives us a method of measuring the spontaneous strain e within a material due to magnetic ordering along a particular direction, by measuring λ_s.

5.3.3 Technical Saturation and Forced Magnetostriction

Can the saturation increase even after the magnetization has reached technical saturation?

As discussed in Section 2.3.2, technical saturation of magnetization occurs when all magnetic domains within a material have been aligned in the same direction to form a single-domain specimen. However, if the magnetic field is increased further, there is still a very slow rise in $\boldsymbol{M}$ and this process is called *forced magnetization*.

Similar behavior is observed in the magnetostriction. Technical saturation magnetostriction is reached when the specimen has been converted to a single domain. However, a very slow change in magnetostriction, called *forced magnetostriction*, is observed as the field is increased further. Forced magnetostriction is a very small effect, being appreciable only at fields of the order of 800 kA m^{-1} (10 kOe).

The phenomenon is caused by an increase in the ordering of individual atomic magnetic moments within the single domain, which is treated in detail in Chapter 9. This is the same mechanism that leads to an increase in the spontaneous magnetization within a domain. The statistical ordering of magnetic moments within a domain is highly temperature dependent [19] and consequently so is the magnitude of the forced magnetostriction.

5.3.4 Magnetostriction at an Angle to the Magnetic Field

How does the saturation magnetostriction vary as a function of angle?

Since we are still considering a completely isotropic medium, we can write down an equation for the saturation magnetostriction $\lambda_s(\theta)$ at any angle θ to the field direction.

$$\lambda_s(\theta) = \frac{3}{2}\lambda_s\left(\cos^2\theta - \frac{1}{3}\right) \tag{5.15}$$

where λ_s is the saturation magnetostriction along the direction of magnetization.

This leads to an explanation of some of the magnetostriction measurements, which you will often find in the literature, in which the magnetostriction with the field parallel to a given $\lambda_{s\parallel}$ direction and the magnetostriction with the field perpendicular to the given direction $\lambda_{s\perp}$ are measured and the difference taken. The difference between them gives the spontaneous strain within a single domain.

$$\lambda_{s\parallel} - \lambda_{s\perp} = \lambda_s + \frac{\lambda_s}{2} = \frac{3}{2}\lambda_s = e \tag{5.16}$$

5.3.5 Anisotropic Materials

Are the magnetic properties identical in all directions in a single crystal?

Although single crystal nickel comes fairly close to having isotropic properties such as magnetostriction, the reality is that all solids are anisotropic to some degree, and therefore the saturation magnetostriction needs to be defined in relation to the crystal axis along which the magnetization lies. An extensive review of magnetostriction in anisotropic materials has been given by Lee [20].

The magnetostrictions or spontaneous strains are defined along each of the principal axes of the crystal. For cubic materials, there are two independent magnetostriction constants λ_{100} and λ_{111} as shown in Table 5.2.

The saturation magnetostriction in single domain, single crystal cubic materials is then given by a generalized version of the equation for isotropic materials above.

TABLE 5.2
Magnetostriction Coefficients of Some Cubic Materials

Material	λ_{100} (10^{-6})	λ_{111} (10^{-6})
Iron	21	–21
Nickel	–46	–24
Terfenol	90	1640

$$\lambda_s = \frac{3}{2}\lambda_{100}\left(\alpha_1^2\beta_1^2 + \alpha_2^2\beta_2^2 + \alpha_3^2\beta_3^2 - \frac{1}{3}\right) + 3\lambda_{111}\left(\alpha_1\alpha_2\beta_1\beta_2 + \alpha_2\alpha_3\beta_2\beta_3 + \alpha_3\alpha_1\beta_3\beta_1\right) \quad (5.17)$$

where:

λ_{100} is the saturation magnetostriction measured along the <100> directions
λ_{111} is the saturation magnetostriction along the <111> directions

In this equation, α_1, α_2, and α_3 are the direction cosines, relative to the field direction, of the axis along which the magnetic moments are aligned and hence along which ***M*** is saturated, while β_1, β_2, and β_3 are the direction cosines relative to the field direction of the axis along which the magnetostriction is measured. These equations of course only apply to the magnetostriction within a domain. The anisotropic magnetostriction equations for cubic and other crystal classes have been given by Lee [21].

Usually, we need to know the saturation magnetostriction in the same direction as the field in which case the above expression reduces to the following:

$$\lambda_s = \frac{3}{2}\lambda_{100} + 3(\lambda_{111} - \lambda_{100})\left(\alpha_1^2\alpha_2^2 + \alpha_2^2\alpha_3^2 + \alpha_3^2\alpha_1^2\right) \quad (5.18)$$

Once again, as in the case of isotropic materials, the two constants λ_{100} and λ_{111} can be determined by saturating the magnetostriction along the axis of interest and then at right angles. The difference in strain remains (3/2) λ_{100} and (3/2) λ_{111} depending on the axis chosen.

The behavior of the magnetostriction of an assembly of domains, a poly-crystal, for example, can only be calculated by averaging the effects. This is not possible in general, and therefore, it is assumed that the material consists of a large number of domains and hence that the strain is uniform in all directions. In a randomly oriented polycrystalline cubic material (i.e., one in which there is no preferred grain orientation), this formula simplifies further to become

$$\lambda_s = \frac{2}{5}\lambda_{100} + \frac{3}{5}\lambda_{111} \quad (5.19)$$

5.3.6 Field-Induced Bulk Magnetostriction

How does the length of a ferromagnet change with magnetic field?

The field-induced bulk magnetostriction is the variation of λ with ***H*** or ***B*** and is often the most interesting feature of the magnetostrictive properties to the materials scientist. However, the variations $\lambda(\mathbf{H})$ or $\lambda(\boldsymbol{B})$ are very structure sensitive, so that it is not possible to give any general formula for the relation of magnetostriction to field. This may at first seem strange when we have already been able to give equations for the saturation magnetostriction and for the variation of magnetization with field.

The magnetostrictions of polycrystalline iron, cobalt, and nickel are shown in Figure 5.7. This indicates some of the problems in developing a simple theory based

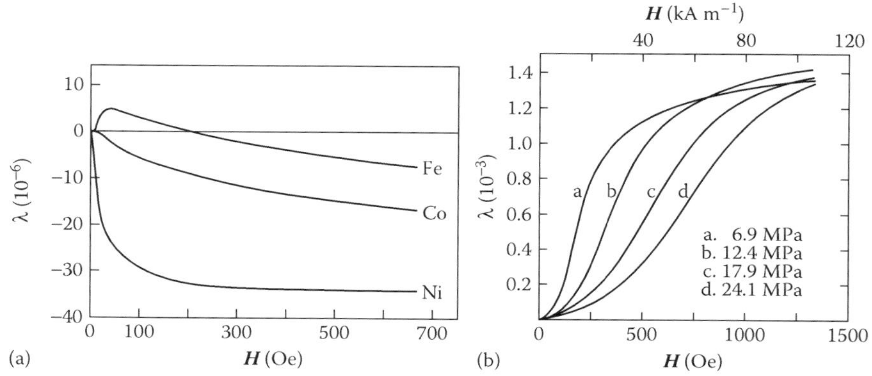

FIGURE 5.7 Dependence of the bulk magnetostriction on applied magnetic field in: (a) iron, nickel, and cobalt and (b) a highly magnetostrictive rare earth iron alloy $Tb_{0.27}Dy_{0.73}Fe_{1.95}$.

on the relationship of magnetostriction λ to magnetization $\boldsymbol{M}$. For example, while magnetization increases monotonically with increase in magnetic field, the bulk magnetostriction of iron actually changes sign from positive at low fields to negative at high fields as both $\boldsymbol{H}$ and $\boldsymbol{M}$ increase. Lee [21] has reported on the magnetostriction curves of other polycrystalline ferromagnets.

There is one case at least where a fairly simple solution occurs. That is when the magnetic field is applied in a direction perpendicular to the easy axis in a single crystal with uniaxial anisotropy, or perpendicular to the axis in which the moments have been completely aligned in a polycrystalline material such as nickel under extreme tension or terfenol under compression.

In this case, magnetization takes place entirely by rotation of magnetization. Therefore, we may make the substitution

$$\boldsymbol{M} = \boldsymbol{M}_s \cos\theta \tag{5.20}$$

which gives the magnetization along the field axis in terms of the angle θ, which the saturation magnetization within the domains makes with this axis. The magnetostriction along the field axis is given by

$$\lambda = \frac{3}{2}\lambda_s \cos^2\theta \tag{5.21}$$

and substituting for $\cos^2\theta$ leaves

$$\lambda = \frac{3}{2}\lambda_s \left(\frac{\boldsymbol{M}}{\boldsymbol{M}_s}\right)^2 \tag{5.22}$$

which gives the variation of the observed magnetostriction with magnetization $\boldsymbol{M}$.

5.3.7 Transverse Magnetostriction

When the length of a ferromagnet changes what happens to the cross section?

Between the demagnetized state and the saturation magnetization, the volume of a ferromagnet remains fairly constant (although there is a very small volume magnetostriction in most materials). Therefore, there is a transverse magnetostriction of one-half the longitudinal magnetostriction and opposite in sign.

$$\lambda_t = \frac{-\lambda}{2} \tag{5.23}$$

5.3.8 Magnetostrictive Materials and Applications

Where do such materials find uses?

Highly magnetostrictive materials, that is, materials with saturation magnetostrictions of beyond a few hundred parts per million, are invariably alloys of rare earth metals with transition metals such as iron, cobalt, and nickel. The rare earth metals provide the high magnetostriction and the transition metals provide the high Curie temperatures, so that the materials can be used under ambient temperature conditions. A review of magnetostrictive materials for applications has been given by Greenough and Schulze [22]. These materials find practical applications today mostly in actuators (e.g., transducers and positioning devices) and in magnetic sensors. These applications have been summarized by Greenough, Reed, and Schulze [23].

5.4 MAGNETORESISTANCE

Does electrical resistance change with magnetization in materials?

The electrical resistivity of magnetic materials changes when they are magnetized. In most cases, the change is an increase in resistivity in bulk magnetic materials when current and magnetization are parallel, and a decrease when they are orthogonal. In the classical free electron model of metals, the magnetic contribution ρ_{mag} to the resistivity ρ is given by [24]

$$\rho = \rho_0 + \rho_{mag} = \rho_0 \left[1 + \left(\frac{eB}{m_e} \right)^2 \tau \right] \tag{5.24}$$

$$\rho_{mag} = \rho_0 \left(\frac{eB}{m_e} \right)^2 \tau \tag{5.25}$$

where:

ρ_0 is the resistivity in the absence of a magnetic field
m_e is the mass of the electrons
e is the charge on the electrons
B is the magnetic flux density
τ is the mean free time of the electrons between collisions

The magnetoresistance is usually quoted in terms of the fractional change in resistance $\Delta R/R$ and is typically a few percent, although as mentioned in Chapter 3, it can be as high as 300% in some materials, and in the so-called colossal magnetoresistive materials, it can be even higher.

Investigation of magnetoresistance in magnetic multilayers led to a rapid increase in the number of materials available, which show substantial magnetoresistance (i.e., more than a few percent). Giant magnetoresistance was first observed in layered structures consisting of iron and chromium by Fert et al. [25] and by Grunberg et al. [26]. In these multilayers, the resistance is controlled by changing the relative orientation of the magnetization in subsequent magnetic layers from antiparallel to parallel by the application of a magnetic field. In order to understand this phenomenon, the ideas of exchange coupling and antiferromagnetic order need to be developed, and therefore, the discussion of the mechanism needs to be deferred until later in the book.

However, the essential idea is that the scattering of the conduction electrons is altered depending on whether the subsequent magnetic layers are aligned parallel, with consequently the same available energy states for the electrons, or antiparallel in which case, the conduction electrons from one layer may find that equivalent energy states are not available in the layers with magnetization oriented antiparallel. Fractional changes in resistance of 50%–90% are achievable in this way, and bearing in mind that this is a reduction, the final resistance may be as low as one-tenth of the zero field resistance [27]. These materials have been used to develop magnetic sensors, with one of the largest areas of application being magnetic read heads for magnetic data storage. A review of developments in magnetoresistance in multilayers was given by Levy [28].

REFERENCES

1. Fröhlich, O. *Electrotech Z.*, 2, 134, 1881.
2. Kennelly, A. E. *Trans. Am. IEE.*, 8, 485, 1891.
3. Weiss, P. *J. Phys.*, 9, 373, 1910.
4. Becker, R., and Doring, W. *Ferromagnetismus*, Springer, Berlin, Germany, 1938.
5. Mee, C. D. *The Physics of Magnetic Recording*, North Holland, Amsterdam, the Netherlands, 1964.
6. Rayleigh, L. *Phil. Magn.*, 23, 225, 1887.
7. Bozorth, R. M. *Ferromagnetism*, Van Nostrand, New York, 1951, p. 484.
8. Barkhausen, H. *Phys. Z.*, 29, 401, 1919.
9. Schroeder, K., and McClure, J. C. The Barkhausen effect, *Crit. Rev. Solid State.*, 6, 45, 1976.
10. Matzkanin, G. A., Beissner, R. E., and Teller, C. M. *The Barkhausen Effect and Its Application to NDE*, SWRI Report No. NTIAC–79–2, Southwest Research Institute, San Antonio, TX, 1979.

11. Alessandro, B., Beatrice, C., Bertotti, G., and Montorsi, A. *J. Appl. Phys.*, 68, 2901, 1990.
12. Jiles, D. C., Sipahi, L. B., and Williams, G. *J. Appl. Phys.*, 73, 5830, 1993.
13. Lord, A. E. Acoustic emission, in *Physical Acoustics*, Vol. XI (eds. W. P. Mason and R. N. Thurston), Academic Press, New York, 1975.
14. Ono, K., and Shibata, M. *Mater. Eval.*, 38, 55, 1980.
15. Kusanagi, H., Kimura, H., and Sasaki, H. *J. Appl. Phys.*, 50, 1989, 1979.
16. Buttle, D. J., Scruby, C. B., Briggs, G. A. D., and Jakubovics, J. P. *Proc. Roy. Soc. Lond. A*, 414, 469, 1987.
17. Joule, J. P. *Ann. Electr. Magn. Chem.*, 8, 219, 1842.
18. Joule, J. P. *Phil. Magn.*, 30, 76, 1847.
19. Kittel, C. *Introduction to Solid State Physics*, 6th edn, Wiley, New York, 1987, p. 248.
20. Lee, E. W. Magnetostriction and magnetomechanical effects, *Rep. Prog. Phys.*, 18, 184, 1955.
21. Lee, E. W. Magnetostriction curves of polycrystalline ferromagnets, *Proc. Phys. Soc.*, 72, 249, 1958.
22. Greenough, R. D., and Schulze, M. P. in *Intermetallic Compounds* (eds. J. H. Westbrook and R. L. Fleischer), John Wiley, New York, 1994.
23. Greenough, R. D., Reed, I. M., and Schulze, M. P. Magnetostrictive actuators, in *Advances in Actuators* (eds. A. P. Dorey and J. H. Moor), Institute of Physics, Bristol, 1995.
24. Jiles, D. C., *Introduction to the Electronic Properties of Materials*, 2nd edn, Nelson-Thornes, Cheltenham, Gloucestershire, 2001, p. 170.
25. Baibich, M. N., Broto, J. M., Fert, A. et al. *Phys. Rev. Letts.*, 61, 2472, 1988.
26. Binash, G., Grunberg, P., Saurenbach, F., and Zinn, W. *Phys. Rev. B*, 39, 4828, 1989.
27. Fert, A., Grunberg, P., Bartholemy, A. et al. *J. Magn. Mater.*, 140, 1, 1995.
28. Levy, P. M. *Solid State Phys.*, 47, 367, 1994.

FURTHER READING

Cullity, B. D. Fundamentals of magnetostriction, *Int. J. Magnetism*, 1, 323, 1971.
Cullity, B. D. *Introduction to Magnetic Materials*, Addison-Wesley, Reading, MA, 1972.
Heck, C. *Magnetic Materials and Their Applications*, Crane Russak, New York, 1974, Ch. 4.

EXERCISES

5.1 *Fröhlich-Kennelly relation.* Show that the Fröhlich equation for the anhysteretic magnetization

$$M - \alpha H\left(1+\beta H\right) \tag{5.1.1}$$

and the Kennelly expression

$$\frac{1}{\left(\mu-\mu_0\right)} = a + bH \tag{5.1.2}$$

are equivalent. Expand these equations in terms of a series and compare the resulting forms with the law of approach to saturation. Do you notice any similarities? Comment on the behavior at high fields.

5.2 *Magnetization of paramagnets.* A piece of aluminum has a permeability of 1.00002 and a piece of copper has a susceptibility of -1×10^{-5}. Determine the strength of the magnetic field required to induce the same level of magnetization in these materials as in a piece of nickel. Are such field strengths practically attainable in the laboratory?

5.3 *Determination of Rayleigh coefficients at low fields.* Using the results for the initial magnetization of iron given in Table 5.3.1, plot the initial magnetization curve. Estimate the extent of the Rayleigh region. Calculate the Rayleigh coefficients and use them to determine the remanence that would be observed if the field were reduced to zero after reaching a maximum value of (1) 10 A m^{-1} and (2) 20 A m^{-1}. Find the hysteresis loss when the field is given one complete cycle at an amplitude of 20 A m^{-1}.

5.4 *Magnetostriction of terfenol without stress.* In polycrystalline terfenol, which is not subjected to stress and does not have a preferred orientation, the domains are aligned randomly. If $\lambda_{111} = 1600 \times 10^{-6}$ and $\lambda_{100} = 90 \times 10^{-6}$, calculate the saturation magnetostriction along the field direction on the basis of initial random alignment of domains.

For the same specimen of terfenol, calculate the *saturation magnetostriction* $\lambda_{\parallel} - \lambda_{\perp}$ obtained by applying a saturating field at right angles to a given direction and setting the strain at an arbitrary value of zero, then rotating the field into the direction of interest. How does this value of magnetostriction compare with the saturation magnetostriction obtained by the first method?

5.5 *Magnetostriction of terfenol under compressive load.* By applying a uniaxial compressive stress to terfenol, the moments can also be made to line

TABLE 5.3.1
Magnetic Properties of Annealed Iron

H (A m^{-1})	*B* (T)
0	0
5	0.0019
10	0.0042
15	0.0069
20	0.010
40	0.028
50	0.043
60	0.095
80	0.45
100	0.67
150	1.01
200	1.18
500	1.44
1,000	1.58
10,000	1.72

up perpendicular to the stress axis. Using the saturation magnetostriction calculated above for the nonoriented terfenol, determine how the magnetostriction of stressed terfenol varies with magnetization (assume the stress is sufficient to align all domains at right angles to the stress axis) and hence find the magnetostriction at $\boldsymbol{M} - \boldsymbol{M}_s/2$ and at $\boldsymbol{M} = \boldsymbol{M}_s$. How does this compare with $\lambda_\parallel - \lambda_\perp$, which was calculated in Exercise 5.4?

Discuss how the magnetostriction along the field axis would be changed if instead of having a nonoriented sample we had one with the <111> directions aligned preferentially along the field axis.

5.6 *Hysteresis loss.* The hysteresis curve shown in Figure 5.6.1 is for a material with coercivity 50 A m^{-1}, field amplitude 250 A m^{-1}, induction amplitude 1.6 T, and remanence 1 T. Estimate the hysteresis loss per cycle in J m^{-3}. The material is to be used as a transformer core with a cross-sectional area of 4×10^{-4} m^2 and an average path length of 300 mm. Estimate the hysteresis power dissipation at 60 Hz as the material is driven round this hysteresis curve. Estimate the eddy current losses if the resistivity is 0.57 μΩ m, the thickness of the laminations in the transformer is 0.002 m, and the shape factor is 6. (You will need to find the expression for classical eddy current losses in a magnetic material.)

5.7 *Mathematical models for magnetization processes in ferromagnets.* Describe the various mathematical models that have been developed for (a) low field, (b) high field, and (c) anhysteretic magnetization curves (***M*** versus ***H***) and

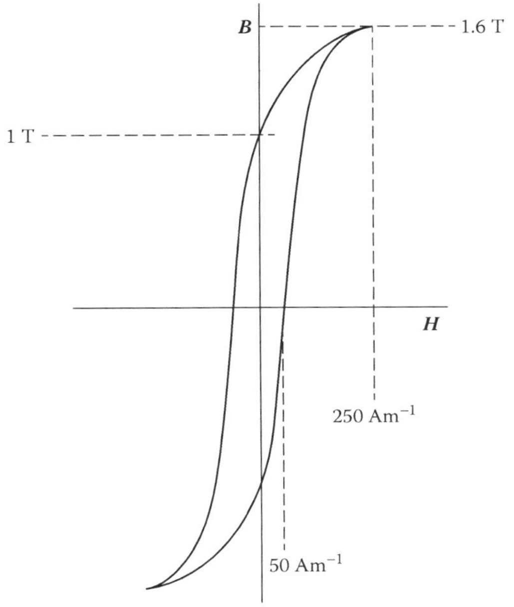

FIGURE 5.6.1 The hysteresis curve for a material with coercivity 50 A m^{-1}, field amplitude 250 A m^{-1}, induction amplitude 1.6 T, and remanence 1 T.

TABLE 5.9.2
Magnetic Properties of Nickel

H (A/m)	M (10^6 A m^{-1})
323	0.049
430	0.092
646	0.196
1075	0.298
1398	0.330
1720	0.349
2152	0.365
2712	0.379
3232	0.388

explain their limitations. In the law of approach to saturation, prove mathematically that the second-order term (the b/H^2 term) is determined by the magnetocrystalline anisotropy. In particular, show how b is related to the anisotropy coefficient K, and explain any assumptions that you have made to reach this conclusion.

5.8 *Magnetostriction along different directions relative to the magnetization.* Describe what is meant by magnetostriction. Derive an equation for the dependence of strain along the <110> direction in a cubic material as a function of the angle of $\boldsymbol{M}$ relative to the <100> direction as the individual magnetic moments (and hence the domain magnetization vector $\boldsymbol{M}_s$) rotate away from the <100> direction toward the <110> direction, assuming that they always lie in the (001) plane.

5.9 *Differences in magnetization curves for the same material in a toroid and in a sphere.* These values of $\boldsymbol{H}$ and $\boldsymbol{M}$ were measured on a toroidal sample of nickel (Table 5.9.2).

Plot the $\boldsymbol{M}$ versus $\boldsymbol{H}$ curve of the material under these conditions. Then plot the expected $\boldsymbol{M}$ versus applied field $\boldsymbol{H}_{app}$ curve for the material in the form of a sphere. What is the value of the reciprocal slope ($d\boldsymbol{H}_{app}/d\boldsymbol{M}$) at the origin ($\boldsymbol{H}_{app} = 0$, $\boldsymbol{M} = 0$) for the sphere? How does this relate to the demagnetizing factor N_d?

5.10 *Classical dipole magnetic field of an array of atoms.* Assuming that each atom behaves as if it has a fixed, small (meaning short, so that you are calculating $\boldsymbol{H}$ in the *far field* and not in the *near field*), classical magnetic dipole moment of 1 Bohr magneton, calculate the magnetic field at a distance of 0.3 nm from the magnetic dipole moment: (1) along the axis of the moment and (2) in the plane perpendicular to the moment. Discuss whether this field can be responsible for ferromagnetic ordering in the material and/or a Curie temperature of typically a few hundred to several hundred degrees Kelvin.

6 Magnetic Domains

We now consider the organization of the magnetic moments within ferromagnets. Two questions arise: Are the magnetic moments permanent or field induced, and are they randomly aligned or ordered? It will be shown that the moments are permanent and are aligned parallel within volumes containing large numbers of atoms, but these regions, known as *magnetic domains*, are on the microscopic scale in most cases.

6.1 DEVELOPMENT OF DOMAIN THEORY

What microscopic theories are needed to account for the observed macroscopic properties of ferromagnets?

On the macroscopic scale, the bulk magnetization $\boldsymbol{M}$ is clearly field induced. On the microscopic scale, we need to find how the magnetization varies within a ferromagnet. In the demagnetized state, for example, is the magnetization everywhere zero or are there large local values of magnetic moment that sum to zero on the macroscopic scale? If there are large local magnetic moments, we need to find how these are arranged and what happens to them when subjected to a field.

6.1.1 Atomic Magnetic Moments

Do the atoms of a ferromagnet have permanent magnetic moments or are the moments induced by the magnetic field?

Since atoms are the units from which solids are composed, it is reasonable to suppose that when a ferromagnetic solid is magnetized, there is a net magnetic moment per atom. For example, in Chapter 4, we have calculated the net magnetic moment per atom in saturated iron. There are two possible origins for the atomic magnetic moments in ferromagnets. The material could already have small atomic magnetic moments within the solid that are randomly aligned (or at least give a zero vector sum over the whole solid) in the demagnetized state but which became ordered (or aligned) under the action of a magnetic field. This was first suggested by Weber [1]. Alternatively, the atomic magnetic moments may not exist at all in the demagnetized state but could be induced on the application of a magnetic field as suggested by Poisson [2].

The existence of saturation magnetization and remanence support the former idea, and in fact, it has been established beyond doubt that in ferromagnets, permanent magnetic moments exist on the atomic scale and that they do not rely on the presence of field for their existence. The origin of the atomic moments was first suggested by Ampère [3], who, with extraordinary insight, suggested that they were due to electrical currents continually circulating within the atom. This was some 75 years before Thomson discovered the electron and at a time when it was not known whether charge separation existed within an atom.

We should note that both paramagnets and ferromagnets have permanent atomic magnetic moments. Next, it is necessary to distinguish between them on a microscopic scale knowing that on the macroscopic scale, the permeabilities of ferromagnets are much higher.

6.1.2 Magnetic Order in Ferromagnets

Are the ferromagnets already in an ordered state before a magnetic field is applied or is the order induced by the field?

Ewing [4] followed the earlier ideas of Weber in explaining the difference between a magnetized and demagnetized ferromagnet as due to the atomic moments (or *molecular magnets*, as they called them in those days) being randomly oriented in demagnetized iron but aligned in the magnetized material. Ewing was particularly interested in explaining hysteresis on the basis of interactions between the atomic dipole moments of the type envisaged by Weber. As we shall see, this was inevitably to fail because he did not realize that the demagnetized iron was actually already in an ordered state with large numbers of atomic magnetic moments aligned locally in parallel.

6.1.3 Permeability of Ferromagnets

Can the properties of ferromagnets be explained best by assuming that a magnetic field rearranges existing ordered-volume magnetic moments or by assuming that the field aligns disordered (randomly oriented) atomic magnetic moments?

One of the problems in the field of magnetism that needed to be addressed was the very large permeabilities and susceptibilities of ferromagnets. In their original state, the bulk magnetization of ferromagnets is zero, but on application of a magnetic field, they become *magnetically polarized*; that is, they acquire a magnetization. However, the magnetizations of ferromagnets are mostly orders of magnitude greater than the field strengths that produce them.

There are two possible explanations. The atomic magnetic moments are randomly oriented on the interatomic scale and the field gradually aligns them as in the case of paramagnets. Alternatively, the moments could be aligned on the interatomic scale but at some larger scale, the magnetizations of whole aligned regions could be randomly aligned from one domain to the next.

The properties of ferromagnets can be explained if long-range magnetic order exists within the solid but the volumes, known as *domains*, containing the magnetic moments are randomly aligned in the demagnetized state. Magnetization is then simply the process of rearranging these volumes so that their magnetizations are aligned parallel. The paramagnets, which also have permanent atomic magnetic moments, can then be distinguished from the ferromagnets because the paramagnets do not exhibit long-range order such as is found in ferromagnetic domains. In fact, in paramagnets, the atomic magnetic moments are randomly aligned in the absence of a field due to the thermal energy, also known as *Boltzmann energy*.

6.1.4 Domain Theory

If the atomic magnetic moments are ordered, then how do we explain the demagnetized state?

Some years after Ewing's work, one of the most important advances in the understanding of ferromagnetism was made by Weiss in two works [5,6]. In these works, Weiss built on the earlier work of Ampère, Weber, and Ewing and suggested the existence of magnetic domains in ferromagnets, in which the atomic magnetic moments were aligned parallel over much larger volumes of the solid than had previously been suspected. In these domains, large numbers of atomic moments are aligned parallel so that the magnetization within the domain is almost saturated. However, the direction of alignment varies from domain to domain in a more or less random manner, although certain crystallographic axes are preferred by the magnetic moments, which in the absence of a magnetic field will align along one of these equivalent *magnetic easy axes*.

The immediate consequences of this were (1) atomic magnetic moments were in permanent existence (Weber's hypothesis), (2) the atomic moments were ordered (aligned) even in the demagnetized state, (3) it was the domains only that were randomly aligned in the demagnetized state, and (4) the magnetization process consisted of reorienting the domains, so that either more domains were aligned with the field or the volumes of domains aligned with the field were greater than the volume of domains aligned against the field.

6.1.5 Mean Field Theory

What is the underlying cause of the alignment of atomic magnetic moments?

If the atomic moments are aligned within the domains of ferromagnets, it is necessary to explain this ordering and if possible to explain why when a ferromagnet is heated up it eventually undergoes a transition to a paramagnet at the Curie temperature.

In order to explain these observations, Weiss further developed the statistical thermodynamic ideas of Boltzmann and Langevin as they applied to magnetic materials. Some years previously, Langevin [7] had produced a theory of paramagnetism based on classical Boltzmann statistics. Weiss used the basic Langevin model but added an extra term, the so-called Weiss mean field, which was in effect an interatomic interaction that caused neighboring atomic magnetic moments to align parallel because their energy was lower if they did so.

In the original Weiss theory, the mean (or *molecular*) field was proportional to the bulk magnetization $\boldsymbol{M}$ so that

$$\boldsymbol{H}_{\mathrm{e}} = \alpha \boldsymbol{M} \tag{6.1}$$

where α is the mean field coefficient that depends on the strength of the field. This mean field can be proved to be equivalent assuming that each atomic moment interacts equally with every other atomic moment within the solid. This was found to be a viable assumption in the paramagnetic phase because, due to the homogeneous distribution of magnetic moment directions, the local value of magnetization, obtained by considering a microscopic volume of the material surrounding a given atomic magnetic moment, is equal to the bulk magnetization.

However, in the ferromagnetic phase, the magnetization is locally inhomogeneous on a scale larger than the domain size due to the variation in the direction of magnetization from domain to domain. Therefore, subsequent authors preferred to apply the idea of a Weiss mean field only within a domain, arguing that the interaction between the atomic moments decayed with distance and that therefore such an interaction was unlikely to extend beyond the domain. It is generally considered that the Weiss field is a good approximation to the real situation within a given domain because within the domain, the magnetization is homogeneous and has a known $\boldsymbol{M}_s$. So, the interaction field that is responsible for the ordering of moments within domains can be expressed as

$$\boldsymbol{H}_e = \alpha \boldsymbol{M}_s \tag{6.2}$$

where $\boldsymbol{M}_s$ is the spontaneous magnetization within the domain, which has been discussed in Section 2.3.2; it is equal to the saturation magnetization at 0 K but decreases as the temperature is increased, becoming zero at the Curie point.

It will be seen later that this equivalent field $\boldsymbol{H}_e$ is not a magnetic field in the normal sense. This *magnetic field* is a useful abstraction that can be used to represent the additional energy that forces the strong alignment of magnetic moments inside the domains of a ferromagnetic material. The molecular or mean field energy of a single magnetic moment $\boldsymbol{m}$ is then

$$E = -\mu_0 \boldsymbol{m} \cdot \boldsymbol{H}_e = -\mu_0 \alpha \boldsymbol{m} \cdot \boldsymbol{M}_s \tag{6.3}$$

A positive value of α gives a minimum energy when the alignment of the magnetic moments is parallel resulting in ferromagnetism.

This energy per magnetic moment arising from this mean field is an additional energy, known as the *exchange energy*, which causes alignment of the magnetic moments within a domain. The existence of a Curie temperature is a direct consequence of this energy and the exchange energy can be calculated from the Curie temperature T_c.

When the thermal energy per magnetic moment equals or exceeds the exchange energy per magnetic moment, the magnetic alignment or order is destroyed. This is the Curie temperature T_c, which marks the transition between the magnetically ordered state of ferromagnetism and disordered state of paramagnetism. The Curie temperature can therefore be related directly to the exchange energy and the mean field coupling coefficient α. According to the three-dimensional classical theory, the thermal energy per moment at T_c will be $3k_BT_c$, and therefore

$$E_{ex} = 3k_BT_c \tag{6.4}$$

and consequently

$$T_c = \frac{\mu_0 \alpha N \boldsymbol{m}^2}{3k_B} \tag{6.5}$$

From these equations, we can obtain an estimate of the exchange energy per moment for a material with a known Curie temperature. For iron, $T_c = 1043$ K and therefore $E_{ex} = 4.3 \times 10^{-20}$ J; for cobalt, $T_c = 1404$ K and therefore $E_{ex} = 5.8 \times 10^{-20}$ J; and for nickel $T_c = 631$ K and therefore $E_{ex} = 2.6 \times 10^{-20}$ J. From these values, and assuming that only exchange between nearest-neighbor pairs is significant, the exchange energy is 5×10^{-21} J per pair in iron, 5×10^{-21} J per pair in cobalt, and 2×10^{-21} J per pair in nickel, since these have $n = 8$, 12, and 12 nearest neighbors, respectively.

The spontaneous magnetization $\boldsymbol{M}_s$ in a domain is the vector sum of all the individual magnetic moments m_j in the domain divided by the volume of the domain V.

$$\boldsymbol{M}_s = \frac{1}{V}\sum_j \boldsymbol{m}_j \tag{6.6}$$

Therefore, the energy of the magnetic moment $\boldsymbol{m}$ shown in Equation 6.3 can also be written in terms of microscopic quantities

$$E = -\mu_0 \frac{\alpha}{V}\boldsymbol{m}\cdot\sum_j \boldsymbol{m}_j \tag{6.7}$$

This is the summation of all pair interaction energies between $\boldsymbol{m}$ and all other magnetic moments $\boldsymbol{m}_j$ in the domain. These pair interaction energies can be written as

$$E_{\text{pair}} = -\mu_0 \frac{\alpha}{V}\boldsymbol{m}_i\cdot\boldsymbol{m}_j \tag{6.8}$$

Equation 6.8 assumes, in accordance with the mean field approach, that the interaction coefficient α is identical for all pairs of moments. This is unlikely in general but could be true in specific cases. More generally, the interaction coefficient α will differ from pair to pair, so that if the interaction coefficient between $\boldsymbol{m}_i$ and $\boldsymbol{m}_j$ is α_{ij}, then their pair interaction energy will be as follows:

$$E_{\text{pair}} = -\mu_0 \frac{\alpha_{ij}}{V}\boldsymbol{m}_i\cdot\boldsymbol{m}_j \tag{6.9}$$

When $\alpha_{ij} > 0$, the minimum energy state results in ordering of the magnetic moments that is parallel, leading to ferromagnetism, but when $\alpha_{ij} < 0$, the minimum energy state results in ordering of the magnetic moments that is antiparallel, leading to simple antiferromagnetism. The resulting ordering of moments within a domain is shown in Figure 6.1.

Subsequent models, such as the Ising model [8] applied to ferromagnets, have been based on interaction fields only between nearest neighbors. A range of different types of magnetic order is possible depending on the nature of the interaction parameter α. Some of these configurations are shown in Figure 6.2.

In Chapter 9, we will discuss both the Langevin and the Weiss theories in detail, showing derivation of the equations and how the Curie temperature is determined by

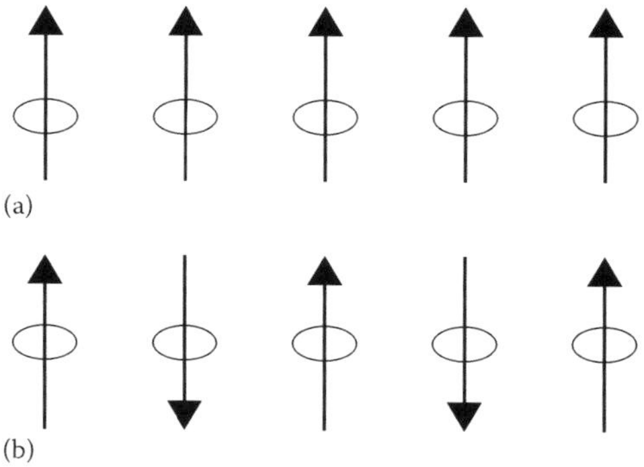

FIGURE 6.1 Ordered arrangement of a linear array of atomic magnetic moments (a) with $\alpha > 0$ leading to ferromagnetism and (b) with $\alpha < 0$ leading to simple antiferromagnetism.

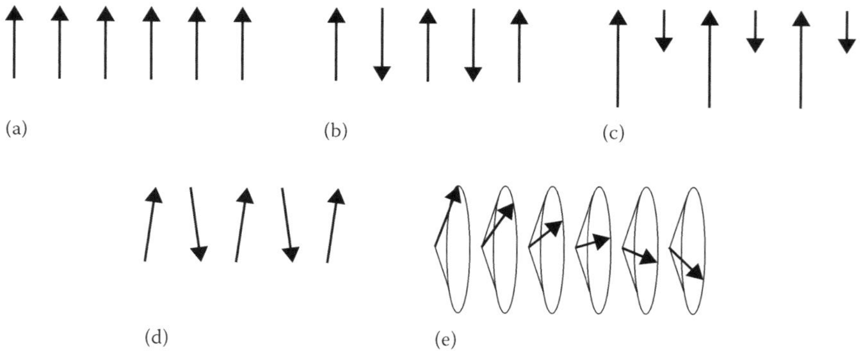

FIGURE 6.2 Examples of different types of magnetic order using a linear array of localized moments. These include (a) ferromagnetism, (b) simple antiferromagnetism, (c) ferrimagnetism, (d) canted antiferromagnetism, and (e) helical spin array.

the mean field. We will also leave the explanation of the Weiss mean field until later, although we should note in passing that it cannot be explained in classical terms and depends entirely on quantum mechanical considerations. In quantum mechanics, it is known as the *exchange interaction* and we shall refer to it as such occasionally.

As an example, we can calculate the equivalent field strength of the Weiss mean field. Suppose the field experienced by any magnetic moment $\boldsymbol{m}_i$ within a domain due to its interaction with any other moment $\boldsymbol{m}_j$ is $\boldsymbol{H}_e = \alpha_{ij}\boldsymbol{m}_j$. (1) Find the field experienced by this moment as a result of its interactions with all other moments, (2) find the energy of the moment as a result of this interaction, (3) if each moment interacts equally with all other moments within the domain, show that this is equivalent to a mean field, (4) calculate the field strength if the mean field parameter in iron is $\alpha = 400$ in SI units ($= 400 \times 4\pi$ in CGS units).

It is known that in iron $M_0 = 1.7 \times 10^6$ A m^{-1}, and at room temperature, the spontaneous magnetization M_s in iron may be taken as almost equal to the saturation magnetization M_0.

If $\alpha = 400$ in iron, it follows that the mean field is

$$H_e = \alpha M_s = 6.8 \times 10^8 \text{ A m}^{-1} \quad (6.10)$$

which is equivalent to a magnetic induction in free space of $B = \mu_0 H = 855$ T.

Notice in particular the enormous size of this field. Remember that a standard laboratory electromagnet generates an induction of typically 2 T. The expected field based on simple classical dipole interactions between the moments in a solid is of the order of 8×10^4 A m^{-1} or equivalently 0.1 T. The strength of the exchange field is therefore several orders of magnitude greater than the strength of the expected classical magnetostatic field.

6.1.6 Energy States of Different Arrangements of Moments

If bar magnets in a linear chain prefer to align antiparallel, why do atomic magnetic moments align parallel?

Consider the two configurations of magnetic moments in Figure 6.3. It can easily be shown that for $\alpha > 0$, the configuration with all moments aligned parallel is a lower energy state than the configuration with one moment antiparallel. If we consider the exchange energy of the six-moment system, the energy of any moment m_i, following from Equation 6.7, is simply the sum of all the pair interaction energies for moment m_i with all the other moments m_j.

$$E_i = -\mu_0 m_i \cdot \sum_j \frac{\alpha_{ij}}{V} m_j \quad (6.11)$$

In the mean field approximation, for which $\alpha_{ij} = \alpha$, for all i and j, this becomes

$$E_i = -\mu_0 \frac{\alpha}{V} m_i \cdot \sum_j m_j \quad (6.12)$$

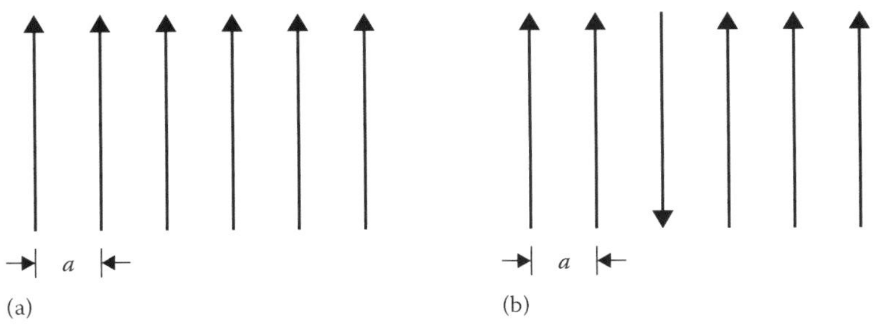

FIGURE 6.3 Two possible configurations of a linear array of magnetic moments. When $\alpha > 0$, the parallel alignment is the ground state.

The total energy obtained by summing over all magnetic moments $\boldsymbol{m}_i$ is therefore

$$E = -\mu_0 \frac{\alpha}{V} \sum_i \boldsymbol{m}_i \cdot \sum_{j \neq i} \boldsymbol{m}_j \tag{6.13}$$

for $j \neq i$. When all moments are parallel, this energy is

$$E = -\frac{\mu_0 \alpha \, (6\boldsymbol{m})(5\boldsymbol{m})}{V} = -\frac{30\boldsymbol{m}^2 \mu_0 \alpha}{V} \tag{6.14}$$

With one moment aligned antiparallel, this energy is

$$E = -\frac{\mu_0 \alpha \left(5\boldsymbol{m}3\boldsymbol{m} - \boldsymbol{m}5\boldsymbol{m}\right)}{V} = -\frac{10\boldsymbol{m}^2 \mu_0 \alpha}{V} \tag{6.15}$$

Therefore, it can be seen that, as a result of the positive exchange interaction α, the energy is lower when all moments are aligned parallel within the domain and hence the parallel aligned ferromagnetic state is preferred.

6.1.7 Early Observational Evidence of Domains

If these magnetic domains exist, how can we see them?

There were two important experimental observations after Weiss' work that served to confirm the essential correctness of his theories. There have subsequently been innumerable observations of ferromagnetic domains, both direct and indirect. The first confirmation was the indirect detection of domains by the Barkhausen effect [9], in which the reorientation of domains caused discrete changes in magnetic induction within a ferromagnet that could be detected by suitable amplification of the signals from a search coil wound around the specimen. The Barkhausen effect leads to discontinuities in other bulk properties if they are measured accurately enough. Recently, discontinuities in magnetoresistance have been reported, for example [10], and of course, acoustic emissions can be generated as discussed in Section 5.2.3.

The second confirmation was by direct observations of domain patterns on the surfaces of ferromagnetic materials made by Bitter [11]. He used a very fine magnetic powder suspended in a carrier liquid, which was spread on the surface of the material. Patterns were observed in the particle accumulations when viewed under a microscope. The particles accumulate at positions where the magnetic field gradient is greatest. This occurs where domain walls intersect the surface. It seems likely that the idea of doing this had come from the magnetic particle inspection techniques of Hoke and DeForest [12], which had been developed a few years before.

More recently, colloidal solutions of ferromagnetic particles in a carrier liquid (*ferrofluids*) have been used. The particles are usually Fe_3O_4 and the commercially available ferrofluids usually have to be diluted for the best observation of domains. In order to produce optimum surface conditions for domain observations, the surface of the material should be electropolished to remove strains, which would otherwise reduce the size of domains. Some Bitter patterns produced by magnetic colloids

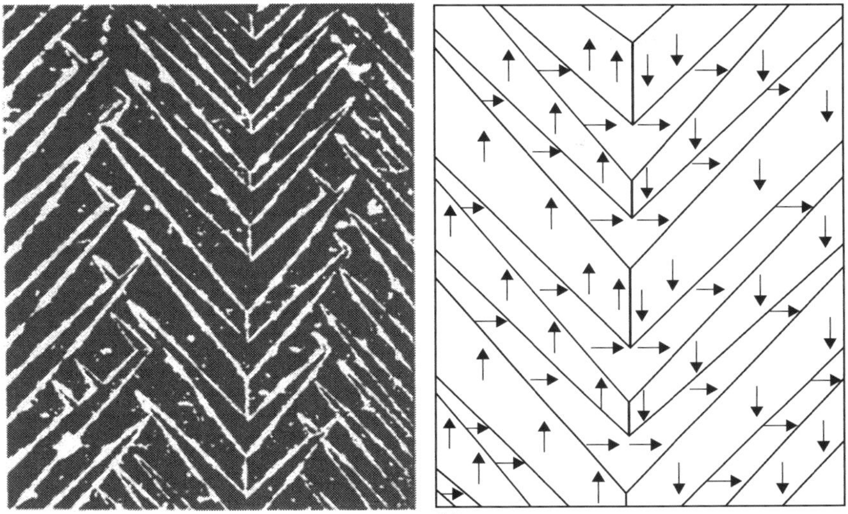

(a) (b)

FIGURE 6.4 (a) Magnetic domains in the surface of iron observed using the Bitter method (magnification × 120); (b) interpretation of domain pattern in (a) showing the direction of the spontaneous magnetizations within the domains. (Data from Williams, H. J. et al., *Phys. Rev.*, 75, 155, 1949.)

such as ferrofluid on the surface of iron from the work of Williams, Bozorth, and Shockley [13] are shown in Figure 6.4.

A variation of the Bitter method has been developed [14], in which the conventional Bitter colloid pattern is observed in polarized light in order to study stray field-induced birefringence on the surfaces of magnetic materials.

6.1.8 Techniques for Domain Observation

What other methods are available for looking at domains?

Several other techniques are regularly used for the observation of domains. Two of these are related optical methods, the Faraday and Kerr effects, in which the axis of polarization of a linearly polarized light beam is rotated by the action of a magnetic field.

The rotation of the direction of polarization of a polarized light beam reflected from the surface of magnetic material is known as the *Kerr effect* [15]. The angle of rotation of the axis of polarization is dependent upon the magnitude and direction of the magnetization ***M*** at the surface of the material. This is determined by the domain configuration in the surface and hence an image of the domain structure at the surface can be formed. One of the difficulties with the Kerr effect is that the angle of rotation is usually very small, so that there is little contrast between the different domains.

There are three types of Kerr effects that can be observed depending on the relative orientation of the magnetization with respect to the plane of incidence of the

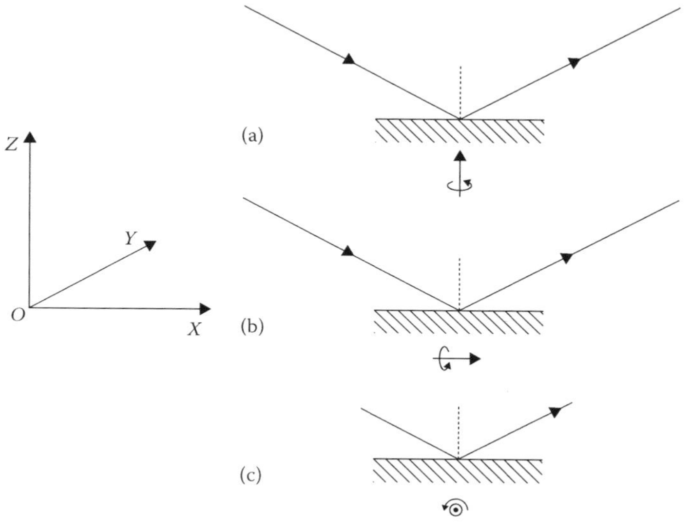

FIGURE 6.5 Arrangement for (a) polar, (b) longitudinal, and (c) transverse Kerr effect observation of surface domain structures showing the relative orientation of the magnetization ***M*** with respect to the incident linearly polarized light beam. The plane of incidence is the *X*-*Z* plane and the surface is in the *X*-*Y* plane.

light beam and the plane of the reflecting surface. These are known as the *polar*, *longitudinal*, and *transverse* Kerr effects [16] and are shown in Figure 6.5.

In the polar Kerr effect, Figure 6.5a, the domain magnetization ***M*** has a component perpendicular to the surface of the specimen. In this case, the angle of rotation is largest and may be up to 20 min of arc. However, the demagnetizing energy strongly favors the alignment of magnetization in the plane of the surface, so unless there is sufficient anisotropy to maintain a component of ***M*** perpendicular to the surface, this method cannot be used. In the longitudinal Kerr effect, Figure 6.5b, the domain magnetization ***M*** lies in the plane of incidence and also in the plane of the surface. In this case, the rotation of the direction of polarization is much smaller, typically being up to 4 min of arc. The maximum rotation in the longitudinal Kerr effect is obtained at an angle of incidence 60°. In the transverse Kerr effect (Figure 6.5c), ***M*** lies in the surface plane but is perpendicular to the plane of incidence. This leads to an effective rotation of the direction of polarization, as described originally by Ingersoll [17]. This is comparable in magnitude to the rotation observed in the longitudinal Kerr effect.

Hubert and coworkers [18] developed a technique based on the Kerr effect that uses digital image processing using a combination of longitudinal and transverse Kerr effect measurements. This enables quantitative determination of the magnetization and its direction in the domain pattern of soft magnetic materials.

The Faraday effect [19], which is less useful for domain observations, is similar except that the rotation of the axis of polarization is caused during transmission of a polarized light beam through a ferromagnetic solid. This can therefore only be applied to a thin transparent sample of a ferromagnetic material and is hence restricted to thin slices of ferromagnetic oxide or to metal films. In both cases, the resulting beam is analyzed using a second polarizing filter which then reveals the location of various domains as either light or dark regions in the polarized image.

A combination of magneto-optic Faraday and Kerr effects with the Bitter pattern technique has been reported by Hartmann [20]. This provides a high contrast image of the surface domain pattern and can be used for both transparent and opaque magnetic materials. It is known as the *interference contrast colloid technique.*

A development in the general area of magneto-optic methods for domain observations is the laser magneto-optic microscope of Argyle and Herman [21]. This is a derivative of the Kerr effect that has been used for domain studies in read/write heads for magnetic recording devices [22].

Another method of domain observation is the transmission electron microscopy [23], also known as *Lorentz microscopy*, in which the specimens are usually in the form of thin films. The transmitted electrons are deflected by the local magnetic field gradients in the material and this can be used to produce an image of the domain structure [24,25]. The angular deflection of the electrons in ferromagnetic materials such as iron is typically 0.01°. In order to obtain images, normal bright-field imaging conditions cannot be used. Instead, the image must be either under- or over-focused; as the focus is changed, the images of domains can be obtained with different directions. The domain walls then appear as either dark or bright lines. The transmission electron micrographs obtained under these conditions reveal only the domain walls.

The force $\boldsymbol{F}$ on an electron as it passes through the material is

$$\boldsymbol{F} = -\mu_0 e \boldsymbol{v} \times \boldsymbol{M} \tag{6.16}$$

where:

$\boldsymbol{M}$ is the local magnetization
$\boldsymbol{v}$ is the velocity

For 100 keV, the maximum thickness of the ferromagnetic material that can usefully be observed is about 200 nm. This method is capable of a spatial resolution of 5 nm; however, one problem is that the electron beam is focused with a strong magnetic field, and therefore, a ferromagnetic specimen cannot be placed in the usual location in the microscope without some magnetic shielding, otherwise the domain structure will be disturbed. An alternative is to move the specimen from the normal position inside the objective lens.

Scanning electron microscopy can also be used for domain imaging [26]. This has the added advantage over transmission electron microscopy that it can be used to image domains in thick specimens. Two methods are used in scanning electron microscopy: one that depends on the deflection of incident electrons after they enter the material, so that the number of backscattered electrons at a given angle is dependent on the direction of magnetization within a domain [27]; the second depends on the deflection of secondary electrons by stray magnetic fields near the specimen surface [28].

X-ray topography is another technique of domain imaging [29]. It is well known that the diffraction of X-rays by a solid is dependent on the lattice spacing. In ferromagnets, because of spontaneous magnetostriction, the lattice parameter is dependent upon the direction of magnetization within the domain. This means that domains with magnetic moments aligned at angles other than 180° can be distinguished on the basis of Bragg diffraction of X-rays. This method can be used on thick specimens but the diffraction of X-rays is affected by dislocations and grain boundaries, and therefore, the method is best suited to fairly pure single-crystal materials with few dislocations or other defects.

Another method that has been reported in recent years is neutron diffraction topography [30]. The neutron carries a magnetic moment but no charge therefore can be affected by a magnetic field or the magnetization within a domain. In this method, a beam of polarized neutrons passes through a specimen and the angle of rotation of polarization is detected. The results are then used to reconstruct an image of the domain pattern in the material.

Magnetic domains can also be imaged by the method known as *magnetic force microscopy* [31,32]. This technique is a derivative of atomic force microscopy, in which a small (~10–100 nm radius) magnetized probe tip is scanned across the surface of a material. The magnetic force of attraction or repulsion can then be measured, and from the application of the magnetic analog of Coulomb's law, the magnetization on the surface can be deduced. The sharp probe tips produce high field gradients, and for measurements on soft magnetic materials, a ball-shaped tip is preferred. The variation of magnetization with position can then be used to provide a magnetic map of the surface with resolution down to 50 nm. This method is widely used.

The use of magnetic force microscopy for scanning the magnetic image at the surface of very soft magnetic materials can be problematic because the magnetization of the probe tip can perturb the magnetization in the surface of the material. This effect has been studied by, among others, Matteucci, Muccini, and Hartmann [33]. An excellent overview of recent developments has been given by Gomez et al. [34].

6.2 ENERGY CONSIDERATIONS AND DOMAIN PATTERNS

6.2.1 Existence of Domains as a Result of Energy Minimization

Why do domains exist at all? Surely the moments should simply be aligned throughout the solid?

Since Weiss showed that there exists an interaction field between the magnetic moments within a ferromagnet that causes alignment of the magnetic moments, we are left once again with the question, Why a ferromagnet is not spontaneously magnetized, or if you wish, why the domains themselves are not aligned throughout the whole volume of the solid?

In postulating the existence of domains, Weiss had sought to provide an empirical explanation of why the mean field did not lead to spontaneous magnetization of the material. It was left to Landau and Lifschitz in 1935 [35] to show that the existence of domains is a consequence of energy minimization. A single-domain specimen has associated with it a large magnetostatic energy, but the breakup of the magnetization into localized regions (domains), providing for flux closure at the ends of the specimen, reduces the magnetostatic energy. Providing that the decrease in magnetostatic energy is greater than the energy needed to form magnetic domain walls, then multidomain specimens will arise.

6.2.2 Magnetostatic Energy of Single-Domain Specimens

What is the self-energy of a single-domain particle?

The energy per unit volume of a dipole of magnetization $\boldsymbol{M}$ in a magnetic field $\boldsymbol{H}$ is given by

$$E = -\mu_0 \int \boldsymbol{M} \cdot d\boldsymbol{H} \tag{6.17}$$

as given earlier in Section 1.2.4. When it is subjected only to its own demagnetizing field $\boldsymbol{H}_d$, which is generated by $\boldsymbol{M}$ anyway, we can put $\boldsymbol{H}_d = -N_d\boldsymbol{M}$ in the integral where N_d is the demagnetizing factor so that the energy becomes

$$E = \mu_0 N_d \int \boldsymbol{M} \cdot d\boldsymbol{M} \tag{6.18}$$

$$E = \frac{\mu_0}{2} N_d \boldsymbol{M}^2 \tag{6.19}$$

This then is the magnetostatic energy of a single-domain magnetic material with magnetization $\boldsymbol{M}$ subjected only to its own demagnetizing field $N_d\boldsymbol{M}$. The calculation of the energy of a multidomain specimen is more complex and we shall consider it later.

It should be noted that the overall $\boldsymbol{M}$ will be reduced by the emergence of two antiparallel domains in the particle and so the magnetostatic self-energy will be reduced. It may then be energetically favorable for the particle to split into more than one domain. However, if the particle is small enough that the energy needed to create a domain boundary between two domains in the particle is greater than the magnetostatic energy of the single-domain particle, the particle will remain a single domain.

This means that there is a single-domain size for each material, below which size it will be a single domain and above which size it will be a multidomain. The single-domain sizes for various ferromagnetic materials are shown in Figure 6.6. This single-domain size is also determined by another factor, which we have not considered yet: the domain-wall energy.

Furthermore, if the particle size gets much smaller, a situation can be reached at which the internal coupling energy between the magnetic dipoles in the particle is not sufficient to hold the dipoles in alignment against thermal energy. This critical particles size is known as the *super-paramagnetic limit*. The super-paramagnetic limit is of course dependent on the prevailing temperature, being smaller as the temperature is reduced. That means the ferromagnetic order can exist in smaller particles at lower temperatures.

6.2.3 Domain Patterns and Configurations

How do the domains arise from a single-domain specimen as the magnetic field is reduced?

Many direct observations of domain patterns have been made by Bitter pattern techniques or by other methods such as the magneto-optic Kerr and Faraday effects.

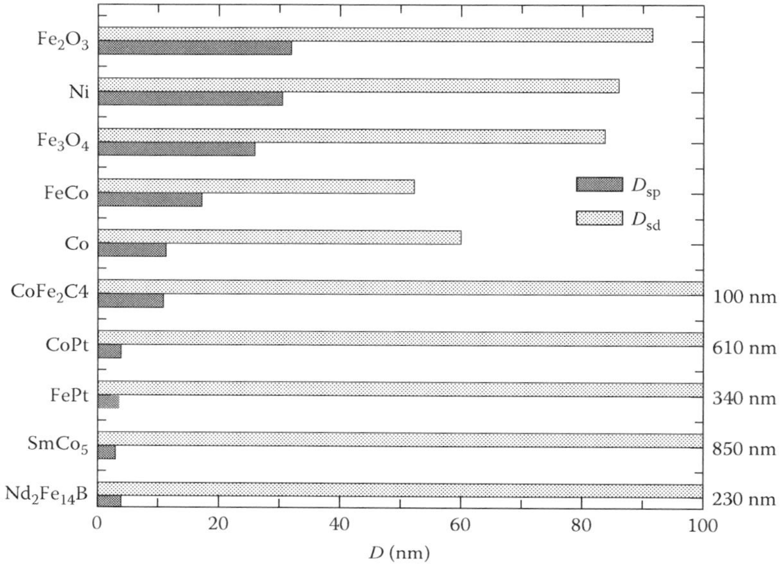

FIGURE 6.6 Single-domain particle diameter size (D_{sd}) and super-paramagnetic transition diameter size (D_{sp}) of selected ferromagnetic materials.

Figure 6.7 shows diagrammatically the emergence of domains as a sample that is originally in the saturated condition is demagnetized. The lower diagram shows a closure domain at the end of a single crystal of iron. Closure domains usually emerge fairly early in the demagnetizing process since they provide return paths for the magnetic flux within the solid. They are nucleated by defects including the boundary of the material. They are also usually the last domains to be swept out at higher fields.

6.2.4 Magnetization Process in Terms of Domain Theory

What changes occur within domains as a magnetic field is applied and gradually increased?

Since according to the domain theory the atomic magnetic moments are ordered even in the demagnetized state in a ferromagnet, the difference between the demagnetized state and the magnetized state must be due to the configuration of the domains.

When a magnetic field is applied to a demagnetized ferromagnetic material, the changes in magnetic induction $\boldsymbol{B}$ when traced on the $\boldsymbol{B}$-$\boldsymbol{H}$ plane generate the initial magnetization curve. At low fields, the first domain process occurs, which is a growth of domains that are aligned favorably with respect to the field according to the minimization of the field energy $E = -\mu_0 \boldsymbol{M}_s \cdot \boldsymbol{H}$ and a consequent reduction in size of domains, which are aligned in directions opposing the field, as shown in Figure 6.8.

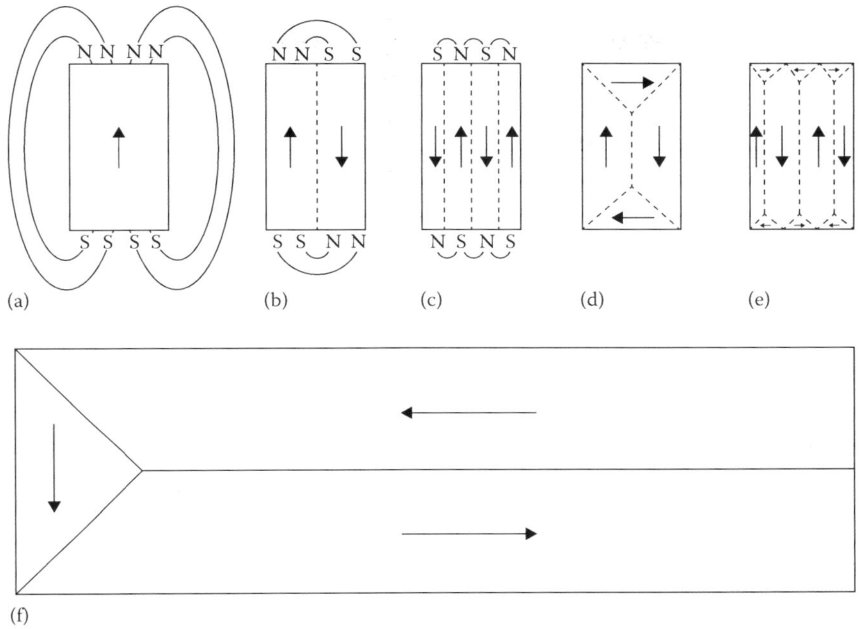

FIGURE 6.7 Changing domain patterns (a)–(e) as a sample of single-crystal iron is demagnetized. Closure domain at end of sample (f).

At moderate field strengths, a second mechanism becomes significant; this is domain rotation, in which the atomic magnetic moments within an unfavorably aligned domain overcome the anisotropy energy and suddenly rotate from their original direction of magnetization into one of the crystallographic *easy* axes that is nearest to the field direction.

The final domain process that occurs at high fields is coherent rotation. In this process, the magnetic moments, which are all aligned along the preferred magnetic crystallographic easy axes lying close to the field direction, are gradually rotated into the field direction as the magnitude of the field is increased. This results in a single-domain sample.

6.2.5 Technical Saturation Magnetization

Why is the magnetization within the domains not equal in magnitude to the saturation magnetization?

When all domains have been aligned with their spontaneous magnetization vectors parallel to the field, the material consists of a single magnetic domain and is said to have reached technical saturation magnetization. If the magnetic field is increased further, it is noticed, however, that the magnetization does continue to increase very slowly. This is due to an increase in the spontaneous magnetization $\boldsymbol{M}_s$ within the domain as the atomic magnetic moments within the single domain, which are not perfectly aligned with the field because of thermal activation, are brought into complete alignment.

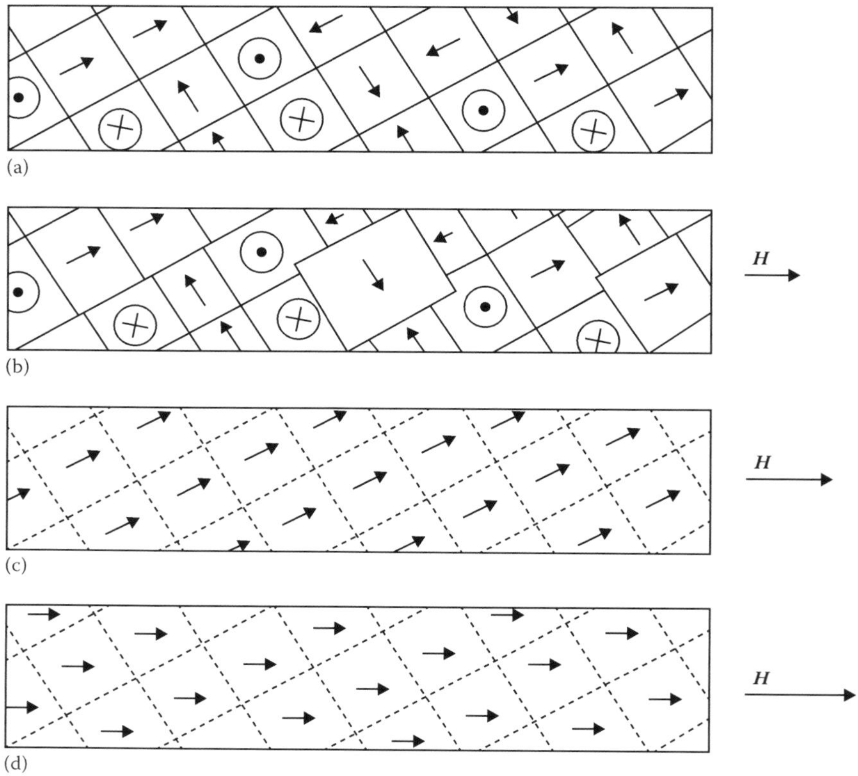

FIGURE 6.8 Domain processes occurring as a material is magnetized to saturation, from the demagnetized state (a) to partial magnetization; (b) by domain-wall movement, from partial magnetization to the knee of the magnetization curve; (c) by irreversible rotation of domain magnetization, from the knee of the magnetization curve to technical saturation; and (d) by reversible rotation.

The spontaneous magnetization is temperature dependent as we have noted earlier. At 0 K, it is equal to the saturation magnetization but decays to zero at the Curie point. At temperatures above 0 K, the individual magnetic moments have thermal energy that causes them to precess about the field direction as shown in Figure 6.9. The precession becomes greater as the temperature increases. It is the precession that causes the spontaneous magnetization to be smaller than the saturation magnetization. Ultimately, when all the magnetic moments within the domain are completely aligned parallel due to a very high magnetic field, the magnetization reaches saturation magnetization $\boldsymbol{M}_0$.

6.2.6 Domain Rotation and Anisotropy

How does anisotropy affect the ability of domains to change the direction of their magnetization?

Although the domain growth process is not much affected by anisotropy, both the reversible rotation and irreversible rotation of magnetic moment within a domain

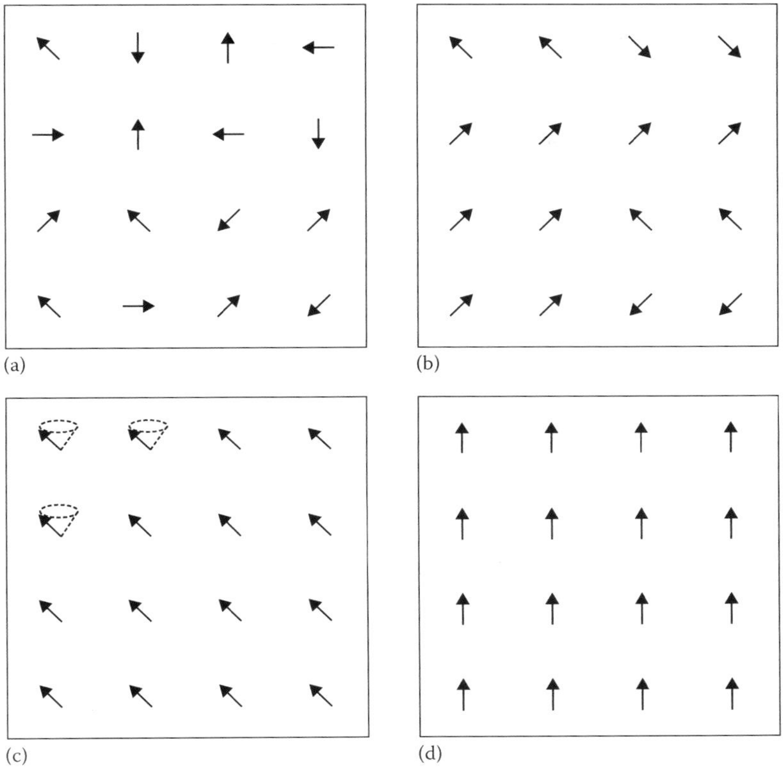

FIGURE 6.9 Alignment of individual magnetic moments within a domain at various temperatures: (a) above the Curie point showing random alignment; (b) below the Curie point; (c) at low temperatures in which the magnetic moments precess about the field direction in low-level excited states; and (d) perfect alignment at 0 K where there is no thermal energy for precession.

are determined principally by the magnetocrystalline anisotropy. The rotation of the domains can be considered as a competition between the anisotropy energy $E_a(\theta, \phi)$ and the field energy E_H giving a total energy E_{Tot} of

$$E_{Tot} = E_a(\theta, \phi) + E_H \tag{6.20}$$

where

$$E_H = -\mu_0 \boldsymbol{M}_s \cdot \boldsymbol{H} \tag{6.21}$$

$\boldsymbol{M}_s$ is the spontaneous magnetization within a domain.

6.2.7 Axial Anisotropy

What is the simplest representation of anisotropy?

In this case, the anisotropy can be represented by the one-constant approximation

$$E_a = K_{u1} \sin^2 \phi \tag{6.22}$$

where:
K is the anisotropy coefficient
ϕ is the angle of the magnetization $\boldsymbol{M}$ with respect to the unique axis

For $K > 0$, the unique axis is the easy axis, (i.e., the preferred direction for the magnetic moments to align along) while for $K < 0$, the unique axis is the hard axis (i.e., the least preferred direction for the magnetic moments to align along). In the case of single-crystal cobalt, for example, which is hexagonal, $K_{u1} = 4.1 \times 10^5$ J m^{-3}. Table 6.1 provides anisotropy constants for various ferromagnetic materials.

6.2.8 Anisotropy as an Equivalent Magnetic Field

Just as in the case of the mean field energy (Section 6.1.5), an equivalent field description is often used to describe anisotropy. This is not a real field in the same sense of a magnetic field generated by a solenoid, because it does not have the normal vector characteristics of a magnetic field. The *anisotropy field* is the equivalent field strength $\boldsymbol{H}_{an}$ that gives the same amount of energy to the individual magnetic moments as the anisotropy, approximately $\mu_0 \boldsymbol{M}_s \boldsymbol{H}_{an} = K$.

We calculate the *anisotropy field* for a cobalt crystal whose magnetization is close to the hexagonal axis. Using $\boldsymbol{M}_s = \boldsymbol{M}_0 = 1.42 \times 10^6$ A m^{-1} and $K_{u1} = 4.1 \times 10^5$ J m^{-3}.

The anisotropy energy for a uniaxial material such as cobalt is given by Equation 6.22 and for small angles ϕ, we can put $\sin^2 \phi \approx \phi^2$

$$E_a = K_{u1} \cdot \phi^2 \tag{6.23}$$

TABLE 6.1
Anisotropy Constants for Various Ferromagnetic Materials

Material	Crystal Structure	K_1	K_{u1} (10^5 J m^{-3})	K_2	K_{u2}
Iron	Cubic	0.480	–	0.050	–
Nickel	Cubic	−0.045	–	0.023	–
Cobalt	Hexagonal	–	4.1	–	1.0
$SmCo_5$	Hexagonal	–	110	–	–
NdFeB	Tetragonal	–	94	–	–
Barium ferrite	Hexagonal	–	3.2	–	–

If a magnetic field is applied along the unique axis, then the field energy is given by

$$E_{\mathrm{H}} = -\mu_0 \boldsymbol{M}_{\mathrm{s}} \boldsymbol{H} \cos\phi \tag{6.24}$$

and for small angles ϕ, $\cos\phi = 1-\phi^2/2$

$$E_{\mathrm{H}} = -\mu_0 \boldsymbol{M}_{\mathrm{s}} \boldsymbol{H} \left(1 - \frac{\phi^2}{2}\right) \tag{6.25}$$

Removing the constant term, which has no bearing on the equilibrium condition

$$E_{\mathrm{H}} = \frac{\mu_0 \boldsymbol{M}_{\mathrm{s}} \boldsymbol{H} \phi^2}{2} \tag{6.26}$$

Equating these energies will give the magnetic field along the unique axis, which is equivalent to the anisotropy

$$K_{\mathrm{u1}} = \frac{\mu_0 \boldsymbol{M}_{\mathrm{s}} \boldsymbol{H}}{2} \tag{6.27}$$

and hence the effective field due to anisotropy is

$$\boldsymbol{H}_{\mathrm{an}} = \frac{2K_{\mathrm{u1}}}{\mu_0 \boldsymbol{M}_{\mathrm{s}}} \tag{6.28}$$

Substituting in the value of anisotropy for cobalt, $K_{\mathrm{u1}} = 8.2 \times 10^5$ J m^{-3} gives

$$\boldsymbol{H} = 4.6 \times 10^5 \ \mathrm{A\,m^{-1}} \tag{6.29}$$

6.2.9 Cubic Anisotropy

What is the simplest representation of cubic anisotropy?

In the case of cubic anisotropy, the following one-constant anisotropy equation can also be used as a first approximation:

$$E_{\mathrm{a}} = K_1 \left(\cos^2\theta_1 \cos^2\theta_2 + \cos^2\theta_2 \cos^2\theta_3 + \cos^2\theta_3 \cos^2\theta_1\right) \tag{6.30}$$

where θ_1, θ_2, and θ_3 are angles that the magnetization makes relative to the three principal crystal axes. Therefore, for example, if we consider the anisotropy in, say, the (001) plane only, we will have $\theta_3 = \pi/2$, and consequently $\cos\theta_3 = 0$

$$E_{(001)} = K_1 \left(\cos^2\theta_1 \cos^2\theta_2\right) \tag{6.31}$$

Once we have established that we are only considering moments in this plane, it follows that $\theta_1 = 90° - \theta_2$

$$E_{(001)} = \left(\frac{K_1}{4}\right)\sin^2 2\theta \qquad (6.32)$$

where now $\theta = 0$ is the direction of the magnetic easy axis within the plane (001) when $K_1 > 0$, that is, the <100> direction, as in the case of iron for which $K_1 = 4.8 \times 10^4$ J m^{-3}. When $K_1 < 0$, the magnetic easy axes are the <111> direction, as in nickel for which $K_1 = -4.5 \times 10^3$ J m^{-3}.

6.2.10 Domain Magnetization Reversal in Isolated Single Domains

How does the magnetization change direction within a single domain when subjected to a magnetic field?

The simplest case to describe when considering domain magnetization changes is the case of an isolated single-domain particle. In this case, one only needs to consider the anisotropy energy and the energy of the magnetic moment at different directions from the applied field in order to determine how the magnetization will change with field strength. In fact, this particular case, which seems at first to be highly idealized, has practical applications in particulate magnetic recording media, which normally consist of elongated isolated domains of magnetic material in a substrate. So, for example, if one considers an isolated particle with axial anisotropy and applies a magnetic field in the direction antiparallel to the magnetization, the equilibrium orientation of the magnetization as a function of applied field strength can be calculated using the principles discussed in this chapter.

For uniaxial anisotropy, as shown in Section 6.2.7, $E_a = K_{u1}\sin^2\phi$ and, if the magnetic field is along the direction $\phi = 180°$, we obtain $E_H = -\mu_0 \boldsymbol{M}_s \boldsymbol{H}\cos(180 - \phi) = \mu_0 \boldsymbol{M}_s \boldsymbol{H}\cos\phi$, where ϕ is the angle of orientation of the moment relative to its original, zero field direction. The total energy is then simply the sum of these two energy terms

$$E_{Tot} = K_{u1}\sin^2\phi + \mu_0 \boldsymbol{M}_s \boldsymbol{H}\cos\phi \qquad (6.33)$$

and equilibrium locations for the magnetic moment arise when $dE_{Tot}/d\phi = 0$ and $d^2E_{Tot}/d\phi^2 < 0$.

$$\frac{dE_{Tot}}{d\phi} = (2K_{u1}\cos\phi - \mu_0 \boldsymbol{M}_s \boldsymbol{H})\sin\phi = 0 \qquad (6.34)$$

where clearly $\phi = 0$ is an unstable equilibrium, while $\phi = 180°$ is the trivial case of alignment along the field direction. Other solutions occur when $\cos\phi = \mu_0 \boldsymbol{M}_s \boldsymbol{H}/2K_{u1}$ and these are of most interest because they show how the angle of magnetization changes with the strength of the applied field $\boldsymbol{H}$. Once $\boldsymbol{H} > 2K_{u1}/\mu_0 \boldsymbol{M}_s$, the magnetization switches discontinuously into the field direction along the *antiparallel* direction to the easy axis. This critical field for switching has characteristics of the bulk coercivity.

The concept of single-domain magnetization reversal forms the basis of some of the simplest descriptions of the magnetization mechanism such as those of the Stoner-Wohlfarth model, which is described in more detail in Chapter 13.

REFERENCES

1. Weber, W. *Pogg. Ann.*, LXXXVII, 167, 1852.
2. Poisson, S. D. in *Magnetic Induction in Iron and Other Methods* (ed. J. A. Ewing), The Electrician, London, 1893, p. 282.
3. Ampère, A. M. *Theorie Mathematique des Phenomenes Electrodynamiques Uniquement Deduite de l'Experience.* Reprinted by Blanchard, Paris, France, 1958.
4. Ewing, J. A. *Magnetic Induction in Iron and Other Metals*, The Electrician, London, 1893.
5. Weiss, P. *Compt. Rend.*, 143, 1136, 1906.
6. Weiss, P. *J. Phys.*, 6, 661, 1907.
7. Langevin, P. *Ann. Chem. et Phys.*, 5, 70, 1905.
8. Ising, E. *Z. Phys.*, 31, 253, 1925.
9. Barkhausen, H. *Physik Z.*, 20, 401, 1919.
10. Tsang, C., and Decker, S. K. *J. Appl. Phys.*, 52, 2465, 1981.
11. Bitter, F. *Phys. Rev.*, 38, 1903, 1931.
12. Betz, C. E. *Principles of Magnetic Particle Testing*, Magnaflux, Chicago, IL, 1967, p. 48.
13. Williams, H. J., Bozorth, R. M., and Shockley, W. *Phys. Rev.*, 75, 155, 1949.
14. Jones, G. A., and Puchalska, I. *Phys. Stat. Sol. A*, 51, 549, 1979.
15. Kerr, J. *Rep. Brit. Assoc.*, 5, 1876.
16. Carey, R., and Isaac, E. D. *Magnetic Domains and Techniques for Their Observation*, Academic Press, London, 1966, Ch. 5.
17. Ingersoll, L. P. *Phys. Rev.*, 35, 315, 1912.
18. Rave, W., Schafer, R., and Hubert, A. *J. Magn. Magn. Mater.*, 65, 7, 1987.
19. Faraday, M. *Phil. Trans. Roy. Soc.*, 136, 1, 1846.
20. Hartmann, U. *Appl. Phys. Letts.*, 51, 374, 1987.
21. Argyle, B., and Herman, D. *IEEE Trans. Magn.*, MAG-22, 772, 1986.
22. Herman, D., and Argyle, B. *J. Appl. Phys.*, 61, 4200, 1987.
23. Hale, M. E., Fuller, H. W., and Rubinstein, H. *J. Appl. Phys.*, 30, 789, 1959; *J. Appl. Phys.*, 31, 238, 1960.
24. Hwang, C., Laughlin, D. E., Mitchell, P. V., Layadi, A., Mountfield, K., Snyder, J. E., and Artman, J. O. *J. Magn. Magn. Mater.*, 54–57, 1676, 1986.
25. Lodder, J. C. *Thin Solid Films*, 101, 61, 1983.
26. Mayer, L. *J. Appl. Phys.*, 28, 975, 1957.
27. Koike, K., Matsuyama, H., Todokoro, H., and Hayakawa, K. *Jap. J. Appl. Phys.*, 24, 1078, 1985.
28. Unguris, J., Hembree, G., Celotta, R. J., and Pierce, D. T. *J. Magn. Magn. Mater.*, 54–57, 1629, 1986.
29. Polcarova, M., and Lang, A. R. *Appl. Phys. Letts.*, 1, 13, 1962.
30. Baruchel, J., Palmer, S. B., and Schlenker, M. *J. Phys.*, 42, 1279, 1981.
31. Saenz, J. J., and Garcia, N. *J. Appl. Phys.*, 63, 4293, 1987.
32. Wickramsinghe, H. K., and Martin, Y. *J. Appl. Phys.*, 63, 2948, 1988.
33. Matteucci, G., Muccini, M., and Hartman, U. *JMMM*, 133, 422, 1994.
34. Gomez, R., Adly, A. A., Mayergoyz, I. D., and Burke, E. R. *IEEE Trans. Magn.*, 29, 2494, 1993.
35. Landau, I. D., and Lifschitz, E. M. *Physik. Z. Sowjetunion*, 8, 153, 1935.

FURTHER READING

Argyle, B., and Herman, D. *IEEE Trans. Magn.*, MAG–22, 772, 1986.

Carey, R., and Isaac, E. D. *Magnetic Domains and Techniques for Their Observation*, Academic Press, London, 1966.

Craik, D. J., and Tebble, R. S. *Ferromagnetism and Ferromagnetic Domains*, North Holland, Amsterdam, the Netherlands, 1965.

Crangle, J. *The Magnetic Properties of Solids*, Arnold, London, 1977, Ch. 6.

Cullity, B. D. *Introduction to Magnetic Materials*, Addison-Wesley, Reading, MA, 1972, Ch. 7.

Jakubovics, J. P. *Magnetism and Magnetic Materials*, Institute of Metals, London, 1994.

Kittel, C., and Galt, J. K. Ferromagnetic domains, *Solid State Phys.*, 3, 437, 1956.

EXERCISES

6.1 *Magnetic domains.* Write a brief essay on magnetic domains. Discuss the reasons why magnetic ordering occurs in magnetic materials such as ferromagnets, and why a magnetic material does not just spontaneously magnetize throughout its volume. What evidence is there for the existence of internal coupling between the magnetic moments located on the atomic sites in a magnetic material such as iron? What evidence do we have for the existence of domains and what techniques are now widely used for their observation? If we observe the magnetic domains on the surface of a piece of iron, does this tell us anything about the magnetization processes taking place inside the material?

6.2 *Magnetocrystalline anisotropy.* Explain what is meant by magnetocrystalline anisotropy. Using a model of cubic anisotropy with only one constant, K, prove that the magnetic moments will point along one of the $<100>$ directions when K is positive and find the direction that they will point when K is negative.

6.3 *Effects of anisotropy on rotation of magnetization.* Derive an expression for the magnetization of a specimen with cubic anisotropy when a field is applied along the specific $<110>$ direction and the easy axes are the $<100>$ directions in general. You may assume that the material is a perfect single crystal so that the domain-wall motion all occurs at negligibly small fields, leaving the magnetizations within domains aligned along one of the $<100>$ directions closest to the applied field direction. Also, you may use the one-constant approximation for the anisotropy.

6.4 *Critical fields as determined by anisotropy.* A magnetic field is applied in the base plane of cobalt. The easy direction is the unique axis. Using a one-constant model for the anisotropy with $K_{u1} = 4.1 \times 10^5$ J m^{-3} and $\boldsymbol{M}_s = 1.42 \times 10^6$ A m^{-1}, calculate the field needed to rotate the magnetic moments (1) into the base plane (i.e., perpendicular to the unique axis) and (2) 45° from the hexagonal axis.

6.5 *Stress-induced anisotropy.* If the saturation magnetostriction of a randomly oriented rod of polycrystalline terfenol is $\lambda_s = 1067 \times 10^{-6}$, and assuming the anisotropy constant is $K_1 = -2 \times 10^4$ J m^{-3}*, determine the stress needed completely to align the moments in the plane perpendicular to the stress. You may use the one-constant approximation to the cubic anisotropy and since the material has no preferred orientation in the bulk sense, you may use the expression $E_\sigma = -(3/2)\lambda_s\sigma \cos^2 \theta$ for the stress-induced anisotropy, where σ is the stress and θ is the angle from the stress axis.

6.6 *Simple Ising lattice.* Consider a square array of nine magnetic moments located on atoms in a lattice. If the interactions between these decay so rapidly with distance that only the nearest-neighbor interactions need to be considered, show that the lowest energy state obtained when the interaction coefficient α is less than zero, in which the neighboring moments are aligned antiparallel, and when the interaction coefficient is positive, the lowest state is one, in which the neighboring moments are aligned parallel. Derive a relationship between the Curie temperature T_c (or Néel temperature T_n) and α for this 3×3 array.

6.7 *Exchange energy and domains.* Explain what magnetic domains are and discuss why domains are necessary to account for the differences in permeabilities between ferromagnets and paramagnets. What evidence is there for the existence of domains? Calculate the strength of the interaction (Weiss) field in cobalt (T_c = 1140 C, $m = 1.59 \times 10^{-23}$ A m^2 per atom) and iron (T_c = 770 C, $m = 2.04 \times 10^{-23}$ A m^2 per atom). From this information calculate the exchange energy in joules for both iron and cobalt. Determine the nearest neighbor coupling coefficients between atoms for hexagonal cobalt (12 nearest neighbors) and body-centered cubic iron (8 nearest neighbors), assuming that only nearest neighbor interactions are significant. The lattice parameters for hexagonal cobalt are $a = 0.25 \times 10^{-9}$ m and $c = 0.41 \times 10^{-9}$ m, while for cubic iron, the lattice parameter is $a = 0.29 \times 10^{-9}$ m.

6.8 *Magnetocrystalline anisotropy.* What is magnetocrystalline anisotropy and how does it determine the direction of magnetization within a domain? What forms of anisotropy, magnitudes, and signs of anisotropy coefficients would be best for (1) soft magnetic materials and (2) hard magnetic materials?

Nickel orders in the face-centered cubic structure. If the first-order cubic anisotropy coefficient for nickel is -0.05×10^5 J m^{-3}, show how to determine the crystallographic easy axis and calculate the effective *anisotropy field* for small deviations of moments from the easy axis.

6.9 *Magnetocrystalline anisotropy, domain-wall energy, and domain-wall thickness.* Barium ferrite is a hexagonal material and has K = 330 kJ m^{-3}, $\boldsymbol{M}_S = 0.4 \times 10^6$ A m^{-1}, T_c = 470 C, and lattice parameters 3 nm along

* The anisotropy coefficient of this alloy varies significantly with temperature and alloy composition. Therefore, this value, although in the correct range, should not be assumed to be generally valid for terfenol.

the unique axis and 0.6 nm in the base plane perpendicular to the unique axis. Calculate the equivalent anisotropy field $\boldsymbol{H}_a$, the domain-wall surface energy γ, and the domain-wall thickness δ for barium ferrite.

6.10 *Critical radius for single-domain particles.* Calculate the critical radius (maximum radius) for a spherical particle of cobalt to remain single domain if the effective energy terms that determine this are the magnetostatic (demagnetizing) energy and the domain-wall surface energy. For cobalt, $\boldsymbol{M}_s = 1.4 \times 10^6$ A m^{-1}, $T_C = 1404$ K, and $K = 5 \times 10^5$ J m^{-3}. You may assume the exchange stiffness is $A = k_B T_c / a$ (uniaxial), where a is the lattice spacing for which you may also assume a value of 0.3×10^{-9} m.

7 Domain Walls

In this chapter, we shall investigate the local magnetic properties in the vicinity of domain boundaries. These regions are called *domain walls* and have properties that are very different from the rest of the domain. Most of the magnetic changes under the action of weak and moderate magnetic fields occur at the domain walls and hence an understanding of domain-wall behavior is essential for describing the magnetization processes in many materials, particularly soft magnetic materials.

7.1 PROPERTIES OF DOMAIN BOUNDARIES

How do the magnetic moments behave in the domain boundaries?

Once we have accepted the idea of domains within ferromagnets, there arises the question of how the moments change direction in the regions between domains. There are two possibilities to consider. Either the domain boundaries are infinitesimal in width with nearest-neighbor moments belonging either to one domain or to the other or there could be a transition region of finite width over which the magnetic moments are aligned in different directions between the domains and therefore belong to neither domain, as shown for example in Figure 7.1 [1].

7.1.1 Domain Walls

How thick are the domain walls?

The existence of these transition layers between domains, in which the magnetic moments undergo a reorientation, was first suggested by Bloch [2]. The transition layers are commonly referred to as *domain walls* or *Bloch walls*, although we should note immediately that not all domain walls are necessarily Bloch walls. The total angular displacement across a domain wall is often 180° or 90°, particularly in cubic materials because of the anisotropy and, as we shall see, the change in direction of the moments takes place gradually over many atomic planes.

7.1.2 Domain-Wall Energy

Is there any additional energy associated with the domain wall?

We may define the domain-wall energy as the difference in energy of the magnetic moments when they are part of the domain wall and when they are within the main body of the domain. This is usually expressed as the energy per unit area of the wall and typically has a value in the range of 10^{-3} J m^{-3}.

It can be calculated by considering a combination of the Weiss-type interaction energy between the atomic magnetic moments and the anisotropy energy. These energies have

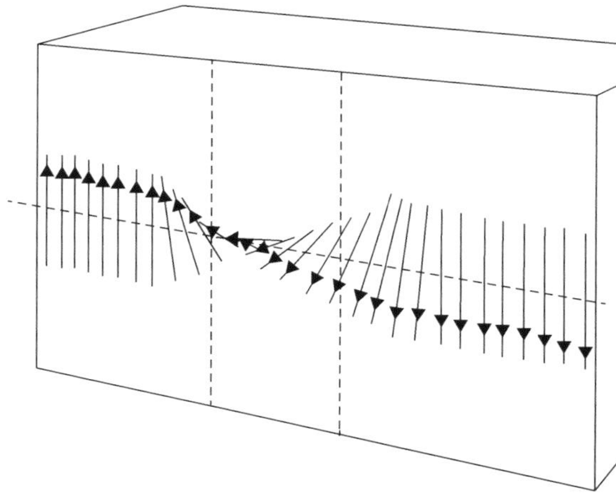

FIGURE 7.1 Alignment of individual magnetic moments within a 180° domain wall. (After Kittel, C., *Revs. Mod. Phys.*, 21, 541, 1949.)

already been described in Chapter 6. The anisotropy energy tends to make the domain wall thinner because anisotropy energy is lowest with all moments aligned along crystallographically equivalent axes known as the "easy" axes. The interaction energy, also known as the *exchange energy*, tends to make the walls thicker because this energy in a ferromagnet is minimized when neighboring moments are aligned parallel.

7.1.2.1 Exchange Energy

If we consider only the interactions between the magnetic moments, these give rise to the interaction or exchange energy as described in Section 6.1.5.

$$E_{ex} = -\mu_0 \boldsymbol{m} \cdot \boldsymbol{H}_e \tag{7.1}$$

where $\boldsymbol{H}_e$ is the interaction field. In the Weiss mean field model, we have said that $\boldsymbol{H}_e$ is proportional to the magnetization within the domain, that is, $\boldsymbol{H}_e = \alpha \boldsymbol{M}_s$, which at temperatures well below the Curie temperature is almost equal to $\boldsymbol{M}_0$. Consequently,

$$\boldsymbol{H}_e = \alpha \boldsymbol{M}_0 \tag{7.2}$$

where α is the mean field parameter. If we consider this in terms of an interaction with the individual magnetic moments $\boldsymbol{m}$, then, since $\boldsymbol{M}_0 = N\boldsymbol{m}$, where N is the number of atoms per unit volume

$$\boldsymbol{H}_e = \alpha N \boldsymbol{m} \tag{7.3}$$

Therefore,

$$E_{ex} = -\mu_0 \alpha N \boldsymbol{m}^2 \tag{7.4}$$

This energy per magnetic moment is an additional energy, known as the *exchange energy*, which causes alignment of the magnetic moments within a domain.

In order to consider the energy in domain walls, we need look at situations where only the nearest-neighbor interactions are significant. We need to do this because the directions of the magnetic moments within a domain wall vary with position, and therefore, the mean field approximation, which can work within the body of a domain due to uniform magnetization within the domain, is not valid for those moments located in the domain wall.

We therefore consider nearest-neighbor interactions in order to derive an equation for the domain-wall energy. The interaction or exchange energy in joules between any pair of moments $\boldsymbol{m}_i$ and $\boldsymbol{m}_j$ is given by Equation 6.7.

$$E_{\text{pair}} = -\mu_0 \frac{\alpha_{ij}}{V} \boldsymbol{m}_i \cdot \boldsymbol{m}_j \tag{7.5}$$

If the interaction coefficient α_{ij} is the same for all pairs of moments, then it will have a single value α and the pair interaction energy for any pair of moments $\boldsymbol{m}_i$ and $\boldsymbol{m}_j$ will be

$$E_{\text{pair}} = -\mu_0 \frac{\alpha}{V} \boldsymbol{m}_i \cdot \boldsymbol{m}_j \tag{7.6}$$

Each moment $\boldsymbol{m}_i$ then interacts with the nearest moments like $\boldsymbol{m}_j$ around it. The sum of these pair interaction energies is the interaction or exchange energy for the individual magnetic moment in joules.

$$E_{\text{ex}} = -\mu_0 \frac{\alpha}{V} \boldsymbol{m}_i \cdot \sum_j \boldsymbol{m}_j \tag{7.7}$$

If on average each moment $\boldsymbol{m}_i$ interacts with only n magnetic moments like $\boldsymbol{m}_j$ around it, then the sum $\Sigma_j \boldsymbol{m}_j$ can be replaced with $n\boldsymbol{m}_j$, α/V needs to be rescaled accordingly. We therefore will replace it with an interaction between nearest neighbors only represented by a nearest-neighbor coupling coefficient ϑ such that the interaction energy per moment becomes

$$E_{\text{ex}} = -\mu_0 n \vartheta \boldsymbol{m}_i \cdot \boldsymbol{m}_j \tag{7.8}$$

where n is the number of nearest neighbors. The approach used here in defining the exchange interaction in terms of magnetic moment will be useful in introducing the quantum mechanical exchange in terms of electron spins in Section 7.1.4. The term $-\mu_0 \vartheta \boldsymbol{m}_i \cdot \boldsymbol{m}_j$ is the exchange energy between any pair of nearest neighbors, assuming that all interactions other than nearest-neighbor exchange interactions are negligible. It also follows from Equation 7.8 that

$$T_{\text{c}} = \frac{\mu_0 n \vartheta \boldsymbol{m}^2}{3k_{\text{B}}} \tag{7.9}$$

Since the magnetic moments $\boldsymbol{m}_i$ and $\boldsymbol{m}_j$ are not necessarily parallel in the domain wall, the interaction energy per moment becomes dependent on the angle between neighboring moments. If ϕ is the angle between the neighboring moments $\boldsymbol{m}_i$ and $\boldsymbol{m}_j$, the interaction energy per moment becomes

$$E_{\text{ex}} = \mu_0 n \vartheta \boldsymbol{m}^2 \cos\phi \tag{7.10}$$

If we consider a linear chain, each moment has two nearest neighbors. Making the substitution $\cos\phi = 1 - \phi^2/2$ for small ϕ, the exchange energy per moment becomes

$$E_{\text{ex}} = -\mu_0 \vartheta \boldsymbol{m}^2 \left(\phi^2 - 2\right) \tag{7.11}$$

The extra interaction energy due to the rotation of the angle of the moments within the wall is therefore the sum of the exchange energies of all magnetic moments in the wall

$$E_{\text{ex}} = -\mu_0 \vartheta \boldsymbol{m}^2 \phi^2 n_{\text{wall}} \tag{7.12}$$

where n_{wall} is the number of moments that span the thickness of the domain wall. From this result, it is clear that the interaction energy is minimized when ϕ is very small, corresponding to a very wide domain wall [3]. Therefore, the exchange interaction energy favors wide domain walls.

For a 180° domain wall of thickness equal to n_{wall} lattice parameters, each of size a, the angle between the orientations of the neighboring moments is $\phi = \pi/n_{\text{wall}}$, and the additional energy per unit area due to exchange is then

$$\frac{E_{\text{ex}}}{a^2} = \frac{\mu_0 \vartheta \boldsymbol{m}^2 \pi^2}{n_{\text{wall}} a^2} \tag{7.13}$$

The term $\mu_0 \vartheta \boldsymbol{m}^2/a$ in Equation 7.13 is often replaced by the symbol A called the *exchange stiffness*, which has units of joules per meter.

$$A = \frac{\mu_0 \vartheta \boldsymbol{m}^2}{a} \tag{7.14}$$

As will be shown in the next section, the domain-wall surface energy γ and domain-wall width δ (= $n_{\text{wall}}a$) can be easily expressed in terms of this exchange stiffness A and the anisotropy coefficient K. The values of A can be estimated from $A = 3k_B T_c/na = E_{\text{ex}}/na$, where a is the lattice parameter and n is the number of nearest-neighbor atoms. For iron, the exchange stiffness is typically $A = 1.75 \times 10^{-11}$ J m^{-1}; for nickel, $A = 0.6 \times 10^{-11}$ J m^{-1}; and for cobalt, $A = 1.5 \times 10^{-11}$ J m^{-1}.

7.1.2.2 Anisotropy Energy

The anisotropy energy of a cubic crystal expressed using the single coefficient K_1 approximation gives the energy of the pth moment in the walls as

$$E_a = \left(\frac{K_1}{4}\right)\sin^2 2p\phi \tag{7.15}$$

and summing this over the width of the domain wall gives [4] the energy per unit area as

$$E_a = K_1 a n_{wall} \tag{7.16}$$

where:

a is the lattice *spacing*

n_{wall} is the number of layers of atoms in the domain wall

If $\delta = a\, n_{wall}$

$$E_a = K_1\delta \tag{7.17}$$

Therefore, the anisotropy energy is minimized for a very thin wall.

Energy per unit area of the wall

If the anisotropy and exchange energies are summed, this gives the domain-wall energy per unit area γ as determined by these two main components of the domain-wall energy.

$$\gamma = \frac{\mu_0 \vartheta \boldsymbol{m}^2 \pi^2}{\delta a} + K_1\delta \tag{7.18}$$

If the anisotropy is the dominant term, then energy is minimized at small δ, whereas if the exchange is dominant, then energy is minimized at large δ. The domain-wall thickness is determined by competing influence of these two factors. The domain-wall surface energy shown here is usually represented by the symbol γ, which has units of joules per square meter.

7.1.3 Width of Domain Walls

How wide are the domain walls and what determines the width?

The width of the domain walls is determined by minimizing energy of the wall with respect to its width. Therefore, differentiating the energy with respect to δ to find the equilibrium, remembering that $\phi = \pi/n_{wall}$ and $\delta = a\, n_{wall}$

$$\frac{d\gamma}{d\delta} = \frac{-\mu_0 \vartheta \boldsymbol{m}^2 \pi^2}{\delta^2 a} + K_1 = 0 \tag{7.19}$$

This leads to the result

$$\delta = \pi\sqrt{\frac{\mu_0 \vartheta \boldsymbol{m}^2}{K_1 a}} = \pi\sqrt{\frac{3k_B T_c}{K_1 a n}} \tag{7.20}$$

The width of the domain wall therefore increases with increasing exchange energy but decreases with increasing anisotropy. The anisotropy energy favors alignment of moments only along the crystallographic easy axes, such as the <100> axes in iron or the <111> axes in nickel.

In the equation for the domain-wall thickness δ, the term $\mu_0 \vartheta \boldsymbol{m}^2/a$ can now be replaced by A, giving the particularly simple relation

$$\delta = \pi\sqrt{\frac{A}{K}} \tag{7.21}$$

from which it is clear that exchange stiffness and anisotropy have opposite effects on the wall thickness.

As an example, we can calculate the domain-wall width of a 180° domain wall in iron, assuming that for iron, the anisotropy is $K_1 = 4.8 \times 10^4$ J m^{-3}, the lattice parameter is $a = 0.29$ nm, the magnetic moment per atom is $\boldsymbol{m} = 2.14$ Bohr magnetons ($= 1.98 \times 10^{-23}$ A m^2), and the exchange energy, or interaction energy, between any nearest neighbor pair of magnetic moments is $E_{ex} = 24 \times 10^{-20}$ J, giving an exchange energy per nearest neighbor pair of $E_{ex}/n = 2 \times 10^{-21}$ J.
The domain-wall thickness δ is given by

$$\delta = \pi\sqrt{\frac{\mu_0 \vartheta \boldsymbol{m}^2}{Ka}} = \pi\sqrt{\frac{E_{ex}}{nKa}} \tag{7.22}$$

Substituting in the various values

$$\delta = \pi\sqrt{\frac{2\times10^{-21}}{\left(4.8\times10^{4}\right)\left(2.9\times10^{-10}\right)}} \tag{7.23}$$

$$= 38\times10^{-9}\,\text{m} = 130 \text{ lattice parameters (or atomic layers)} \tag{7.24}$$

This is an approximate value for the domain-wall width in iron. Actually the width depends on the type of domain wall. Typical values of domain-wall thicknesses in iron, nickel, and cobalt along different crystallographic directions have been given by Lilley [5]. Table 7.1 lists the magnetic properties of iron, cobalt, and nickel.

Substituting the expression for domain-wall thickness δ into the equation for domain-wall energy gives

$$\gamma = 2K\delta \tag{7.25}$$

and consequently,

$$\gamma = 2\pi\sqrt{AK} \tag{7.26}$$

which gives a particularly simple form for the domain-wall energy in terms of the exchange coefficient A and the anisotropy coefficient K. In this case, both A and K have similar effects on domain-wall energy.

TABLE 7.1
Magnetic Properties of Iron, Cobalt, and Nickel

	Fe	Co	Ni
Magnetic moment per atom [6] at 0 K (in Bohr magnetons)	2.22	1.72	0.62
Saturation magnetization (in 10^6 A m^{-1})			
(at 0 K)	1.74	1.43	0.52
(at 300 K)	1.71	1.40	0.48
Exchange energy *J* [3,7]			
(in J)	2.5×10^{-21}	4.5×10^{-21}	2×10^{-21}
(in meV)	0.015	0.03	0.020
Curie temperature [6]			
(in °C)	770	1131	358
(in K)	1043	1404	631
Anisotropy energy K_1 [8,9,10,11] (in J m^{-3})			
(at 0 K)	5.7×10^4	68×10^4	-5.7×10^4
(at 300 K)	4.8×10^4	45×10^4	-0.5×10^4
Lattice spacing [12] (in nm)			
A	0.29	0.25	0.35
C		0.41	
Domain-wall thickness [3,4]			
(in nm)	40	15	100
(in lattice parameters)	138	36	285
Domain-wall energy [6,11] (in J m^{-2})	3×10^{-3}	8×10^{-3}	1×10^{-3}

Source: For further details consult the paper by Lilley, B.A., *Phil. Magn.*, 41, 792, 1950.
Note: Domain-wall thicknesses and energies are approximate values only, since they will depend on the crystallographic direction of the moments in the domains on either side of the wall.

For the example of iron given above, we get

$$\gamma = 2\pi\sqrt{AK} = 2\pi\sqrt{\frac{E_{\text{ex}}K}{na}} = 0.0036 \text{ J m}^{-2} \tag{7.27}$$

7.1.4 180° and Non-180° Domain Walls

What different types of domain wall exist and how may we classify them?

Generally, domain walls can be classified into 180° and non-180° walls. The 180° domain walls occur in virtually all materials and are distinct from all other non-180° walls in that they are not affected by stress [13]. In 180° walls, the directions of magnetization in the neighboring domains are antiparallel and consequently the moments in the two domains always lie in equivalent crystallographic directions.

In cubic materials with $K > 0$, the non-180° walls are all 90°, so that the direction of the moments in neighboring domains are at right angles. Therefore, in iron, the easy axes are all the <100> directions, and domain walls between the <100> and <$\bar{1}$00> directions are 180° walls, while those between <100> and <010> are 90° walls. In nickel, for which $K < 0$, the easy axes are all the <111> directions. Consequently, the non-180° domain walls will be either 71° or 109°.

Often, all non-180° domain walls are loosely and incorrectly referred to collectively as *90° walls* to distinguish the stress-sensitive from the non-stress-sensitive walls.

7.1.5 Effects of Stress on 180° and Non-180° Domain Walls

Why are 90° domain walls stress-sensitive but 180° domain walls are not?

Suppose, for example, a uniaxial stress is applied along the <100> direction in iron. If this is a tensile stress, it will make the <100> direction lower in energy than the <010> and the <001> directions, which were degenerate in energy in the unstressed state. However, by symmetry, the energy of the <$\bar{1}$00> direction will be reduced by the same amount as the <100> direction. Consequently, a 180° wall separating domains aligned along these two directions will not be affected by stress.

Consider now a 90° wall between domains along the <100> and <010> directions in iron that are energetically equal. Application of tensile stress along the <100> direction causes the <100> oriented domain to be energetically favored over the <010> direction, so the 90° wall will move under the action of such a stress to increase the volume of the <100> domain at the expense of the <010> domain. It can be seen therefore that the 90° domain walls are stress-sensitive, whereas the 180° domain walls are not.

7.1.6 Closure Domains

Where do these different types of domain wall occur?

Closure domains occur more often in cubic materials than in hexagonal materials because the cubic anisotropy ensures that the directions at right angles to the magnetization in a given domain are also magnetically easy axes. Therefore, it is energetically favorable to have 90° as well as 180° domain walls. In a material such as cobalt, in which the unique axis is the easy direction, it is difficult to form 90° walls since the magnetic moments in one of the domains must then lie in the hexagonal basal plane that is energetically unfavorable.

One example of where 90° domain walls occur is in the closure domains of grain-oriented silicon-iron, Figure 7.2. The domain boundaries between the neighboring longitudinal domains are 180° walls, while those between the closure domains at the end of the material and the main longitudinal domains are 90° walls. Of course,

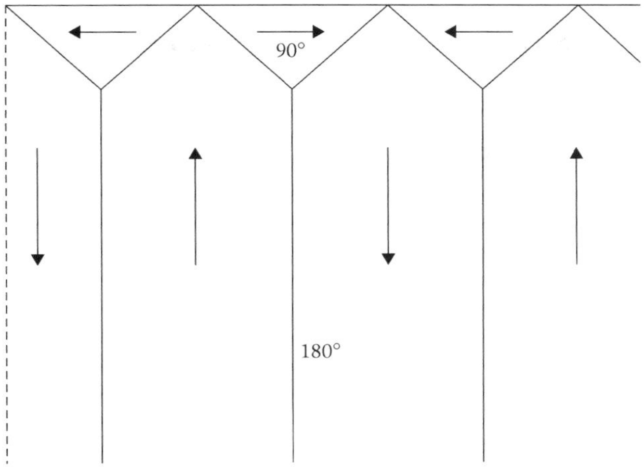

FIGURE 7.2 Closure domains at the surface in a material with cubic crystal structure, in this case grain-oriented silicon-iron.

there are many other instances of 90° walls, and they certainly are not dependent on high anisotropy. In fact, high anisotropy can impede the formation of closure domains if the change in anisotropy energy is larger than the change in magnetostatic energy that is driving the formation of the closure domains.

7.1.7 Néel Walls

Are there other types of domain wall apart from Bloch walls?

If a specimen is in the form of thin film, the ferromagnetic domains can extend across the whole width of the specimen. In this case, the Bloch wall, which would have its magnetization normal to the plane of the material, as shown in Figure 7.3, causes a large

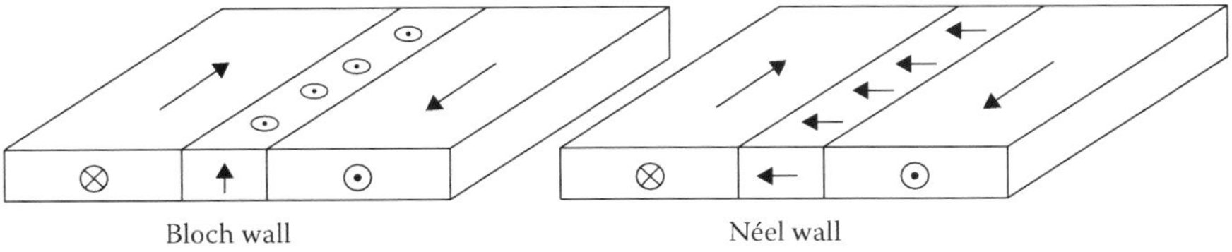

FIGURE 7.3 A conventional Bloch and Néel walls in a thin film of ferromagnetic material. The Bloch wall requires that some of the magnetic moments be oriented normal to the plane of the film. This leads to a demagnetization energy associated with the Bloch wall. The Néel wall has all moments oriented in the plane. The Néel wall is energetically favored once the film thickness decreases below a certain critical value.

increase in magnetostatic energy, whereas the Néel wall [14], in which the moments rotate within the plane of the specimen, results in a lower magnetostatic energy.

Néel walls do not occur in bulk specimens because they generate a higher magnetostatic energy within the volume of the domain wall. It is only in thin films that this energy becomes lower than the magnetostatic energy of the Bloch wall, leading therefore to the appearance of Néel walls.

7.1.8 Antiferromagnetic Domain Walls

What about domain walls in other ordered magnetic materials?

Magnetic domain walls occur in all forms of ordered magnetic materials and so there are correspondingly many different forms of domain walls. Domain walls in simple antiferromagnets and ferrimagnets are similar to domain walls in ferromagnets. However, helical antiferromagnetism, as occurs in dysprosium and terbium, presents a very interesting case. Domains in these materials were first suggested by Palmer [15] in which it was proposed that the domains consisted of helices of different chirality. These were subsequently observed experimentally [16]. The domain walls are transition regions between helical domains and are regions in which the neighboring moments are aligned nearly parallel, or at least the turn angle between successive moments ϕ is very small compared to within the domain.

Generally, the two senses of helix are energetically equal so that either may occur. Application of a field perpendicular to the unique axis of the helix can cause favorably oriented domain walls to grow. Therefore, the helical antiferromagnetic to ferromagnetic transition, which can be induced by application of a field perpendicular to the unique axis, is caused by growth of domain walls in this type of magnetic structure.

7.2 DOMAIN-WALL MOTION

What changes occur at the domain walls when the magnetic field increases?

In discussing the magnetizing process, we have mentioned the rotation of magnetic moments within a domain as a result of the competing effects of anisotropy and field energy. We have also touched very briefly on the idea of the growth of favorably oriented domains without really explaining what is meant by this.

In fact, on application of a field, it is the moments within domain walls that can most easily be rotated. This is because the resulting directions of the moments within the walls are a fine balance between the exchange and anisotropy energies. Therefore, a change in the field energy $E = -\mu_0 \boldsymbol{m} \cdot H$ can alter this balance causing the moments to rotate. Within the main body of the domain, the moments are locked into a particular direction by the exchange interaction, so that an applied field cannot immediately alter the balance of the energy in favor of another direction.

7.2.1 Effect of Magnetic Field on the Energy Balance in Domain Walls

Why do magnetic changes under the action of weak and moderate fields occur at domain boundaries?

Consider, for example, Figure 7.1. If a weak field is applied in the *up* direction, the moments within the *down* domain will not change direction because they are at the bottom of a deep energy well caused by their mutual interactions through the exchange field. However, in the domain wall, the energy introduced by the magnetic field will alter the energy balance slightly in favor of the *up* direction. The net result is that the moments within the walls rotate slightly into the field direction as the field is increased. To an observer, it therefore appears as though the domain wall moves toward the right. This process is called *domain-wall motion*, although in truth there is no translational motion at all. The movement is more like that of a wave, and in fact, Kittel [17] and others have considered the movement of Bloch walls to be an example of the wave motion of a soliton, that is, a solitary wave.

In discussing domain-wall motion, it is conventional to treat the domain wall as an entity in itself and to discuss its motion through the material in terms similar to those used for interfaces such as elastic membranes.

7.2.2 Domain Walls as Elastic Membranes

Is there a simple model that we can use to envisage the behavior of domain walls?

Since domain walls have an energy associated with them, which is proportional to the area, the walls behave in such a way as to minimize their area. We may therefore consider them to be analogous to the surfaces of liquids where the domain-wall energy is equivalent to the surface tension of the liquid. This analogy works well in trying to understand the behavior of domain walls, particularly the reversible bowing of domain walls [18,19] under the action of a magnetic field, and their tendency to become attached to nonmagnetic particles, impurities, second-phase materials, or other inhomogeneities within the solid [20] and to regions of inhomogeneous microstrain [21,22].

7.2.3 Forces on Domain Walls

What is the force exerted on a domain wall by a magnetic field?

If a magnetic field is applied to a ferromagnetic material, this results first in the movement of domain walls, in such a way that domains aligned favorably with the field direction grow at the expense of those aligned unfavorably. The energy per unit volume of such a domain with magnetization $\boldsymbol{M}_s$ when subjected to field $\boldsymbol{H}$ is

$$E_H = -\mu_0 \boldsymbol{M}_s \cdot \boldsymbol{H} \tag{7.28}$$

Consequently, the energy change caused by displacement of a 180° domain wall through a distance x is

$$E = -2\mu_0 A\boldsymbol{M}_s \cdot \boldsymbol{H}x \tag{7.29}$$

where A is the area of the wall. Therefore, the force per unit area on such a wall is

$$\boldsymbol{F} = -\left(\frac{1}{A}\right)\left(\frac{dE}{dx}\right) = 2\mu_0 \boldsymbol{M}_s \cdot \boldsymbol{H} \tag{7.30}$$

7.2.4 Planar Displacement of Rigid, High-Energy Domain Walls: Potential Approximation

How do walls with high surface energy compared to pinning energy behave?

Walls with high surface energy tend to remain planar. Planar domain-wall motion has been considered by Kersten [23,24]. The movement of planar domain walls under the action of a magnetic field in a specimen of high-purity iron is shown in Figure 7.4.

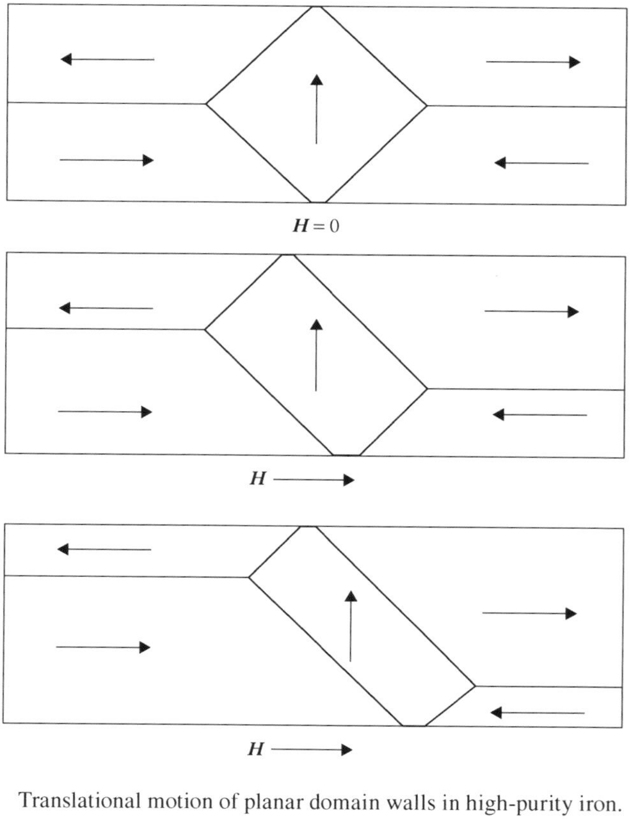

FIGURE 7.4 Translational motion of planar domain walls in high-purity iron.

The energy supplied by a magnetic field $\boldsymbol{H}$ to a ferromagnet is given by

$$\Delta E = -\mu_0 \int \boldsymbol{H} \cdot d\boldsymbol{M} \tag{7.31}$$

and consequently if a 180° domain wall with a unit cross-sectional area is moved through a distance dx, the change in magnetization is

$$d\boldsymbol{M} = 2\boldsymbol{M}_s dx \tag{7.32}$$

Therefore, the change in field energy is given by

$$\Delta E_H = -2\mu_0 \int \boldsymbol{M}_s \cdot \boldsymbol{H} dx = -2\mu_0 \boldsymbol{M}_s \cdot \boldsymbol{H} x \tag{7.33}$$

where x is the displacement of the wall. We assume in this analysis that the $\boldsymbol{H}$ field is parallel to the direction of magnetization in one of the domains and antiparallel to the other domain.

If the domain wall is subject to a potential energy E_p within the material, then the total energy is [25,26]

$$E_{tot} = E_p - 2\mu_0 \boldsymbol{M}_s \cdot \boldsymbol{H} x \tag{7.34}$$

The potential E_p will vary throughout the solid since defects such as inclusions or voids within the solid cause local minima in E_p, while regions of inhomogeneous microstress, associated with dislocations, can provide either energy minima or maxima depending on the sign of the stress and the magnetostriction coefficient. In general, the form of this potential will vary in a random manner as shown in Figure 7.5, where the energy is plotted against wall displacement.

The displacement x of the 180° wall can then be calculated from the condition

$$\frac{dE_{tot}}{dx} = 0 = \frac{dE_p}{dx} - 2\mu_0 \boldsymbol{M}_s \cdot \boldsymbol{H} \tag{7.35}$$

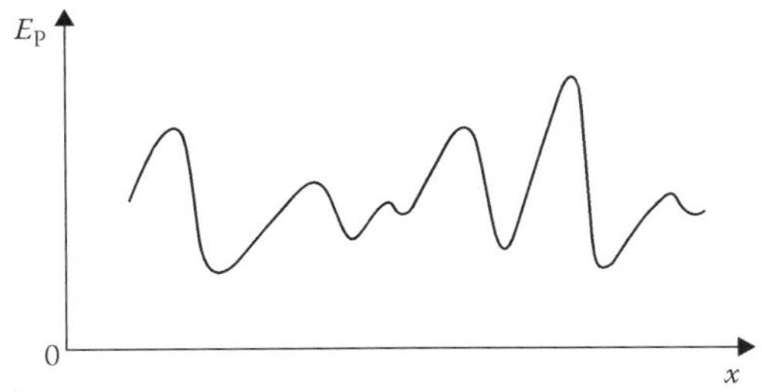

FIGURE 7.5 Potential energy seen by domain wall as a function of position. This variation of energy with displacement is used in models based on the rigid wall approximation.

Specific solutions may be attempted if the form of E_p is known or can be approximated. As an example, consider the simplest form of potential well

$$E_p = \frac{1}{2}ax^2 \tag{7.36}$$

In this case, the solution for the displacement x of the wall is

$$x = \frac{2\mu_0 \boldsymbol{M}_s \cdot \boldsymbol{H}}{a} \tag{7.37}$$

This refers to a small displacement, and hence to a completely reversible process since if the field is removed, then x returns to zero. As might be expected, the wall displacement under a given field strength increases as the potential well becomes flatter and decreases as it becomes steeper.

7.2.5 Magnetization and Initial Susceptibility in the Rigid Wall Approximation

What value of initial reversible susceptibility is expected on the basis of the rigid wall model?

This enables us to calculate the initial reversible change in magnetization. Suppose $x = 0$ corresponds to the demagnetized state $\boldsymbol{M} = 0$, then for a 180° wall, the magnetization is given by

$$\boldsymbol{M} = 2\boldsymbol{M}_s Ax\cos\theta \tag{7.38}$$

where:

A is the cross-sectional area of the wall

θ is the angle between the magnetization vector in the domain and the direction of displacement

Substituting for x from Equation 7.37

$$\boldsymbol{M} = \frac{4\mu_0 A\boldsymbol{H}\boldsymbol{M}_s^2\cos^2\theta}{a} \tag{7.39}$$

and consequently, the reversible initial susceptibility is

$$\chi_{in} = \frac{d\boldsymbol{M}}{d\boldsymbol{H}} = \frac{4\mu_0 A\boldsymbol{M}_s^2\cos^2\theta}{a} \tag{7.40}$$

In a cubic material such as iron, the average of $\cos^2\theta$ over the three easy axes is 1/3, so for a three-dimensional multidomain specimen

$$\chi_{in} = \frac{4\mu_0 A M_s^2}{3a} \quad (7.41)$$

For a 90° domain wall, the analogous result is

$$\chi_{in} = \frac{2\mu_0 A M_s^2}{3a} \quad (7.42)$$

The initial reversible susceptibility calculated on the basis of the rigid wall approximation, therefore depends on the saturation magnetization and on the potential seen by the domain wall as expressed by the parameter a. As the potential well becomes steeper, the initial susceptibility decreases. Equation 7.42 shows how a macroscopic measurable magnetic quantity, the initial susceptibility, is linked to a microstructural feature, the internal potential experienced by the domain walls.

7.2.6 Bending of Flexible, Low-Energy Domain Walls: Wall Bowing Approximation

How do walls with low surface energy behave compared to those with pinning energy?

Walls with low surface energy show a tendency to bend. In practice, domain walls exhibit both reversible bending and reversible translation under the action of a magnetic field. Wall bending is shown in Figure 7.6. Of course, the amount of bending that the walls undergo depends on many factors, including the field strength, but most particularly on the domain-wall energy γ [27–31]. Walls with high energy do not bend easily, while those with a low energy are more flexible.

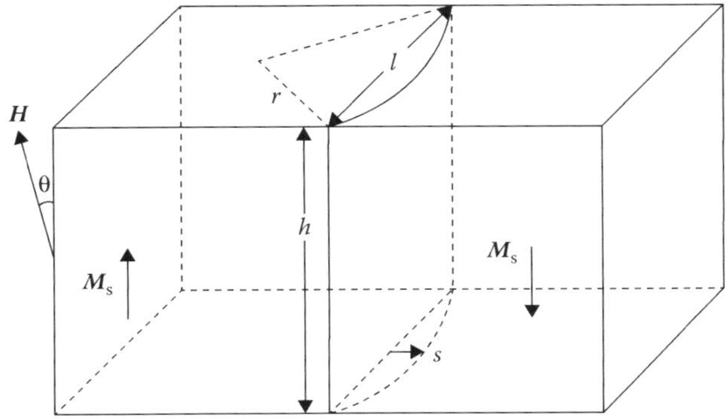

FIGURE 7.6 Bending of a domain wall under the action of a field. The domain wall is pinned at the boundaries and expands in the manner of an elastic membrane. (Data from Néel, L., *Ann. Univ. Grenoble*, 22, 299, 1946; Kersten, M., *Z. Angewandte Phys.*, 7, 313, 1956.)

To give some indication, the calculated wall energies, based on a one-constant anisotropy model, suggest that 180° domain walls have surface energies of about 2.9×10^{-3} J m^{-2}, 0.7×10^{-3} J m^{-2}, and 7.6×10^{-3} J m^{-2} in iron, nickel, and cobalt, respectively. The domain walls in cobalt are therefore more rigid than the domain walls in nickel. However, if a specimen is very pure and is largely free of defects, the low surface energy domain walls such as those of nickel can still move in a planar manner, because there are so few pinning sites to cause domain-wall bowing.

Suppose that a 180° domain wall extends throughout a single grain of a relatively pure specimen. The grain boundary is then the principal microstructural impediment to domain-wall motion. The domain wall will attach itself to the grain boundary in much the same way that it becomes attached to any other defect in the material. If a field is applied along the direction of one of the domains, that domain will grow by domain-wall motion. However, as the domain wall is attached to the grain boundary, that motion will initially occur by bending.

The force per unit area on the domain wall is, as before

$$\boldsymbol{F} = 2\mu_0 \boldsymbol{M}_s \cdot \boldsymbol{H} \tag{7.43}$$

The difference in wall energy caused by bending will be

$$E = \gamma\left[A(\boldsymbol{H}) - A(0)\right] \tag{7.44}$$

where:

γ is the wall surface energy

$A(0)$ is the area of the domain wall under zero magnetic field, that is, before deformation

$A(\boldsymbol{H})$ is the area of the domain wall under a field $\boldsymbol{H}$

Assuming for simplicity a cylindrical deformation, this leads to an expression for the force per unit area on the wall in terms of its radius of curvature

$$\boldsymbol{F} = \frac{\gamma}{r} \tag{7.45}$$

as in the expression for excess pressure across an elastic membrane such as a liquid interface, where γ is the surface tension.

7.2.7 Magnetization and Initial Susceptibility in the Flexible Approximation

What initial reversible susceptibility is expected on the basis of the wall-bending model?

Once again, it is possible to calculate the initial reversible susceptibility. The change in magnetic moment as a result of bending of the domain wall is

$$m = 2\boldsymbol{M}_s dV \tag{7.46}$$

where $dV = (2/3)lhx$, x being the displacement of the wall at its center, and h and l represent the spatial extent of the undeformed wall.

$$\boldsymbol{m} = \frac{4}{3}\boldsymbol{M}_s lhx \tag{7.47}$$

Since for small x, $x \approx l^2/8r$

$$\boldsymbol{m} = \frac{1}{6}\frac{\boldsymbol{M}_s l^3 h}{r} \tag{7.48}$$

and at equilibrium, the force due to the field trying to move the wall must equal the force due to the surface tension trying to restrain the movement of the wall. Equating the pressures, we get

$$\frac{\gamma}{r} = 2\mu_0 \boldsymbol{M}_s \boldsymbol{H} \tag{7.49}$$

This allows a substitution to be made for r, since now $r = \gamma/(2\mu_0 \boldsymbol{M}_s \boldsymbol{H})$

$$m = \frac{\mu_0 \boldsymbol{M}_s^2 \boldsymbol{H} l^3 h}{3\gamma} \tag{7.50}$$

and if the typical domain volume over which this change occurs is V, then

$$\boldsymbol{M} = \frac{\mu_0 \boldsymbol{M}_s^2 \boldsymbol{H}}{3\gamma}\frac{hl^3}{V} \tag{7.51}$$

The initial susceptibility is thus

$$\chi_{\text{in}} = \frac{\mu_0 \boldsymbol{M}_s^2}{3\gamma}\frac{hl^3}{V} \tag{7.52}$$

This is the initial susceptibility, which depends on the saturation magnetization and on the domain-wall surface energy γ. For high domain-wall surface, energy the wall is more rigid and the initial susceptibility is small. For low domain-wall surface energy, the wall is more flexible and the initial susceptibility is large. For a higher density of defects or pinning sites (meaning low values of l and h), the initial susceptibility will be lower. Here, again it has been shown how a macroscopic measurable magnetic quantity, the initial susceptibility, is linked to a microstructural feature, the separation of the defects of pinning sites in the material and the domain-wall surface energy.

We see that in general two types of wall motion can occur: wall displacement and wall bending. The strength of the domain-wall pinning and the surface energy of the wall determine which of these occurs in a particular case.

REFERENCES

1. Kittel, C. *Revs. Mod. Phys.*, 21, 541, 1949.
2. Bloch, F. *Z. Physik*, 74, 295, 1932.
3. Kittel, C., and Galt, J. K. Ferromagnetic domain theory, *Solid State Phys.*, 3, 437, 1956.
4. Chen, C. W. *Magnetism and Metallurgy of Soft Magnetic Materials*, North Holland, Amsterdam, the Netherlands, 1977, p. 84.
5. Lilley, B. A. *Phil. Magn.*, 41, 792, 1950.
6. Cullity, B. D. *Introduction to Magnetic Materials*, Addison-Wesley, Reading, MA, 1972, p. 617.
7. Hoffmann, J. A., Paskin, A., Tauer, K. J., and Weiss, R. J. *J. Phys. Chem. Sol.*, 1, 45, 1956.
8. Graham, C. D. *Phys. Rev.*, 112, 1117, 1958.
9. Sato, H., and Chandrasekar, B. S. *J. Phys. Chem. Sol.*, 1, 228, 1957.
10. Pauthenet, R., Barnier, Y., and Rimet, G. Proceedings of the international conference on magnetism and crystallography, Kyoto, Japan, 1961, *J. Phys. Soc. Japan*, 17 (Suppl. B-l), 309, 1962.
11. Chikazumi, S. *Physics of Magnetism*, Wiley, New York, 1964, p. 130.
12. Kittel, C. *Introduction of Solid State Physics*, 6th edn, Wiley, New York, 1986, p. 23.
13. Chikazumi, S. *Physics of Magnetism*, Wiley, New York, 1964, p. 192.
14. Néel, L. *Comptes Rendus*, 241, 533, 1955.
15. Palmer, S. B. *J. Phys. F.*, 5, 2370, 1975.
16. Palmer, S. B., Baruchel, J., Drillat, A., Patterson, C., and Fort, D. *J. Magn. Magn. Mater.*, 53–57, 1626, 1986.
17. Kittel, C. *Introduction to Solid State Physics*, 6th edn, Wiley, New York, 1986.
18. Kersten M. *Z. Angewandte Phys.*, 8, 496, 1956.
19. Néel, L. *Cahiers Phys.*, 25, 21, 1944.
20. Kersten. M. *Grundlagen einer Theorie der Ferromagnetischen Hysterese und der Koerzitivkraft*, Leipzig, Germany, 1943.
21. Becker, R. *Phys. Zeits.*, 33, 905, 1932.
22. Kondorsky, E. *Phys. Z. Sowjetunion*, 11, 597, 1937.
23. Kersten, M. *Phys. Zeits.*, 39, 860, 1938.
24. Kersten, M. in *Problem der Technischen Magnetisierungskurve* (ed. R. Becker), Springer, Berlin, Germany, 1938, p. 42.
25. Hoselitz, K. *Ferromagnetic Properties of Metals and Alloys*, Oxford University Press, Oxford, 1952, Ch. 2, p. 79.
26. Chikazumi, S. *Physics of Magnetism*, Wiley, New York, 1964, p. 267.
27. Néel, L. *Ann. Univ. Grenoble*, 22, 299, 1946.
28. Néel, L. *Cahiers Phys.*, 25, 21, 1944.
29. Kersten, M. *Z. Angewandte Phys.*, 7, 313, 1956.
30. Kersten, M. *Z. Angewandte Phys.*, 8, 382, 1956.
31. Kersten, M. *Z. Angewandte Phys.*, 8, 496, 1956.

FURTHER READING

Chikazumi, S. *Physics of Magnetism*, John Wiley, New York, 1964, Ch. 9.
Crangle, J. *The Magnetic Properties of Solid*, Arnold, London, 1977, Ch. 6.

Cullity, B. D. *Introduction of Magnetic Materials*, Addition-Wesley, Reading, MA, 1972, Ch. 9.
Kittel, C. Physical theory of ferromagnetic domains, *Rev. Mod. Phys.*, 21, 541, 1949.
Kittel, C., and Galt, J. K. Ferromagnetic domain theory, *Solid State Phys.*, 3, 437, 1956.

EXERCISES

7.1 *Magnetostatics.* A specimen of magnetic material is in the shape of a cube of side L and consists of an array of slab-like antiparallel domains of width d separated by 180° domains walls. The magnetostatic energy per unit volume of a magnetized body is given by

$$E = \left(\frac{1}{2}\right)\mu_0 N \boldsymbol{M}_s^2 \left(\frac{d}{L}\right) \quad (7.1.1)$$

where:

N is the demagnetizing factor
$\boldsymbol{M}_s$ is the saturation magnetization within the domains

If $N = 0.5$, $\boldsymbol{M}_s = 1 \times 10^6$ A m^{-1}, and $L = 10$ mm, calculate the domain-wall surface energy that is necessary to give a domain spacing of $d = 5 \times 10^{-6}$ m.

7.2 *Magnetostatics.* If in the previous example the material has a cubic lattice with atomic spacing of 0.3 nm and the anisotropy is 1×10^5 J m^{-3}, calculate the exchange energy between each pair of moments located on the atomic sites needed to get the above result.

7.3 *Critical dimensions for single-domain particles in nickel.* Calculate the magnetostatic energy of a small spherical particle of nickel of radius r and magnetized to saturation $\boldsymbol{M}_s = 5.1 \times 10^5$ A m^{-1}. (You can assume the magnetostatic energy of a sphere with magnetization $\boldsymbol{M}$ and volume V is $E_{ms} = \mu_0 \boldsymbol{M}^2 V/6$ and that this energy is halved when the particle is divided into two domains.) The energy required to form a Bloch wall is $\gamma = 0.7 \times 10^{-3}$ J m^{-2} in nickel. Hence find the critical radius for single-domain particles, which will be such that the reduction in magnetostatic energy obtained by division into two domains is less than the energy needed to form a wall.

7.4 *Calculation of wall energy and thickness from anisotropy energy, saturation magnetization, and exchange energy.* Find the domain-wall energy γ, wall thickness δ, and critical radius r_c for single-domain particles of a material with anisotropy energy $K = 33 \times 10^4$ J m^{-3}, spontaneous magnetization within a domain of $\boldsymbol{M}_s = 0.38 \times 10^6$ A m^{-1}, lattice spacing 3×10^{-10} m, and exchange energy $J = 3 \times 10^{-21}$ J. You may assume that this material has a spin of $S = 1/2$, or equivalently a magnetic moment per atom of one Bohr magneton.

7.5 *Estimation of domain spacing in cobalt.* Estimate the domain spacing d in a specimen of single-crystal cobalt in the form of a plate of infinite extent in two dimensions and of height l and assume that the magnetostatic energy per unit area is $0.135\mu_0 M_s^2 d$. The domain-wall energy for cobalt is 7.6×10^{-3} J m^{-2} and $M_s = 1.42 \times 10^6$ A m^{-1}. Calculate the domain spacing when $l = 0.01$ m.

7.6 *Exchange energy.* The anisotropy constant of cobalt is 4.5×10^5 J m^{-3} and of neodymium-iron-boron is 4.0×10^6 J m^{-3}. Both materials have uniaxial anisotropy and therefore only 180° domain walls exist. The Curie temperatures are 1400 and 585 K, respectively, and you can assume an average lattice parameter $a = 0.3$ and 0.1 nm, respectively, for the purposes of the calculation. Estimate the exchange energy that corresponds to these Curie temperatures. Then calculate the domain boundary width and energy per unit area of the domain boundaries in the two materials.

7.7 *Existence of domain walls.* Explain the reasons for the existence of domain boundaries and the factors determining the thickness δ. Explain why a domain boundary has a surface energy γ and specify which factors determine the magnetic domain-wall surface energy.

7.8 *Exchange energy.* Determine (1) the exchange energy per *atomic* spin, (2) the exchange energy between two nearest neighbor spins, (3) the domain-wall thickness, and (4) domain-wall surface energy for a simple cubic material with Curie temperature $T_c = 700$°C, anisotropy $K_1 = 4 \times 10^4$ J m^{-3}, lattice spacing $a = 0.25$ nm, and magnetic moment per atom $\boldsymbol{m} = 2$ Bohr magneton.

7.9 *Exchange energy, anisotropy, and domain-wall properties.* For a simple cubic (sc lattice) material state the following expressions:

a. The exchange energy per *atomic* spin E_{ex}
b. The exchange stiffness A
c. The exchange energy between any two nearest neighbor pairs of spins E_{ex}/N_{nn}
d. The domain-wall thickness δ
e. The domain-wall surface energy γ

Calculate the values of each of the above (a)–(e) if the material has Curie temperature $T_c = 750$°C, anisotropy $K_1 = 5 \times 10^4$ J m^{-3}, and lattice spacing $a = 0.3$ nm.

7.10 *Stress and magnetoelastic energy.* A ferromagnetic crystal with negligible magnetocrystalline anisotropy has a magnetostriction coefficient λ_s that is isotropic. If this material is subjected to a uniform tensile stress σ throughout, and has an exchange stiffness A, derive an expression for the surface energy γ of a 180° domain wall in this material under stress. Find the magnetostriction λ_s and determine the exchange energy per *atomic* spin E_{ex} for nickel if $T_c = 630$ K. From this, calculate the exchange stiffness A for nickel, assuming a lattice spacing of $a = 0.3$ nm and thereby determine the energy of a 180° wall in nickel under a tensile stress of 100 MPa.

8 Domain Processes

In this chapter, we will look at the behavior of domains under the action of a magnetic field that varies with time. We are interested principally in the changes that occur in the domain structure and the mechanisms by which these are brought about. We will look at reversible and irreversible changes in magnetization and describe them in terms of the two main mechanisms: (1) domain-wall motion and (2) domain rotation. We will use some of these ideas to formulate a model of hysteresis, and finally, we will look at time-dependent magnetization processes.

8.1 REVERSIBLE AND IRREVERSIBLE DOMAIN PROCESSES

Are the changes in magnetization reversible or irreversible or both?

The changes in magnetization arising from the application of a magnetic field to a ferromagnet can be either reversible or irreversible depending on the domain processes involved. A reversible change in magnetization is one in which after application and removal of a magnetic field, the magnetization returns to its original value. In ferromagnetic materials, this only occurs for small field increments.

More often both reversible and irreversible changes occur together, so that on removal of the field the magnetization does not return to its initial value. If the magnetic field is cycled under these conditions, hysteresis is observed in $\boldsymbol{M}$. The next task is to interpret such changes in magnetization in terms of domain mechanisms, so that both reversible and irreversible changes can be explained.

8.1.1 Domain Rotation and Wall Motion

What mechanisms are available for interpreting and describing the changes in magnetization with field?

The domain mechanisms that we have discussed so far are rotation and wall motion. Both of these processes can be manifested as either reversible or irreversible mechanisms, and the transition from reversible to irreversible is in both cases dependent on the amplitude of the magnetic field.

Wall motion incorporates two distinct effects: (1) bowing of domain walls and (2) translation. Domain-wall bowing is a reversible process at low-field amplitudes. The domain wall expands like an elastic membrane under the action of a magnetic field. When the field is removed, the wall returns reversibly to its original position. Wall bowing becomes irreversible in two ways. The deformation of the domain wall can become irreversible if during this bowing process, the wall encounters further pinning sites that prevent it from relaxing once the field is removed. Alternatively, if the domain wall is sufficiently deformed, the expansion can continue without further increase of field until the wall encounters further pinning sites that prevent it from relaxing once the field is removed.

The translation of domain walls is usually irreversible unless the material is sufficiently pure that the domain wall can exist in a region of the material that is free from defects. The displacement of planar walls can be modeled using a potential energy, which fluctuates as a function of distance, as in Figure 8.1. There are two possible origins for the short-range fluctuations in potential experienced by the domain walls. These are short-range variations in strain associated with dislocations within the material and the microstructural inhomogeneties such as the presence of particles of a second phase within the matrix material. There is no general way of treating these energy fluctuations since they are random in nature; however, we can consider a simple case of a sinusoidal variation of potential energy, which will have many of the essential features of the real solid.

Rotation of magnetic moments within a domain has been discussed in Chapter 6. At low-field amplitudes, the direction of alignment of the magnetic moments, which corresponds to minimum energy, is displaced slightly from the crystallographic easy axes toward the field direction. This results in a reversible rotation of the magnetic moments within a domain.

At intermediate to high-field amplitudes, there is an irreversible mechanism within the domain when the moments rotate from their original easy axis to the easy axis closest to the field direction. This occurs when the field energy overcomes the anisotropy energy. In this case, once the magnetic moments within the domain have rotated into a different easy axis, the moments remain within the potential well surrounding this easy axis if the field is reduced.

At high fields, the energy minimum of the easy axis closest to the field is perturbed by the field energy until the minimum lies in the field direction. This results in a reversible rotation of the moments into the field direction and hence a reversible change in magnetization at high fields. Finally, at very high fields there is a reversible change in which the magnetic moments within the specimen, which is already a single domain, are aligned more closely with the field direction. This occurs because the individual magnetic moments precess about the field direction

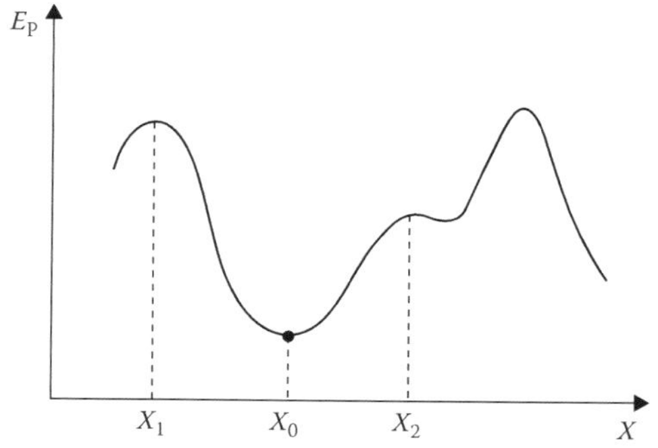

FIGURE 8.1 Magnetic energy potential as a function of distance seen by a magnetic domain wall.

due to thermal energy. This precession does not give any net moment in other directions but does reduce the component of magnetization along the field direction. As the field strength increases, the angle of precession is reduced. Similarly, if the temperature is lowered, the angle of precession is reduced due to a reduction in thermal energy.

8.1.2 Strain Theory: Pinning of Domain Walls by Strains

What causes domain-wall motion to be irreversible?

Even before Bloch's prediction of the existence of a finite transition region between domains, which he called *domain walls*, it was realized that the principal domain process occurring at low fields was caused by domain boundary motion. Early experiments on domain boundary motion were reported by Sixtus and Tonks [1]. In view of the hysteresis loss, most of which occurs at low fields, it was necessary to explain the loss mechanism in terms of wall motion.

One of the earliest suggestions was by Becker [2] that the domain walls were impeded in their movements by regions of inhomogeneous strain, which interacted via the magnetostriction with the magnetic moments to provide local energy barriers that the domain walls needed to overcome. A survey of this early work has been given by Hoselitz [3]. The first calculations were by Becker and Kersten [4] and by Kondorsky [5]. Becker and Kersten determined the initial susceptibility and the coercivity as a function of internal stress. The results of Kersten's investigations are shown in Figure 8.2.

The strain theory of coercivity was treated in detail by Becker and Doring [6] and by Kersten [7]. During the time in which the effects of stress on magnetization were being investigated, the existence of dislocations within crystals was first suggested by Orowan [8], Taylor [9], and Polanyi [10]. Dislocations have an associated local stress field, which gives rise to highly inhomogeneous microstrains within a solid [11].

Through the magnetoelastic coupling, the dislocations pin domain walls, and therefore, the higher the dislocation density within a ferromagnet, the greater the impedance to domain-wall motion. This explains why cold-worked specimens have higher coercivity and lower initial susceptibility than the same material in a well-annealed state. This result can be exploited for nondestructive evaluation of the mechanical state of a ferromagnetic material using magnetic measurements.

We can calculate the susceptibility as a result of planar wall displacement in a rather simplified example using a sinusoidal internal potential. Suppose the stress σ varies from location to location within a material and can be approximately represented as a periodic function of distance x as shown in Figure 8.3

$$\sigma = \sigma_0 \left[1 - \cos\left(\frac{2\pi x}{l} \right) \right] \tag{8.1}$$

where:

l is the spatial periodicity of the stresses
σ_0 is the maximum level of stress

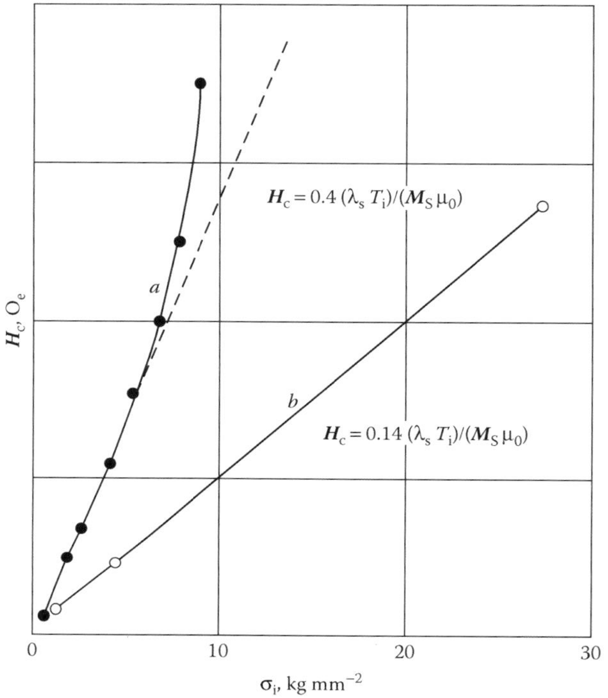

FIGURE 8.2 Variation of coercivity of nickel with internal stress calculated by Kersten: (a) recrystallized nickel and (b) hard-drawn nickel. This result confirmed the strain theory of coercivity.

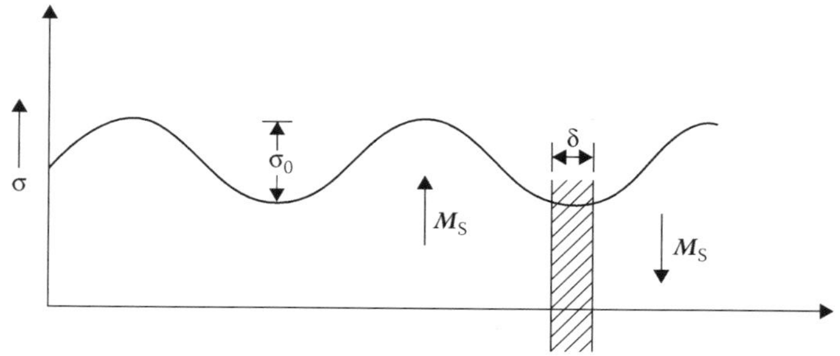

FIGURE 8.3 Idealized variation of internal micro stress with displacement, approximated by a sinusoidal function.

This has a minimum energy at $x = 0$ so any small movement of the domain wall away from $x = 0$, caused, for example, by a magnetic field to reduce the magnetostatic energy in the applied field, will result in an increase of stress of the domain wall and hence to an increase of the magnetoelastic energy. These two energy terms will then have different effects on the domain wall and will compete to determine the equilibrium position of the domain wall.

The magnetoelastic energy E_σ for a domain wall of area A and thickness δ is, assuming that domain-wall thickness is much less than the periodicity l,

$$E_\sigma = \left(\frac{3}{2}\right)\lambda_s \sigma A\delta = \left(\frac{3}{2}\right)\lambda_s \sigma_0 \left[1 - \cos\left(\frac{2\pi x}{l}\right)\right] A\delta \tag{8.2}$$

where λ_s is the magnetostriction coefficient. Then consider a 90° domain wall of area A with the domain on one side of the boundary aligned parallel with an applied $\boldsymbol{H}$ field, and the domain on the other side of the boundary aligned perpendicular to the field. On change of the applied field $\boldsymbol{H}$, the domain wall moves through a distance x aligning moments from the unfavorably oriented 90° domain into the field direction. The magnetization $\boldsymbol{M}_s$ is changed in part of the 90° domain, specifically in the volume Ax swept out by the domain wall as it moves under the action of the field $\boldsymbol{H}$. The change in energy E_H as the magnetization is reversed in the 90° domain is

$$E_H = -\mu_0 \boldsymbol{M}_s \boldsymbol{H} A\, x \tag{8.3}$$

The total energy will be the sum of the field energy and the stress energy.

$$E_{tot} = E_\sigma + E_H \tag{8.4}$$

$$E_{tot} = \left(\frac{3}{2}\right)\lambda_s \sigma_0 \left[1 - \cos\left(\frac{2\pi x}{l}\right)\right] A\delta - \mu_0 \boldsymbol{M}_s \boldsymbol{H} A x \tag{8.5}$$

Since these energy terms vary with position x, the equilibrium position of the domain wall is obtained when the derivative is zero.

$$\frac{dE_{tot}}{dx} = \left(\frac{3\pi}{l}\right)\lambda_s \sigma_0 \sin\left(\frac{2\pi x}{l}\right) A\delta - \mu_0 \boldsymbol{M}_s \boldsymbol{H} A = 0 \tag{8.6}$$

Therefore, at equilibrium,

$$\mu_0 \boldsymbol{M}_s \boldsymbol{H} A = \left(\frac{3\pi}{l}\right)\lambda_s \sigma_0 \sin\left(\frac{2\pi x}{l}\right) A\delta \tag{8.7}$$

which can be solved for x, the displacement of the domain wall. For small x, we can make the approximation $\sin(2\pi x/l) \approx 2\pi x/l$.

$$\mu_0 \boldsymbol{M}_s \boldsymbol{H} = \frac{6\pi^2 \lambda_s \sigma_0}{l^2} x\delta \tag{8.8}$$

The equilibrium displacement of the domain wall is therefore

$$x = \frac{\mu_0 \boldsymbol{M}_s \boldsymbol{H} l^2}{6\pi^2 \lambda_s \sigma_0 \delta} \tag{8.9}$$

The initial susceptibility χ_{in} for small displacement x can be obtained by noting that for a 90° domain wall the change in magnetic moment along the field direction $\Delta \boldsymbol{m}$ is

$$\Delta \boldsymbol{m} = A\boldsymbol{M}_s x \tag{8.10}$$

$$\chi_{in} = \frac{d\boldsymbol{M}}{d\boldsymbol{H}} = \frac{1}{V}\left(\frac{dm}{dx}\right)\left(\frac{dx}{d\boldsymbol{H}}\right) = \frac{1}{V}\frac{A\mu_0 \boldsymbol{M}_s^{\,2} l^2}{6\pi^2 \lambda_s \sigma_0 \delta} \tag{8.11}$$

The initial susceptibility χ_{in} decreases with the amplitude of the stress σ_0, and decreases with the magnetostriction λ_s, both results are well known in practice, and increases with the separation distance l of the peaks in stress. Therefore, the lower the stress amplitude (pinning strength) and the greater the distance between the stress maxima (pinning site density), the less are the impediments to domain-wall motion and the higher the susceptibility.

8.1.3 Inclusion Theory: Pinning of Domain Walls by Impurities

What other factors cause irreversibility in domain-wall motion?

Isolated regions of second-phase materials with magnetic properties different from those of the matrix material are known as *magnetic inclusions*. These reduce the energy of domain walls when the domain walls intersect them and consequently the domain walls are attracted to the inclusions, which effectively impede wall motion.

The inclusion theory of domain-wall pinning was suggested by Kersten [12,13]. He assumed that the magnetic domain walls move in a planar manner through the solid and that the energy of the walls is reduced when they intersect inclusions. The inclusions themselves may take many forms, such as insoluble second-phase material; they may be oxides or carbides, or they may be pores, voids, cracks, or other mechanical inhomogeneities. Well-known examples of magnetic inclusions are cementite (Fe_3C) particles in iron and steel.

Néel [14] criticized the assumption of planar wall motion and further indicated that Kersten's interpretation ignored the fact that the magnetic free poles associated with a defect such as a void and a crack would be a greater source of energy. Thus, a naked inclusion totally enclosed within the body of a domain would have free poles attached, as shown in Figure 8.4a, with an attendant magnetostatic energy of $2\pi\mu_0 \boldsymbol{M}_s^2 r^3/9$. In practice, the inclusion will instead have closure domains (also known as *spike domains*) attached to it, as shown in Figure 8.4b, because these reduce the magnetostatic energy of the inclusion compared with the naked inclusion. However, when a domain wall bisects the inclusion, there occurs a change in distribution of the free poles on the inclusion, as shown in Figure 8.4c. This results in a further reduction of the magnetostatic energy compared with the configuration with spike domains, to $\pi\mu_0 \boldsymbol{M}_s^2 r^3/9$. Néel's model became to be known as the *disperse field theory* or the *variable field theory*.

The energy reduction of the domain wall using the simple elastic membrane model is the reduction in area of the wall times the wall energy per unit area. This can be calculated as $\pi r^2 \gamma$, where γ is the domain surface energy (wall energy per unit area)

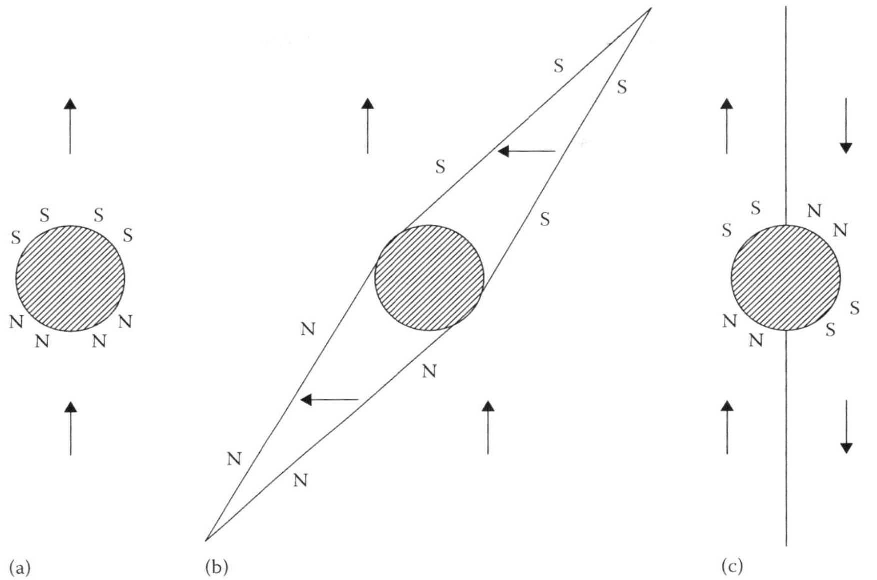

FIGURE 8.4 (a) Free-pole distribution on a naked inclusion; (b) spike domains attached to an inclusion to reduce the magnetostatic energy associated with the inclusion compared with a naked inclusion; and (c) reduction of magnetostatic energy associated with an inclusion when intersected by a domain wall.

in J m^{-2}. For a 1 μm diameter nonmagnetic inclusion in iron with $\gamma = 3 \times 10^{-3}$ J m^{-2}, the energy reduction due to free poles is about 70 times the energy reduction due to wall area. Figure 8.5 shows the results of a comparison of calculated and observed coercivities in steels by Néel [15].

An attempt to describe the effect of carbide inclusions on the coercivity of iron was made by Dijkstra and Wert [16]. In their model, the domain wall was considered to be an elastic surface, so that the difference in energy between the wall when it intersected the defect and when it was free was determined simply from the difference in area of the wall. In reality, an inclusion would not be entirely free from spike domains when within the body of a domain as shown in Figure 8.4b. These spike domains themselves lower the magnetostatic energy of the inclusion. Nevertheless, the energy of the inclusion and spike domains is still reduced when intersected by a domain wall and the energy reduction is still considerably greater than the energy obtained from reducing the wall area.

8.1.4 Critical Field When a Domain Wall Is Strongly Pinned

What is the field strength needed to break a domain wall away from a pinning site in the situation where the wall bends?

Kersten [17] modified the assumptions of his previous model and calculated the initial susceptibility of a solid with flexible domain walls. Consider, for

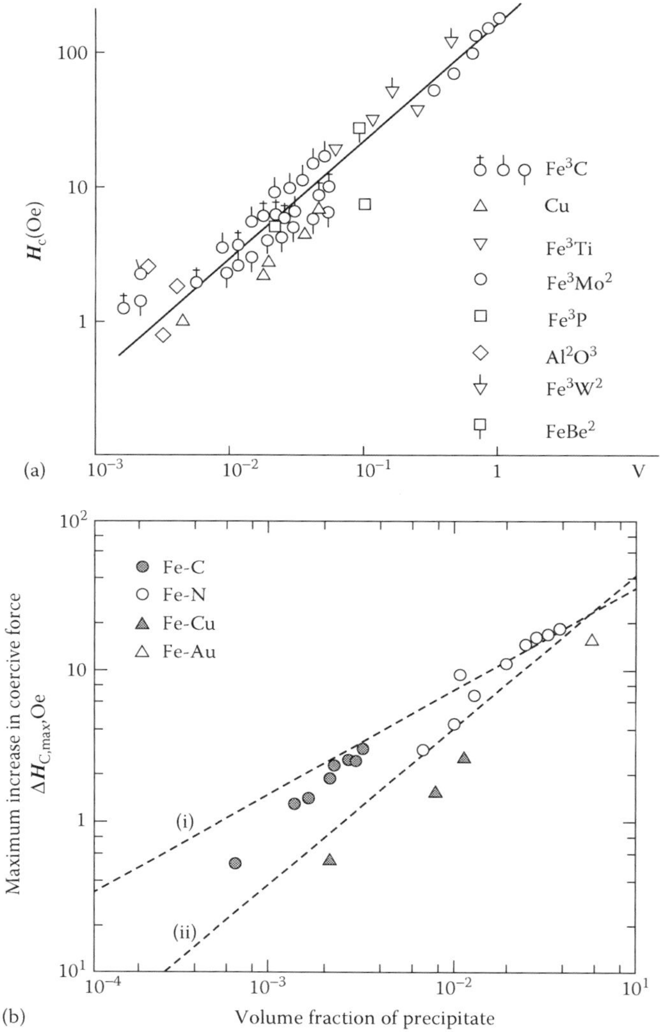

FIGURE 8.5 (a) Coercivity of various steels as a function of the total volume fraction of inclusions. Experimental results of Kersten and calculations by Néel. (Data from Néel, L., *Ann. Univ. Grenoble*, 22, 299, 1946.) (b) Increase in coercivity of iron caused by precipitates of interstitial or substitutional solutes. Theoretical calculations are represented by the two dashed lines: (i) due to Kersten and (ii) due to Néel.

example, the case where the pinning strength of the defect sites is relatively high and the wall energy is relatively low. In this case, the domain wall will bend under the action of a field. The process will be reversible until the radius of curvature reaches a critical value, after which the wall expands discontinuously and irreversibly.

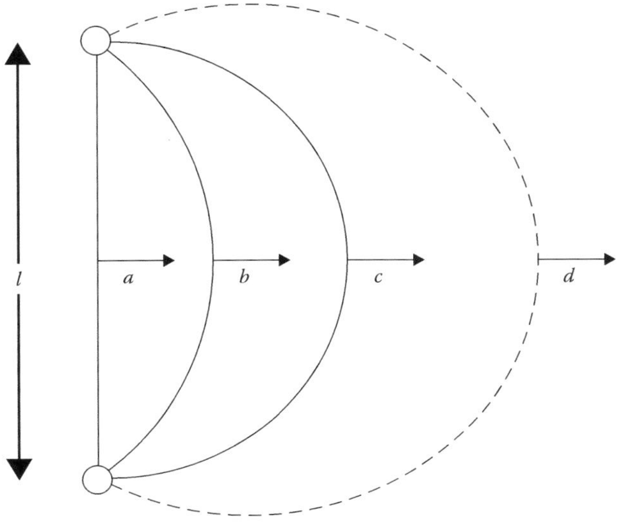

FIGURE 8.6 Reversible and irreversible expansion of a domain wall according to Kersten.

Consider the situation depicted in Figure 8.6. The radius of curvature r of the domain wall as it bends under the influence of an applied field $\boldsymbol{H}$ is such that the excess pressure on the wall γ/r exactly balances the force per unit area due to the field. For a 180° wall,

$$\frac{\gamma}{r} = 2\mu_0 \boldsymbol{M}_s \boldsymbol{H} \cos\theta \quad (8.12)$$

where θ is the angle between the direction of the $\boldsymbol{H}$ field and the magnetization within one of the domains. The critical condition arises when the radius of curvature of the wall reaches $l/2$, where l is the distance between the pinning sites.

$$\boldsymbol{H}_{c180^\circ} = \frac{\gamma}{\mu_0 \boldsymbol{M}_s l \cos\theta} \quad (8.13)$$

and similarly for a 90° wall,

$$\boldsymbol{H}_{c90^\circ} = \frac{\gamma\sqrt{2}}{\mu_0 \boldsymbol{M}_s l \cos\theta} \quad (8.14)$$

Kersten used these equations to determine the coercivity of nickel-iron alloys. This tells us that for strong pinning, the critical field needed to break the domain wall from its pinning sites, and hence the bulk coercivity, is dependent on the number density of pinning sites, expressed through the coefficient l and the domain-wall energy γ.

8.1.5 Critical Field When a Domain Wall Is Weakly Pinned

What is the field strength needed to break a domain wall away from its pinning site before it bends?

In this case, the walls move in a planar manner because they break away from the pinning sites before they have a chance to bend. The walls experience a certain potential as a result of the distribution of defects and dislocations. The force exerted on unit area of a 180° domain wall by a field $\boldsymbol{H}$ is given by

$$\boldsymbol{F} = 2\mu_0 \boldsymbol{M}_s \boldsymbol{H} \cos\theta \tag{8.15}$$

where θ is the angle between the magnetization and the magnetic field. If the potential energy is given by

$$E_p = E_p(x) \tag{8.16}$$

and the maximum slope of this is $\{[dE_p(x)]/dx\}_{max}$, then at the critical field, we must have the maximum force per unit area exerted on the wall.

$$\boldsymbol{F}_{max} = \left[\frac{dE_p(x)}{dx}\right]_{max} \tag{8.17}$$

Therefore, the critical field, beyond which the domain wall breaks away from its pinning site, is given by

$$\boldsymbol{H}_{crit} = \frac{\boldsymbol{F}_{max}}{2\mu_0 \boldsymbol{M}_s \cos\theta} = \frac{1}{2\mu_0 \boldsymbol{M}_s \cos\theta}\left[\frac{dE_p(x)}{dx}\right]_{max} \tag{8.18}$$

In the case of a simplified sinusoidal variation of the potential energy E_p,

$$E_p(x) = E_{max} \sin\left(\frac{2\pi x}{l}\right) \tag{8.19}$$

and therefore

$$\left[\frac{dE_p(x)}{dx}\right]_{max} = \frac{2\pi E_{max}}{l} \tag{8.20}$$

this leads to

$$\boldsymbol{H}_{crit} = \frac{\pi E_{max}}{l \mu_0 \boldsymbol{M}_s \cos\theta} \tag{8.21}$$

In this case, therefore, the critical field $\boldsymbol{H}_{crit}$ and hence the bulk coercivity $\boldsymbol{H}_c$ are dependent on the maximum pinning energy E_{max} and the number density of pinning sites, expressed in terms of their spatial periodicity or separation l.

8.2 DETERMINATION OF MAGNETIZATION CURVES FROM PINNING MODELS

Can realistic models of the magnetization process be devised from the concept of domain-wall pinning?

One of the most difficult problems in the field of magnetism is to describe the magnetization curves of a ferromagnet in terms of materials properties. The complexity of domain-wall interactions with distributed structural features is further compounded by the possibility of magnetization changes by domain rotation. Therefore, attempts at deriving the underlying theory have dealt only with the simplest situations.

8.2.1 Effects of Microstructural Features on Magnetization

How do dislocations and other defects affect coercivity and initial susceptibility?

The investigations of Becker, Kersten, and Néel served to clarify the processes taking place on the domain level when a domain wall interacts with defects. To be useful, such a theory must be related to the observed bulk properties. As we have shown, calculations were made by Kersten of the initial susceptibility based on a single Bloch wall model. The next development in understanding the magnetization process was a generalization of the earlier models to include the interaction of domain walls with defects such as dislocations. Dislocations have an associated stress field, which impedes the motion of magnetic domain walls. The suggestion that dislocations affect the coercive force was first made by Vicena [18]. Subsequently, there were a number of works published by Kronmuller and coworkers on this subject [19–23].

Seeger et al. [19] considered the influence of lattice defects on the magnetization curve. They noted that the internal stresses in the earlier work of Becker were of an unspecified origin and were characterized by the average stress amplitude. Their theoretical calculations, based on a rigid wall approximation, suggested that the product of coercivity and initial susceptibility should be a constant.

$$\chi_{in}\boldsymbol{H}_c = \text{constant} \tag{8.22}$$

In addition, it was shown that in the rigid wall approximation, the coercivity was dependent on the square root of dislocation density ρ, which in a face-centered cubic material is proportional to applied stress.

$$\frac{\boldsymbol{H}_c}{\rho^{1/2}} = \text{constant} \tag{8.23}$$

Kronmuller [20] derived a statistical theory of Rayleigh's law in the low-field magnetization region based on a rigid-wall approximation model of domain-wall motion. The Rayleigh law (see Section 5.1.6) gives the relation between magnetization $\boldsymbol{M}$ and magnetic field $\boldsymbol{H}$ as

$$M = \chi_{in} H + \nu H^2 \tag{8.24}$$

where:

ν is the Rayleigh coefficient

χ_{in} is the initial susceptibility

The coefficients were shown to be determined by the domain-wall potential, and it was found that they were related to the dislocation density ρ by the following:

$$\chi_{in} \rho^{1/2} = \text{constant} \tag{8.25}$$

$$\nu \rho = \text{constant} \tag{8.26}$$

Hilzinger and Kronmuller [21] also investigated theoretically the coercive field of hard ferromagnetic materials. In particular, they found that the interaction of domain walls in hard magnetic materials with defects led to the experimentally observed temperature dependence of coercivity of $SmCo_5$. They have shown that high coercivities can be obtained by (1) pinning of domain walls by defects and (2) elimination of domain walls using single-domain particles. For multidomain materials, the first mechanism is more important and it is known that in many materials, the most significant mechanism for pinning is the interaction between dislocations and domain walls. However, in $SmCo_5$, they considered that point defects are more important pinning sites because the domain walls in this material are so thin, being typically 3 nm in width.

Further investigation of the Bloch wall pinning in rare earth cobalt alloys [22] suggested that antiphase boundaries are the dominant mechanism determining coercive force in these materials. Antiphase boundaries occur when the A and B sublattices within an ordered superlattice become out of phase, so that the type of atoms that were on sublattice A switch to sublattice B and vice versa. The nearest-neighbor coordination of the superlattice is therefore disrupted locally.

Hilzinger and Kronmuller [23] developed and generalized these theories further for rigid (planar) domain-wall motion, considering the pinning of Bloch walls by randomly distributed defects. As they have indicated, the theories of domain-wall pinning basically still fall into two categories: (1) potential theories with rigid domain walls and (2) wall-bowing theories with flexible walls. The problem of domain-wall defect interactions has been treated previously for dislocations, point defects, and antiphase boundaries. They have shown that the coercivity depends upon defect density, the interaction (or pinning force), and the area and flexibility of domain walls. These concepts have also been considered by Jiles and Atherton [24].

In the case of potential theories, Hilzinger and Kronmuller have shown further that the coercive force H_c depends on the square root of the total domain-wall area A, as well as the defect density ρ

$$H_c \sim \frac{\rho^{1/2}}{A^{1/2}} \tag{8.27}$$

In the case of domain-wall bowing models, the coercive field was found to be dependent on $\rho^{2/3}$

$$H_c \sim \rho^{2/3} \quad (8.28)$$

according to Labusch [25]. The question of how much domain-wall bowing takes place is still slightly controversial. In Kronmuller's statistical theory of Rayleigh's law, the domain walls were assumed to remain planar, but the works of Kersten [17] and Dietze [26] assumed that bowing took place. It seems that the amount of domain-wall bowing that occurs depends on the strength of the pinning sites and the domain-wall surface energy. Chikazumi [27] considered that the domain walls remain planar along the axis of magnetization within the domain, such as the <100> direction in iron, in order to avoid the appearance of free poles at the curved portion of the walls; however, the walls are free to bend in other directions. For example, magnetizations along the <100> and <$\bar{1}$00> directions on either side of a domain wall lead to a planar wall when viewed from either the (001) or (010) plane. But when viewed from the (100) plane, the wall can bend, as in Figure 8.7.

Finally, after considering only planar walls in their earlier works, Hilzinger and Kronmuller [28] considered the pinning of curved domain walls by randomly distributed defects. The two theories, the potential theory for planar walls and the bowing theory for flexible walls, were shown to be merely extreme cases of a more general theory. In this more general theory, the coercive force was found to depend on $\rho^{1/2}$ for weak defect domain-wall interactions and on $\rho^{2/3}$ for strong defect domain-wall interactions. Therefore, the two models provided limiting cases of a single theoretical model. The result was important because it brought together the two principal wall mechanisms under a unified theory of domain-wall motion.

8.2.2 Domain-Wall Defect Interactions in Metals

What do experimental observations of domain-wall defect interactions show?

The problem of describing the magnetization process in terms of Bloch wall motion was also discussed by Porteseil, Astie, and coworkers. In these works, a number of observations of domain-wall motion were made and the results were interpreted in terms of theoretical models. Porteseil and Vergne in a theoretical study [29] calculated the expected magnetization curve of a polycrystal based on a single isolated Bloch wall model. They investigated conditions under which a single irreversible

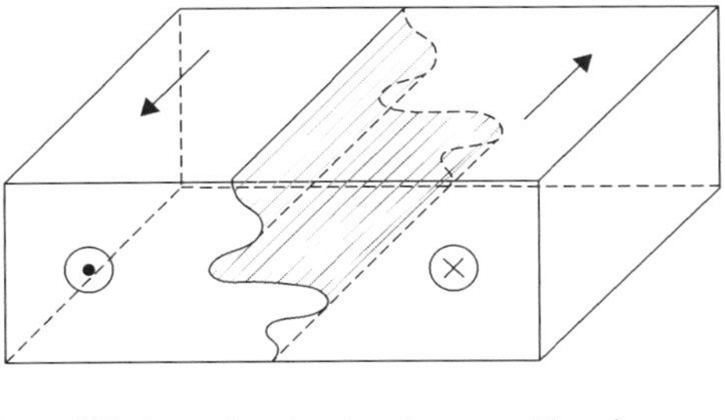

FIGURE 8.7 Domain wall in iron showing its planar and bowing aspects simultaneously in different crystallographic planes.

event remained independent and when it led to an avalanche effect with numerous subsequent irreversible wall jumps or Barkhausen discontinuities.

Astie et al. [30] studied the influence of dislocations on the magnetic and magnetomechanical properties of high-purity iron. The dislocation structures were formed by strain hardening of polycrystalline iron. They criticized earlier work on domain-wall defect interactions because earlier studies failed to make precise observations of lattice defects. Their investigation involved the dependence of the initial susceptibility χ_{in}, the Rayleigh coefficient ν, and the coercivity $\boldsymbol{H}_c$ on the strain hardening $\Delta\sigma = \sigma - \sigma_0$, where σ_0 is the yield strength and σ is the maximum stress applied. The authors concluded that there were three distinct regions on the strain hardening curve between $\Delta\sigma = 0$ and $\Delta\sigma = 100$ MPa and these were identified both from transmission electron microscopy and from the magnetic parameters.

Interestingly in the low-stress region (<30 MPa), it was found that the coercivity and initial permeability were almost independent of stress. However, at higher stresses, which resulted in larger dislocation densities in the form of tangles of dislocations separated by relatively large distances with lower dislocation densities, $\boldsymbol{H}_c$ increased rapidly, while χ_{in} was less affected. Further studies of the interaction of magnetic domain walls with defects in high-purity iron were reported by Degauque and Astie [31].

The effect of grain size on the magnetic properties of high-purity iron was reported by Degauque et al. [32]. The coercivity $\boldsymbol{H}_c$, the initial susceptibility χ_{in}, and the Rayleigh coefficient ν were studied together with the remanent magnetic induction $\boldsymbol{B}_R$. The study concluded that the initial susceptibility χ_{in} was independent of grain size but the Rayleigh coefficient ν was proportional to grain diameter d as shown in Figure 8.8.

$$\frac{\nu}{d} = \text{constant} \tag{8.29}$$

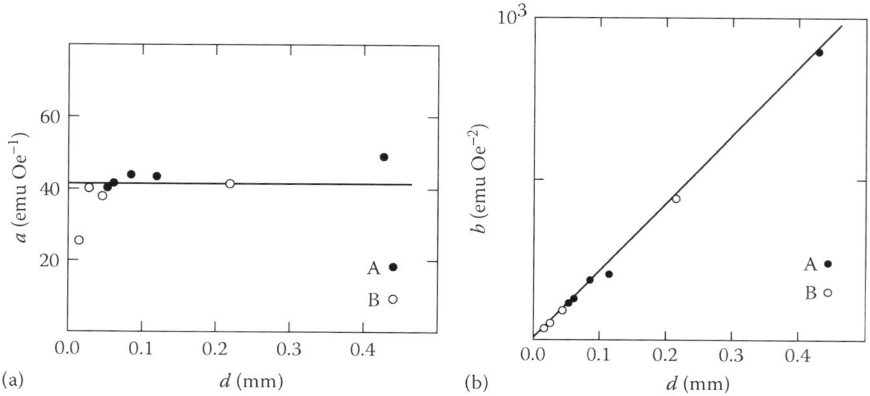

FIGURE 8.8 Dependence of Rayleigh coefficients on the grain diameter d in high-purity iron (see Equation 8.23). (After Degauque, J. et al., *J. Magn. Magn. Mater.*, 26, 261, 1982.)

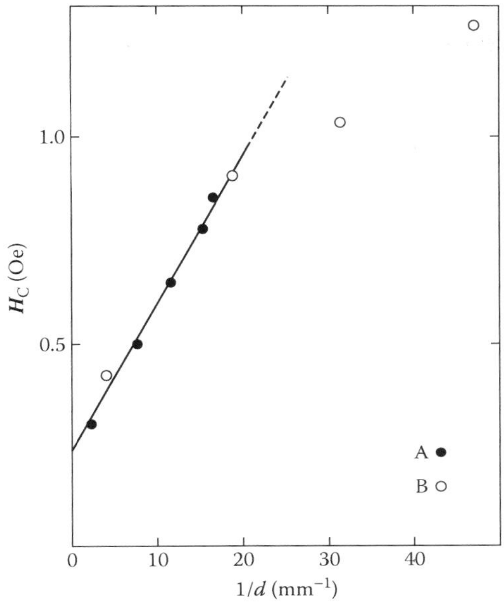

FIGURE 8.9 Dependence of the coercivity $\boldsymbol{H}_c$ on the grain diameter d in high-purity iron. (After Degauque, J. et al., *J. Magn. Magn. Mater.*, 26, 261, 1982.)

They also found that coercivity $\boldsymbol{H}_c$ was linearly dependent on $1/d$ as shown in Figure 8.9.

$$\boldsymbol{H}_c d = \text{constant} \quad (8.30)$$

The potential model for domain-wall motion was used by Astie et al. [33] to interpret results obtained on polycrystalline iron. From the work of Hilzinger and Kronmuller reported above, it was known that the theory predicted

$$\chi_{in}\boldsymbol{H}_c = \text{constant} \quad (8.31)$$

and

$$v\boldsymbol{H}_c^2 = \text{constant} \quad (8.32)$$

However, it was found from experimental measurements on polycrystalline iron, in which the density of defects was altered, that these values did change in contradiction of the predictions of the potential model of Hilzinger and Kronmuller. They found that the initial susceptibility χ_{in} was independent of grain size in iron. This is because χ_{in} is determined by the short-range reversible motion of domain walls in low fields. Under these conditions, the dislocations are the predominant factor contributing to χ_{in}. The Rayleigh coefficient ν was found to be dependent on grain size d and increased with grain size, whereas the coercivity decreased with grain size.

In summary, the predictions of invariance for both $\chi_{in}\boldsymbol{H}_c$ and $v\boldsymbol{H}_c^2$ follow from the model only if the assumption is valid that the potential curve $V(x)$ is only altered by scaling. In that case, V maintains a similar dependence on x even when the magnitude of V changes. However, Astie et al. indicate that such an assumption is unlikely to be valid in most cases, and as observed in their results and those of Jiles et al. [34], these products do not remain constant.

8.2.3 Magnetization Processes in Materials with Few Defects

Which factors influence the magnetization curve in materials with few defects?

The investigation of domain-wall motion in materials, such as ferrites, garnets, and spinels, which are relatively free of defects compared with iron, cobalt, and nickel, was conducted by Globus et al. The effects of grain boundaries on domain-wall motion in these materials proved to be the most significant microstructural factor affecting the bulk magnetization curves. In fact, domain-wall motion in these materials can be modeled with relatively few parameters, due to the simplicity of the situation, with grain size being the dominant microstructural feature.

In the Globus model, which was suggested in the earliest work of this series [35], it was assumed that domain walls are pinned at the grain boundaries and that the walls deform by bending like an elastic membrane under the action of a magnetic field as was first described by Kersten. In experimental studies, Globus and Duplex [36] prepared specimens that had only one variable parameter: the grain size. They investigated ferrites such as nickel ferrite $NiO{\cdot}Fe_2O_3$, spinels such as nickel-zinc ferrite $Ni_xZn_{1-x}O{\cdot}Fe_2O_3$, and yttrium-iron-garnet $5Fe_2O_3{\cdot}3Y_2O_3$. Precautions were taken to avoid other features such as nonmagnetic inclusions, pores, and dislocations within the grains, since these would strongly influence the domain-wall motion and hence the bulk magnetization curve.

From the model, it is possible to calculate the wall curvature that corresponds to the initial susceptibility χ_{in}. It was found that χ_{in} depended linearly on the grain diameter d, as shown in Figure 8.10 and Equation 8.33

$$\chi_{in} = \chi_{ss} + \frac{\mu_0 \boldsymbol{M}_s^2 d}{\gamma} \tag{8.33}$$

where χ_{ss} is the spin susceptibility, which is the value of susceptibility in the limit as d tends to zero.

It was also found that the initial susceptibility of these materials is due almost entirely to wall motion. From the results, it was therefore possible to separate the influence of wall motion (proportional to d) and rotation (independent of d) on the initial susceptibility. These contributions can be found in Figure 8.10b from the slope of the line, which is due to wall pinning, and the intercept at $d = 0$, which is due to rotation of the magnetic moments. The rotational contribution to χ_{in} is, according

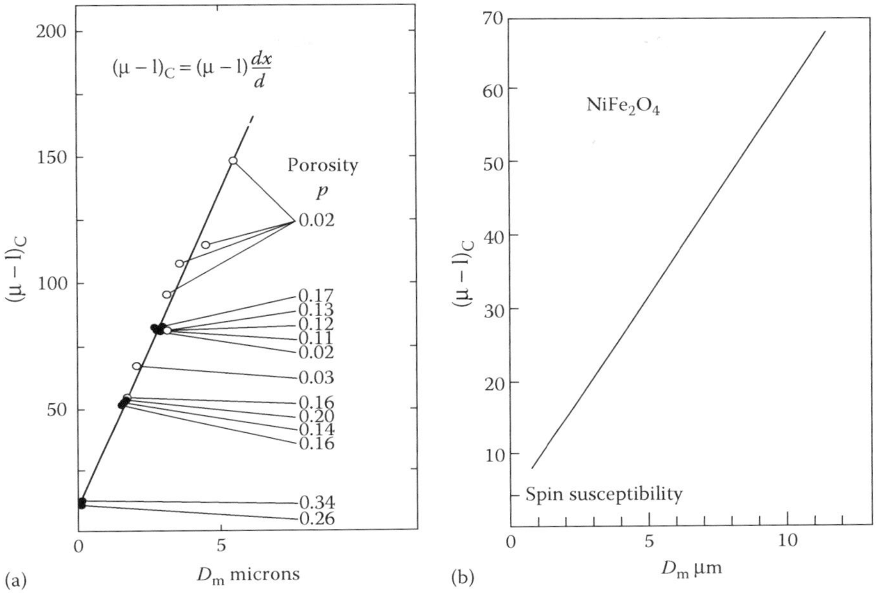

FIGURE 8.10 Dependence of the initial susceptibility χ_{in} on grain diameter d in yttrium-iron garnet (a) and nickel ferrite (b). (Reproduced with permission from Globus, A., and Duplex, P., *IEEE Trans. Magn.*, 2, 441, 1966. Copyright 1966 IEEE; Globus, A. et al., *IEEE Trans. Magn.*, 7, 617, 1971. Copyright 1971 IEEE.)

to Globus and Duplex [37], dependent on anisotropy but independent of structure, whereas the wall motion contribution is highly structure-sensitive. Subsequently, Globus and Duplex [38] applied the same model to spinels and garnets.

Globus, Duplex, and Guyot [39] have considered the magnetization process in yttrium-iron-garnet in terms of domain-wall processes. In their work, the reversible component of magnetization was considered to be due to wall bulging, while the irreversible component of magnetization was due to domain-wall displacement, since the fields used were well below those required for domain rotation. In this study, an explicit formula for the susceptibility in terms of domain-wall bending was given.

The extended Globus-Guyot model [40] included both wall bulging and displacement. From this, a model was provided for both coercivity and remanence. In the extended model, the authors determined a critical field strength $\boldsymbol{H}_{cr}$ beyond which wall translation began and hence irreversible magnetization processes began. This according to the model was dependent on the maximum restraining force $\boldsymbol{F}$ imposed by the grain boundary on the domain wall, the grain diameter d, and the spontaneous magnetization within the domain $\boldsymbol{M}_s$. In the model, the critical field strength was given in Equation 8.34:

$$\boldsymbol{H}_{cr} = \frac{2F}{\mu_0 \boldsymbol{M}_s d} \tag{8.34}$$

Further work by Guyot and Globus [41,42] attempted to relate hysteresis loss to the creation and annihilation of domain-wall surface area and the energy loss by continual pinning and unpinning of magnetic domain walls. They concluded that the losses in these materials cannot be explained on the basis of pinning losses alone. The first contribution to losses arises from a solid friction term due to the pinning and unpinning of domain walls. The second contribution to losses comes from the variations in magnetic domain surface area. Both of these lead to hysteresis loss. It was concluded that the magnetization curves of these materials all had the same general form that was dependent upon grain size d, saturation magnetization $\boldsymbol{M}_s$, and the magnetic anisotropy field $\boldsymbol{H}_{an}$. In general, the critical field for wall motion was much smaller than the anisotropy field ($\boldsymbol{H}_{cr}/\boldsymbol{H}_{an} = 0.003$) so that domain rotation mechanisms could not play a significant role.

Finally, Globus [43] has summarized the earlier work giving a universal curve for the initial magnetization and hysteresis loss of spinels, ferrites, and garnets. This universal curve depends upon $\boldsymbol{M}_s$, the anisotropy K, and the grain size d. The model seems to work well for these materials because they have very few intragranular defects, and therefore, the factors influencing the magnetization curve are relatively few and can be expressed in an analytic form of the magnetization curve. In metals such as iron and nickel, the presence of other defects such as nonmagnetic inclusions and dislocations play a significant role, and consequently, the magnetization mechanism is more complicated.

8.2.4 Barkhausen Effect and Domain-Wall Motion

How can other magnetic phenomena be interpreted in terms of wall motion?

The related phenomena of the Barkhausen effect and magnetoacoustic emission are both due to discontinuous irreversible changes in magnetization. These irreversible changes can occur as a result of irreversible domain-wall motion, either as the unpinning of planar or nonplanar domain walls from their pinning sites or as the discontinuous expansion of a domain wall once its curvature has exceeded the critical value. Barkhausen emissions can also result from the discontinuous rotation of moments within a domain from one of the easy axes into the easy axis aligned closest to the field direction.

The Barkhausen effect can be caused by motion of either 180° walls or non-180° walls. However, magnetoacoustic emission, which relies on the generation of changes in stress, is caused by discontinuous motion of non-180° walls. This is because no additional stresses are generated as a result of 180° wall motion, where the strain is independent of the direction of the magnetic moments, whether parallel or antiparallel, on either side of the wall.

The theoretical description of the Barkhausen effect has proved quite difficult because of its random nature. However, Alessandro, Bertotti, and coworkers have developed a stochastic process model of the Barkhausen effect in which the magnetic domain walls move through a randomly fluctuating internal potential [44]. At various locations, the domain wall is forced over pinning sites by the applied field and this leads to discontinuous motion or Barkhausen events. The predictions of the theory have been tested against an experiment [45] with good results. Subsequently, Pust et al. [46] developed these theoretical ideas further and showed that the local coercivity, or pinning field, for a domain wall is dependent on the gradient of the internal potential experienced by the wall as it moves through the material. In the work of Pust et al., the variations in this internal field were also expressed as a stochastic function of the position of the domain wall. This stochastic approach seems to provide a useful mathematical way of describing discontinuous domain-wall motion in soft magnetic materials.

8.2.5 Magnetostriction and Domain-Wall Motion

Are there differences in the contribution to magnetostriction from different types of domain wall?

There is no change in the bulk magnetostriction as a result of 180° wall motion or rotation. So in the case of terfenol and other highly magnetostrictive materials, it is important to arrange for as much non-180° activity as possible to optimize the performance of the material, that is, to maximize the bulk magnetostriction. This is achieved by inducing stress anisotropy via an applied compressive stress to align the domains at right angles to the field direction.

8.3 THEORY OF FERROMAGNETIC HYSTERESIS

How can the bulk magnetic properties of ferromagnets be described with a minimum number of parameters?

It is useful to be able to describe hysteresis mathematically in order to develop a model to describe magnetic properties of ferromagnetic materials. Therefore, we will now consider how the ideas developed so far can be brought together to provide a theoretical model of hysteresis in ferromagnets. A few models are in use at present; these include the Preisach model [47], which is widely used in the magnetic recording industry for describing the magnetization characteristics of recording tapes, and the Stoner-Wohlfarth model [48] of rotational hysteresis, which really only applies to single-domain particles, but has been used for modeling properties of hard magnetic materials. We will consider a model of hysteresis [24] based originally on domain-wall motion that, as we have mentioned previously, is the principal cause of hysteresis in multidomain specimens. This model can also take into account the effects of domain rotation since the essence of the model is the establishment of a relationship between energy dissipation (*losses*) and change in magnetization.

8.3.1 Energy Loss through Wall Pinning

Can we describe the energy loss in pinning of a domain wall?

Consider a 180° domain wall of area A between two domains aligned parallel and antiparallel to a magnetic field $\boldsymbol{H}$. If the wall moves through a distance dx under the action of a field, the change in energy in joules due to this movement is

$$dE = -2\mu_0 \boldsymbol{M}_s \cdot \boldsymbol{H} A dx \tag{8.35}$$

Suppose the pinning energy of a given pinning site is $\mu_0\varepsilon_\pi$ for a 180° domain wall, and that in general the pinning energy ε_{pin} is proportional to the change in energy per unit volume caused by moving the wall

$$\varepsilon_{pin} = \frac{1}{2}\varepsilon_\pi\left(1-\cos\phi\right) \tag{8.36}$$

where ϕ is the angle between the moments in the neighboring domains. This means that for any pinning site, the energy dissipation caused by unpinning is proportional to the change in magnetization $\Delta\boldsymbol{M} = \boldsymbol{M}_s(1 - \cos\phi)$. Equation 8.36 gives an expression for the pinning energy of a given site as a function of angle ϕ. When $\phi = 0$, the pinning energy must go to zero since the wall no longer exists. In the case of 180° wall, $\varepsilon_{pin} = \varepsilon_\pi$.

If there are n pinning sites per unit volume, the energy lost by moving a 180° domain wall depends on the pinning energy ε_π, the number of pinning sites per unit volume n and the volume Adx swept out by the wall as it moves. In this case, the energy loss in joules is

$$dE_{loss} = \mu_0 n\varepsilon_\pi A dx \tag{8.37}$$

where A is the cross-sectional area of the wall. The change in magnetic moment will be

$$d\boldsymbol{m} = 2\boldsymbol{M}_s A dx \tag{8.38}$$

where Adx represents the volume swept out by the motion of the domain wall. Therefore, the energy loss in joules per cubic meter will be

$$dE_{loss} = \frac{\mu_0 n\varepsilon_\pi d\boldsymbol{m}}{2\boldsymbol{M}_s V} = \frac{\mu_0 n\varepsilon_\pi d\mathbf{M}}{2\boldsymbol{M}_s} \tag{8.39}$$

where V is the volume, so that $d\boldsymbol{m}/V = d\boldsymbol{M}$. Replacing $(n\varepsilon_\pi/2M_s)$ by the loss coefficient k, which represents the pinning, gives the energy loss per unit volume as

$$dE_{loss} = \mu_0 k d\boldsymbol{M} \tag{8.40}$$

which gives the reasonable result that the energy loss per unit volume is proportional to the change in magnetization.

8.3.2 Irreversible Magnetization Changes

If we know the energy loss, can we write down an energy equation for the magnetization process?

Suppose the change in energy of a ferromagnet is manifested either as a change in magnetization or as hysteresis loss. Then we can write the energy equation as follows:

$$\begin{pmatrix}\text{Energy supplied to}\\ \text{material}\end{pmatrix} = \begin{pmatrix}\text{Change in}\\ \text{magnetostatic}\\ \text{energy}\end{pmatrix} + \begin{pmatrix}\text{Hysteresis}\\ \text{loss}\end{pmatrix} \tag{8.41}$$

In the case where there is no hysteresis loss, then the change in magnetostatic energy must equal the total energy supplied. When there is no hysteresis, the magnetization follows the anhysteretic curve $\boldsymbol{M}_{\text{an}}(\boldsymbol{H})$

$$\mu_0 \int \boldsymbol{M}_{\text{an}}(\boldsymbol{H}) d\boldsymbol{H} = \mu_0 \int \boldsymbol{M}(\boldsymbol{H}) d\boldsymbol{H} + \mu_0 \int k \left(\frac{d\boldsymbol{M}}{d\boldsymbol{H}} \right) d\boldsymbol{H} \tag{8.42}$$

and differentiating with respect to $\boldsymbol{H}$ gives

$$\boldsymbol{M}_{\text{an}}(\boldsymbol{H}) = \boldsymbol{M}(\boldsymbol{H}) + k \left(\frac{d\boldsymbol{M}}{d\boldsymbol{H}} \right) \tag{8.43}$$

$$\frac{d\boldsymbol{M}}{d\boldsymbol{H}} = \frac{\boldsymbol{M}_{\text{an}}(\boldsymbol{H}) - \boldsymbol{M}(\boldsymbol{H})}{k} \tag{8.44}$$

This simple result states that the rate of change of magnetization $\boldsymbol{M}$ with field $\boldsymbol{H}$ is proportional to the displacement of the magnetization from the anhysteretic magnetization. That is the bulk magnetization $\boldsymbol{M}$ experiences a harmonic potential about $\boldsymbol{M}_{\text{an}}(\boldsymbol{H})$.

In fact, the situation is a little more complex than this in reality due to the internal coupling between the magnetic domains, which is of the form envisaged by Weiss, so that the effective field becomes $\boldsymbol{H}_{\text{e}} = \boldsymbol{H} + \alpha \boldsymbol{M}$, resulting in the following equation, which we should note represents only the irreversible component of magnetization:

$$\frac{d\boldsymbol{M}_{\text{irr}}}{d\boldsymbol{H}} = \frac{\boldsymbol{M}_{\text{an}}(\boldsymbol{H}) - \boldsymbol{M}_{\text{irr}}(\boldsymbol{H})}{k\delta - \alpha\left[\boldsymbol{M}_{\text{an}}(\boldsymbol{H}) - \boldsymbol{M}_{\text{irr}}(\boldsymbol{H})\right]} \tag{8.45}$$

Here, δ is a directional term that has the value +1 for increasing magnetic field ($d\boldsymbol{H}/dt > 0$), or −1 for decreasing magnetic field ($d\boldsymbol{H}/dt < 0$), to ensure that the dissipation of energy due to hysteresis is always a loss, irrespective of the direction of change of the field.

8.3.3 Reversible Magnetization Changes

Can we incorporate reversible changes in magnetization into the differential equation of hysteresis?

In most cases, both reversible and irreversible magnetization changes occur simultaneously. The reversible component of magnetization is due to reversible domain-wall bowing, reversible translation, and reversible rotation. In the model, this has the form

$$\boldsymbol{M}_{\text{rev}} = c\left(\boldsymbol{M}_{\text{an}} - \boldsymbol{M}_{\text{irr}}\right) \tag{8.46}$$

and since magnetization changes must be either reversible or irreversible, changes in the total magnetization $\boldsymbol{M}_{\text{tot}}$ may be expected to be given by

$$\boldsymbol{M}_{\text{tot}} = \boldsymbol{M}_{\text{rev}} + \boldsymbol{M}_{\text{irr}} \tag{8.47}$$

In fact, the form of Equation 8.47 is not very helpful since magnetization changes that begin as reversible can become locked-in and end up as irreversible. A much more useful expression is the differential equation that expresses the change in magnetization with field. In this case, we can distinguish between a reversible contribution to the differential susceptibility and an irreversible contribution to the differential susceptibility, so that at any instant, it is possible to say how much of the change in magnetization is being contributed by reversible or irreversible magnetization changes.

$$\frac{d\boldsymbol{M}_{\text{tot}}}{d\boldsymbol{H}} = \frac{d\boldsymbol{M}_{\text{irr}}}{d\boldsymbol{H}} + \frac{d\boldsymbol{M}_{\text{rev}}}{d\boldsymbol{H}} \tag{8.48}$$

Removing the subscript and assuming that whenever we talk about a change in magnetization without further qualification, we mean the total magnetization

$$\frac{d\boldsymbol{M}}{d\boldsymbol{H}} = \frac{\left(\boldsymbol{M}_{\text{an}} - \boldsymbol{M}_{\text{irr}}\right)}{k\delta - \alpha\left(\boldsymbol{M}_{\text{an}} - \boldsymbol{M}_{\text{irr}}\right)} + c\left(\frac{d\boldsymbol{M}_{\text{an}}}{d\boldsymbol{H}} - \frac{d\boldsymbol{M}_{\text{irr}}}{d\boldsymbol{H}}\right) \tag{8.49}$$

It is clear from this model that if $k \to 0$, then $\boldsymbol{M} \to \boldsymbol{M}_{\text{an}}(\boldsymbol{H})$, which conforms with earlier comments that if there are no pinning sites, the magnetization will follow the anhysteretic magnetization curve. Solutions of this differential equation for various values of the parameters are shown in Figure 8.11.

8.3.4 Relationship between Hysteresis Coefficients and Measurable Magnetic Properties

How can these theoretical parameters be calculated from the conventional magnetization curve?

It is clearly important from the viewpoint of applications to be able to calculate the values of the various model parameters governing hysteresis from a measured magnetization curve [49]. From Equations 8.48 and 8.49, it is easily shown that the initial susceptibility of the normal magnetization curve is given by

$$\chi_{\text{in}} = c\chi_{\text{an}} \tag{8.50}$$

where χ_{an} is the differential susceptibility of the anhysteretic curve at the origin. This concept agrees well with Rayleigh's idea that χ_{in} represents the reversible component of magnetization at the origin of the initial magnetization curve.

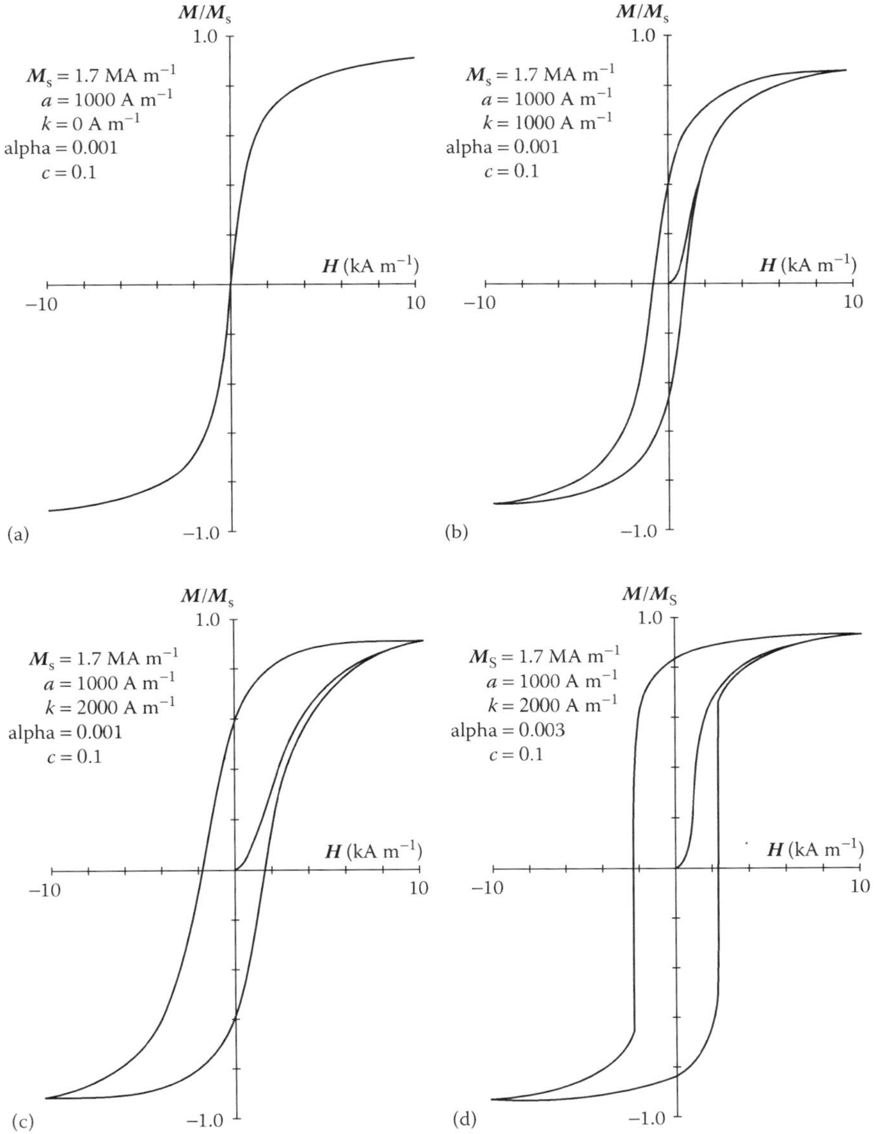

FIGURE 8.11 Theoretical hysteresis loops calculated using the equations derived in the theory of hysteresis for different values of model parameters (a–d) by Jiles and Atherton.

In the case where k is constant and the reversible component is negligible, that is, $c = 0$, we have the following very simple solution for k in terms of the coercivity $\boldsymbol{H}_c$ and the slope χ'_{H_c} of the hysteresis loop at $\boldsymbol{H}_c$:

$$k = \boldsymbol{M}_{an}(\boldsymbol{H}_c)\left(\alpha + \frac{1}{\chi'_{H_c}}\right) \tag{8.51}$$

when c $\neq$ 0, this becomes

$$k = \boldsymbol{M}_{an}(\boldsymbol{H}_c)\left(\frac{\alpha}{(1-c)} + \left\{\frac{1}{(1-c)\chi'_{H_c} - c\left[d\boldsymbol{M}_{an}(\boldsymbol{H}_c)/d\boldsymbol{H}\right]}\right\}\right) \tag{8.52}$$

In the case of soft magnetic materials that have a low coercivity, we can make some approximations that lead to an interesting relation between k and $\boldsymbol{H}_c$. For low coercivity materials, the slope of the hysteresis loop at the coercive point is approximately equal to the slope of the anhysteretic curve at the origin. Setting these equal

$$\chi'_{an} = \chi'_{in} \tag{8.53}$$

$$\frac{\boldsymbol{M}_s}{3a - \alpha\boldsymbol{M}_s} = \frac{\boldsymbol{M}_{an}(\boldsymbol{H}_c)}{k - \alpha\boldsymbol{M}_{an}(\boldsymbol{H}_c)} \tag{8.54}$$

which leads to the following equation for k:

$$k = \frac{3a}{\boldsymbol{M}_s}\boldsymbol{M}_{an}(\boldsymbol{H}_c) \tag{8.55}$$

Second, the slope of the anhysteretic curve at the origin is quite linear and therefore, for small $\boldsymbol{H}_c$, we can write

$$\boldsymbol{M}_{an}(\boldsymbol{H}_c) = \chi'_{an}\boldsymbol{H}_c = \left(\frac{\boldsymbol{M}_s}{3a - \alpha\boldsymbol{M}_s}\right)\boldsymbol{H}_c \tag{8.56}$$

and substituting this result into the expression for k gives

$$k = \frac{\boldsymbol{H}_c}{\left[1 - (\alpha\boldsymbol{M}_s/3a)\right]} \tag{8.57}$$

This brings us to the important result that the coercivity $\boldsymbol{H}_c$ of a soft ferromagnetic material is determined principally by the pinning of domain-wall motion. In fact, $k \simeq \boldsymbol{H}_c$ for soft magnetic materials where $\alpha\boldsymbol{M}_s << 3a$.

8.3.5 Effects of Microstructure and Deformation on Hysteresis

How does the hysteresis depend on the details of the material microstructure?

Changes in microstructure, in the form of additional magnetic inclusions such as second-phase particles with different magnetic properties from those of the matrix material, cause changes in the hysteresis properties by introducing more pinning sites that impede domain-wall motion and thereby lead to increased coercivity and hysteresis loss [50–52]. The same is also true of dislocations when their number density is increased by plastic deformation, either in tension or compression [53]. So, for example, the addition of carbon in the form of iron carbide particles increases coercivity and hysteresis loss. Cold-working of the material has a similar effect. The effect of these pinning sites is expressed via the coefficient k in the theory of ferromagnetic hysteresis. Clearly, as their numbers increase, k will increase proportionally and this results in an increase in the coercivity $\boldsymbol{H}_c$ as given in Equation 8.57.

As shown above, in the low-field limit the coercivity is $\boldsymbol{H}_c = k$ and consequently the coercivity is proportional to the product of number density and average pinning energy per site.

8.3.6 Effects of Stress on Bulk Magnetization

How does applied stress alter the hysteresis properties?

Following classical thermodynamics of reversible systems, the Gibbs energy per unit volume (in joules per cubic meter) is as follows:

$$G = U - TS + \frac{3}{2}\sigma\lambda \tag{8.58}$$

where:

- λ is the bulk magnetostriction
- σ is the stress
- U is the internal energy per unit volume
- T is the thermodynamic temperature
- S is the entropy

The Helmholtz energy per unit volume is

$$A = G + \mu_0 \boldsymbol{H}\boldsymbol{M} \tag{8.59}$$

where:

- H is the field
- M is the magnetization

The internal energy U due to magnetization is

$$U = \frac{1}{2}\alpha\mu_0 \boldsymbol{M}^2 \tag{8.60}$$

The total effective field is given by [54]

$$\boldsymbol{H}_{\text{eff}} = \left(\frac{1}{\mu_0}\right)\left(\frac{dA}{d\boldsymbol{M}}\right)_T \tag{8.61}$$

assuming the material is under a constant stress σ, we can write the effective field as

$$\boldsymbol{H}_{\text{eff}} = \boldsymbol{H} + \alpha\boldsymbol{M} + \frac{3\sigma}{2\mu_0}\left(\frac{d\lambda}{d\boldsymbol{M}}\right)_T \tag{8.62}$$

where:

H is the magnetic field
$\alpha\boldsymbol{M}$ is the mean-field coupling to the magnetization
$\boldsymbol{H}_\sigma$ is an equivalent stress field [54]

$$\boldsymbol{H}_\sigma = \frac{3\sigma}{2\mu_0}\left(\frac{d\lambda}{d\boldsymbol{M}}\right)_T \tag{8.63}$$

This can be used to determine the reversible magnetization under the action of stress. Therefore, for example, the anhysteretic magnetization curve, which is a reversible magnetization curve, can be determined by adding the stress equivalent field $\boldsymbol{H}_\sigma$ to the sum of the true field $\boldsymbol{H}$ and the internal field due to coupling between the magnetic moments $\alpha\boldsymbol{M}$. Using the Fröhlich-Kennelly relation with $\alpha = 0$ (the Fröhlich-Kennelly equation does not contain any coupling to the magnetization), this would become under the action of a stress σ

$$\boldsymbol{M}_{\text{an}}(\boldsymbol{H}) = \frac{\alpha\left[\boldsymbol{H} + (3\sigma/2\mu_0)(d\lambda/d\boldsymbol{M})_T\right]}{1+\beta\left[\boldsymbol{H} + (3\sigma/2\mu_0)(d\lambda/d\boldsymbol{M})_T\right]} \tag{8.64}$$

This result shows how the anhysteretic magnetization curve is altered under the action of a constant stress σ for a material with magnetostriction λ. Note that this analysis applies only to a reversible process. In the case of irreversible processes, the thermodynamic relationships become more complicated.

8.4 DYNAMICS OF DOMAIN MAGNETIZATION PROCESSES

What happens to the magnetic moments in the time between application of a magnetic field and the final orientation in the equilibrium positions?

So far we have discussed the magnetization processes only in terms of equilibrium conditions, in which the domain magnetizations and the domain walls have reached their final positions. However, when a magnetic material is subjected to a change in the applied magnetic field, there are time-dependent effects to consider, as the magnetic moments reorient and the domain walls move. These time-dependent effects have been discussed by Chikazumi [55] and Chen [56] and essentially our current understanding of these phenomena remains much the same, so only an outline of the main results will be presented here.

8.4.1 Domain Rotational Processes

How do the individual magnetic moments in a domain behave when subjected to an applied field?

We know from the previous discussion that a magnetic dipole, when subjected to an applied magnetic field, experiences a torque, which tends to align it with the field direction. Under strongly damped motion, when a magnetic moment is strongly coupled to other moments in the material, the magnetic moment will slowly rotate into the field direction and thus within a period of time, it will occupy an equilibrium orientation where it is aligned with the magnetic field. If on the other hand the motion is completely undamped, then the magnetic moment may go into precession, in which its direction oscillates about the field direction at an angle that is dependent upon its initial orientation. In the absence of damping, therefore, we expect the rate of change of magnetization to be proportional to the torque τ

$$\frac{\partial \boldsymbol{M}}{\partial t} = -\gamma\tau \tag{8.65}$$

where the torque is given by $\mu_0 \boldsymbol{M} \times \boldsymbol{H}$. The coefficient of proportionality γ is the gyromagnetic ratio, which we encountered in Section 3.3.6. The solutions of Equation 8.65 yield a constant component along the direction of the applied field and a sinusoidal component with fixed frequency ω_0 in the plane normal to the direction of the field

$$\omega_0 = \gamma\mu_0 \boldsymbol{H} \tag{8.66}$$

This shows that for undamped conditions, there is a spin resonance at a frequency, which increases linearly with the applied field strength. When there is damping, the equation of motion must be modified. Landau and Lifschitz have treated this problem [57] and gave the following equation known as the *Landau-Lifschitz equation*:

$$\frac{\partial \boldsymbol{M}}{\partial t} = -\gamma\mu_0 \boldsymbol{M} \times \boldsymbol{H} - \frac{4\pi\lambda_d}{\boldsymbol{M}^2}\left(\boldsymbol{M} \times \boldsymbol{M} \times \boldsymbol{H}\right) \tag{8.67}$$

where the term $4\pi\lambda_d$ on the right-hand side of the equation is a damping term [58]. It has dimensions of seconds^{-1} and is known as the *damping coefficient* or *relaxation frequency*.

Subsequently, Gilbert [59] derived a modified form of this equation in which he considered that the damping of the motion should depend on resultant motion $\partial \boldsymbol{M}/\partial t$. He therefore proposed the following alternative equation:

$$\frac{\partial \boldsymbol{M}}{\partial t} = -\gamma\mu_0 \boldsymbol{M} \times \boldsymbol{H} + \frac{4\pi\lambda_d}{\gamma\mu_0 \boldsymbol{M}^2}\left(\boldsymbol{M} \times \frac{\partial \boldsymbol{M}}{\partial t}\right) \tag{8.68}$$

This equation is now widely used in the description of spin dynamics, and in fact, is a generalization of the Landau-Lifschitz equation, since the Landau-Lifschitz equation can be derived from the Gilbert equation by neglecting the higher-order terms in the expression for $\partial \boldsymbol{M}/\partial t$ on the right-hand side of the equation.

Critical damping of the motion of the magnetic moments occurs when $4\pi\lambda_d = \gamma\mu_0\boldsymbol{M}$. For values of $\lambda_d << \gamma\mu_0\boldsymbol{M}/4\pi$, the spins perform a number of precessions before finally reaching the field direction, while for $\lambda_d >> \gamma\mu_0\boldsymbol{M}/4\pi$, the spins will rotate slowly but without oscillations toward the field direction. The value of this relaxation frequency λ_d for materials such as nickel-zinc ferrite and manganese-zinc ferrite is typically 10^7–10^8 s^{-1}, and in other materials is generally in the range of 10^6–10^9 s^{-1}. This relaxation frequency is insufficient to reach critical damping, which from the above inequalities occurs at typically $\lambda_d \approx 10^{10}$ s^{-1} for values of $\boldsymbol{M} = 0.8 \times 10^6$ A m^{-1}.

8.4.2 Wall Motion Processes

How can we describe the actual motion of the magnetic domain walls before they reach their final equilibrium positions?

The movement of domain walls under carefully controlled conditions was studied by Williams and coworkers [60,61]. Their results confirmed an earlier prediction by Sixtus and Tonks that the velocity of a domain wall at any given time is given by

$$v = \xi(\boldsymbol{H} - \boldsymbol{H}_t) \tag{8.69}$$

where:

- $\boldsymbol{H}$ is the prevailing magnetic field
- $\boldsymbol{H}_t$ is the local *threshold* field needed to activate the domain wall at its present location
- ξ is the mobility of the domain wall

The motion of magnetic domain walls is subjected to various damping mechanisms including eddy current effects and spin relaxation damping, as discussed in Section 8.4.1. The movement of a domain wall obeys a damped simple harmonic oscillator equation of the following form:

$$m\frac{d^2x}{dt^2} + \beta\frac{dx}{dt} + \alpha x = \boldsymbol{F}(t) \tag{8.70}$$

where:

$\boldsymbol{F}(t)$ is the force on the domain wall, which is determined principally by the magnetic field and the nature of the domain wall (e.g., 90° and 180°)

m is the inertia of the domain wall, which simply relates its acceleration to the force causing acceleration. Typical values of m are 10^{-9} kg m^{-2}

β is the damping coefficient for wall motion, which is determined by the combined effect of all energy dissipation mechanisms

α is the stiffness or restoring force coefficient

The force term $\boldsymbol{F}(t)$ is normally expressed as $\boldsymbol{F}(t) = b\mu_0\boldsymbol{M}_s\boldsymbol{H}(t)$, where $b = 2$ for 180° walls and $b = \sqrt{2}$ for 90° walls.

Under conditions where the motion of the domain walls is lightly damped, so that $\beta^2 < 4m\alpha$, the motion of the domain walls can exhibit resonance. When damping is heavy, so that $\beta^2 > 4m\alpha$, the domain walls will simply exhibit relaxation behavior. Critical damping occurs when $\beta^2 = 4m\alpha$, leading to the fastest approach of the magnetic moment to equilibrium. For a sinusoidal external field,

$$\boldsymbol{H} = \boldsymbol{H}_0\exp(i\omega t) \tag{8.71}$$

and if the wall is not accelerating, $d^2x/dt^2 = 0$, then

$$\frac{dx}{dt} = \frac{1}{\beta}\left(b\boldsymbol{M}_s\mu_0\boldsymbol{H} - \alpha x\right) = \frac{b\boldsymbol{M}_s\mu_0}{\beta}\left(H - \frac{\alpha x}{b\boldsymbol{M}_s\mu_0}\right) \tag{8.72}$$

where the term $\alpha x/b\boldsymbol{M}_s\mu_0$ has dimensions of the magnetic field. Therefore, replacing $\boldsymbol{H}_c = \alpha x/b\boldsymbol{M}_s\mu_0$, we obtain the velocity as

$$v = \frac{b\boldsymbol{M}_s\mu_0}{\beta}\left(\boldsymbol{H} - \boldsymbol{H}_c\right) \tag{8.73}$$

which is identical to the form of equation obtained by Williams, Shockley, and Kittel [61], where now the mobility ξ is $b\boldsymbol{M}_s\mu_0/\beta$. If the domain wall experiences an internal potential of the form $E_p(x) = \alpha x^2/2$, then the force per unit area on the domain wall, $\boldsymbol{F}$, is

$$\boldsymbol{F} = \frac{dE_p}{dx} = -\alpha x = b\mu_0\boldsymbol{M}_s\boldsymbol{H} \tag{8.74}$$

and therefore the effective magnetic field that the wall experiences as a result of this internal potential is

$$\boldsymbol{H}_c = \frac{\alpha x}{b\mu_0\boldsymbol{M}_s} \tag{8.75}$$

This explains the origin of the local pinning field, or threshold field $\boldsymbol{H}_c$, needed to activate the domain wall. This local pining field is the derivative of the internal potential, normalized with respect to $b\mu_0\boldsymbol{M}_s$. The domain wall also has a natural frequency ω_0 and a resonant frequency ω_r, which can be determined from the structural factors contained in the coefficients m, β, and α of the equation of motion.

8.4.3 Ferromagnetic Resonance

What causes resonance when a ferromagnet is subjected to a time-dependent magnetic field?

When a ferromagnet is subjected to a time-dependent magnetic field, two types of resonance may occur. These are a spin resonance based on the Landau-Lifschitz-Gilbert equation for spin dynamics, and a domain-wall resonance based on the equation of motion of the domain walls.

8.4.3.1 Spin Resonance

For spin resonance starting from the Landau-Lifschitz-Gilbert equation, if we assume that the damping term λ_d is small, compared with $\gamma\mu_0 \boldsymbol{M}/4\pi$, then it is clear that the resonance frequency ω_0 is

$$\omega_0 = \gamma\mu_0 \boldsymbol{H} \tag{8.76}$$

In practice, the value of $\boldsymbol{H}$ in Equation 8.76 is often strongly affected by demagnetizing effects and internal fields such as exchange and anisotropy. Ferromagnetic spin resonance has been observed by many, including in the work of Snoek [62]. His interpretation of the observed resonance of the rotation of magnetization vectors was that it was controlled by the anisotropy field, $\boldsymbol{H}_a$, so that

$$\omega_0 = \gamma\mu_0 \left(\boldsymbol{H} + \boldsymbol{H}_a\right) \tag{8.77}$$

and if $\boldsymbol{H} = 0$, this gives a relationship between the anisotropy field and the resonance frequency ω_0. A resonance frequency of typically $f_0 = \omega_0/2\pi = 100$ MHz gives $\boldsymbol{H}_a = 2.8 \times 10^3$ A m^{-1} for a material such as nickel-zinc ferrite. In other materials, the resonance can be at higher fields, for example 10^9 Hz, giving a value of anisotropy field in the range of 3×10^4 A m^{-1}.

The above relationship between anisotropy field $\boldsymbol{H}_a$ and ferromagnetic resonance frequency ω_0 can be extended to include the anisotropy coefficient and initial susceptibility. Assuming $\omega_0 = \gamma\mu_0\boldsymbol{H}_a$, then for cubic anisotropy with the one-constant approximation, the anisotropy field is determined by the anisotropy coefficient K_1 and the spontaneous magnetization within a domain $\boldsymbol{M}_s$

$$\boldsymbol{H}_a = \frac{2K_1}{\mu_0 \boldsymbol{M}_s} \tag{8.78}$$

and in the limit as the $\boldsymbol{H}$ field tends to zero the initial susceptibility χ_{in} is related to the anisotropy by

$$\chi_{in} = \frac{\mu_0 \boldsymbol{M}_s^2}{3K_1} \tag{8.79}$$

so that

$$\omega_0 = \frac{2\gamma K_1}{\boldsymbol{M}_s} = \frac{2}{3}\frac{\gamma\mu_0 \boldsymbol{M}_s}{\chi_{in}} \tag{8.80}$$

which leads to the interesting conclusion that the resonance frequency is inversely proportional to the initial susceptibility $\chi_{in.}$ A similar result holds true for axially anisotropic materials with anisotropy constant K_{u1}, because the anisotropy field is again $\boldsymbol{H}_a = 2K_{u1}/\mu_0\boldsymbol{M}_s$ and the initial susceptibility $\chi_{in} = \mu_0 \boldsymbol{M}_s^2/3K_{u1}$.

8.4.3.2 Domain-Wall Resonance

Domain-wall resonance occurs at much lower frequencies, typically in the range 50 kHz. The first observation of domain-wall resonance was by Rado [63] in manganese ferrite. For wall motion resonance, we can begin from the damped simple harmonic motion (Equation 8.66) and observe that in the absence of damping effects ($\beta = 0$) but with a restoring coefficient α, the walls will oscillate at a natural frequency ω_0 given by

$$\omega_0 = \sqrt{\frac{\alpha}{m}} \tag{8.81}$$

where m is the effective mass or inertia of the domain walls. In the presence of damping, the resonant frequency ω_r is related to the natural (undamped) resonant frequency ω_0 by the following equation:

$$\omega_r = \omega_0\sqrt{1-2\left(\frac{\beta}{\beta_{crit}}\right)^2} = \omega_0\sqrt{1-\frac{\beta^2}{2m\alpha}} \tag{8.82}$$

In the case of negligible damping, the resonant frequency depends largely on the restoring coefficient α and this can be related to the internal potential $E_p(x)$ experienced by the domain wall, as discussed in Section 7.2.4. The force per unit area on the domain wall is $dE_p/dx = b\mu_0\boldsymbol{M}_s d\boldsymbol{H}_c/dx$ ($b = 2$ for 180° walls, $b = \sqrt{2}$ for 90° walls), where $\boldsymbol{H}_c$ is the equivalent internal magnetic field, like a local coercivity, which needs to be overcome to move the domain wall as discussed in Section 8.4.2. These terms can now be equated with the restoring pressure on the domain wall

$$\frac{dE_p}{dx} = b\mu_0\boldsymbol{M}_s\frac{d\boldsymbol{H}_c}{dx} = \alpha x \tag{8.83}$$

Therefore, the restoring coefficient α is seen to be a representation of the internal potential seen by the domain wall at its current location $E_p = \alpha x^2/2$. As shown in Section 7.2.5, this is directly related to the initial susceptibility $\chi_{in} = 4\mu_0\boldsymbol{M}_s^2 A/3\alpha$ for 180° domain walls and $\chi_{in} = 2\mu_0\boldsymbol{M}_s^2 A/3\alpha$ for 90° walls.

Therefore, the domain-wall resonance frequency can be directly connected to the initial susceptibility χ in

$$\omega_{0\ 180} = \sqrt{\frac{4\mu_0 \boldsymbol{M}_s^2 A}{3m\chi_{in}}} \tag{8.84}$$

$$\omega_{0\ 90} = \sqrt{\frac{2\mu_0 \boldsymbol{M}_s^2 A}{3m\chi_{in}}} \tag{8.85}$$

Derivation of an explicit expression for the domain-wall mass per unit area in terms of the domain-wall energy and width is somewhat lengthy, but has been discussed by Chen [56] and Chikazumi [55].

Ferromagnetic resonance has been treated in detail by Slichter [64], Wigen [65], and Heinrich [66]. Other forms of magnetic resonance, which we shall not cover here, include nuclear magnetic resonance and electron spin resonance (also known as *electron paramagnetic resonance*). These are discussed in detail by Slichter [64] and other authors [67].

8.4.4 Damping and Relaxation Effects

How long does it take the magnetization to change after a field is applied?

When a step change in the applied magnetic field is experienced by a magnetic material, the resulting change in magnetization $d\boldsymbol{M}/dt$ is delayed, depending on factors such as spin reorientation, domain-wall motion, damping effects, and eddy currents. In its simplest form, the step change in field causes the magnetization to approach its new value at a rate that decays exponentially with time

$$\boldsymbol{M}(t) = \boldsymbol{M}_\infty - \left(\boldsymbol{M}_\infty - \boldsymbol{M}_0\right)\exp\left(\frac{-t}{\tau}\right) \tag{8.86}$$

where:

$\boldsymbol{M}_0$ is the initial magnetization before the field is applied ($t = 0$)

$\boldsymbol{M}_\infty$ is the final magnetization (the limit as $t \to \infty$)

$\boldsymbol{M}(t)$ is the magnetization at any time t

τ is the relaxation time coefficient, which describes how fast the magnetization approaches its final value. This relaxation time coefficient is temperature-dependent

Equation 8.86 can give a first approximation for time-dependent effects in magnetization, but in reality, there are different mechanisms by which the magnetization changes and therefore in general, the behavior is not necessarily described by this simple exponential function. One of the important works on time-dependent effects in magnetization was by Street and Woolley [68]. In particular, when the relaxation processes are thermally activated with a range of pinning strengths, the relationship between change in magnetization $\boldsymbol{M}$ and time t is logarithmic, so that

$$\boldsymbol{M}(t) = S \log_e t + \text{constant} \tag{8.87}$$

where S is a coefficient of proportionality [69], the value of which depends on the material, the temperature, the point of measurement on the hysteresis curve, and the demagnetizing factor for the specimen. In this case, t is the time interval since the sudden step change in the applied field $\boldsymbol{H}$.

There has been a resurgence of interest in these time-dependent effects as the dynamic behavior of magnetization in materials has begun to impact the performance of magnetic materials in applications where they are exposed to time-dependent fields such as in magnetic disk drives, actuators, and sensors. The dynamic behavior of magnetization in magneto-optic materials [70] and in permanent magnet materials [71] are examples.

8.4.5 Micromagnetic Modeling

How can the arrangement of individual magnetic moments within a domain be determined?

The rapid advances in computer capabilities, coupled with the ease of access to such facilities, have led to an increase in interest in numerical modeling of magnetic structures within domains. At this level, the subject is known as *micromagnetic modeling*, to distinguish it from magnetic modeling on the multidomain scale such as Preisach modeling, or on the macroscopic scale such as finite-element modeling of engineering components.

The seminal text in this field is the work of Brown [72]. The general approach is to assume an array of localized magnetic moments of fixed magnitude, which are allowed to point in any direction. The time needed to make these calculations was previously prohibitive, so that early work focused on two-dimensional arrays to ease the computational burden. However, with the advances in computer capabilities, both in terms of speed and data handling capacity, three-dimensional arrays are now readily modeled.

The usual approach taken is to calculate a local minimum in the free energy of the moment array. An exchange coupling is assumed, which can be treated as an effective internal magnetic field that varies locally throughout the array. Then the equation of motion for the individual magnetic moments is used to determine the equilibrium state of the array. The equation of motion that is normally chosen is the Landau-Lifschitz-Gilbert equation described in Section 8.4.1. The calculation is an integrative one because changes in the orientations of the magnetic moments result in local changes of the effective exchange field, which is the dominant field term. The energy minimization problem is complicated because of this, but providing the damping term is sufficiently large (the precession term in the Landau-Lifschitz equation is often omitted or ignored) an equilibrium configuration will eventually be reached.

Many of the early ideas were developed by Brown [73] and Aharoni [74]. A comprehensive review by Berkov, Ramstock, and Hubert [75] has dealt with subsequent developments, and also outlined some of the problems, while Muller-Pfeiffer, Schneider, and Zinn [76] have used micromagnetic modeling to interpret the results of magnetic force microscopy images. This last study is particularly interesting since it is only through the development of the magnetic force microscope that experimental methods have been available on a suitable scale for verifying the results of micromagnetic calculations. Micromagnetic methods were used by Aharoni and Jakubovics [77] to calculate the structure of moving domain walls in thin permalloy films.

Aharoni [78] has provided a summary of the progress in micromagnetic modeling, both in terms of the analytical equations and the numerical or computational aspects of the subject. Finally, in the work by Antropov, Harmon, and Stocks [79], a modified version of the traditional Landau-Lifschitz approach has been used to determine noncollinear equilibrium orientations of three-dimensional arrays of magnetic moments, including variations in the magnitude of the local magnetic moments from electron band structure calculations.

REFERENCES

1. Sixtus, K. J., and Tonks, L. *Phys. Rev.*, 37, 930, 1931.
2. Becker, R. *Phys. Zeits.*, 33, 905, 1932.
3. Hoselitz, K. *Ferromagnetic Properties of Metals and Alloys*, Oxford University Press, Oxford, 1952.
4. Becker, R., and Kersten, M. *Z. Phys.*, 64, 660, 1930.
5. Kondorsky, E. *Physik. Z. Sowjetunion*, 11, 597, 1937.
6. Becker, R., and Doring, W. *Ferromagnetismus*, Springer-Verlag, Berlin, Germany, 1938.
7. Kersten, M. in *Problems of the Technical Magnetisation Curve* (ed. R. Becker), Springer, Berlin, Germany, 1938.
8. Orowan, E. *Z. Phys.*, 89, 605, 1934.
9. Taylor, G. I. *Proc. Roy. Soc. Lond. A*, 145, 362, 1934.
10. Polanyi, M. *Z. Phys.*, 89, 660, 1934.
11. Hull, D., and Bacon, D. J. *Introduction to Dislocations*, 3rd edn, Pergamon, Oxford, 1964.
12. Kersten, M. *Underlying Theory of Ferromagnetic Hysteresis and Coercivity*, Leipzig, Germany, 1943.
13. Kersten, M. *Phys. Z.*, 44, 63, 1943.
14. Néel, L. *Cahiers de Phys.*, 25, 21, 1944.
15. Néel, L. Basis of a new general theory of the coercive field, *Ann. Univ. Grenoble*, 22, 299, 1946.
16. Dijkstra, L. J., and Wert, C. *Phys. Rev.*, 79, 979, 1950.
17. Kersten, M. *Z. Angew. Phys.*, 7, 313; 8, 382; 8, 496, 1956.
18. Vicena, F. *Czech. Phys.*, 5, 480, 1955.
19. Seeger, A., Kronmuller, H., Rieger, H., and Trauble, H. *J. Appl. Phys.*, 35, 740, 1964.
20. Kronmuller, H. *Z. Angew. Phys.*, 30, 9, 1970.
21. Kronmuller, H., and Hilzinger, H. R. *Int. J. Magn.*, 5, 27, 1973.
22. Hilzinger, H. R., and Kronmuller, H. *Phys. Letts. A*, 51, 59, 1975.
23. Hilzinger, H. R., and Kronmuller, H. *J. Magn. Magn. Mater.*, 2, 11, 1976.
24. Jiles, D. C., and Atherton, D. L. *J. Magn. Magn. Mater.*, 61, 48, 1986.
25. Labusch, R. *Cryst. Latt. Def.*, 1, 1, 1969.
26. Dietze, H. D. *Z. Phys.*, 149, 276, 1957.
27. Chikazumi, S. *Physics of Magnetism*, Wiley, New York, 1964, p. 193.
28. Hilzinger, H. R., and Kronmuller, H. *Physica*, 86–8B, 1365, 1977.
29. Porteseil, J. L., and Vergne, R. *J. Phys.*, 40, 871, 1981.

30. Astie, B., Degauque, J., Porteseil, J. L., and Vergne, R. *IEEE Trans. Magn.*, 17, 2929, 1981.
31. Degauque, J., and Astie, B. *Phys. Stat. Sol. A*, 74, 201, 1982.
32. Degauque, J., Astie, B., Porteseil, J. L., and Vergne, R. *J. Magn. Magn. Mater.*, 26, 261, 1982.
33. Astie, B., Degauque, J., Porteseil, J. L., and Vergne, R. *J. Magn. Magn. Mater.*, 28, 149, 1982.
34. Jiles, D. C., Chang, T. T., Hougen, D. R., and Ranjan, R. *J. Appl. Phys.*, 64, 3620, 1988.
35. Globus, A. *Comptes Rendus Acad. Seances*, 255, 1709, 1962.
36. Globus, A., and Duplex, P. *IEEE Trans. Magn.*, 2, 441, 1966.
37. Globus, A., and Duplex, P. *Phys. Stat. Sol.*, 31, 765, 1969.
38. Globus, A., and Duplex, P. *Phys. Stat. Sol. A*, 3, 53, 1970.
39. Globus, A., Duplex, P., and Guyot, M. *IEEE Trans. Magn.*, 7, 617, 1971.
40. Globus, A., and Guyot, M. *Phys. Stat. Sol. B*, 52, 427, 1972.
41. Guyot, M., and Globus, A. *Phys. Stat. Sol. B*, 59, 447, 1973.
42. Guyot, M., and Globus, A. *J. Phys.*, 38, Cl-157, 1977.
43. Globus, A. *Physica*, 86-8B, 943, 1977.
44. Alessandro, B., Beatrice, C., Bertotti, G., and Montorsi, A. *J. Appl. Phys.*, 68, 2901, 1990.
45. Alessandro, B., Beatrice, C., Bertotti, G., and Montorsi, A. *J. Appl. Phys.*, 68, 2908, 1990.
46. Pust, L., Bertotti, G., Tomas, I., and Vertesy, G. *Phys. Rev. B*, 54, 12262, 1996.
47. Preisach, F. *Z. Phys.*, 94, 277, 1935.
48. Stoner, E. C., and Wohlfarth, E. P. *Phil. Trans. Roy. Soc. A*, 240, 599, 1948.
49. Jiles, D. C., and Thoelke, J. B. Proceedings of the 1989 intermag conference, *IEEE Trans. Magn.*, 25, 3928, 1989.
50. Leslie, W. C., and Stevens, D. W. *Trans. ASM*, 57, 261, 1964.
51. English, A. T. *Acta. Metall.*, 15, 1573, 1967.
52. Jiles, D. C. *J. Phys. D. (Appl. Phys.)*, 21, 1186, 1988.
53. Jiles, D. C. *J. Phys. D. (Appl. Phys.)*, 21, 1196, 1988.
54. Sablik, M. J., Kwun, H., Burkhardt, G. L., and Jiles, D. C. *J. Appl. Phys.*, 63, 3930, 1988.
55. Chikazumi, S. *Physics of Magnetism*, Wiley, New York, 1964, Ch. 16, p. 353.
56. Chen, C. W. *Magnetism and Metallurgy of Soft Magnetic Materials*, North Holland, Amsterdam, the Netherlands, 1977, pp. 151 and 159.
57. Landau, L. D., and Lifschitz, E. M. *Phys. Z. Sovietunion*, 8, 153, 1935.
58. Cullity, B. D., and Graham, C.D. *Introduction to Magnetic Materials*, 2nd edn, Wiley, Hoboken, NJ, 2009, p. 436.
59. Gilbert, T. L. *Phys. Rev.*, 100, 1243, 1955.
60. Williams, H. J., and Shockley, W. *Phys. Rev.* 75, 178, 1949.
61. Williams, H. J., Shockley, W., and Kittel, C. *Phys. Rev.*, 80, 1090, 1950.
62. Snoek, J. *Physica*, 14, 207, 1948.
63. Rado, G. T. *Phys. Rev.*, 80, 273, 1950.
64. Slichter, C. *Principles of Magnetic Resonance*, 3rd edn, Springer-Verlag, Berlin, Germany, 1990, Ch. 10.
65. Wigen, P. E. *Thin Solid Films*, 114, 135, 1984.
66. Heinrich, B. *Ultrathin Magnetic Structures II*, Springer-Verlag, Berlin, Germany, 1994, Ch. 3.
67. Oxford Instruments. *NMR and EPR Spectroscopy*, Pergamon, Oxford, 1960.
68. Street, R., and Woolley, J. C. *Proc. Phys. Soc. Lond. A*, 62, 562, 1949.
69. Street, R., and Woolley, J. C. *Proc. Phys. Soc. Lond. A*, 69, 1189, 1956.
70. Street, R., and Brown, S. D. *J. Appl. Phys.*, 76, 6386, 1994.

71. Folks, L., and Street, R. *J. Appl. Phys.*, 76, 6391, 1994.
72. Brown, W. F. *Micromagnetics*, Robert Krieger, Malabar, FL, 1978.
73. Brown, W. F., and LaBonte, A. E. *J. Appl. Phys.*, 36, 1380, 1965.
74. Aharoni, A. *Phys. Stat. Sol. A*, 16, 1, 1966.
75. Berkov, D. V., Ramstock, K., and Hubert, A. *Phys. Stat. Sol. A*, 137, 207, 1993.
76. Muller-Pfeiffer, S., Schneider, M., and Zinn, W. *Phys. Rev. B*, 49, 15745, 1994.
77. Aharoni, A., and Jakubovics, J. P. *JMMM*, 140, 1893, 1995.
78. Aharoni, A. *Introduction to the Theory of Ferromagnetism*, Clarendon Press, Oxford, 1996.
79. Antropov, V., Harmon, B. N., and Stocks, G. M. *Phys. Rev. B*, 54, 1019, 1996.

FURTHER READING

Aharoni, A. *Introduction to the Theory of Ferromagnetism*, Oxford Science, Oxford, 1996.
Becker, R., and Doring, W. *Ferromagnetismus*, Springer-Verlag, Berlin, Germany, 1939.
Chikazumi, S. *Physics of Magnetism*, Wiley, New York, 1964, Chs. 11, 13, 14.
Crangle, J. *The Magnetic Properties of Solids*, Arnold, London, 1977, Ch. 6.
Cullity, B. D. *Introduction to Magnetic Materials*, Addison-Wesley, Reading, MA, 1972, Ch. 9.
Hoselitz, K. *Ferromagnetic Properties of Metals and Alloys*, Oxford University Press, Oxford, 1952, Ch. 2.
Stoner, E. C. *Phil. Trans. Roy. Soc. A*, 240, 599, 1948.

EXERCISES

8.1 *Magnetization mechanisms.* Write a short essay describing the principal mechanisms by which the magnetization of a magnetic material changes under the action of an applied magnetic field. Be sure to describe all processes and how these are affected by factors such as anisotropy, domain-wall surface energy, and the presence and strength of pinning sites in the material. Are the initial permeability and coercivity of a magnetic material related in any way?

8.2 *Magnetostatic energy associated with a void.* Show that the magnetostatic energy of a spherical void enclosed entirely within a domain as shown in Figure 8.4 is $2\mu_0 \boldsymbol{M}_s^2 \pi r^3/9$, where $\boldsymbol{M}_s$ is the spontaneous magnetization of the material and r is the radius of the void.

Assuming that the reduction in energy when a 180° domain wall intersects such a void is $\mu_0 \boldsymbol{M}_s^2 \pi r^3/9$, calculate the energy reduction when a domain wall in iron intersects spherical voids of $r = 5 \times 10^{-8}$ m and $r = 10^{-6}$ m. Compare this to the difference in wall energy caused by the intersection with the void, assuming the domain-wall energy in iron is 2×10^{-3} J m^{-2} and $\boldsymbol{M}_s = 1.7 \times 10^6$ A m^{-1}.

8.3 *Reduction of domain-wall energy by voids.* If the reduction in energy when a domain wall intersects a void is given by $\mu_0 \boldsymbol{M}_s^2 \pi r^3/9$, estimate the reduction in energy per unit volume in a material that has N voids per unit volume each of radius r. (Assume all domain walls form parallel planes of separation d within a cube of unit volume as shown in Figure 8.3.1.)

If the wall energy is to be completely compensated for by the decrease in energy associated with the voids, estimate the number density of voids of radius r that can generate domain walls in a crystal.

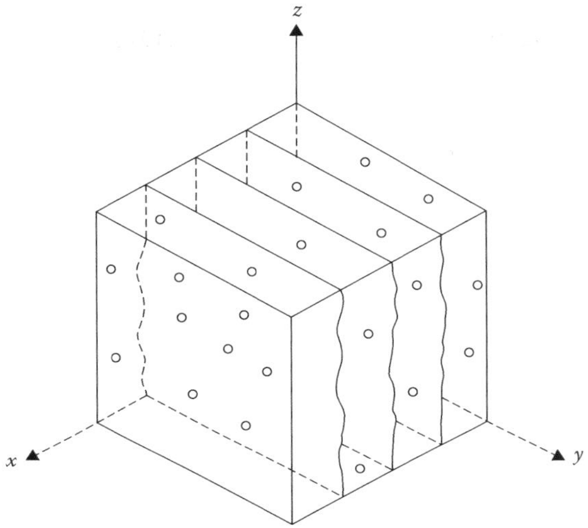

FIGURE 8.3.1 Domain walls within a ferromagnetic material stabilized by inclusions distributed within the material. In this case, the inclusions are voids.

Calculate this number for iron with $\boldsymbol{M}_s = 1.7 \times 10^6$ A m^{-1}, $r = 0.01$ mm, and a wall energy of 2×10^{-3} J m^{-3}.

8.4 *Effect of stress on anhysteretic susceptibility.* Derive an expression for the susceptibility of a 180° domain wall moving in a linear stress field $\sigma = ax^2$, where a is a constant. Assume there is no change in wall energy as it moves in the stress field.

Determine the anhysteretic susceptibility at the origin $\chi'_{an}(\sigma)$ of a magnetic material under a compressive stress of −20 MPa, if zero stress value of $\chi'_{an}(0)$ is 1000, and the low-field magnetostriction increases with magnetization $\boldsymbol{M}$ according to $\lambda = b(\sigma)\boldsymbol{M}^2$, where $b(\sigma) = 2.3 \times 10^{-18}$ (A m^{-1})$^{-2}$ when $\sigma = -20$ MPa. The saturation magnetization of this material $\boldsymbol{M}_s = 0.9 \times 10^6$ A m^{-1}.

8.5 *Susceptibility from domain-wall motion.* Derive an expression for the initial susceptibility of a ferromagnet in terms of relevant materials properties, assuming axial anisotropy, with a single easy axis, so that only 180° domain walls exist, low domain-wall energy γ, and a uniform density of strong pinning sites a distance L apart. Discuss any additional assumptions that you have made to reach this expression.

8.6 *Rotation against anisotropy.* If a thin film of iron can be treated as a material with *in plane* two-dimensional anisotropy only, with an anisotropy constant of $K = 0.5 \times 10^5$ J m^{-3}, calculate the field strength along the <010> direction needed to rotate the magnetic moments from the <100> direction into the field direction.

8.7 *Effects of stress on domain-wall motion and domain rotation.* Explain the various ways in which the presence of stress alters the magnetic properties of ferromagnetic materials. Do long-range (~50 μm) and short-range (~50 nm) stresses have the same or different effects on magnetic-properties such as coercivity and differential susceptibility? Explain why. Is high magnetostriction advantageous or disadvantageous in soft magnetic materials? Justify your answer in terms of residual stresses in the material.

8.8 *Saturation magnetostriction along different crystallographic directions.* Calculate the saturation magnetostriction of a cubic crystal of nickel in the <110> direction when the magnetization is saturated along (1) the <100>, (2) the <111>, and (3) the <110> directions. (For nickel $\lambda_{100} = -46 \times 10^{-6}$ and $\lambda_{111} = -24 \times 10^{-6}$.)

8.9 *Magnetocrystalline anisotropy and domain magnetization rotation.* A piece of single crystal cobalt with uniaxial anisotropy coefficient $K_{u1} = 5 \times 10^5$ J m^{-3} and spontaneous magnetization $\boldsymbol{M}_s = 1.44 \times 10^6$ A m^{-1} is in the form of a sphere (so there is no shape anisotropy to compete with the magnetocrystalline anisotropy). If the particle has been magnetized to saturation along a particular direction and remains a single domain after removal of the field, derive an expression for and calculate the value of:

a. The magnetic field strength $\boldsymbol{H}_c$ needed to reverse the direction of magnetization (the *coercivity*) when an external field is applied antiparallel to the initial direction of magnetization
b. The differential susceptibility χ when an external field is applied perpendicular to the direction of magnetization
c. The field strength $\boldsymbol{H}_s$ that is needed when applied perpendicular to the direction of magnetization to rotate the magnetization into the field direction (the *saturation field*)

8.10 *Exchange energy, anisotropy energy, thermal energy, and super-paramagnetic particle size.* Explain how the magnetic order of a material is affected by exchange energy, anisotropy energy, and temperature. Assuming a temperature of 300 K and an anisotropy energy of 4.8×10^4 J m^{-3}, calculate the *super-paramagnetic critical radius* for a spherical particle of iron ($T_c = 1043$ K in bulk), so that above this radius, it will be ferromagnetic and below this radius, it will be paramagnetic. How might this critical radius be affected as the temperature changes?

9 Magnetic Order and Critical Phenomena

In this chapter, we discuss theories of magnetic behavior of materials, particularly the alignment of magnetic moments within the material. These theories can provide useful phenomenological models of the magnetic properties, including the order-disorder transition that occurs at the Curie temperature. These models assume that each atom in a paramagnet or ferromagnet has a magnetic moment of fixed magnitude that is localized at the atomic site and that can change its orientation.

9.1 THEORIES OF PARAMAGNETISM AND DIAMAGNETISM

What atomic scale theories do we have to account for the properties of diamagnets and paramagnets?

Diamagnets are solids with no permanent net magnetic moment per atom. Diamagnetic susceptibility is assumed to arise from the realignment of electron orbitals under the action of a magnetic field. Therefore, all materials exhibit a diamagnetic susceptibility, although not all are classified as diamagnets. Some materials have a net magnetic moment per atom, actually due to an unpaired electron spin in each atom, which leads to paramagnetism or even to ordered magnetic states such as ferromagnetism. In either case, the paramagnetic or ferromagnetic susceptibility is much greater than the diamagnetic susceptibility and therefore is the dominant effect. In paramagnetism, the atomic magnetic moments are randomly oriented but can be aligned by a magnetic field.

9.1.1 DIAMAGNETISM

What causes the negative susceptibility observed in some materials?

The magnetic moments associated with atoms in magnetic materials have three origins. These are the electron spins, the electron orbital motion, and the change in orbital motion of the electrons caused by an applied magnetic field. Only the change in orbital motion gives rise to a diamagnetic susceptibility. Diamagnetism leads to a very weak magnetization, which opposes the applied field. The diamagnetic susceptibility is therefore negative and has an order of magnitude of 10^{-5} or 10^{-6}. It is also found to be independent of temperature. Most elements in the periodic table are diamagnetic. Examples of diamagnets are copper, gold, silver, and bismuth.

9.1.2 Langevin Theory of Diamagnetism

How can we explain the negative susceptibility of diamagnets in terms of the motion of electrons?

The susceptibility of diamagnets was first explained by Langevin [1], who combined some of the earlier ideas of Ampère, Weber, and Lenz on the effect of a magnetic field on a current-carrying conductor. This gave a description of the change in orbital motion of electrons within an atom that arises as a result of exposure to an applied magnetic field. An electron in orbit about an atomic nucleus can be compared with current passing through a closed loop of conductor, and it will therefore have an orbital magnetic moment $\boldsymbol{m}_0$ since, as we already know from Chapter 1, electric charge in closed loop motion generates a magnetic moment.

In the case of a current loop, the magnetic moment is

$$\boldsymbol{m}_0 = iA \tag{9.1}$$

where:

i is the current
A is the area of the loop

For an electron in orbital motion,

$$\boldsymbol{m}_0 = \frac{eA}{\tau} \tag{9.2}$$

where:

e is the charge on the electron
τ is the orbital period

If we have circular orbital of area $A = \pi r^2$ and $\tau = 2\pi r/v$, where v is the instantaneous tangential velocity of the electron and r is the radius of the orbital, then

$$\boldsymbol{m}_0 = \frac{evr}{2}$$

This is the magnetic moment obtained as a result of the orbital motion of an electron. In the absence of a magnetic field, the orbital moments of paired electrons within an atom will cancel. An applied magnetic field $\boldsymbol{H}$ will accelerate or decelerate the orbital motion of the electron and thereby contribute to a change in the orbital magnetic moment. Once a magnetic field has been applied, the perturbation of the electron velocity can be determined. The change in magnetic flux through the current loop described by the electron about the nucleus gives rise to an electromotive force or voltage V_e in the current loop. This leads to an electric field $\boldsymbol{E}$ given by

$$\boldsymbol{E} = \frac{V_e}{L} = -\left(\frac{1}{L}\right)\frac{d\phi}{dt} \tag{9.3}$$

where L is the orbit length ($L = 2\pi r$).

$$\boldsymbol{E} = \left(\frac{-1}{L}\right)\frac{d(\boldsymbol{B}A)}{dt} = -\left(\frac{-A}{L}\right)\frac{d\boldsymbol{B}}{dt} \tag{9.4}$$

The acceleration of the electron is

$$a = \frac{dv}{dt} = \frac{e\boldsymbol{E}}{m_e} \tag{9.5}$$

where:

$\boldsymbol{F} = e\boldsymbol{E}$ is the force on the electron due to the field $\boldsymbol{E}$

m_e is the mass of the electron

$$\frac{dv}{dt} = \frac{e\boldsymbol{E}}{m_e} = -\left(\frac{eA}{m_e L}\right)\frac{d\boldsymbol{B}}{dt} = -\left(\frac{er}{2m_e}\right)\frac{d\boldsymbol{B}}{dt} = -\left(\frac{\mu_0 er}{2m_e}\right)\frac{d\boldsymbol{H}}{dt} \tag{9.6}$$

Now integrating from zero field strength to an arbitrary field strength $\boldsymbol{H}$ gives

$$\int_{v_1}^{v_2} dv = -\left(\frac{\mu_0 er}{2m_e}\right)\int_0^H d\boldsymbol{H} \tag{9.7}$$

$$v_2 - v_1 = -\left(\frac{\mu_0 er}{2m_e}\right)\boldsymbol{H} \tag{9.8}$$

The change in magnetic moment arising from this is

$$\Delta\boldsymbol{m}_0 = \left(\frac{er}{2}\right)(v_2 - v_1) = -\frac{\mu_0 e^2 r^2 \boldsymbol{H}}{4m_e} \tag{9.9}$$

This result only applies in the case where the magnetic field is perpendicular to the plane of motion of the electron. In the case, where the field $\boldsymbol{H}$ is in the plane of motion, the net change is zero. In the general case, therefore, we have to take into account the projection R of the orbit radius r on a plane normal to the field

$$R = r\sin\theta \tag{9.10}$$

where:

$\theta = 0$ corresponds to a field in the plane of orbit

$\theta = \pi/2$ corresponds to a field perpendicular to the plane

$$\Delta\boldsymbol{m}_0 = -\left(\frac{\mu_0 e^2}{4m_e}\right)\int\left(\frac{R^2}{A}\right)\sin^2\theta\, dA \tag{9.11}$$

where A is the area of a hemisphere and $dA = 2\pi R^2 \sin\theta\, d\theta$, as shown in Figure 9.1.

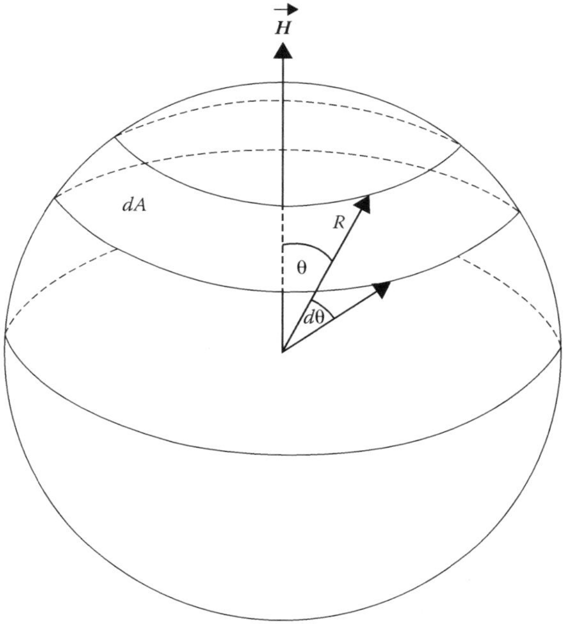

FIGURE 9.1 Unit sphere defining the parameters R, θ, and A used to describe the orbital motion of an electron about the nucleus.

The average value of R^2 over all possible orientations is

$$\langle R^2 \rangle = \left(\frac{2}{3}\right) r^2 \tag{9.12}$$

and consequently in cases where the orbit can have any orientation

$$\Delta \boldsymbol{m}_0 = -\frac{\mu_0 e^2 r^2 \boldsymbol{H}}{6m_e} \tag{9.13}$$

If we consider Z outer electrons in the atom then the change in magnetic moment per atom is

$$\Delta \boldsymbol{m}_0 = -\frac{\mu_0 Z e^2 r^2 \boldsymbol{H}}{6m_e} \tag{9.14}$$

and converting this to a bulk magnetization

$$\Delta \boldsymbol{M} = -\left(\frac{N_0 \rho}{W_a}\right) \frac{\mu_0 Z e^2 r^2 \boldsymbol{H}}{6m_e} \tag{9.15}$$

where:

N_0 is Avogadro's number
ρ is the density
W_a is the relative atomic mass

$$\chi = \frac{\boldsymbol{M}}{\boldsymbol{H}} = -\left(\frac{N_0 \rho}{W_a}\right)\frac{\mu_0 Z e^2 r^2}{6 m_e} \tag{9.16}$$

This classical model shows that in the case of a diamagnet, which has no net atomic magnetic moment in the absence of a field, the action of a magnetic field causes changes in the orbits of electrons in the atom in such a way that the induced moment opposes the field producing it. The above expression (Equation 9.16) for diamagnetic susceptibility is independent of temperature, which is in accordance with experimental observations.

A result that is sometimes quoted as confirmation of the Langevin model of diamagnetism is the susceptibility of carbon, for which $N_0 = 6.02 \times 10^{-26}$ kg mol^{-1}, $\rho = 2220$ kg m^{-3}, $e = 1.6 \times 10^{-19}$ C, and $Z = 6$.

$$\left\langle R^2 \right\rangle = \left(0.7 \times 10^{-10}\right)^2 \text{ m}^2 \tag{9.17}$$

The expected value of susceptibility on the basis of the Langevin model equation is

$$\chi = -18.85 \times 10^{-6} \tag{9.18}$$

The measured value is

$$\chi = -13.82 \times 10^{-6} \tag{9.19}$$

9.1.3 Paramagnetism

How can we explain the paramagnetic susceptibility of solids that have a permanent magnetic moment per atom?

In atoms for which the spin and orbital moments do not cancel out, there is a net permanent magnetic moment per atom, and this results in contributions to the magnetization, which lead to positive susceptibility. The susceptibilities of paramagnets are typically of the order of $\chi \approx 10^{-3}$–10^{-5}, and at low fields $\boldsymbol{M}$ is proportional to $\boldsymbol{H}$, although deviations from proportionality occur at high fields, where the magnetization begins to saturate. Examples of paramagnets are aluminum, platinum, and manganese above its Néel temperature of 100 K. Iron, nickel, and cobalt, although classified primarily as ferromagnets, each has a high temperature paramagnetic phase above their Curie temperature, as do all ferromagnets.

There are a number of possible explanations of paramagnetic behavior in solids. These range from the localized moments model of Langevin [1], in which the noninteracting electronic magnetic moments on the atomic sites are randomly oriented as a result of their thermal energy, to the Van Vleck model [2] of localized moment paramagnetism, which leads to a temperature-independent susceptibility under certain circumstances. Finally, there is the Pauli paramagnetism model [3], which depends on the weak spin paramagnetism of the conduction-band electrons in metals. In this model, the conduction electrons are considered essentially to be free and so nonlocalized. The Pauli model like the Van Vleck model also leads to a temperature-independent paramagnetic susceptibility.

9.1.4 Curie's Law

How does the susceptibility of a paramagnet depend on temperature?

The susceptibilities of a number of paramagnetic solids were measured over a wide temperature range by Curie [4]. He found that the susceptibility varied inversely with temperature, as shown in Figure 9.2.

$$\chi = \frac{C}{T} \tag{9.20}$$

where C is the Curie constant.

The materials that obey this law are those in which the magnetic moments are localized at the atomic or ionic sites. Examples are *dilute* magnetic materials, in

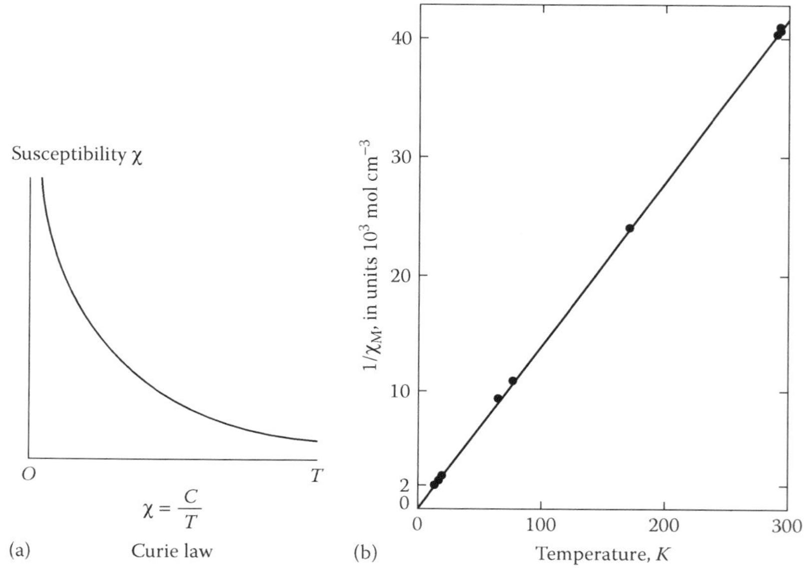

FIGURE 9.2 Temperature dependence of paramagnetic susceptibility: (a) a schematic of the variation of χ with T and (b) the variation of $1/\chi$ with temperature for the paramagnetic salt Gd $(C_2H_5{\cdot}SO_4){\cdot}9H_2O$. The circles are experimental points; the straight line is the Curie law prediction.

which the magnetic atoms are surrounded by a number of nonmagnetic atoms. Hydrated salts of transition metals such as $CuSO_4{\cdot}5H_2O$ and $CrK(SO_4){\cdot}12H_2O$ obey the Curie law.

9.1.5 Langevin Theory of Paramagnetism

If the electronic magnetic moments are localized on the atom, how does the susceptibility depend on magnetic field and temperature?

In materials with unpaired electrons, and consequently in which the orbital magnetic moments are not balanced, there is a net permanent magnetic moment per atom. If this net atomic magnetic moment is $\boldsymbol{m}$, which will be the vector sum of the spin $\boldsymbol{m}_s$ and orbital $\boldsymbol{m}_o$ components, then the energy of the moment in a magnetic field $\boldsymbol{H}$ will be

$$E = -\mu_0 \boldsymbol{m} \cdot \boldsymbol{H} \tag{9.21}$$

Thermal energy tends to randomize the alignment of the moments. Langevin [1] supposed that the moments are noninteracting in which case we can use Boltzmann statistics to express the probability of any given electron occupying an energy state E. If $k_B T$ is the thermal energy per moment, then

$$p(E) = \exp\left(\frac{-E}{k_B T}\right) \tag{9.22}$$

We evaluate the probability function for the case of an isotropic material. The number of moments lying between angles θ and $\theta + d\theta$ is dn, which will be proportional to the surface area dA

$$dA = 2\pi r^2 \sin\theta \, d\theta \tag{9.23}$$

$$dn = C 2\pi r^2 \sin\theta \, d\theta \tag{9.24}$$

where C is here a normalizing constant, which gives the total number of moments per unit area.

Now incorporating the probability of occupation of any given state

$$dn = C 2\pi \sin\theta \, d\theta \exp\left(\frac{\mu_0 \boldsymbol{mH} \cos\theta}{k_B T}\right) \tag{9.25}$$

Integrating this expression over the entire range of possible direction gives the resultant total number of moments per unit volume N

$$N = 2\pi C \int_0^{\pi} \sin\theta \exp\left(\frac{\mu_0 \boldsymbol{mH}}{k_B T}\right) d\theta \tag{9.26}$$

$$C = \frac{N}{2\pi \int_0^\pi \sin\theta \exp(\mu_0 \boldsymbol{mH}/k_B T)\, d\theta} \tag{9.27}$$

The magnetization is then given by the expression

$$\boldsymbol{M} = \int_0^N m \cos\theta \, dn = \frac{N\boldsymbol{m}\int_0^\pi \cos\theta \sin\theta \exp(\mu_0 \boldsymbol{mH}\cos\theta/k_B T)\, d\theta}{\int_0^\pi \sin\theta \exp(\mu_0 \boldsymbol{mH}\cos\theta/k_B T)\, d\theta} \tag{9.28}$$

If we put $x = \cos\theta$, $dx = -\sin\theta \, d\theta$ and integrate, this gives

$$\boldsymbol{M} = N\boldsymbol{m}\left[\coth\left(\frac{\mu_0 \boldsymbol{mH}}{k_B T}\right) - \left(\frac{k_B T}{\mu_0 \boldsymbol{mH}}\right)\right] = L\left(\frac{\mu_0 \boldsymbol{mH}}{k_B T}\right) \tag{9.29}$$

This is the Langevin equation for the magnetization of a paramagnet. The term $\left[\coth(\mu_0 \boldsymbol{mH}/k_B T) - (k_B T/\mu_0 \boldsymbol{mH})\right]$ is called the *Langevin function* and always lies in the range from −1 to +1. The Langevin function can be expressed as an infinite power series in $\mu_0 \boldsymbol{mH}/k_B T$.

In many cases, $\mu_0 \boldsymbol{mH}/k_B T \ll 1$ so the expression for $\boldsymbol{M}$ becomes approximately equal to the first term in the series

$$\boldsymbol{M} = \frac{N\mu_0 \boldsymbol{m}^2 \boldsymbol{H}}{3k_B T} \tag{9.30}$$

which leads immediately to the Curie law, since

$$\chi = \frac{\boldsymbol{M}}{\boldsymbol{H}} = \frac{N\mu_0 \boldsymbol{m}^2}{3k_B T} = \frac{C}{T} \tag{9.31}$$

This demonstrates that the Langevin model leads to a paramagnetic susceptibility that varies inversely with temperature in accordance with the Curie law.

9.1.6 Curie-Weiss Law

Is there a more general law for the dependence of paramagnetic susceptibility on temperature?

It was found that the susceptibilities of a number of paramagnetic metals obey a modified or generalized law known as the *Curie-Weiss law* [5]. In metals such as nickel and the lanthanides, susceptibility was found to vary as shown in Figure 9.3, which can be represented in the paramagnetic region by an equation of the following form:

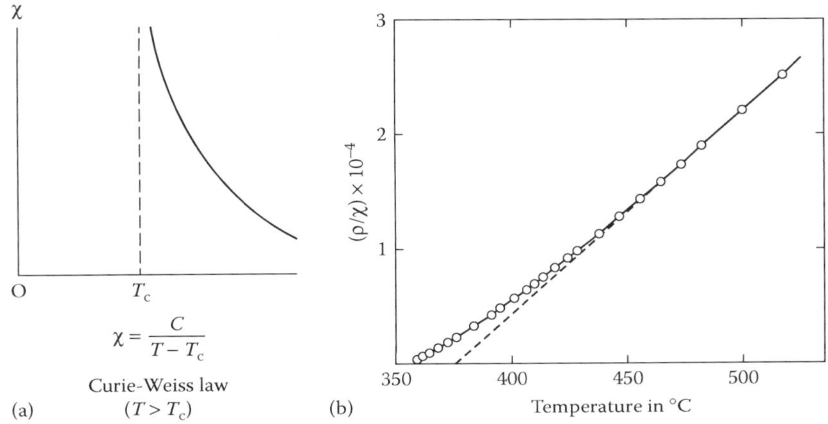

FIGURE 9.3 Variation of χ with temperature for paramagnetic materials which undergo a transition to ferromagnetism at the Curie temperature T_c: (a) a schematic of the variation of χ with T and (b) the variation of $1/\chi$ with temperature for nickel.

$$\chi = \frac{C}{(T - T_c)} \tag{9.32}$$

where:

C is again the Curie constant
T_c is another constant with dimensions of temperature

T_c can be either positive, negative, or zero. $T_c = 0$ corresponds of course to the earlier Curie law. For materials that undergo a paramagnetic to ferromagnetic transition, $T_c > 0$. For materials that undergo a paramagnetic to antiferromagnetic transition, the term T_c is less than zero, although in practice, the transition temperature between the paramagnetic and antiferromagnetic phases occurs at a positive temperature T_N known as the *Néel temperature.*

It should be remembered that the susceptibility only follows the Curie-Weiss law in the paramagnetic region. Once the material becomes ordered, the susceptibility behaves in a very complicated way and no longer has a unique value for a given field strength.

9.1.7 Weiss Theory of Paramagnetism

What does the Curie-Weiss law tell us about the interactions between the individual electronic magnetic moments?

Weiss [6] showed that the variation of paramagnetic susceptibility with temperature of materials that obeyed the Curie-Weiss law could be explained if the individual atomic magnetic moments interacted with each other via an interaction field $\boldsymbol{H}_e$, which was called the *molecular field* but more accurately should be called the *atomic field*.

Since in paramagnets the magnetization is locally homogeneous, the magnetic moment per unit volume will be everywhere equal to the bulk magnetization $\boldsymbol{M}$ (unlike in ferromagnets where the local magnetic moment per unit volume, the *spontaneous magnetization*, is unrelated to the bulk magnetization because of the existence of domains). Therefore, interactions between any individual magnetic moment and other moments within a localized volume can be expressed as an interaction between the given moment and the bulk magnetization $\boldsymbol{M}$ as a mean field approximation. This is represented as the interaction field $\boldsymbol{H}_e$, which can be expressed as

$$\boldsymbol{H}_e = \alpha \boldsymbol{M} \tag{9.33}$$

where so far we have made no assumptions about the nature of α. The total magnetic field experienced by a magnetic moment then becomes the sum of the applied field $\boldsymbol{H}$ and the interaction field $\boldsymbol{H}_e$

$$\boldsymbol{H}_{tot} = \boldsymbol{H} + \boldsymbol{H}_e \tag{9.34}$$

and hence

$$\boldsymbol{H}_{tot} = \boldsymbol{H} + \alpha \boldsymbol{M} \tag{9.35}$$

Consider the paramagnetic susceptibility of a material in which such a field operates. This is a perturbation of the Langevin model, so a Curie-type law should still be obeyed, providing that the orientation of the magnetic moments is in thermal equilibrium and obeys Boltzmann statistics.

$$\frac{\boldsymbol{M}}{\boldsymbol{H}_{tot}} = \frac{C}{T} \tag{9.36}$$

The susceptibility is still given by

$$\chi = \frac{\boldsymbol{M}}{\boldsymbol{H}} \tag{9.37}$$

and substituting $\boldsymbol{H} = \boldsymbol{H}_{tot} - \alpha \boldsymbol{M}$ leads to

$$\chi = \frac{C}{T - \alpha C} = \frac{C}{T - T_c} \tag{9.38}$$

This is the Curie-Weiss law. The derivation shows that a paramagnetic solid with localized but interacting atomic moments will have a susceptibility that obeys the Curie-Weiss law and the susceptibility is larger than it would be if there were no interaction field present. The critical temperature T_c is the Curie temperature and marks the boundary between the paramagnetic and ferromagnetic states of the material.

9.1.8 Consequences of Weiss Theory

What is the magnetization equation for a material with a Weiss interaction field?

Having established the concept of an interatomic coupling in the form of a mean field interaction, it is possible to provide an equation for the paramagnetic magnetization as a function of the applied magnetic field using a perturbation of the Langevin equation. In this case, the energy of a magnetic moment within a magnetic field needs to be modified to

$$E = -\mu_0 \boldsymbol{m} \cdot (\boldsymbol{H} + \alpha \boldsymbol{M}) \tag{9.39}$$

and consequently the magnetization as a function of field becomes

$$\boldsymbol{M} = N\boldsymbol{m}\left\{ \cot h \left[\frac{\mu_0 \boldsymbol{m}(\boldsymbol{H} + \alpha \boldsymbol{M})}{k_B T} \right] - \left[\frac{k_B T}{\mu_0 \boldsymbol{m}(\boldsymbol{H} + \alpha \boldsymbol{M})} \right] \right\} \tag{9.40}$$

This means that the paramagnetic susceptibility is greater in the case of the interacting moment system. It is also apparent however that at the critical temperature T_c, there arises a discontinuity in the function on the right-hand side of the equation, so that below T_c, the behavior is very different.

9.1.9 Critique of Langevin-Weiss Theory

In what way does the classical Langevin-Weiss theory of paramagnetism fail?

The Langevin model requires that the magnetic moments be localized on the atomic sites. The model does not apply to a material in which the moments are not localized and such materials should not follow the Curie or Curie-Weiss law. Although a number of paramagnetic materials obey the Curie law, most transition metals do not. In many of these cases, the susceptibility is independent of temperature.

The Langevin theory does not work for most transition metals because the magnetic electrons are usually the outer electrons of the atom, which are not localized at the atomic core. The unpaired electrons must exist in unfilled shells and for many elements this means that the magnetic electrons are the outer electrons, which are only loosely bound. These are unlikely to remain localized at the atomic sites.

Despite this, the Curie-Weiss law still works very well for some metals such as nickel, for which it is unlikely that the magnetic electrons are tied to the ionic sites. One possible explanation is that the electrons do migrate but that they spend a large amount of their time close to the ionic sites, which results in behavior that is similar to that predicted by a localized model. Also, in the rare earth metals and their alloys and compounds, the 4f electrons, which determine the magnetic properties, are closely bound to the atomic core. Hence, these materials do exhibit the Curie-Weiss type behavior.

9.2 THEORIES OF ORDERED MAGNETISM

What types of ordered magnetic structures exist and how do they differ?

There are a number of different types of magnetic order in solids including ferromagnetism, antiferromagnetism, ferrimagnetism, and helimagnetism. Some materials, such as the heavy rare earths, exhibit more than one ordered magnetic state. These ordered states undergo transitions at critical temperatures, so that every solid that exhibits one of these types of magnetic order will become paramagnetic at higher temperatures. For example, in a ferromagnet, the Curie point is the transition temperature above which the material becomes paramagnetic and below which an ordered ferromagnetic state exists. The Néel point is the temperature below which an ordered antiferromagnetic state exists. As the temperature of a ferromagnet is increased, the thermal energy increases while the interaction energy is unaffected. Depending on whether the material is a ferromagnet or an antiferromagnet, at the Curie or Néel temperature, the randomizing effect of thermal energy overcomes the aligning effect of the interaction energy and above these temperature the magnetic state becomes disordered.

Examples of familiar ferromagnetic elements are the three transition metals: iron, $T_c = 770°C$; nickel, $T_c = 358°C$; and cobalt, $T_c = 1131°C$. Several of the rare earth metals also exhibit ferromagnetism, mostly only at temperatures below ambient including dysprosium, $T_c = 85$ K; terbium, $T_c = 219$ K; holmium, $T_c = 19$ K; erbium, $T_c = 19.5$ K; and thulium, $T_c = 32$ K. Some solids such as terbium, dysprosium, and holmium have both Curie and Néel temperatures. Gadolinium is a ferromagnetic rare earth with a Curie temperature at an ambient temperature $T_c = 293$ K (20°C).

9.2.1 FERROMAGNETISM

What causes the transition from paramagnetism to ferromagnetism?

In ferromagnetic solids at temperatures well below the Curie temperature, the magnetic moments are aligned parallel within domains. This can be explained phenomenologically by the Weiss interaction field, which was originally suggested in order to explain the dependence of paramagnetic susceptibility on temperature in certain materials. The magnetic moments also have a preference for certain crystallographic directions a phenomenon known as *magnetocrystalline anisotropy*. This preferred direction varies from material to material. The alignment of magnetic moments in some ferromagnetic solids is shown in Figure 9.4.

9.2.2 WEISS THEORY OF FERROMAGNETISM

How can the Weiss interaction be used to explain magnetic order in ferromagnets?

If the unpaired electronic magnetic moments, which are responsible for the magnetic properties, are localized on the atomic sites, then we can consider an interaction between the unpaired moments of the form as discussed above in Section 9.1.7.

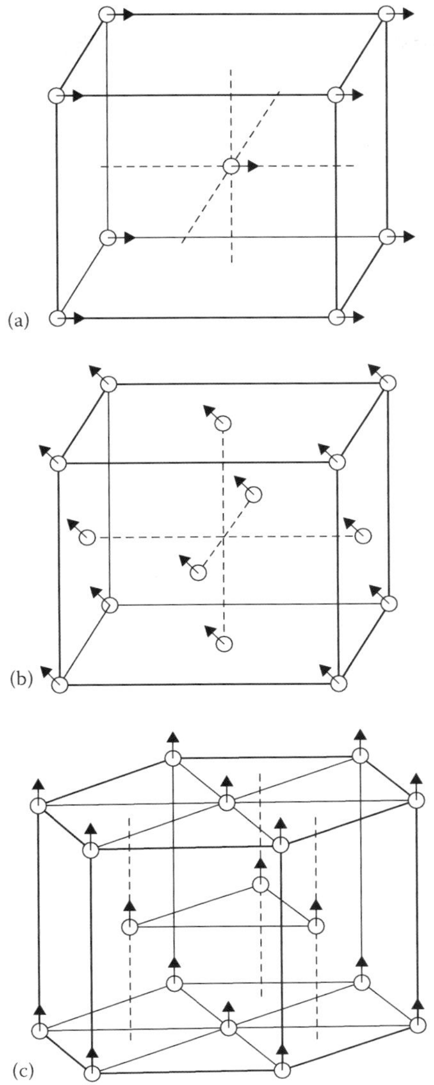

FIGURE 9.4 Crystallographic alignment of the magnetic moments in various ferromagnetic solids: (a) Fe, (b) Ni, and (c) Co.

This interaction, which Weiss introduced to explain the paramagnetic susceptibilities of certain materials, leads to the existence of a critical temperature below which the thermal energy of the electronic moments is insufficient to cause random paramagnetic alignment. This means that the effective field $\boldsymbol{H}_e$ can be used to explain the alignment of magnetic moments within domains for temperatures below T_c.

There are a number of possible variations on the theme of the interatomic interaction or exchange field. We will look at two of these: (1) the mean-field approximation, which was used successfully in the paramagnetic region, and (2) a nearest-neighbor-only coupling, which is more appropriate in the ferromagnetic regime particularly within domains. We will begin by considering the interaction between just two magnetic moments.

As in Section 7.1.2, an atomic magnetic moment $\boldsymbol{m}_i$ experiences an effective field $\boldsymbol{H}_{\text{pair}}$ due to another moment $\boldsymbol{m}_j$. If we assume that this field is also in the direction of $\boldsymbol{m}_j$, we can write

$$\boldsymbol{H}_{\text{pair}} = \frac{\alpha_{ij}}{V} \cdot \boldsymbol{m}_j \tag{9.41}$$

The total exchange interaction field at the moment $\boldsymbol{m}_i$ will then simply be the vector sum of all the interactions with other moments

$$\boldsymbol{H}_{\text{ex}} = \sum_j \frac{\alpha_{ij}}{V} \cdot \boldsymbol{m}_j \tag{9.42}$$

9.2.3 Mean-Field Approximation

Is there a simple explanation of the Weiss interaction?

So far we have made no assumption about the form of α_{ij}. We can show that if the interactions between all moments are identical and hence independent of displacement between the moments, then all of the α_{ij} are equal. Let these be α/V,

$$\boldsymbol{H}_{\text{ex}} = \frac{\alpha}{V} \sum_j \boldsymbol{m}_j \tag{9.43}$$

so that, as seen in Section 6.1.5, within a domain, the mean-field interaction is

$$\boldsymbol{H}_{\text{ex}} = \alpha \boldsymbol{M}_{\text{s}} \tag{9.44}$$

The interaction energy of the moment under these conditions is

$$E_{\text{e}} = -\mu_0 \boldsymbol{m}_i \cdot \boldsymbol{H}_{\text{ex}} = -\mu_0 \alpha \boldsymbol{m}_i \cdot \boldsymbol{M}_{\text{s}} \tag{9.45}$$

This was the original formulation of the Weiss theory. The mean-field approximation gives an explanation for the spontaneous magnetization within domains but ultimately is not very realistic because each moment does not interact equally with all other moments. However, for moments within the body of a domain, it works out reasonably well in practice simply because each moment will experience the same exchange field as its neighbors, and this will be in the direction of the spontaneous magnetization $\boldsymbol{M}_{\text{s}}$ in the domain. Therefore, if the mean-field parameter is treated completely empirically, it can give a reasonable explanation of the behavior of the moments within a domain.

If we consider the case of a zero external field, then the only field operating within a domain will be the Weiss field.

$$\boldsymbol{H}_{\text{tot}} = \boldsymbol{H}_{\text{ex}} \tag{9.46}$$

When considering the magnetic moments within the body of a domain, if we apply the mean-field model, the interaction field will be proportional to the spontaneous magnetization $\boldsymbol{M}_s$ within the domain. Following an analogous argument to that given by Langevin for paramagnetism, if there are no constraints on the possible direction of $\boldsymbol{m}$, we arrive at

$$\frac{\boldsymbol{M}_s}{\boldsymbol{M}_0} = \frac{\boldsymbol{M}_s}{N\boldsymbol{m}} = \cot h\left(\frac{\mu_0 \boldsymbol{m} \alpha \boldsymbol{M}_s}{k_B T}\right) - \left(\frac{k_B T}{\mu_0 \boldsymbol{m} \alpha \boldsymbol{M}_s}\right) \tag{9.47}$$

The solution of this equation leads to perfect alignment of magnetic moments within a domain as the temperature approaches absolute zero. As T increases, the spontaneous magnetization within a domain decreases as shown in Figure 9.5.

At a finite temperature, which corresponds to the Curie point, the spontaneous magnetization tends rapidly to zero, representing the transition from ferromagnetism

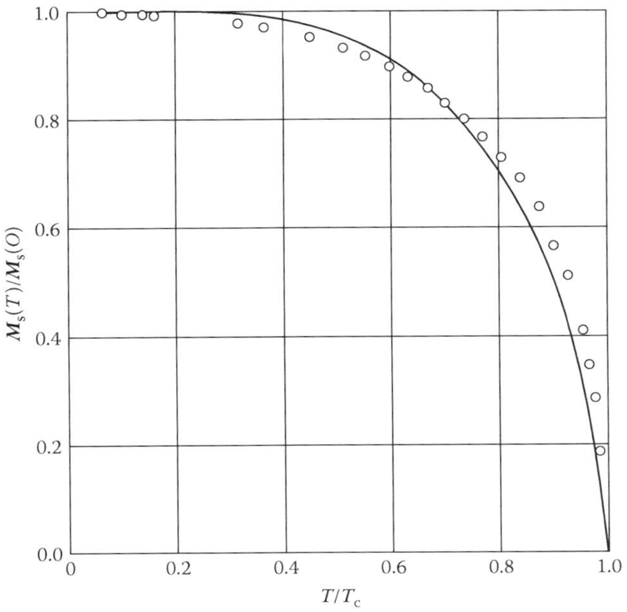

FIGURE 9.5 Variation of the spontaneous magnetization $\boldsymbol{M}_s$ of nickel within a domain as a function of temperature according to the Weiss mean-field model. (After Weiss, P., and Forrer, R., *Ann Phys.*, 12, 279, 1929.)

at lower temperatures to paramagnetism at higher temperatures. The above expression (Equation 9.47) can be generalized to include the effects of a magnetic field $\boldsymbol{H}$, so that the energy of a moment within a domain becomes

$$E = -\mu_0 \boldsymbol{m}(\boldsymbol{H} + \alpha \boldsymbol{M}_s) \tag{9.48}$$

which leads immediately to the following magnetization within a domain

$$\frac{\boldsymbol{M}_s}{\boldsymbol{M}_0} = \frac{\boldsymbol{M}_s}{N\boldsymbol{m}} = \coth\left[\frac{\mu_0 \boldsymbol{m}(\boldsymbol{H} + \alpha \boldsymbol{M}_s)}{k_B T}\right] - \left[\frac{k_B T}{\mu_0 \boldsymbol{m}(\boldsymbol{H} + \alpha \boldsymbol{M}_s)}\right] \tag{9.49}$$

Equation 9.49 is not encountered very often however, because $\alpha \boldsymbol{M}_s >> \boldsymbol{H}$ in ferromagnets (e.g., in iron, $\boldsymbol{M}_s = 1.7 \times 10^6$ A m^{-1}, and so $\alpha \boldsymbol{M}_s$ can be up to 6.8×10^8 A m^{-1}, while $\boldsymbol{H}$ will rarely exceed 2×10^6 A m^{-1}). Consequently, within the body of a domain, the action of the $\boldsymbol{H}$ field is not very significant when compared with the exchange interaction field.

However, it is well known that moderate magnetic fields ($\boldsymbol{H} \approx 8 \times 10^3$ A m^{-1}) can cause significant changes in the bulk magnetization $\boldsymbol{M}$ in ferromagnets. Therefore, these changes occur principally at the domain boundaries where the exchange interaction is competing with the anisotropy energy to give an energy balance and the magnetic moments in the domain boundary are not so strongly held in the direction as the magnetic moments in the body of the domain. Under these conditions, the additional field energy provided by the magnetic field $\boldsymbol{H}$ can just tip the balance and result in changes in the direction of magnetic moments within the domain wall. This mechanism manifests itself as domain-wall motion, as described in Chapter 8.

The magnetic moments in the domain wall, being on the periphery of the domain, do not couple so strongly to the spontaneous magnetization of the domains, although the Weiss-type coupling is still strong between nearest neighbors. The net interaction field per moment is different in this case because the domains on either side are aligned in different directions. Therefore, the energy is very finely balanced in the domain wall and slight perturbations due to an applied magnetic field can cause changes in the direction of alignment of the moments, which would not be possible within the body of the domain.

9.2.4 Nearest-Neighbor Interactions

Can the Weiss model be interpreted on the basis of localized interactions only?

Another variation of the Weiss model that provides mathematically tractable solutions is the nearest-neighbor approximation [8] as discussed in terms of pair interactions in Section 7.1.2. In this case, the electronic moments are assumed to interact only with those of its z nearest neighbors. This is a simplification of course, but it is reasonable to expect that the magnetic moment on an atom will interact more strongly with atomic magnetic moments that are nearby. The nearest-neighbor model just takes this to an extreme and ignores any interactions that are further apart than just between nearest neighbors.

For a simple cubic lattice $z = 6$, for body-centered cubic $z = 8$, for face-centered cubic $z = 12$, and for a hexagonal lattice $z = 12$. The nearest-neighbor approach is particularly useful for considering magnetic moments in the domain walls since in this case, the moments do not couple to the magnetization within the body of the domain simply because they lie between domains with different magnetic directions and the direction of magnetization changes within the wall.

In this approximation, we can write the exchange interaction field as

$$\boldsymbol{H}_e = \sum_{\substack{\text{nearest}\\\text{neighbors}}} \vartheta_{ij}\boldsymbol{m}_j \tag{9.50}$$

We assume that each nearest-neighbor interaction is identical in strength and has the same value of ϑ, where once again $\vartheta = 0$ corresponds to the noninteracting limit described by Langevin theory. When ϑ is nonzero, it is usually convenient to consider that every moment interacts equally with each of its nearest neighbors.

$$\boldsymbol{H}_e = \sum_{\substack{\text{nearest}\\\text{neighbors}}} \vartheta\boldsymbol{m}_j = \vartheta \sum_{\substack{\text{nearest}\\\text{neighbors}}} \boldsymbol{m}_j \tag{9.51}$$

On the basis of this nearest-neighbor interaction, we find that $\vartheta > 0$ corresponds to ferromagnetic alignment, while $\vartheta < 0$ corresponds to antiferromagnetic alignment. This can easily be seen by considering the configuration of moments that leads to a minimum in the interaction energy.

$$E_e = -\mu_0 \boldsymbol{m}\vartheta \sum_{\substack{\text{nearest}\\\text{neighbors}}} \boldsymbol{m}_j \tag{9.52}$$

and summing over n nearest neighbors

$$E_e = -\mu_0 n\vartheta \boldsymbol{m}^2 \tag{9.53}$$

Having established the existence of the Weiss-type interaction, it is possible to provide a description of ferromagnets that is similar to the Langevin model of paramagnetism. Such a model is strictly only correct when the moments are localized on the atomic cores. Thus, it applies to the lanthanide series because the 4f electrons, which determine the magnetic properties, are tightly bound to the nuclei.

9.2.5 Curie Temperature on the Basis of Mean-Field Model

How does the Weiss interaction explain the existence of a critical temperature?

In the original works by Weiss [5,6], it was shown that the existence of an internal coupling field proportional to the magnetization $\boldsymbol{M}$ led to a modified form of Curie's law known as the *Curie-Weiss law*

$$\chi = \frac{C}{T - T_c} \tag{9.54}$$

From the Curie-Weiss law, the Curie constant is given by

$$C = \frac{N\mu_0 \boldsymbol{m}^2}{3k_B} \tag{9.55}$$

and the Curie temperature $T_c = \alpha C$, where α is the mean field coefficient, is therefore given by

$$T_c = \frac{\mu_0 N \alpha \boldsymbol{m}^2}{3k_B} \tag{9.56}$$

We see from this that it is possible to determine the strength of the mean-field coupling or Weiss constant α from the Curie temperature providing the magnetic moment per atom is known. Similarly, for a nearest-neighbor coupling, the interaction parameter ϑ can be found from the Curie temperature using the following equation:

$$T_c = \frac{\mu_0 n \vartheta \boldsymbol{m}^2}{3k_B} \tag{9.57}$$

so that the mean-field coefficient and the nearest-neighbor coupling coefficient are related by $N\alpha = n\vartheta$.

9.2.6 Antiferromagnetism

Is it also possible to explain antiferromagnetic order by a Weiss interaction?

Simple antiferromagnetism [9] in which nearest-neighbor moments are aligned antiparallel can also be interpreted on the basis of the Weiss model. There are two ways of considering this. One is to divide the material into two sublattices A and B, with the moments on one sublattice interacting with the moments on the other with a negative coupling coefficient, but interacting with the moments on their own sublattice with a positive coupling coefficient. This ensures that the magnetic moments on the two sublattices point in different directions. Another way to envisage the problem is on the basis of nearest-neighbor interactions. With a negative interaction between nearest neighbors, this leads to simple antiferromagnetism.

Consequently, if we return to the simplified situation of the linear chain of electron spins as depicted in Figure 6.3 and require that the exchange interaction only be significant between nearest neighbors and have a negative coefficient $-\alpha$ between nearest neighbors, then the energy associated with the parallel alignment of spins is

$$E_{\text{par}} = 10\boldsymbol{m}^2\mu_0\alpha \tag{9.58}$$

while for antiparallel nearest neighbors

$$E_{\text{anti}} = -10\boldsymbol{m}^2\mu_0\alpha \tag{9.59}$$

which shows immediately that for a negative nearest-neighbor interaction coefficient, an antiparallel alignment of magnetic moments is energetically favored.

The Curie-Weiss law also applies to antiferromagnets above their ordering temperatures. However, the sign of the term T_c in the denominator is positive so that the law becomes

$$\chi = \frac{C}{T + T_c} \tag{9.60}$$

It therefore appears superficially as though the critical temperature in this case is below 0 K. In fact, the plot of $1/\chi$ against temperature for these antiferromagnets does appear to be a straight line intercepting the temperature axis at $-T_c$, but at a temperature above 0 K, known as the *Néel temperature*, the materials undergo an order-disorder transition below which they cease to obey the Curie-Weiss law, as shown in Figure 9.6.

The Néel temperature of an antiferromagnet is analogous to the Curie temperature of a ferromagnet. Both mark the borderline temperature above which the material is disordered and below which it is ordered. In both cases, the thermal energy per spin equals the coupling energy per spin at these transition temperatures. However, it should be remembered that these two definitions of T_N and T_c are not quite equivalent. For example, dysprosium and terbium have both a Curie temperature that marks the upper limit of temperature of the ferromagnet regime (85 and 219 K, respectively) and a temperature that marks the lower temperature of the disordered paramagnetic regime (180 and 230 K, respectively). In this respect, one could say that a simple ferromagnet, such as iron, nickel, cobalt, and gadolinium, has coincident Néel and Curie temperatures.

Examples of antiferromagnetic materials are chromium below its Néel temperature of 37°C and manganese below its Néel temperature of 100 K. In these materials, the magnetic moments on neighboring atomic sites are antiparallel so that the net magnetization within a domain in the absence of an applied magnetic field is $\boldsymbol{M}_s = 0$. This kind of magnetic structure is common in the transition metal oxides, MnO,

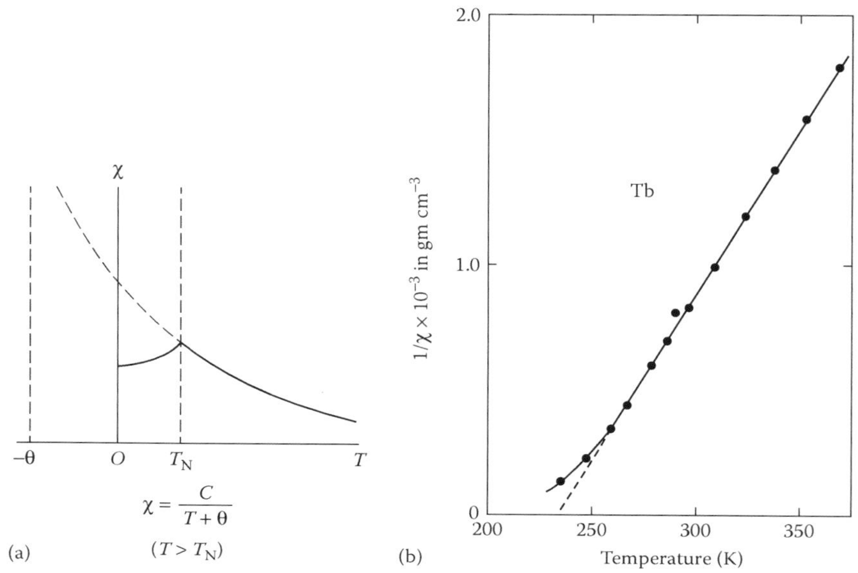

FIGURE 9.6 A schematic variation of χ with temperature in the paramagnetic regime of materials that undergo a transformation to antiferromagnetism (a) and $1/\chi$ versus data for terbium (b).

FeO, CoO, and NiO to name but a few, which have Néel temperatures of 122, 198, 291, and 523 K, respectively.

Despite the fact that the spontaneous magnetization within the domains is zero in these simple linear antiferromagnets, it is possible to determine the spontaneous magnetization on the sublattices using magnetic resonance as described by Feldmann et al. [10].

As yet, we have not really discussed the origin of the coupling between the electron spins that leads to the Weiss mean field, and this must be left until Chapter 11. Therefore, it is not yet apparent from our discussions that the nature of the exchange interaction in these materials, which are mostly electrical insulators, is different from the exchange interaction in ferromagnets, which are metals. In the metals, the exchange interaction takes place between the itinerant outer electrons, which form energy bands, while in most antiferromagnets, the exchange is between localized electrons. The mechanism for the interaction in these antiferromagnets is called *superexchange*, in which the transfer of electrons from the oxygen ions to the neighboring metallic ions allow the *magnetic* cations such as Mn^{2+} to interact with each other indirectly via the intervening anions such as O^{2-}.

As a result of the zero spontaneous magnetization within the domains and the consequent low permeability and susceptibility of antiferromagnets, they do not find many technological applications. One important technological application of antiferromagnets is in the magnetoresistive spin valves in which an antiferromagnetic layer is used to pin the direction of magnetization in one of the ferromagnetic layers. This is discussed in Section 11.3.9.

However, an understanding of the ordering in antiferromagnets provides the key to understanding the ordering in ferrimagnets, a class of magnetic materials that does have important technological applications.

9.2.7 Ferrimagnetism

How can the properties of ferrites be explained?

Ferrimagnetism is a particular case of antiferromagnetism in which the magnetic moments on the A and B sublattices while still pointing in opposite directions have different magnitudes. Ferrimagnetic order was first suggested by Néel in 1948 [11] to explain the behavior of ferrites. The ferrimagnets behave on a macroscopic scale very much like ferromagnets, so that it was not realized for many years that there was a distinction. They have a spontaneous magnetization below the Curie temperature and are organized into domains. They also exhibit hysteresis and saturation in their magnetization curves.

The most familiar ferrimagnet is Fe_3O_4, although other magnetic ferrites with the general formula $MO{\cdot}Fe_2O_3$, where M is a transition metal such as manganese; other metals such as nickel, cobalt, zinc, and magnesium are also widely used. These ferrites are cubic and have the *spinel* crystal structure referred to in Chapter 8. The magnetic moments of the two Fe^{3+} ions are aligned antiparallel, so the magnetic properties are determined by the moment on the M^{2+} ion. Another soft ferrimagnetic material is gamma iron oxide γ-Fe_2O_3, which is widely used as a magnetic recording medium. This is obtained by oxidizing magnetite Fe_3O_4. Above 400°C, this transforms to rhombohedral alpha iron oxide, or haematite, which is a canted antiferromagnet. The soft ferrites find applications in high-frequency magnetic devices because of their combination of high permeability and low conductivity, which leads to high energy conversion efficiency at frequencies up to the megahertz range in some cases such as nickel-zinc ferrite, and up to 0.5 MHz in manganese zinc ferrite. These materials are discussed in more detail in Chapter 12.

Another class of ferrites is made up of the hexagonal ferrites such as barium ferrite $BaO{\cdot}6(Fe_2O_3)$ and strontium ferrite $SrO{\cdot}6(Fe_2O_3)$. These are magnetically hard and have been extensively used as permanent magnet materials. They have a high anisotropy with the moments lying along the c-axis. Their critical temperatures are typically in the range T_c = 500°C–800°C.

A third class of ferrimagnets are the garnets, which have the chemical formula $5Fe_2O_3{\cdot}3R_2O_3$, where R is a rare earth ion. The best known of these is yttrium-iron garnet, $5Fe_2O_3{\cdot}3Y_2O_3$. These materials have a complicated cubic crystal structure. Their order-disorder transition temperatures are around 550°C.

The general model that we have for these materials is one with different magnetic atoms on the respective sublattice sites, with different magnetic moments, and with antiferromagnetic coupling via the indirect superexchange mechanism.

The interlattice coupling coefficient ϑ_{AB} is generally much stronger than the intralattice coupling coefficients ϑ_{AA} or ϑ_{BB}. This leads to an expression for the temperature dependence of susceptibility, which is more complicated than the well-known Curie-Weiss law, but is nevertheless analogous.

$$\chi = \frac{(C_A + C_B)T - 2\alpha C_A C_B}{T^2 - \alpha^2 C_A C_B} \tag{9.61}$$

where:
- T is the temperature
- α is the interlattice coupling coefficient
- C_A and C_B are the Curie constants for the respective sublattices

In this expression (Equation 9.61), the Curie temperature is given by

$$T_c = \alpha\sqrt{C_A C_B} \tag{9.62}$$

From Equations 9.61 and 9.62, it can be seen that, unlike ferromagnets, the reciprocal susceptibility of ferrimagnets shows considerable curvature as a function of temperature as T_c is approached from above.

If the two sublattices of a ferrimagnet are treated as independent, then there arises the possibility that the temperature dependence of the spontaneous magnetizations on the two sublattices can be different. This is because even though the dominant exchange mechanism between the two sublattices is necessarily common to both, the moments on the atoms of the sublattices, and hence their energies, are different. Therefore, it is possible to find two temperatures at which the spontaneous magnetization reaches zero, if the magnetization of the sublattice with the larger magnetic moments decays more rapidly as the temperature is raised. The lower temperature at which $\boldsymbol{M}_s$ reaches zero is known as the *compensation temperature*, when the magnitudes of the magnetic moments on the two lattices are equal and the material exhibits, albeit at only a single temperature, antiferromagnetic order. As the temperature is raised further, the spontaneous magnetization within the domain changes sign, as the sublattice with the previously smaller magnetic moments becomes the sublattice with the larger moments. Finally, as the temperature is raised further, the magnetic moments eventually become disordered at the Curie temperature, when once again the spontaneous magnetization becomes zero.

9.2.8 Helimagnetism

Are there other types of ordered magnetic materials?

So far we have discussed the situations where the nearest-neighbor interaction is positive (ferromagnetic) or negative (antiferromagnetic). There is a more general case in which we consider nearest-neighbor and next-nearest-neighbor interactions. As an example, we will look at the case of dysprosium as treated by Enz [12] and later by Nicklow [13]. In the base plane, the moments are all aligned ferromagnetically; however, successive base planes have their moments inclined at an angle θ to the moments in the next base plane. This gives a helical magnetic structure.

If ϑ_1 is the interaction between nearest-neighbor planes and ϑ_2 is the interaction between next-nearest-neighbor planes, the total exchange energy E_{ex} becomes

$$E_{ex} = -\sum_i \sum_j \vartheta_{ij} \cos(\theta_{i+j} - \theta_i) \boldsymbol{m}^2 \tag{9.63}$$

If the turn angle between moments in successive base planes is θ_t

$$j\theta_t = \theta_{i+j} - \theta_i \tag{9.64}$$

and if $\vartheta_{i1} = \vartheta_1$, and $\vartheta_{i2} = \vartheta_2$ for all i, then

$$E_{ex} = -N \sum_j \vartheta_j \cos(j\theta_t) \boldsymbol{m}^2 \tag{9.65}$$

If we assume that the interaction from further than two planes away is negligible by comparison with the interactions between nearest-neighbors and next-nearest-neighbors,

$$E_{ex} = -N\boldsymbol{m}^2 (\vartheta_0 + 2\vartheta_1 \cos\theta_t + 2\vartheta_2 \cos 2\theta_t) \tag{9.66}$$

at equilibrium

$$\frac{dE_{ex}}{d\theta_t} = 0 = Nm^2 (2\vartheta_1 \sin\theta_t + 4\vartheta_2 \sin 2\theta_t) \tag{9.67}$$

$$\cos\theta_t = -\frac{\vartheta_1}{4\vartheta_2} \tag{9.68}$$

This gives the value of the turn angle θ_t between successive base planes, which leads to the minimum exchange energy. For ferromagnetic alignment, we need $\cos\theta_t = 1$, or $\vartheta_1 = -4\vartheta_2$, while for simple antiparallel antiferromagnetic alignment, we need $\cos\theta_t = -1$, or $\vartheta_1 = 4\vartheta_2$. Examples of this form of helimagnetic order occur in terbium, dysprosium, and holmium.

The sinusoidal nature of the exchange interaction in these helimagnetic structures is quite obvious from the above discussion, and this form of exchange is rather more complex than the simple direct exchange mechanism in ferromagnets or even the indirect superexchange in antiferromagnets. An explanation of the nature of the coupling was provided by Ruderman and Kittel [14], Kasuya [15], and Yosida [16], who suggested that because the 4f electrons, which are primarily responsible for the magnetic properties of these metals, are so closely bound to the nuclei they cannot interact via direct coupling. Instead the coupling proceeds via an indirect mechanism, whereby the 4f electron spins are coupled to the spins of conduction electrons. This coupling is strong and therefore leads to a polarization of the spins of the conduction electrons, which is a long-range effect since the conduction electrons migrate throughout the material. In addition to this, the conduction electrons obey a Fermi distribution and this causes the polarization of the conduction electron spins to be oscillatory. The end result is an effective oscillatory indirect coupling between the spin magnetic moments of the localized 4f electrons, with the attendant possibility of helical magnetic structures.

9.3 MAGNETIC STRUCTURE

How is the magnetic structure within a domain determined?

The magnetic structure of materials is usually deduced from neutron diffraction and magnetization/susceptibility measurements. The use of neutron diffraction for investigating the structure of magnetic materials has been discussed by Bacon [17] and by Lovesey [18]. The first material to be studied in this way was MnO and this was followed by other antiferromagnetic oxides. The determination of the magnetic structure of the 3d series metals iron, nickel, and cobalt by neutron diffraction was first made by Koehler et al. The magnetic structures of the 4f series metals, the lanthanides, were also studied by the same group [19].

9.3.1 Neutron Diffraction

How do neutrons interact with magnetic materials?

Although the various different types of magnetic order described in Sections 9.1 and 9.2 can be inferred from measurements of magnetic anisotropy and susceptibility, their existence has only been directly verified by neutron diffraction. In the older technique of X-ray diffraction, a beam of X-rays is diffracted by the distribution of electric charge at the periodic lattice sites in the solid. This is known as *Bragg reflection*. Neutrons, however, are diffracted both by the distribution of the nuclei on the lattice sites and by the magnetic moments associated with the electron distribution on each atom. This leads to both Bragg peaks and magnetic peaks in the resulting neutron diffraction spectrum.

Neutrons have a net magnetic moment of 5.4×10^{-4} Bohr magnetons ($= 5.0 \times 10^{-27}$ A m^{-2}) but have no electric charge. This combination of properties means that neutrons can pass relatively easily through a solid since they are not influenced by the localized electric charge distribution. The neutrons interact with the nuclei to a greater or lesser extent depending on the type of nuclei present as shown in Table 9.1. This gives rise to a nuclear scattering component in the total neutron diffraction spectrum. For the wavelengths used in neutron diffraction, the nuclei act as point scatterers and the nuclear scattering spectrum is therefore isotropic.

The neutrons necessary for neutron diffraction studies of this type must have wavelengths comparable with the atomic dimensions, which are typically 0.1 nm.

TABLE 9.1
Comparison of Nuclear and Magnetic Scattering Amplitudes of Various Atoms

Atom or Ion	Nuclear Scattering Amplitude (10^{-14} m)	Effective Spin Quantum Number S	Magnetic Scattering Amplitude (10^{-14} m)	
			$\theta = 0$	$\sin\theta/\lambda = 0.25$ A^{-1}
Cr^{2+}	0.35	2	1.08	0.45
Mn^{2+}	−0.37	5/2	1.35	0.57
Fe	0.96	1.11	0.60	0.35
Fe^{2+}	0.96	2	1.08	0.45
Fe^{3+}	0.96	5/2	1.35	0.57
Co	0.28	0.87	0.47	0.27
Co^{2+}	0.28	2.2	1.21	0.51
Ni	1.03	0.3	0.16	0.10
Ni^{2+}	1.03	1.0	0.54	0.23

Source: Bacon, G.E., *Neutron Diffraction*, 3rd edn. Clarendon Press, Oxford, 1975.

Neutrons with de Broglie wavelengths of this order of magnitude are produced in a nuclear reactor as thermal neutrons at a temperature of about 300 K and hence an energy of 4×10^{-21} J (25 meV) and a corresponding wavelength of 0.18 nm.

The neutron diffraction spectra of magnetic materials contain at least three different contributions, which can be used to examine different properties of the materials. There is elastic neutron scattering, which has two components: the first of these is the nuclear (or Bragg-type) diffraction peaks, which are determined by the periodicity of the lattice, and the neutron scattering cross-section of the nucleus; the second is the magnetic scattering, which is determined by the magnetic order in the solid and the magnetic scattering cross-section. Finally, there is inelastic neutron scattering, which results in the creation or annihilation of a magnon (spin wave excitation) and from such measurements, it is possible to study the spin wave spectrum of the solid.

The experimental arrangement for neutron diffraction investigations is similar to that used in X-ray diffraction. A collimated, monochromatic (single-energy), and in some cases, polarized beam of neutrons is directed on to the specimen. A neutron detector can be moved around the specimen to any angle in order to measure the angular dependence of the intensity of the diffracted beam. Nuclear scattering patterns have the same criteria as for X-rays and so the Bragg angle can be defined as $\sin\theta = |G|\lambda/2$, where λ is the de Broglie wavelength of the neutrons and $|G|$ is the magnitude of the reciprocal lattice vector. The magnetic reflections are superimposed on the Bragg reflection spectrum. Generally, both are referenced to the crystallographic unit cell, as indicated in Figure 9.7, which is taken from the work of Shull et al. [20].

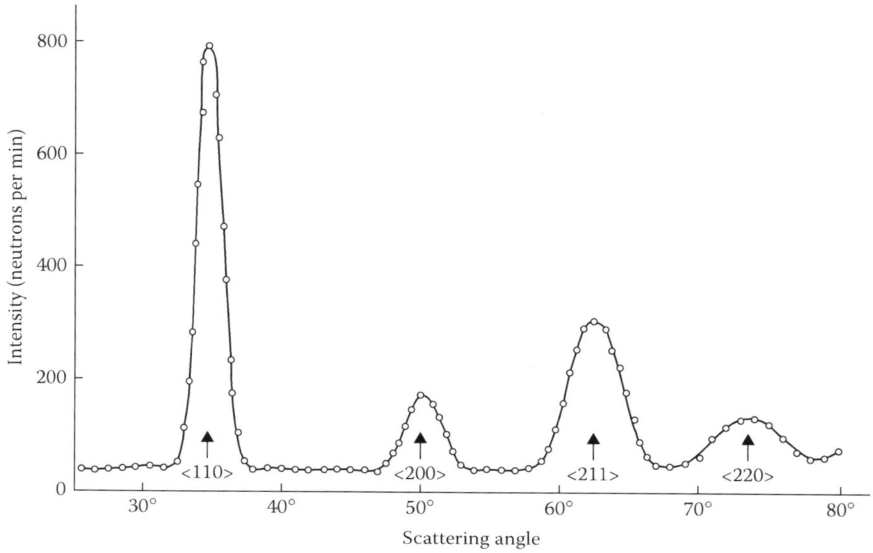

FIGURE 9.7 Neutron diffraction spectrum of iron showing the intensity of diffracted peaks as a function of scattering angle. The peaks have been indexed with the corresponding crystallographic directions. (After Shull, C.G. et al., *Phys. Rev.*, 84, 912, 1951.)

9.3.2 Elastic Neutron Scattering

What can elastic neutron scattering tell us about the magnetic structure of a material?

The idea of using neutrons for investigating directly the magnetic structure of materials was suggested first by Bloch [21] and later by Halpern and Johnson [22]. Elastic neutron scattering gives two types of diffraction peaks. One group is due to nuclear scattering and these are isotropic; that is, they do not depend on the scattering angle, and they also persist above the magnetic ordering temperature as shown Figure 9.8. The second group, the magnetic scattering spectrum, is caused by the presence of localized magnetic moments in the solid. This spectrum is anisotropic and also depends on temperature and the intensity of the peaks decreasing with increasing temperature until at the magnetic ordering temperature, the spectrum disappears to become a diffuse background due to paramagnetic scattering.

The cross sections for neutron-electron and neutron-nuclear scattering are usually comparable in magnitude, with either being larger depending on the particular case. This means that the spectral peaks due to Bragg diffraction and magnetic scattering are usually comparable in magnitude. The two contributions can however be distinguished, for example, by the application of a magnetic field (since the nuclear component remains isotropic) or by raising the temperature above the magnetic ordering temperature (whereupon the magnetic component becomes diffuse due to

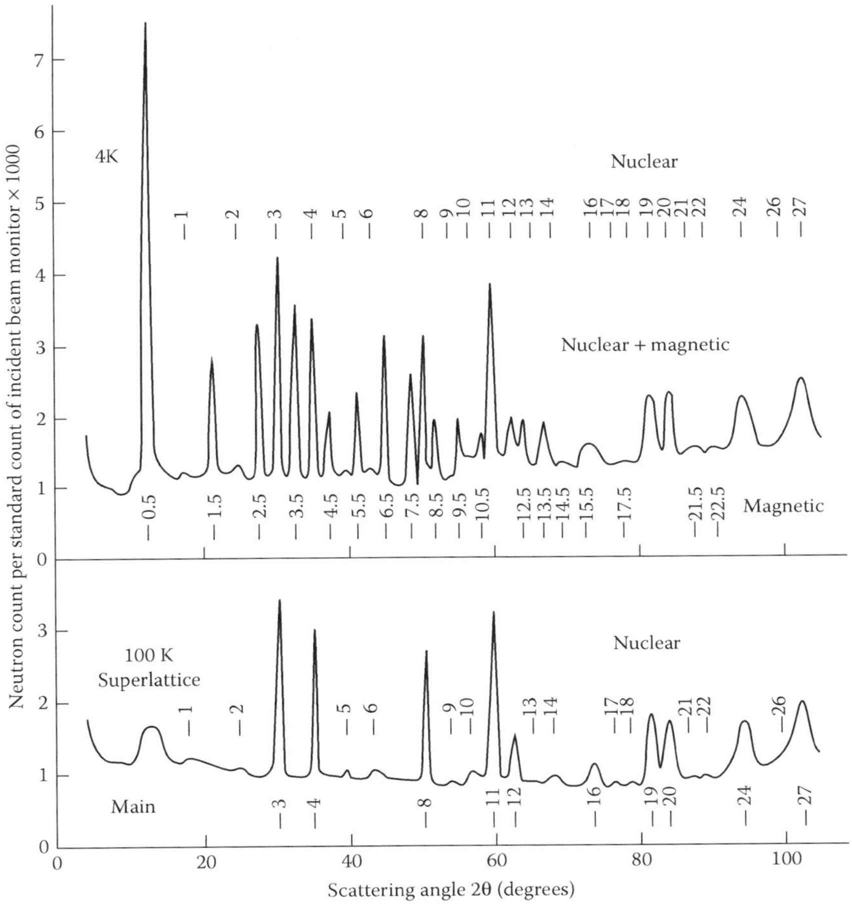

FIGURE 9.8 Neutron diffraction spectra for $TbIn_3$ above and below the magnetic ordering temperature. (After Crangle, J., *The Magnetic Properties of Solids*, Edward Arnold, London, 1977.)

paramagnetic scattering). Once the magnetic scattering contribution to the spectrum has been isolated, the distribution, direction, and ordering of the magnetic moments within the solid can be determined.

9.3.2.1 Paramagnetic Scattering

In the case of a paramagnet, the magnetic scattering is diffuse and appears as a contribution throughout the background, which decreases in intensity with increasing angle of scatter θ. Paramagnetic scattering is found by subtracting all other contributions from the spectrum, providing that the magnetic contribution is sufficiently large compared with the other contributions to render this calculation accurate enough.

9.3.2.2 Ferromagnetic Scattering

The Bragg and magnetic scattering spectra have peaks in the same locations when neutrons are diffracted by a ferromagnet. Therefore, the two components of the spectrum will be completely superposed as shown in Figure 9.9a. The magnetic contribution however decreases with temperature and therefore the two contributions can be distinguished by making measurements above and below the Curie point.

9.3.2.3 Simple Antiferromagnetic Scattering

In the case of simple antiferromagnetism, the magnetic moments on neighboring atoms point in opposite directions. This leads to a doubling of the crystallographic repeat distance for magnetic moments and hence to a halving of the repeat distance in the reciprocal lattice. Therefore, additional magnetic peaks appear midway between the nuclear scattering peaks in the spectrum as shown in Figure 9.9b. The magnetic peaks occur at sin θ values, which are half those of the expected nuclear diffraction peaks. The scattering of neutrons by antiferromagnets has been discussed by Bacon [17].

9.3.2.4 Helical Antiferromagnetic Scattering

In helimagnetic materials, the diffraction peaks consist of the central nuclear peaks each accompanied by a pair of satellite peaks due to the magnetic scattering as shown in Figure 9.10. The displacement of the satellite peaks from the nuclear peaks

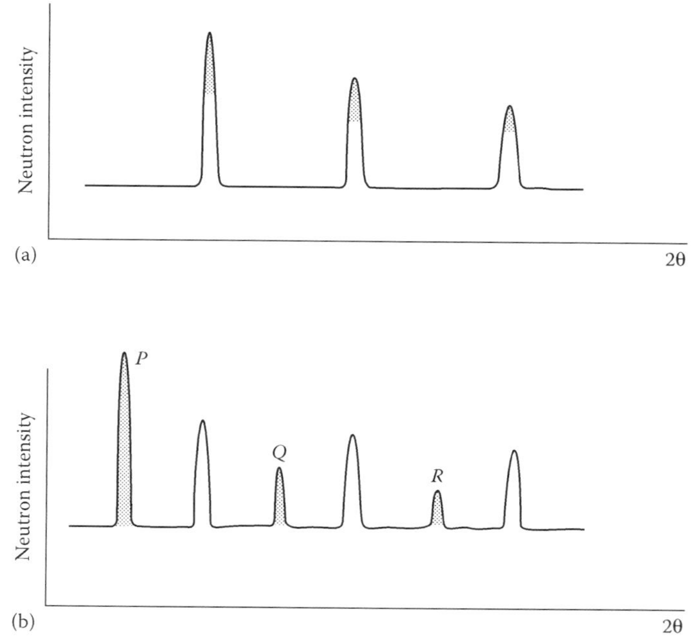

FIGURE 9.9 Neutron diffraction spectra: (a) for a ferromagnetic solid showing superposition of nuclear and magnetic peaks and (b) for a simple antiferromagnetic solid in which the nuclear and magnetic peaks occur at different locations. (After Bacon, G.E., *Neutron Diffraction*, 3rd edn. Clarendon Press, Oxford, 1975.)

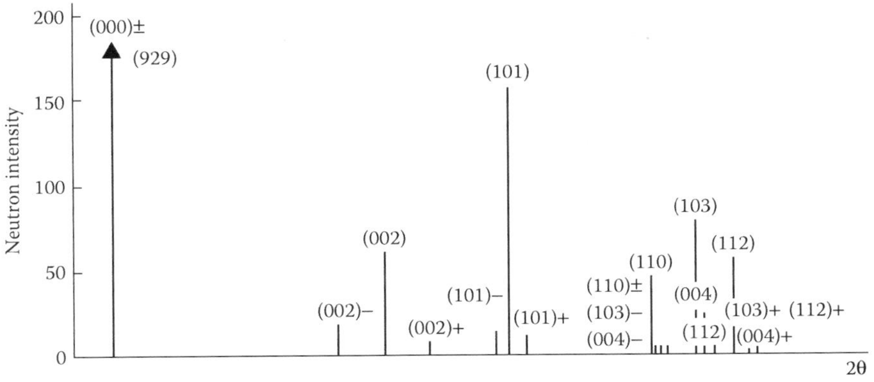

FIGURE 9.10 Neutron diffraction spectrum for a helical antiferromagnetic structure. (After Bacon, G.E., *Neutron Diffraction*, 3rd edn. Clarendon Press, Oxford, 1975.)

can be used to determine both the direction of the helical axis and the magnitude of the turn angle between successive helical planes.

The technique of neutron diffraction finds its greatest use in the investigation of antiferromagnetic (including helical antiferromagnetic) and ferrimagnetic structures. The case of ferromagnetic materials is fairly trivial by comparison since all moments lie parallel within a domain, and therefore, little further information can be obtained, since the crystallographic easy axes can be determined from anisotropic susceptibility measurements. However, information about the spin wave spectrum can be obtained from inelastic neutron scattering data on ferromagnets as they approach the Curie point.

9.3.3 Inelastic Neutron Scattering

How can we study higher energy states (magnetic excitations) in the magnetic structure?

The thermal fluctuations of the individual atomic magnetic moments in a magnetically ordered solid become prevalent as the magnetic ordering temperature is approached. These spin waves can be studied by inelastic neutron scattering. It was observed, for example, by Squires [24] that the scattering cross-section for iron has a peak at the Curie point, as shown in Figure 9.11, which is caused by inelastic scattering of neutrons by spin clusters. This scattering becomes greater as the temperature is increased toward the Curie temperature as the spin fluctuations become greater. A further example for nickel [25] is given in Figure 9.12.

A more detailed discussion of neutron diffraction by magnetic solids is beyond the intended scope of this book. Those interested in this area of magnetism are referred to the two-volume work by Lovesey [18], which provides a comprehensive summary of the subject.

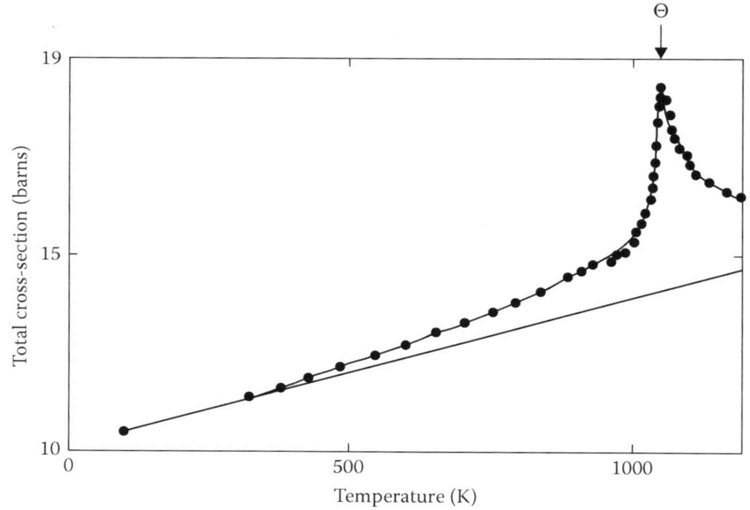

FIGURE 9.11 Total scattering cross-section for inelastic scattering of neutrons by iron as a function of temperature. (After Squires, G.L., *Proc. Phys. Soc. Lond. A*, 67, 248, 1954.)

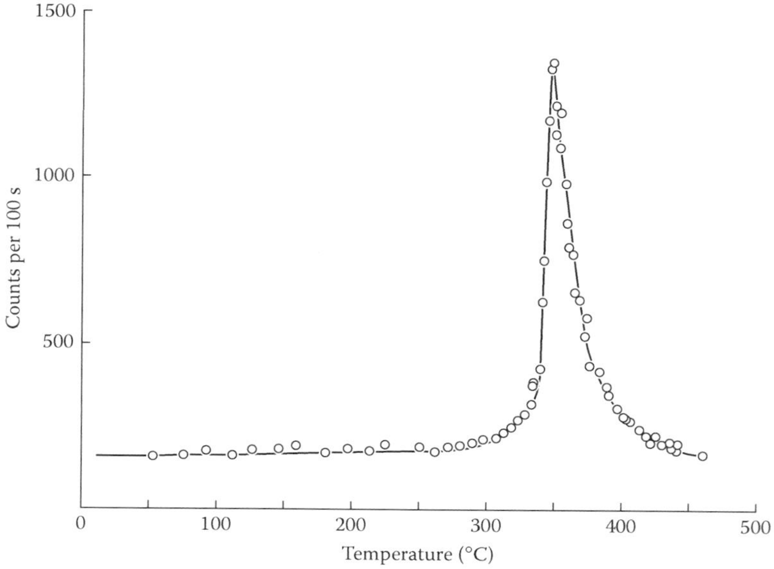

FIGURE 9.12 Magnetic scattering cross-section for inelastic scattering of neutrons by nickel as a function of temperature. (After Cribier, D. et al., *J. Phys. Soc. Jap.*, 17 (Suppl. BIII), 67, 1962.)

9.3.4 Magnetic Order in Various Solids

What examples do we have of these many different types of magnetic order?

For the cubic 3d transition metals such as iron and nickel, the magnetic moments align preferentially along the ⟨100⟩ and ⟨111⟩ axes, respectively [26,27]. The magnetization curves along the various crystal axes in iron and in nickel are shown in Figures 9.13 and 9.14, respectively. From these, it is clear that the initial and low-field susceptibility of iron is highest along the ⟨100⟩ axes, and in nickel, it is highest along the ⟨111⟩ axes. Conversely, it is more difficult to magnetize iron along the ⟨111⟩ axes and nickel along the ⟨100⟩ axes. Cobalt has a hexagonal crystal lattice. In this case, the moments are aligned along the unique [0001] axis, which is the easy direction [28], while the [1010] axis in the base plane is the hard axis. The magnetization curves along these directions are shown in Figure 9.15.

Simple antiparallel antiferromagnetism occurs in chromium and manganese and these ordered structures are shown in Figure 9.16 [29,30]. Antiferromagnetism is actually far more commonly occurring than ferromagnetism, but so far no widespread uses have been found for this type of magnetic order in solids except in magnetoresistive spin values.

In the rare earth metals, the magnetic ordering can be much more complex as shown in Figure 9.17. In gadolinium [31–34], which exhibits the simplest magnetic structure among the rare earth metals, the magnetic moments are aligned along the *c*-axis when the temperature lies between the Curie point of 293 K and the spin reorientation temperature of 240 K. Below 240 K, the moments deviate from

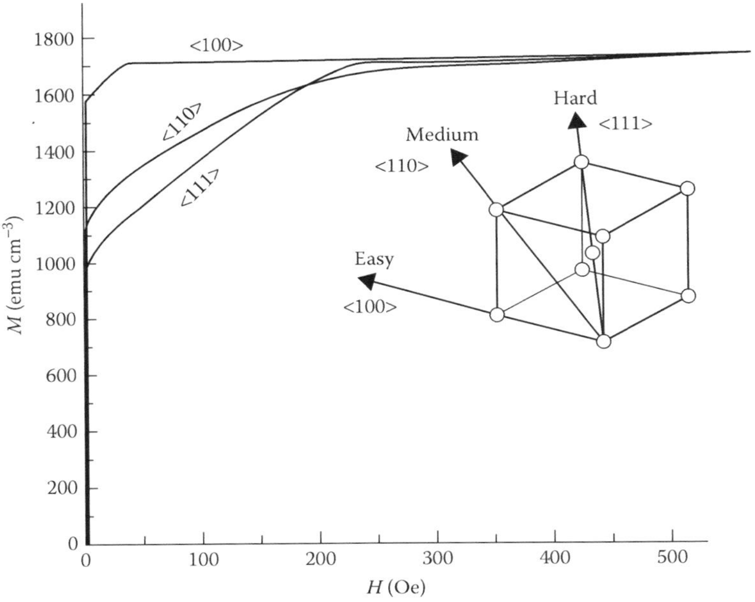

FIGURE 9.13 Magnetization curves for iron along the three axes ⟨100⟩, ⟨110⟩, and ⟨111⟩.

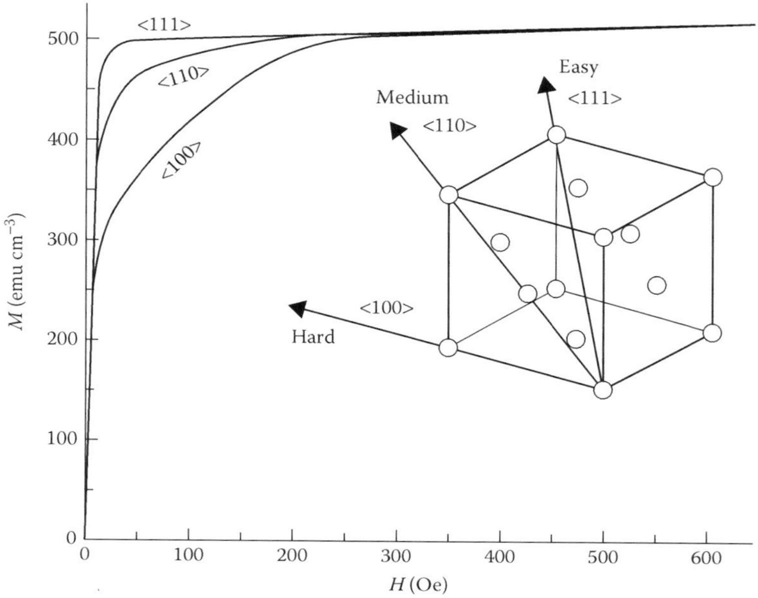

FIGURE 9.14 Magnetization curves for nickel along the three axes ⟨100⟩, ⟨110⟩, and ⟨111⟩.

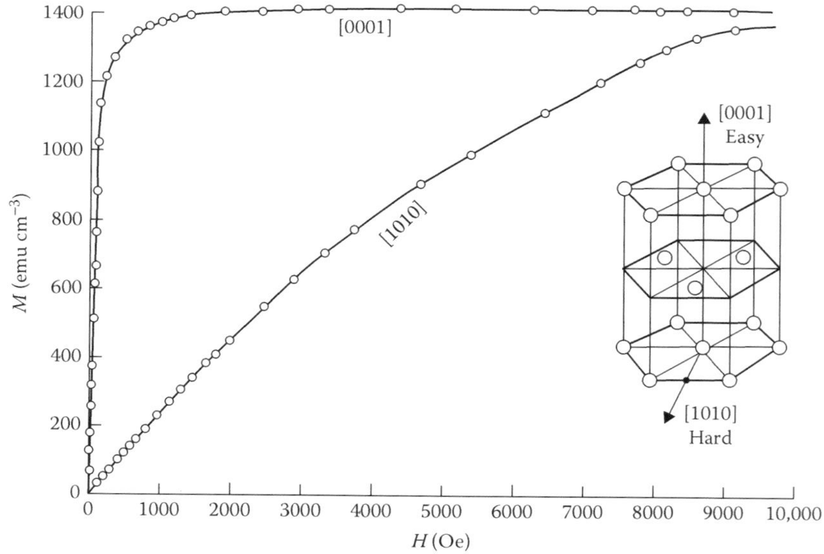

FIGURE 9.15 Magnetization curves for cobalt along the unique axis (0001) and the base plane [1010].

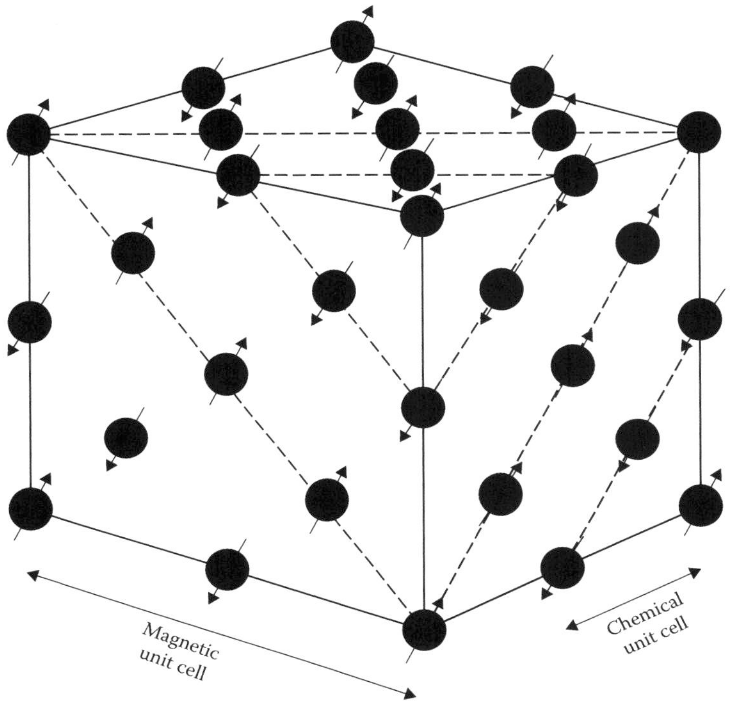

FIGURE 9.16 Simple antiferromagnetism on manganese atoms in manganese oxide.

the [001] axis and the angular deviation increases continuously as the temperature is reduced. In this temperature range, we speak of gadolinium as a *canted* ferromagnet.

Terbium [35] and dysprosium [36–38], the next two elements in the periodic table, behave very differently from gadolinium. Just below the Néel temperature, $T_N = 230$ K for terbium and $T_N = 180$ K for dysprosium, they both form base-plane helical antiferromagnets. In terbium, the easy axis is the [1010] direction (*b*-axis) while in dysprosium, it is the [1000] direction (*a*-axis). At a lower temperature of 219 K in terbium or 85 K in dysprosium, which are their Curie temperatures, they form base-plane ferromagnets. In terbium, the domains have their axes along the *b*-axes, while in dysprosium, the moments align along the *a*-axes.

Holmium has an ordering temperature of 132 K, below which it forms a base-plane antiferromagnet [39] that is similar to dysprosium, with an easy axis along the [1000] direction. This helical structure is retained down to 20 K, the Curie point, below which there appears a ferromagnetic *c*-axis component, while the base plane continues to exhibit helical antiferromagnetic order.

Erbium has a Néel temperature of 85 K, below which the magnetization is sinusoidally modulated along the *c*-axis [39–41]. At 53 K, a further transition occurs when the base plane orders helically, and the *c*-axis structure begins to change to

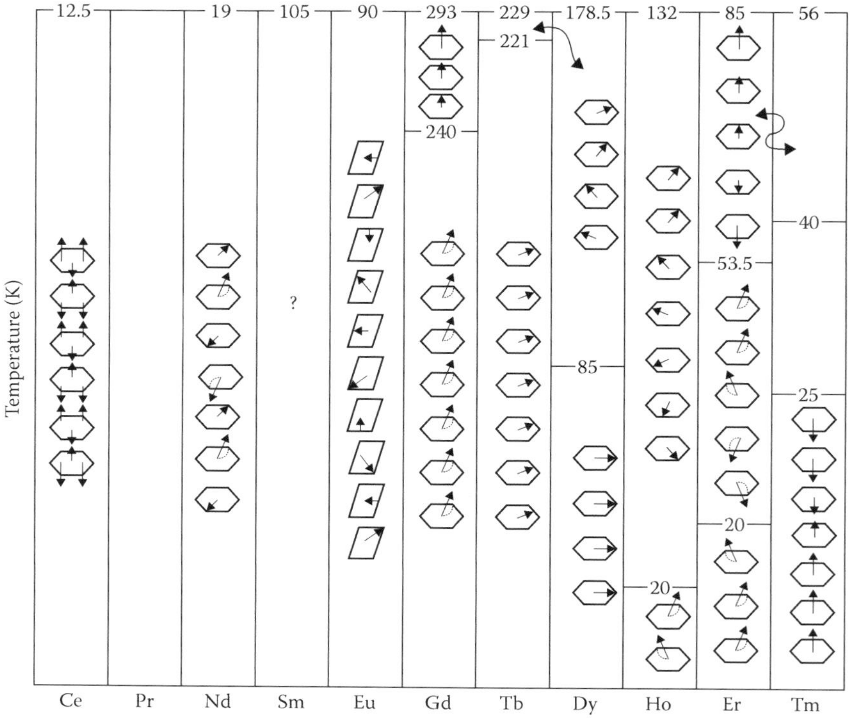

FIGURE 9.17 Magnetic ordering of magnetic moments within the domains of various rare earth metals under zero applied magnetic field.

a square-wave modulation of four moments up followed by four moments down. This square-wave modulation is never completed, however. At the Curie point of 20 K, the *c*-axis components become ferromagnetic, while the base plane continues its helical order. This leads to a *conical* ferromagnetic structure of the type observed also in holmium.

Ferrimagnetism is exhibited by a number of transition metal oxides known as *ferrites*. These fall into three classes: the cubic ferrites, also known as *spinels*, such as $NiO{\cdot}Fe_2O_3$, which are soft magnetic materials with the exception of cobalt ferrite, which is harder. The hexagonal ferrites such as $BaO{\cdot}6(Fe_2O_3)$, which are hard magnetic materials, are used to make ceramic magnets. The garnets such as yttrium-iron garnet $5Fe_2O_3{\cdot}3Y_2O_3$ are also ferrimagnetic.

9.3.5 Excited States and Spin Waves

How does temperature affect the alignment of magnetic moments within a domain?

We have noted in Section 6.2.5 that the spontaneous magnetization within a domain $\boldsymbol{M}_s$ is only equal to the saturation magnetization $\boldsymbol{M}_0 = Nm$ at absolute zero of

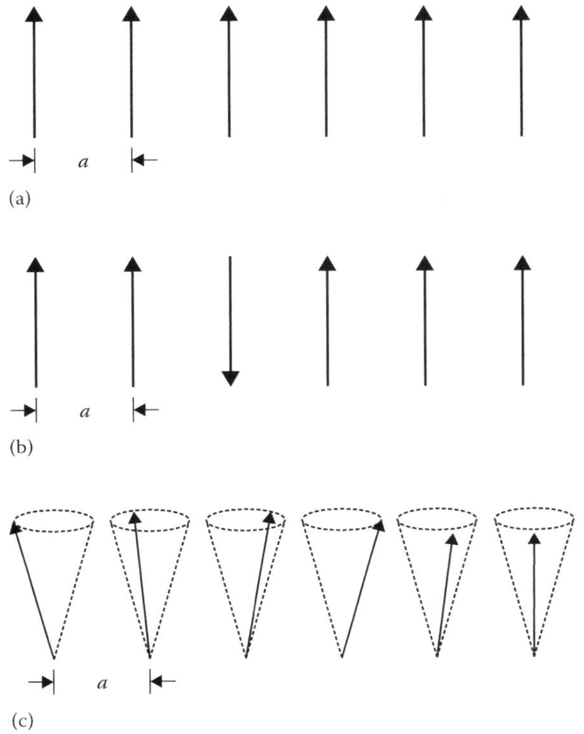

FIGURE 9.18 A six-moment linear ferromagnetic chain: (a) ground state; (b) a state with one moment antiparallel; and (c) a state with each successive moment at an angle 8 to its neighbors.

temperature (0 K). The reason for this is that the thermal energy at temperatures above 0 K causes some misalignment of the directions of magnetic moments within the domains. We shall now explain how this misalignment occurs.

Consider the ground state of a six-moment linear ferromagnet as shown in Figure 9.18a. The ground state occurs when all six moments are aligned parallel. On the basis of nearest-neighbor-only interaction, this has a ground state energy of E_0 given by

$$E_0 = -10\mu_0\vartheta\boldsymbol{m}^2 \tag{9.69}$$

There are several possible candidates for the lowest-level excited state. One possible candidate is the configuration with one moment pointing antiparallel to the rest as shown in Figure 9.18b. This will have an energy of

$$E_1 = -2\mu_0\vartheta\boldsymbol{m}^2 = E_0 + 8\mu_0\vartheta\boldsymbol{m}^2 \tag{9.70}$$

and in fact, for a linear chain with any number of moments greater than 2, the energy of the system with one moment antiparallel is always $8\mu_0\vartheta m^2$ above the ground state.

Another candidate excited state is one in which each moment is aligned at an angle θ to the direction of the previous moment as shown in Figure 9.18c. In this case, the energy of the system is

$$E(\theta) = -10\mu_0\vartheta \boldsymbol{m}^2\cos\theta = E_0 + 10\mu_0\vartheta \boldsymbol{m}^2(1-\cos\theta) \tag{9.71}$$

This allows for much lower-energy excited states, which in the classical approximation form a continuum of allowed energy states above the ground state, each with gradually increasing values of θ.

The spontaneous magnetization within a domain therefore decreases as the temperature increases above 0 K. It can be shown from an analysis of the spin wave structure [42] that

$$\boldsymbol{M}_s = \boldsymbol{M}_0\left(1-aT^{3/2}\right) = \boldsymbol{M}_0\left[1-c\left(\frac{k_BT}{\mu_0\vartheta \boldsymbol{m}^2}\right)^{3/2}\right] \tag{9.72}$$

where:

$\mu_0\vartheta m^2$ is the nearest-neighbor exchange energy

k_BT is the thermal energy

c is a constant, which has values of 0.118, 0.059, and 0.029 for a simple cubic lattice, for a body-centered cubic lattice, and for a face-centered cubic lattice, respectively

9.3.6 Critical Behavior at the Ordering Temperature

What happens to other bulk properties of a ferromagnet at the Curie temperature?

The bulk properties of magnetic materials show anomalous behavior in the vicinity of the transition temperatures such as the Curie and Néel points. The anomalous behavior is due to coupling between the particular bulk property and the magnetic structure. The effects are known as *critical phenomena*. We have already remarked that the bulk susceptibility has anomalous behavior close to T_c, for example. Other properties such as the specific heat, elastic moduli, magnetostriction, magnetoresistance, and thermal expansion all reveal critical behavior at the magnetic transition temperatures.

9.3.7 Susceptibility Anomalies

How does the susceptibility behave at the ordering temperature?

We have shown that the susceptibility of ferromagnets, ferrimagnets, and antiferromagnets behaves in an anomalous way as the temperature is reduced in the paramagnetic regime toward the critical temperature T_c. For materials that obey the Curie-Weiss law, this leads to a dependence of susceptibility on temperature of the form $\chi = C/(T - T_c)$ in the paramagnetic regime. The susceptibility therefore starts to become very large as T_c is approached. Work on the magnetization curves and susceptibility of a number of rare earth metals that obey the Curie-Weiss law was performed by Legvold, Spedding, and coworkers in a series of studies [43–47].

9.3.8 Specific Heat Anomalies

The specific heats of materials that undergo order-disorder transitions show lambda-type anomalies at the critical temperature. Some examples are shown in Figure 9.19 from the work of Hofmann et al. [48]. Measurements of the heat capacities of several heavy rare earth elements were made by Spedding et al. [49–53].

The specific heat of a magnetic material has a magnetic component C_m given by

$$C_m = \left(\frac{dE_{ex}}{dT} \right) \tag{9.73}$$

where E_{ex} is the exchange self-energy of the material per unit volume. In the mean field approximation,

$$E_{ex} = -\mu_0 \alpha \boldsymbol{M}_s^2 \tag{9.74}$$

$$C_m = -2\mu_0 \alpha \boldsymbol{M}_s \frac{d\boldsymbol{M}_s}{dT} \tag{9.75}$$

The dependence of $\boldsymbol{M}_s$ on temperature can be obtained from the equation for magnetization given in Section 9.2.3,

$$\frac{d\boldsymbol{M}_s}{dT} = \boldsymbol{M}_0 \frac{d}{dT}\left[\coth\left(\frac{\mu_0 \boldsymbol{m} \alpha \boldsymbol{M}_s}{k_B T} \right) - \frac{k_B T}{\mu_0 \boldsymbol{m} \alpha \boldsymbol{M}_s} \right] \tag{9.76}$$

The variation of the magnetic component of heat capacity C_m with temperature is plotted for nickel in Figure 9.20. The behavior gives a good fit to the experimental data for such a simple model. The experimental results show some broadening of the lambda anomaly in the paramagnetic region.

9.3.9 Elastic Constant Anomalies

How do the elastic properties behave at the ordering temperature?

The elastic constants of materials show critical behavior close to magnetic phase transitions such as the Curie point as shown in Figure 9.21. Magnetoelastic anomalies are known to occur in rare earth metals as a result of the very strong magnetoelastic coupling. These have been thoroughly investigated by Palmer and coworkers in a series of studies [54–60] and Moran and Luthi [61,62]. Some of the theoretical aspects have been addressed by Tachiki and Maekawa [63]. The dependence of elastic modulus on magnetization, the so-called ΔE effect in iron, is well known and has been reported by Bozorth [64].

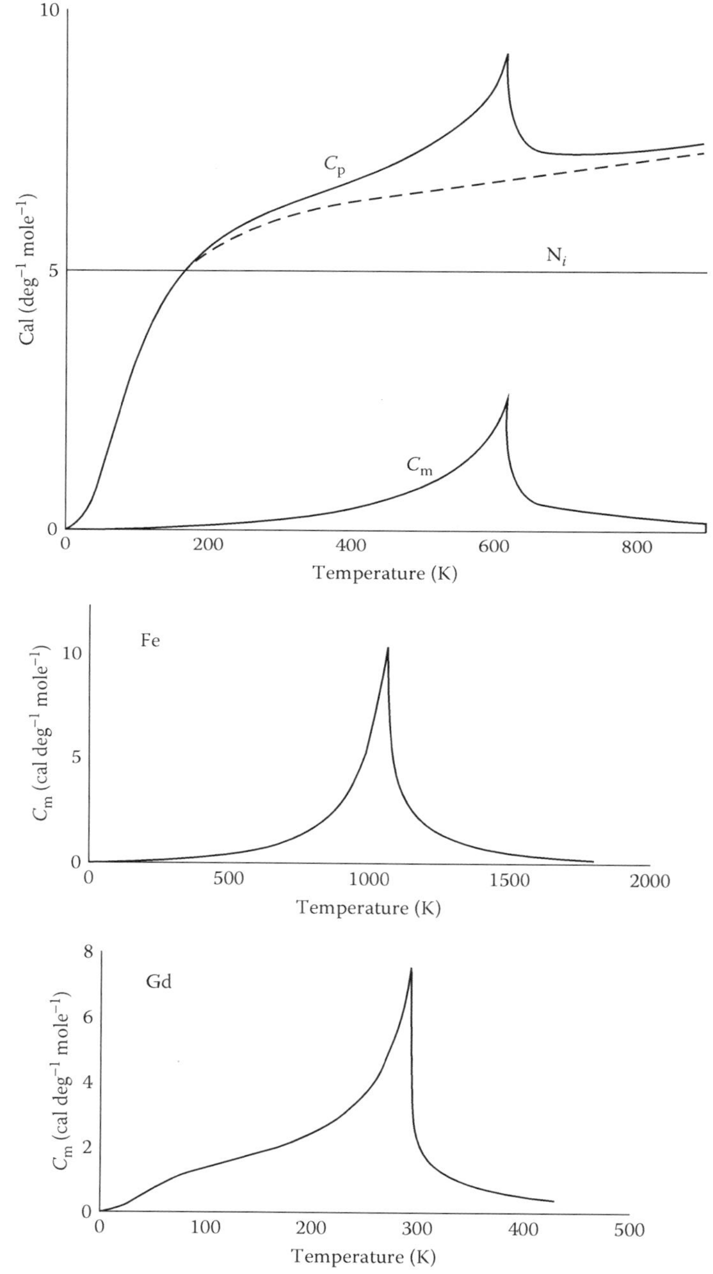

FIGURE 9.19 Examples of specific heat anomalies in nickel, iron, and gadolinium close to their ordering temperatures. (After Hofmann, J.A. et al., *Phys. Chem. Sol.*, 1, 1956.)

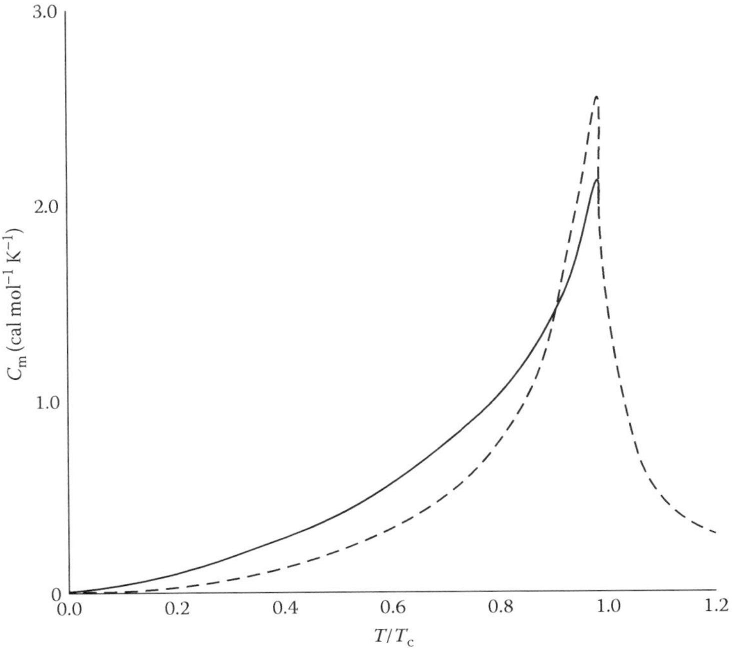

FIGURE 9.20 Specific heat anomaly for nickel at its Curie point compared with the theoretical prediction.

9.3.10 Thermal Expansion Anomalies

How does the thermal expansion behave at the ordering temperature?

Thermal expansion and magnetostriction also undergo unusual behavior at phase transitions such as at the Curie and Néel points as shown in Figure 9.22. This is because there is the sudden appearance of spontaneous magnetostriction at the order-disorder transition temperature. Anomalous thermal expansion and magnetostriction have been investigated by Bozorth and Wakiyama [65] and by Greenough and coworkers in a number of works [66–68].

9.3.11 Ising Model

Is there a thermodynamic model that can be used to describe the behavior of magnetic materials as a function of temperature and field?

The idea of a Weiss-type coupling between electronic moments can be incorporated into a simple classical model that describes the effects of temperature and magnetic field on the magnetization. The Ising model [69] provides a useful mathematical

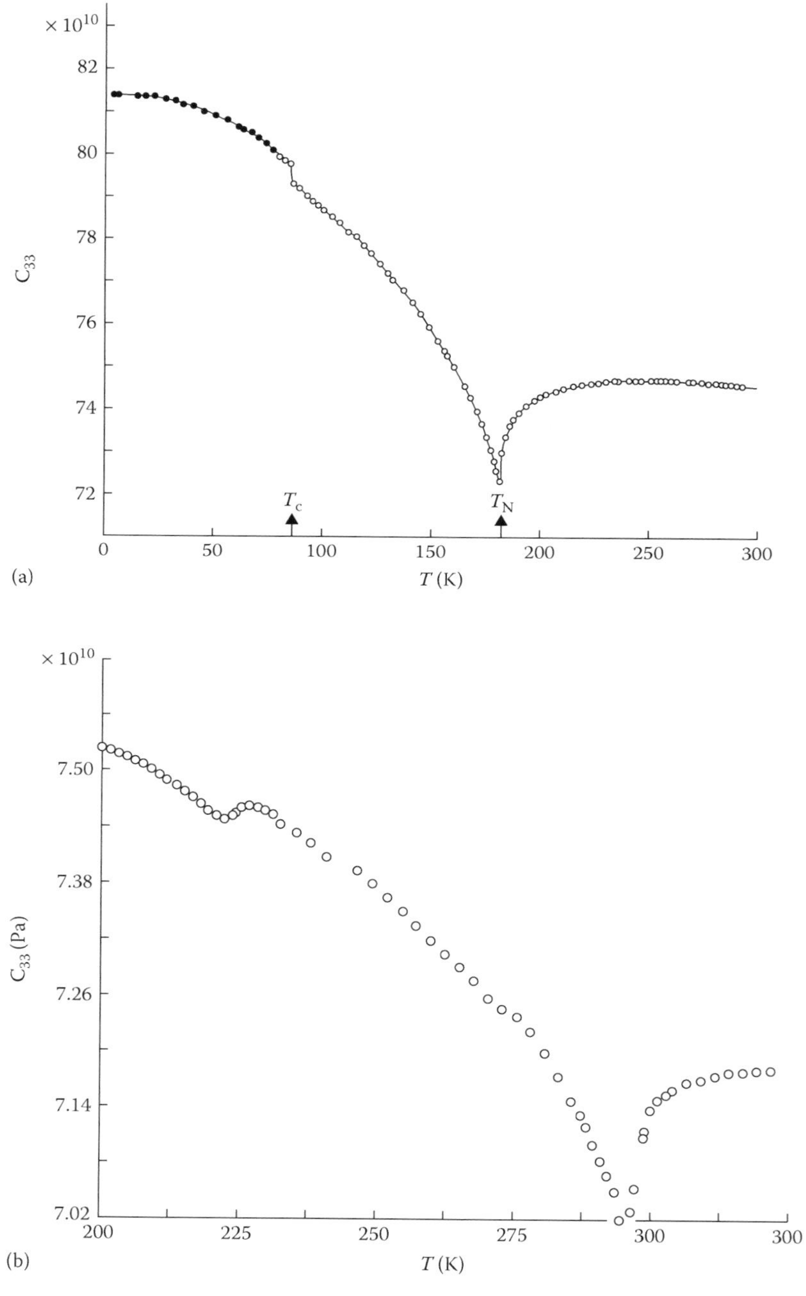

FIGURE 9.21 Critical behavior of the elastic constants of (a) dysprosium and (b) gadolinium.

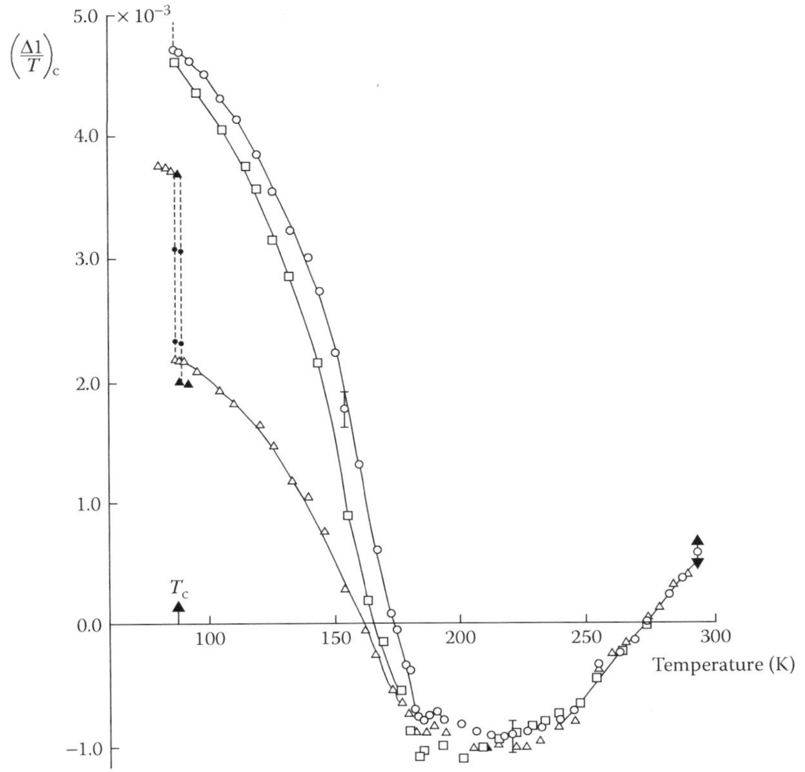

FIGURE 9.22 Anomalous behavior of the *c*-axis thermal expansion of dysprosium at its Néel temperature of 180 K and Curie temperature of 85 K.

description of critical phenomena and is particularly relevant to magnetic materials. It has the unique distinction of being the only model of a second-order phase transition that has so far yielded an analytic solution.

In the Ising model as applied to magnetic materials [70,71], the solid is divided into *cells*, each of which contains one magnetic moment or spin. Each cell is allowed a limited number of possible states and in the simplest models, only two states are allowed, which correspond to *spin up* and *spin down*. At first sight, this restriction may appear to be unrealistic, because classically the spins can point in any direction. But in quantum mechanics, the spins are restricted to certain directions and in particular the directions of the spins can be referenced parallel and antiparallel to an applied magnetic field $\boldsymbol{H}$ to which they are restricted in quantum mechanics.

The Ising model has two imposed restrictions and these are (1) that the spins have two states and (2) that there are interactions between the spins that correlate their orientations. The interactions are restricted to nearest-neighbor, or nearest-neighbor and second-nearest-neighbor, or else all moments interact equally, which is equivalent to the mean-field approximation.

One drawback however is that in the simple Ising model described here, the restriction on orientation of the spins does not allow spin waves. The lowest excited energy state that can be produced is therefore when one spin is antiparallel to the rest and consequently there exists an energy gap of $8\mu_0\vartheta m^2$ between the ground state and the first excited state.

As before, the magnetic interaction has the form

$$E_{\text{ex}} = -\mu_0 \sum \vartheta \boldsymbol{m}_i \cdot \boldsymbol{m}_j \tag{9.77}$$

We can define an order parameter for the Ising model η given by

$$\eta = \frac{p_+ - p_-}{p_+ + p_-} \tag{9.78}$$

where p_+ is the probability of spin up and p_- is the probability of spin down. Under zero applied field when $\eta = 0$, the system is paramagnetic and when $\eta \neq 0$, the system is ferromagnetic. A phase transition therefore occurs when η reaches zero. A mathematical treatment of the probable configuration of magnetic moments on the atomic sites in the two-dimensional Ising lattice has shown that an order-disorder transition occurs at a Curie temperature of

$$T_{\text{c}} = \frac{0.88E_{\text{ex}}}{k_{\text{B}}} = \frac{0.88\mu_0\vartheta \boldsymbol{m}^2}{k_{\text{B}}} \tag{9.79}$$

where E_{ex} is the nearest-neighbor coupling energy between the spins. Similarly, as shown above for the mean-field model, the specific heat has a lambda anomaly.

REFERENCES

1. Langevin, P. *Annates de Chem. et. Phys.*, 5, 70, 1905.
2. Van Vleck, J. H. *The Theory of Electric and Magnetic Susceptibilities*, Oxford University Press, Oxford, 1932.
3. Pauli, W. *Z. Phys.*, 41, 81, 1926.
4. Curie, P. *Ann. Chem. Phys.*, 5, 289, 1895.
5. Weiss, P. *Compt. Rend.*, 143, 1136, 1906.
6. Weiss, P. *J. Phys.*, 6, 661, 1907.
7. Weiss, P., and Forrer, R. *Ann. Phys.*, 12, 279, 1929.
8. Weiss, P. *Phys. Rev.*, 74, 1493, 1948.
9. Néel, L. *Annates de Phys.*, 18, 5, 1932.
10. Feldmann, D., Kirchmayr, H. R., Schmolz, A., and Velicescu, M. *IEEE Trans. Magn.*, 1, 61, 1971.
11. Néel, L. *Annates de Phys.*, 3, 137, 1948.
12. Enz, U. *Physica*, 26, 698, 1960.
13. Nicklow, R. M., Wakabayashi, N., Wilkinson, M. K., and Read, R. E. *Phys. Rev. Letts.*, 26, 140, 1971.
14. Ruderman, M. A., and Kittel, C. *Phys. Rev.*, 96, 99, 1954.

15. Kasuya, T. *Prog. Theor. Phys.*, 16, 45, 1956.
16. Yosida, K. *Phys. Rev.*, 106, 893, 1957.
17. Bacon, G. E. *Neutron Diffraction*, 3rd edn. Clarendon Press, Oxford, 1975, p. 208.
18. Lovesey, S. W. *Theory of Neutron Scattering from Condensed Matter, Vol. 2: Polarization Effects and Magnetic Scattering*, Clarendon Press, Oxford, 1984.
19. Koehler, W. C. *J. Appl. Phys.*, 36, 1078, 1965.
20. Shull, C. G., Wollan, E. O., and Koehler, W. C. *Phys. Rev.*, 84, 912, 1951.
21. Bloch, F. *Phys. Rev.*, 50, 259, 1936.
22. Halpern, O., and Johnson, M. H. *Phys. Rev.*, 55, 898, 1939.
23. Crangle, J. *The Magnetic Properties of Solids*, Edward Arnold, London, 1977.
24. Squires, G. L. *Proc. Phys. Soc. Lond. A*, 67, 248, 1954.
25. Cribier, D., Jacrot B., and Parette, G. *J. Phys. Soc. Jap.*, 17 (Suppl. BIII *Proceedings of the International Conference on Magnetism and Crystallography*, Kyoto, Japan), 67, 1962.
26. Honda, K., and Kaya, S. *S. Sci. Reps.* Tohuku Univ., 15, 721, 1926.
27. Kaya, S. *Sci. Reps.* Tohuku Univ., 17, 639, 1928.
28. Kaya, S. *Sci. Reps.* Tohuku Univ., 17, 1157, 1928.
29. Corliss, L. M., Hastings, J. M., and Weiss, R. J. *Phys. Rev. Letts.*, 3, 211, 1959.
30. Shull, C. G., and Wilkinson, M. K. *Revs. Mod. Phys.*, 25, 100, 1953.
31. Cable, J. W., and Wollan, E. O. *Phys. Rev.*, 165, 733, 1968.
32. Graham, C. D. *J. Appl. Phys.*, 34, 1341, 1963.
33. Corner, W. D., Roe, W. C., and Taylor, K. N. R. *Proc. Phys. Soc.*, 80, 927, 1962.
34. Corner, W. D., and Tanner, B. K. *J. Phys. C.*, 9, 627, 1976.
35. Koehler, W. C., Child, H. R., Wollan, E. O., and Cable, J. W. *J. Appl. Phys.*, 34, 1335, 1963.
36. Bly, P. H., Corner, W. D., and Taylor, K. N. R. *J. Appl. Phys.*, 40, 4787, 1969.
37. Landry, P. C. *Phys. Rev.*, 156, 578, 1967.
38. Wilkinson, M. K., Koehler, W. C., Wollan, E. O., and Cable, J. W. *J. Appl. Phys.*, 32, 48S, 1961.
39. Koehler, W. C., Cable, J. W., Wollan, E. O., and Wilkinson, M. K. *J. Phys. Soc. Jap.*, 17, (Suppl. BIII *Proceedings of the International Conference on Magnetism and Crystallography*, Kyoto, Japan), 32, 1962.
40. Rhyne, J. J., Foner, S., McNiff, E. J., and Doclo, R. *J. Appl. Phys.*, 39, 892, 1968.
41. Cable, J. W., Wollan, E. O., Koehler, W. C., and Wilkinson, M. K. *J. Appl. Phys.*, 32, 49S, 1961.
42. Bloch, F. *Z. Phys.*, 61, 206, 1930.
43. Nigh, H. E., Legvold, S., and Spedding, F. H. *Phys. Rev.*, 132, 1092, 1963.
44. Hegland, D. E., Legvold, S., and Spedding, F. H. *Phys. Rev.*, 131, 158, 1963.
45. Behrendt, D. R., Legvold, S., and Spedding, F. H. *Phys. Rev.*, 109, 1544, 1958.
46. Strandburg, D. L., Legvold, S., and Spedding, F. H. *Phys. Rev.*, 127, 1962.
47. Green, R. W., Legvold, S., and Spedding, F. H. *Phys. Rev.*, 122, 827, 1961.
48. Hofmann, J. A., Paskin, A., Tauer, K. J., and Weiss, R. J. *Phys. Chem. Sol.*, 1, 1956.
49. Jennings, L. D., Stanton, R. M., and Spedding, F. H. *J. Chem. Phys.*, 27, 909, 1957.
50. Skochdopole, R. E., Griffel, M., and Spedding, F. H. *J. Chem. Phys.*, 23, 2258, 1955.
51. Griffel, M., Skochdopole, R. E., and Spedding, F. H. *Phys. Rev.*, 93, 657, 1954.
52. Griffel, M., Skochdopole, R. E., and Spedding, F. H. *Chem. Phys.*, 25, 75, 1956.
53. Gerstein, B. C., Griffel, M., Jennings, L. D., Miller, R. E., Skochdopole, R. E., and Spedding, F. H. *J. Chem. Phys.*, 21, 394, 1957.
54. Palmer, S. B., and Lee, E. W. *Proc. Roy. Soc. A*, 327, 519, 1972.
55. Palmer, S. B., Lee, E. W., and Islam, M. N. *Proc. Roy. Soc. A*, 338, 341, 1974.
56. Isci, C., and Palmer, S. B. *J. Phys. Chem. Sol.*, 38, 1253, 1977.
57. Isci, C., and Palmer, S. B. *J. Phys. F.*, 8, 247, 1978.
58. Jiles, D. C., Blackie, G. N., and Palmer, S. B. *J. Magn. Magn. Mater.*, 24, 75, 1981.

59. Jiles, D. C., and Palmer, S. B. *J. Phys. F.*, 10, 2857, 1980.
60. Jiles, D. C., and Palmer, S. B. *J. Phys. F.*, 11, 45, 1981.
61. Luthi, B., Moran, T. J., and Pollina, R. J. *J. Phys. Chem. Soc.*, 31, 1741, 1970.
62. Moran, T. J., and Luthi, B. *J. Phys. Chem. Sol.*, 31, 1735, 1970.
63. Tachiki, M., and Maekawa, S. *Prog. Theor. Phys. Jap.*, 51, 1, 1974.
64. Bozorth, R. M. *Ferromagnetism*, Van Nostrand, New York, 1951.
65. Bozorth, R. M., and Wakiyama, T. *J. Phys. Soc. Jap.*, 18, 97, 1963.
66. Greenough, R. D., Isci, C., and Palmer, S. B. *Physica*, 86–88B, 61, 1977.
67. Greenough, R. D., and Isci, C. *J. Magn. Magn. Mater.*, 8, 43, 1978.
68. Greenough, R. D. *J. Phys. C.*, 12, 1113, 1979.
69. Ising, E. *Z. Phys.*, 31, 253, 1925.
70. Stanley, H. E. *Introduction to Phase Transition and Critical Phenomena*, Clarendon Press, Oxford, 1971.
71. Green, H. S., and Hurst, C. A. *Order-Disorder Phenomena*, Wiley Interscience, London, 1964.

FURTHER READING

Bacon, G. E. *Neutron Diffraction*, 3rd edn, Clarendon Press, Oxford, 1975.
Chikazumi, S. *Physics of Magnetism*, Wiley, New York, 1964, Chs. 4 and 5.
Coqblin, B. *The Electronic Structure of Rare Earth Solids*, Wiley, New York, 1977.
Craik, D. J. *Magnetism: Principles and Applications*, Wiley, Chichester, West Sussex, 1995.
Cullity, B. D. *Introduction to Magnetic Materials*, Addison-Wesley, Reading, MA, 1977, Chs. 3, 4, and 5.
Elliott, R. J. *Magnetic Properties of Rare Earth Metals*, Plenum, London, 1972.
Lovesey, S. W. *Theory of Neutron Scattering from Condensed Matter*, Clarendon Press, Oxford, 1984.
Mackintosh, A. R. The magnetism of rare earth metals, *Physics Today*, June, 23, 1977.

EXERCISES

9.1 *Paramagnetic susceptibility of oxygen.* Calculate the susceptibility at 0°C of a paramagnetic gas of molecular mass 32 with a magnetic moment of 3 Bohr magnetons per molecule ($S = 3/2$).

9.2 *Magnetic mean interaction field for iron.* Derive the relationship between the Weiss *molecular* field of a ferromagnet and the Curie temperature. Calculate the value of this field for iron that has a Curie temperature of 770°C and an effective magnetic moment per atom of 2.2 Bohr magnetons.

9.3 *Critical behavior of spontaneous magnetization.* In the mean-field approximation for a system with two possible microstates (e.g., magnetic moments either parallel or antiparallel to a unique axis) the spontaneous magnetization within a domain is given by $\boldsymbol{M}_s = \boldsymbol{M}_0 \tanh (\mu_0 m\alpha \boldsymbol{M}_s/k_B T)$. Show that at just below the Curie point the spontaneous magnetization varies with $\sqrt{(T_c - T)}$.

9.4 *Spontaneous magnetization.* Estimate the temperature at which the spontaneous magnetization within the domains of a piece of nickel, and hence its technical saturation magnetization, is (1) 90% and (2) 10% of its value at absolute zero. Calculate the temperature at which the thermal energy of the atoms exactly equals the Weiss mean field (or exchange) coupling energy.

9.5 *Magnetic heat capacity.* Derive an expression for the magnetic contribution C_m to the specific heat capacity of a ferromagnet as a function of temperature. You can use the mean-field approximation and assume only one-dimensional alignment of magnetic moments, so that $\boldsymbol{M}_s = Nm \tanh(\mu_0 m \alpha \boldsymbol{M}_s / k_B T)$ within a domain. Show that C_m has an anomaly at the Curie temperature and sketch the dependence of C_m on temperature.

9.6 A ferromagnet has a Curie temperature of 631 K, a magnetic moment per atom of 0.6 Bohr magnetons, and at 505 K, it has $dm/dT = 1.0 \times 10^{-26}$ A m^2 K^{-1} (or $d\boldsymbol{M}_s/dr = 908$ A m^{-1} K^{-1}). Calculate the magnetic contribution to the specific heat capacity at this temperature in J m^{-3} K^{-1}.

9.7 *Classical theory of paramagnetism.* Explain the differences between paramagnetism and ferromagnetism on the scale of a few atoms in terms of the behavior of the magnetic moments on the atoms. Using the Langevin theory of paramagnetism in three dimensions, plot a graph of the variation of magnetization $\boldsymbol{M}$ versus $\boldsymbol{H}/T$ (magnetic field divided by temperature) for a paramagnetic material with a density of 8000 kg m^{-3}, a relative atomic mass of 56, and a magnetic moment per atom of 2 Bohr magnetons.

9.8 *Magnetization equation for a uniaxial magnet.* Assuming that the magnetic moment on any atom is constrained to lie either parallel or antiparallel to the magnetic field $\boldsymbol{H}$, show that the equation for the probability of an individual magnetic moment $\boldsymbol{m}$ having an energy E in a magnetic field $\boldsymbol{H}$ is $p = \exp(-E/k_B T)/[\exp(-E/k_B T) + \exp(E/k_B T)]$, and from this derive the equation for the bulk magnetization $\boldsymbol{M}$ as a function of field $\boldsymbol{H}$ and temperature T for such a one-dimensional system.

9.9 *Classical theory of ferromagnetism.* The Langevin-Weiss classical theory of ferromagnetism can be considered as a perturbation of paramagnetism. Explain this and show that the solution of the Langevin-Weiss equation for magnetization within a domain in the absence of an applied magnetic field leads to a spontaneous magnetization at temperature below T_c. What happens above T_c?

9.10 *Calculation of spontaneous magnetization within a domain.* Calculate the spontaneous magnetization at room temperature of 20°C in a domain of a magnetic material with Curie temperature 358°C, a magnetic moment of 0.6 Bohr magneton per atom, and 9×10^{28} atoms m^{-3}.

10 Electronic Magnetic Moments

In this chapter, we discuss the prime cause of the magnetic moment within an individual atom. We will look at the properties of electrons, which are of central importance to magnetism and in particular, the origin of the magnetic moment of the electron, which is a result of its angular momentum. We also look at how the magnetic properties of the electrons lead to differences in the available energy states in the presence of a magnetic field. Finally, we show how the magnetic moments of the electrons are combined to give the magnetic moment of the atom.

10.1 CLASSICAL MODEL OF MAGNETIC MOMENTS OF ELECTRONS

Why do electrons have a magnetic moment?

In the classical model, the angular momentum of the electrons can be used to determine the magnetic moments of the electrons by invoking the concept of electrical charge in motion. It is known from Chapter 1, for example, that a current loop behaves as a magnetic dipole and has an associated magnetic moment. There are two contributions to the electronic magnetic moment: an orbital magnetic moment due to orbital angular momentum and a spin magnetic moment due to electron spin.

10.1.1 Electron Orbital Magnetic Moment

How does the angular momentum of an electron lead to a net magnetic moment?

We can envisage an electron moving in an orbit about an atomic nucleus, as shown in Figure 10.1, with orbital area A and period τ. This would then be equivalent to a current i given by

$$i = \frac{e}{\tau} \tag{10.1}$$

From the earlier definitions, in Sections 1.2.6 and 9.1.2, this gives an orbital magnetic moment $\boldsymbol{m}_0$

$$\boldsymbol{m}_0 = iA = -\frac{eA}{\tau} \tag{10.2}$$

The angular momentum of such an orbital $\boldsymbol{p}_0$ will be $\boldsymbol{p}_0 = I\omega$, where I is the moment of inertia of a particle of mass m_e orbiting the nucleus and ω is the angular velocity. $I = m_e r^2$, so that $\boldsymbol{p}_o = m_e r^2 \omega$ and hence

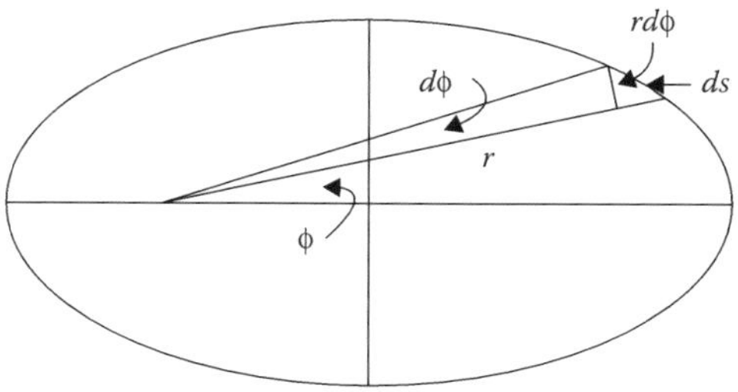

FIGURE 10.1 Classical model of an electron in an elliptical orbit around a nucleus.

$$\boldsymbol{p}_{\mathrm{o}} = m_{\mathrm{e}} r^2 \frac{d\phi}{dt} \tag{10.3}$$

where:
- m_{e} is the mass of an electron
- r is the radius

The area of the orbit is then

$$A = \left(\frac{1}{2}\right) \boldsymbol{p}_{\mathrm{o}} \frac{\tau}{m_{\mathrm{e}}} \tag{10.4}$$

We can therefore write down the orbital magnetic moment $\boldsymbol{m}_{\mathrm{o}}$ of the electron in terms of the orbital angular momentum $\boldsymbol{p}_{\mathrm{o}}$ as

$$\boldsymbol{m}_{\mathrm{o}} = -\left(\frac{e}{2m_e}\right) \boldsymbol{p}_{\mathrm{o}} \tag{10.5}$$

remembering that the magnetic moment vector points in the opposite direction to the angular momentum vector because the electron has negative charge.

10.1.2 Electron Spin Magnetic Moment

How does the electron spin contribute to the magnetic moment?

The electronic spin angular momentum $\boldsymbol{p}_{\mathrm{s}}$ also generates a spin magnetic moment $\boldsymbol{m}_{\mathrm{s}}$. In this case the relation is

$$\boldsymbol{m}_{\mathrm{s}} = -\frac{e\boldsymbol{p}_{\mathrm{s}}}{m_{\mathrm{e}}} \tag{10.6}$$

Notice that for a given angular momentum, the spin gives twice the magnetic moment of the orbit. There is no reason for the relationship between magnetic moment and angular momentum to be the same for spin and orbital components. In fact, the relationship in each case depends on the distribution of the charge. Interestingly, the result of Equation 10.6 can be obtained classically from a rotating disk of mass m_{e} with the charge residing on the circumference. The spin angular momentum is then $\boldsymbol{p}_{\mathrm{s}} = I\omega$, and $I = (1/2)m_{\mathrm{e}}r^2$, so $\boldsymbol{p}_{\mathrm{s}} = (1/2)m_{\mathrm{e}}r^2\omega$ and the magnetic moment is

$$\boldsymbol{m}_{\mathrm{s}} = iA = -\frac{e\omega r^2}{2} = -\frac{e\boldsymbol{p}_{\mathrm{s}}}{m_{\mathrm{e}}} \tag{10.7}$$

10.1.3 Total Electronic Magnetic Moment

How can we combine the spin and orbital angular momentum contributions to get the total electron magnetic moment?

If we consider the total magnetic moment per electron as the vector sum of the orbital and spin magnetic moments, then

$$\boldsymbol{m}_{\mathrm{tot}} = \boldsymbol{m}_{\mathrm{s}} + \boldsymbol{m}_{\mathrm{o}} = -\left(\frac{e}{2m_{\mathrm{e}}}\right)2\boldsymbol{p}_{\mathrm{s}} - \left(\frac{e}{2m_{\mathrm{e}}}\right)\boldsymbol{p}_{\mathrm{o}} \tag{10.8}$$

These terms on the right-hand side can then be combined to give

$$\boldsymbol{m}_{\mathrm{tot}} = -g\left(\frac{e}{2m_{\mathrm{e}}}\right)\boldsymbol{p}_{\mathrm{tot}} \tag{10.9}$$

where now $\boldsymbol{p}_{\mathrm{tot}}$ is the total angular momentum of the electron. The term $-e/2m_{\mathrm{e}}$ is the gyromagnetic ratio, which we already encountered in Section 3.3.6, and the term g is called the *Lande splitting factor*, which has a value of $g = 2$ for spin-only components and $g = 1$ for orbital-only components of magnetic moment. The value of the splitting factor g must therefore change depending on the relative sizes of the contributions from the spin and orbit to the total angular momentum.

Although this equation relating the angular momentum to the magnetic moment of an electron works quite well, allowing for a variable splitting factor, the classical interpretation of the magnetic moment of an electron in terms of a spinning charge leads to unreasonable tangential velocities. This can be shown by a simple calculation using the spin magnetic moment of 1 Bohr magneton, the electronic charge of 1.6×10^{-19} C, and a classical electron radius of 2.8×10^{-15} m. Therefore, the classical interpretation in terms of an effective or *Ampèrian* current has to be abandoned in favor of a quantum mechanical approach. This means that the description of the spin magnetic moment of an electron marks a boundary between the classical and quantum descriptions of magnetic moment, beyond which the classical description cannot be used without encountering some contradictions.

10.2 QUANTUM MECHANICAL MODEL OF MAGNETIC MOMENTS OF ELECTRONS

How are the above definitions modified as a result of quantum mechanics?

We have shown above how the angular momentum of electrons leads to a net magnetic moment. However, in the discussion, we have so far allowed all values of $\boldsymbol{p}_{\mathrm{o}}$ and $\boldsymbol{p}_{\mathrm{s}}$. In fact, this is not realistic, the possible values of angular momentum of electron are restricted by quantum mechanics, and consequently the magnetic moments are also quantized. We begin by defining the quantum numbers needed to fully describe electrons within atoms, then look at the restrictions on magnetic moment, and finally look at experimental evidence confirming the quantization of magnetic moments, specifically the Zeeman effect and the Stern-Gerlach experiment.

We need to use four quantum numbers in order to define uniquely each electron in an atom. These are the principal quantum number defined by Bohr, the angular momentum quantum number l defined by Sommerfeld, the orbital magnetic quantum number $\boldsymbol{m}_l$, and the spin magnetic quantum number m_s.

10.2.1 Principal Quantum Number *n*

What energies can electrons have within the atom?

This n is the quantum number introduced by Bohr in his theory of the atom [1]. It determines the shell of the electron, and hence its energy. In Bohr's original formulation, it also defined the orbital angular momentum but in later models several different orbitals with different angular momenta, but degenerate energy, were found to exist for a given n as shown in Figure 10.2.

The energy E_n of an electron with principal quantum number n is then

$$E_n = -\frac{Z^2 m_e e^4}{8n^2 h^2 \varepsilon_0^2} \tag{10.10}$$

where

$$n = 1, 2, 3, \ldots \tag{10.11}$$

Z is the atomic number, m_e the mass of the electron, e the electronic charge, and ε_0 the permittivity of free space. The maximum number of electrons permitted in the nth shell is $2n^2$.

10.2.2 Orbital Angular Momentum Quantum Number *l*

What values of orbital angular momentum are possible for electrons within the atom?

This l is synonymous with the quantum number k introduced in the Sommerfeld theory of the atom [2] except that $l = k - 1$. It was needed to define the orbital angular momentum $\boldsymbol{p}_o$ of the electron once it was realized that one energy level could have more than one allowed value of angular momentum. It is a measure of the orbital angular momentum of the electron when multiplied by $(h/2\pi)$. In a classical sense, it gives a measure of the eccentricity of the electron orbit.

$$\boldsymbol{p}_o = l\left(\frac{h}{2\pi}\right) \tag{10.12}$$

can have a value of $l = 0, 1, 2, 3, \ldots, (n - 1)$, where n is the principal quantum number.

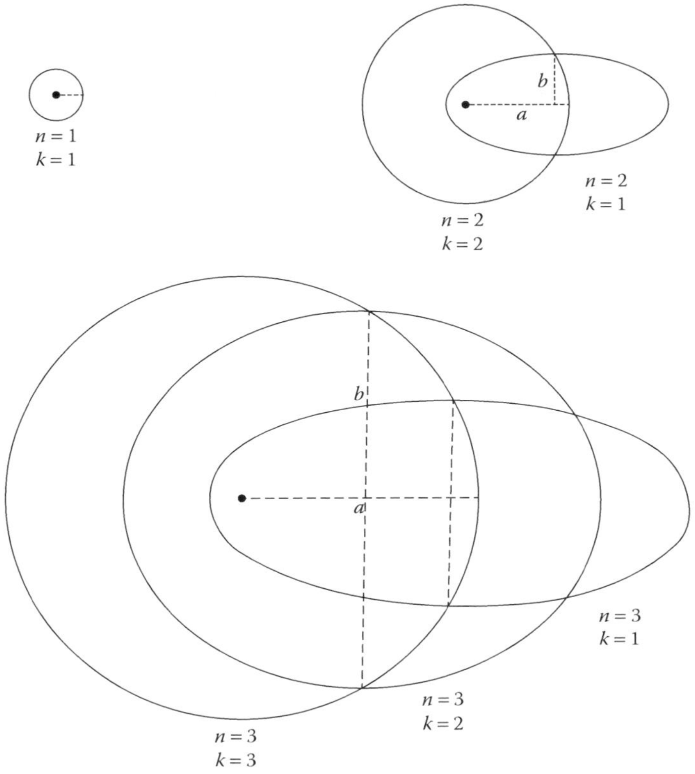

FIGURE 10.2 Electron orbitals with the same energy but with different angular momentum described in the Bohr-Sommerfeld theory.

10.2.3 Spin Quantum Number s

What values of spin angular momentum can electrons have within the atom?

Electrons have spin angular momentum, which can be represented by the spin quantum number s. The value of s is always 1/2 for an electron. The angular momentum due to spin $sh/2\pi$ is therefore always a multiple of $h/4\pi$.

$$\boldsymbol{p}_s = s\left(\frac{h}{2\pi}\right) \tag{10.13}$$

The total angular momentum quantum number j of an electron is not an independent quantum number since it is determined by l and s. However, it is a useful quantity to define, particularly since it gives a measure of the total magnetic moment of an electron. It is the vector sum of the spin and orbital angular momenta and is necessarily quantized.

$$p_j = j\left(\frac{h}{2\pi}\right) = (l+s)\left(\frac{h}{2\pi}\right) \quad (10.14)$$

10.2.4 Magnetic Quantum Numbers m_1 and m_s

What orientations can the spin and orbit angular momentum vectors of an electron take when subjected to a magnetic field?

We can consider that the angular momentum vector precesses about the direction of an applied field as shown in Figure 10.3. This was predicted by Larmor's theorem [3], which states that the effect of a magnetic field on an electron moving in an orbit is to impose on the electron a precession motion about the direction of the magnetic field with an angular velocity given classically by $\omega = \mu_0(e/2m_e)\boldsymbol{H}$. The component of the angular momentum along the field axis is $\boldsymbol{m}_0\cos\theta$, where θ is the angle of precession. The component perpendicular to this direction is time averaged to zero.

The component of orbital angular momentum along the axis of a magnetic field is restricted to discrete values by quantum mechanics. The magnetic quantum number $\boldsymbol{m}_l$, which represents these discrete values, arises from the solution of Schroedinger's equation for a single electron atom. It gives the component l_z of the orbital angular momentum l along the z-axis of a coordinate system, or if you prefer, it gives the orientation of the elliptical orbit.

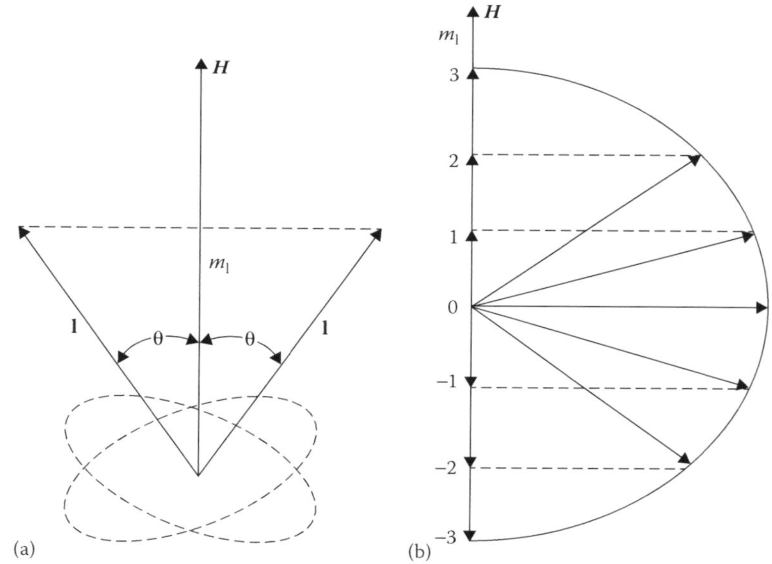

FIGURE 10.3 (a) Precession of the orbital angular momentum vector l about the magnetic field axis and (b) quantization of the allowed projection of angular momentum l along the field direction.

$$l_z = \boldsymbol{m}_l\left(\frac{h}{2\pi}\right) \tag{10.15}$$

where the z-axis is defined as the axis perpendicular to the plane in which the electron orbit lies.

The values of $\boldsymbol{m}_l$ are restricted to $\boldsymbol{m}_l = -l, \ldots, -2, -1, 0, 1, 2, \ldots, +l$. The physical significance of this number is that when a magnetic field is applied to an atom the electron's orbital angular momentum l can only have certain values of the component parallel to the magnetic field direction, and these are given by the magnetic quantum number $\boldsymbol{m}_l$.

Electron spins are constrained to lie either parallel or antiparallel to the direction of the orbital angular momentum. The orientation of the electron can be represented by the spin magnetic quantum number m_s, which is always constrained to have the values $m_s = +(1/2)$ or $-(1/2)$ as shown in Figure 10.4. These are known as the *spin-up* and *spin-down* states.

10.2.5 Quantized Angular Momentum and Magnetic Moments

What restrictions does quantum theory impose on the allowed values of the electronic magnetic moment?

If we express the orbital angular momentum in units of $h/2\pi$, then Equation 10.5 of Section 10.1.1 becomes

$$\boldsymbol{m}_o = -\left(\frac{eh}{4\pi m_e}\right)\left(\frac{2\pi \boldsymbol{p}_o}{h}\right) = -\left(\frac{eh}{4\pi m_e}\right)l \tag{10.16}$$

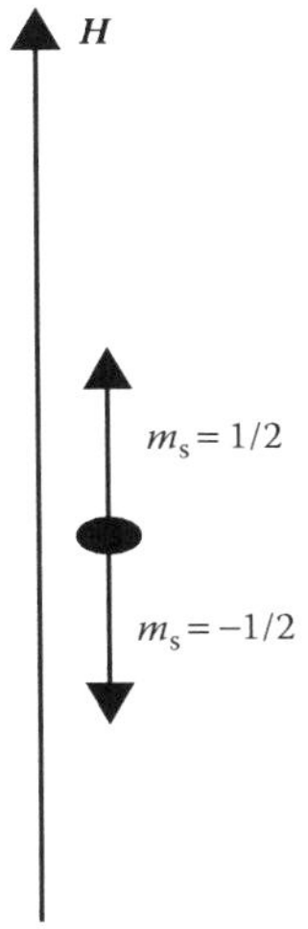

FIGURE 10.4 Quantization of the allowed projection of the spin s along the magnetic field direction.

where l is the orbital angular momentum quantum number, since by the quantum theory we expect the angular momentum of an electron to be an integral multiple of $h/2\pi$.

This means that on this basis we expect the orbital contribution to the magnetic moment to be an integral multiple of $eh/2\pi m_e$.

$$\boldsymbol{m}_o = -\mu_B l \tag{10.17}$$

where l must be an integer.

The spin on an electron is 1/2, and so if we suppose $\boldsymbol{p}_s$ is an integer multiple of $(h/2\pi)$, then $\boldsymbol{p}_s = sh/2\pi$, where s is the spin quantum number. The magnetic moment $\boldsymbol{m}_s$ due to the spin is then

$$\boldsymbol{m}_s = -\frac{eh}{2\pi m_e}\left(\frac{2\pi \boldsymbol{p}_s}{h}\right) \tag{10.18}$$

Replacing $2\pi\boldsymbol{p}_s/h$ by the spin quantum number s

$$\boldsymbol{m}_s = -2\left(\frac{eh}{4\pi m_e}\right)s \tag{10.19}$$

The quantity $eh/4\pi m_e$ is known as the *Bohr magneton*, usually designated μ_B, which has a value of 9.27×10^{-24} A m^2. Replacing $(eh/4\pi m_e)$ by μ_B

$$\boldsymbol{m}_s = -2\mu_B s \tag{10.20}$$

where now because of the quantization of angular momentum $2s = 4\pi\boldsymbol{p}_s/h$. Since s must be $+(1/2)$ or $-(1/2)$, this means that the spin magnetic moment of an electron is 1 Bohr magneton.

As discussed in Section 10.1.2, it is found that the spin gives twice the magnetic moment for a given angular momentum than does the orbit. As described above, there is no fundamental reason why these contributions should be equal; however, a precise treatment of this goes beyond the scope of this book. The situation is discussed by Born [4] and the difference in the relations between the spin and orbital magnetic moments and their respective angular momenta is attributed to relativistic effects.

The total magnetic moment of an electron can be expressed in terms of multiples of the total angular momentum $(h/2\pi)j$ of the electron.

$$\boldsymbol{m}_{tot} = -g\left(\frac{eh}{4\pi m_e}\right)\left(\frac{2\pi \boldsymbol{p}_{tot}}{h}\right) = -g\mu_B\left(\frac{2\pi \boldsymbol{p}_{tot}}{h}\right) = -g\mu_B j \tag{10.21}$$

where j is the total angular momentum quantum number, and for a particular electron

$$\boldsymbol{m}_{tot} = \boldsymbol{m}_o + \boldsymbol{m}_s = -\left(\mu_B l + 2\mu_B s\right) \tag{10.22}$$

The projection $\boldsymbol{m}_j$ of the total angular momentum along the direction of a magnetic field is also quantized as shown in Figure 10.5a.

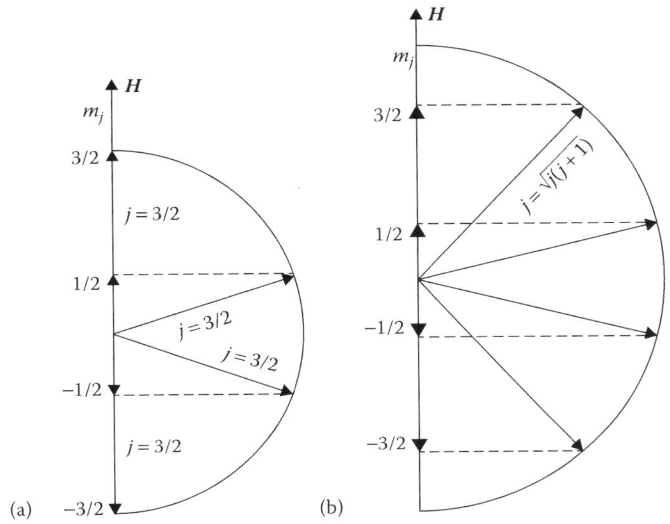

FIGURE 10.5 The projection m_j of the total angular momentum along the magnetic field direction (a) according to the older quantum theory treatment and (b) according to a full wave mechanical treatment.

However, the assumption that the angular momentum is an integral multiple of the orbital quantum number l and the spin quantum number s as shown in Figure 10.5a is not quite valid. More precise values according to wave mechanics are $\boldsymbol{p}_{\mathrm{o}}^2 = (h/2\pi)^2 l(l+1)$ and $\boldsymbol{p}_{\mathrm{s}}^2 = (h/2\pi)^2 s(s+1)$. The respective magnetic moments become $\boldsymbol{m}_{\mathrm{o}} = \mu_{\mathrm{B}}\sqrt{l(l+1)}$ and $\boldsymbol{m}_{\mathrm{s}} = \mu_{\mathrm{B}}\sqrt{s(s+1)}$. Similarly, the total angular momentum needs a correction from wave mechanics as shown in Figure 10.5b because $\boldsymbol{p}_{\mathrm{tot}} \neq (h/2\pi)j$, but is in fact, given by $\boldsymbol{p}_{\mathrm{tot}}^2 = (h/2\pi)^2\, j(j+1)$ so that $\boldsymbol{m}_{\mathrm{tot}} = \mu_{\mathrm{B}}\sqrt{j(j+1)}$. This has been shown, for example, by Sherwin [5].

10.2.6 Wave Mechanical Corrections to Angular Momentum of Electrons

Why is the angular momentum of an electron not simply the quantum number multiplied by (h/2π)?

We may now consider very briefly the quantum theory of angular momentum. This will be needed to provide the expectation value of the angular momentum $\boldsymbol{p}_{\mathrm{o}} = (h/2\pi)\sqrt{l(l+1)}$ for the following discussion of the quantum theories of ferromagnetism and paramagnetism. The treatment given here is necessarily abbreviated and only intended to be a guide to the derivation.

Consider a very simple case of a single electron orbiting an ionic core. We are interested only in the angular momentum of this one electron. We could approximate the situation with the wave function of the single electron in a hydrogen atom. In this case, the wave function can be represented by the expression

$$\Psi(r,\theta,\phi) = R(r)\Theta(\theta)\Phi(\phi) \tag{10.23}$$

which assumes that the dependences on r, θ, and ϕ are independent of each other and hence can be separated. The solutions of the Schroedinger equation in this case are shown in Figures 10.6 and 10.7. These show the possible electron states.

If $\left\langle p_o^2 \right\rangle$ is the operator for the orbital angular momentum squared, and $\bar{p}_o^2$ is the expectation value of the same quantity, then

$$\left\langle p_o^2 \right\rangle \Psi(r,\theta,\phi) = \bar{p}_o^2 \Psi(r,\theta,\phi) \tag{10.24}$$

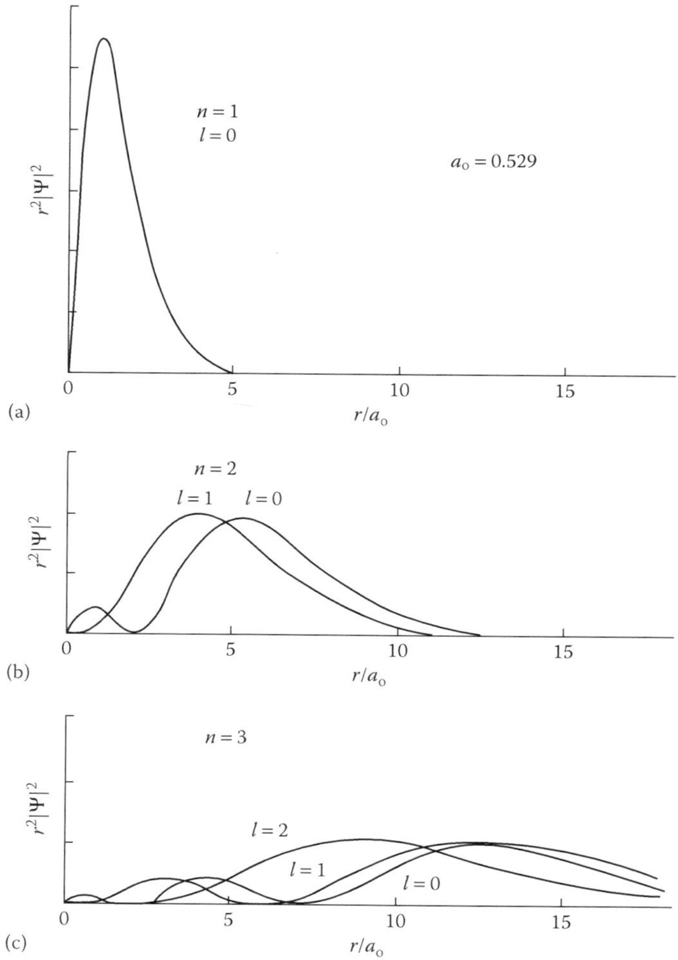

FIGURE 10.6 Probability functions $\psi^*\psi$ for an electron as a function of radial distance from the nucleus in a simple atom such as hydrogen for (a) $n = 1$, (b) $n = 2$, and (c) $n = 3$.

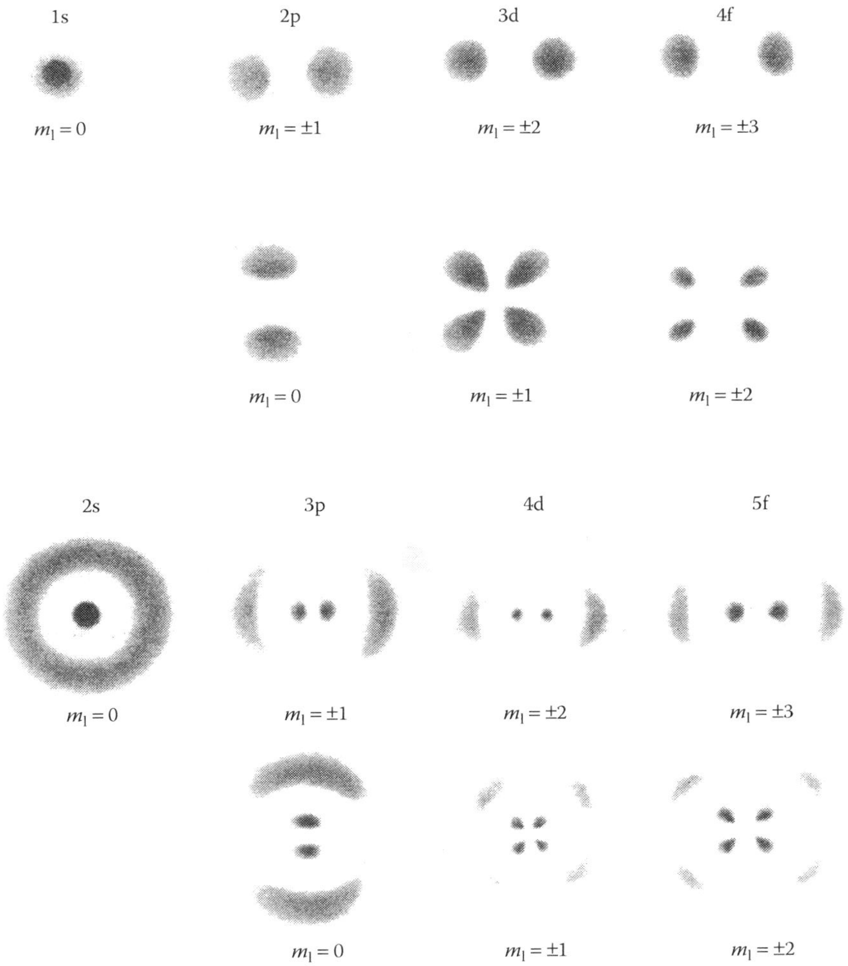

FIGURE 10.7 Probability density functions in real space for various electronic levels in hydrogen.

Replacing $\langle p_o^2 \rangle$ by $\hbar^2 \langle l^2 \rangle$ and $\bar{p}_o^2$ by $\hbar^2 \bar{l}^2$ gives

$$-\hbar^2 \left[\frac{1}{\sin^2\theta} \frac{d^2}{d\phi^2} + \frac{1}{\sin\theta} \frac{d}{d\theta} \left(\sin\theta \frac{d}{d\theta} \right) \right] \Psi(r,\theta,\phi) = \hbar^2 \bar{l}^2 \Psi(r,\theta,\phi) \qquad (10.25)$$

where the first term on the left-hand side in the square brackets is the operator $\langle l^2 \rangle$ and on the right-hand side is the expectation value $\bar{l}^2$. Separating the terms in the wave function, $\Psi(r,\theta,\phi) = R(r)\Theta(\theta)\Phi(\phi)$, as in Equation 10.23, the previous equation becomes

$$-\hbar^2\left[\frac{1}{\sin^2\theta}\frac{d^2}{d\phi^2}+\frac{1}{\sin\theta}\frac{d}{d\theta}\left(\sin\theta\frac{d}{d\theta}\right)\right]R(r)\Theta(\theta)\Phi(\phi)=\hbar^2\bar{l}^2R(r)\Theta(\theta)\Phi(\phi) \quad (10.26)$$

It can be shown by a rather lengthy proof that Equation 10.26 only has well-behaved solutions when $\bar{l}^2 = l(l+1)$, for $l = 0, 1, 2, 3, \ldots$. The expected values of the angular momentum are therefore

$$\bar{p}_o = \frac{h}{2\pi}\bar{l} = \frac{h}{2\pi}\sqrt{l(l+1)} \quad (10.27)$$

The result of this is that $\bar{l} = \sqrt{l(l+1)}$ and therefore an electron with orbital angular momentum quantum number l will have an angular momentum of $\boldsymbol{p}_o = (h/2\pi)\sqrt{l(l+1)}$ and not $\boldsymbol{p}_o = (h/2\pi)l$ as might have been expected. The same argument holds for the spin angular momentum $\boldsymbol{p}_s = (h/2\pi)\sqrt{s(s+1)}$ and the total angular momentum $\boldsymbol{p}_{tot} = (h/2\pi)\sqrt{j(j+1)}$. The vector addition of these quantities is shown in Figure 10.11.

10.2.7 Normal Zeeman Effect

What effect does a magnetic field have on the energy levels of electrons within atoms?

The energy levels of electrons within the atom are altered by the presence of a magnetic field. This is shown in the optical spectra, for example, where the emissions from the atoms in the form of light quanta are the result of electrons moving to lower energy levels. These spectra are altered by the presence of a magnetic field as shown in the Zeeman effect [6,7].

The normal Zeeman effect is exhibited by atoms in which the net spin angular momentum is zero. A spectral line of frequency v_0 in zero field is split into two lines displaced symmetrically about the original zero-field line at frequencies of $v_0 + \Delta v$ and $v_0 - \Delta v$. The displacement in frequency Δv is proportional to the magnetic field strength. When viewed perpendicular to the field direction, all three lines are observed, but when viewed along the direction of the field, only the two displaced spectral lines are observed.

The normal Zeeman effect occurs, for example, in calcium (the singlet at $\lambda = 422.7$ nm), magnesium, and cadmium (the singlet at $\lambda = 643.8$ nm). The displacement in energy is 0.93×10^{-3} J or 0.58×10^{-4} eV in a magnetic induction of 1 T, corresponding in the optical region to a shift in wavelength of typically 0.01 nm.

Under the action of a magnetic field, the allowed changes in the electron energy levels $\Delta E_H = E_0 \pm E_H$ are quantized and are given by

$$\Delta E_H = -\mu_0 \Delta \boldsymbol{m} \cdot \boldsymbol{H} \quad (10.28)$$

where $\Delta \boldsymbol{m}$ is the difference in the component of magnetic moment along the field direction. This can take only the values given by $\boldsymbol{m}_l(eh/4\pi m_e)$, where $\boldsymbol{m}_l = -l, -l+1, -l+2, \ldots 0, \ldots l-2, l-1, l$.

$$\Delta E_{\mathrm{H}} = \boldsymbol{m}_{\mathrm{l}} \left(\frac{eh}{4\pi m_{\mathrm{e}}} \right) \mu_0 \boldsymbol{H} \tag{10.29}$$

Therefore, a P state, which has $l = 1$, splits into three levels, $\boldsymbol{m}_{\mathrm{l}} = -1$, 0, and +1, while a D state, which has $l = 2$, splits into five levels, $\boldsymbol{m}_{\mathrm{l}} = -2$, −1, 0, 1, and 2.

There is also a selection rule that governs allowable transitions between states with different values of $\boldsymbol{m}_{\mathrm{l}}$ [8]. This rule, which is found empirically, states that an electron undergoing a transition from different energy levels cannot change its magnetic quantum number by more than 1. This rule ensures that there are no more than three available transition energies in the normal Zeeman effect.

$$\Delta \boldsymbol{m}_{\mathrm{l}} = 0 \text{ or } \pm 1 \tag{10.30}$$

The shift in energy of the spectral lines is

$$h\nu - h\nu_0 = \left(\frac{eh}{4\pi m_{\mathrm{e}}} \right) \mu_0 \boldsymbol{H} \tag{10.31}$$

By measuring the change in frequency, the ratio of charge to mass of an electron can be found from

$$\nu - \nu_0 = \left(\frac{e}{4\pi m_{\mathrm{e}}} \right) \mu_0 \boldsymbol{H} \tag{10.32}$$

Experimental measurements of the change in frequency yield the value

$$\frac{e}{m_{\mathrm{e}}} = 1.7587 \times 10^{11} \text{ C kg}^{-1} \tag{10.33}$$

Notice that the expression for the change in spectral frequency does not depend on Planck's constant. In fact, an explanation of the normal Zeeman effect of spectral frequency can be given by classical physics [9].

An example of the splitting of the electron energy levels of the D states ($l = 2$) and the P states ($l = 1$) in cadmium is shown in Figure 10.8. In the absence of a net spin, the P levels have three states corresponding to $\boldsymbol{m}_{\mathrm{l}} = 1$, 0, and −1, which are degenerate in zero field. The D levels have five states. Once a magnetic field is applied, the degeneracy is lifted, and on application of the selection rule, this results in three allowed transition energies. The shift in spectral lines is also shown in Figure 10.8.

We should note that the quantization of the various states with $\boldsymbol{m}_{\mathrm{l}} = -l, \ldots, 0, \ldots, +l$ remains even when the field is removed (i.e., it is not the field which induces the quantization), but in zero field the various states are degenerate (i.e., have the same energy) and therefore they all contribute to the same spectral line. All electronic transitions for which $\Delta l = \pm 1$ lead to the same Zeeman effect, that is, the splitting of one spectral line into three.

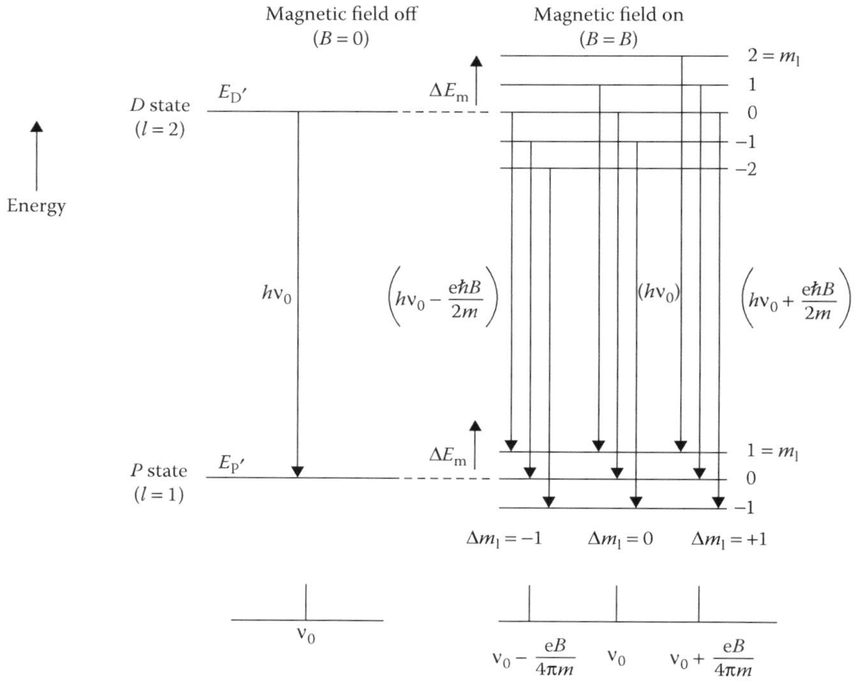

FIGURE 10.8 Splitting of the electronic *D* and *P* states in the normal Zeeman effect.

10.2.8 Anomalous Zeeman Effect

Why do most atoms not exhibit the simple normal *Zeeman effect?*

The spectra of most atoms do not show the simple normal Zeeman effect. In these cases, the spectral lines split into more than three levels described above. The resulting behavior is called the *anomalous Zeeman effect*, even though it occurs more often than the *normal* Zeeman effect.

These spectra cannot be accounted for on the basis of quantization of the orbital angular momentum alone under the action of a magnetic field, and therefore, it is necessary to look for further explanations. In order to account for the anomalous Zeeman effect, Goudsmit and Uhlenbeck [10] suggested that the electrons have a spin angular momentum, which leads to a magnetic moment associated with the spin.

The spectra of most atoms exhibit fine structure even in the absence of a magnetic field and hence the name *anomalous Zeeman effect* since it does not need a magnetic field to cause splitting of the energy states. When such zero-field fine structure is present, it is indicative that the element will exhibit the anomalous Zeeman effect under the action of a field also. The energy splitting between states with spin-up and spin-down in the presence of a magnetic field $\boldsymbol{H}$ is

$$\Delta E_s = 2s\left(\frac{eh}{4\pi m_e}\right)\mu_0 \boldsymbol{H} \tag{10.34}$$

replacing $\mu_B = eh/4\pi m_e$

$$\Delta E_s = 2s\mu_B\mu_0 \boldsymbol{H} \tag{10.35}$$

and $\boldsymbol{m}_s = 2s\mu_B$

$$\Delta E_s = \mu_0 \boldsymbol{m}_s \cdot \boldsymbol{H} \tag{10.36}$$

Therefore, since any energy state can be occupied by electrons in degenerate spin-up and spin-down states, it will be split into two; that is, it has the degeneracy lifted by a magnetic field.

A well-known example of the anomalous Zeeman effect occurs in the sodium spectrum. The zero-field transitions can be resolved into a doublet with wavelengths 589.0 and 589.6 nm. Each of these lines itself exhibits fine structure and can be resolved further into two or three lines. The doublet is due to the 3p–3s transition as shown in Figure 10.9. Application of a magnetic field splits the P states into four new

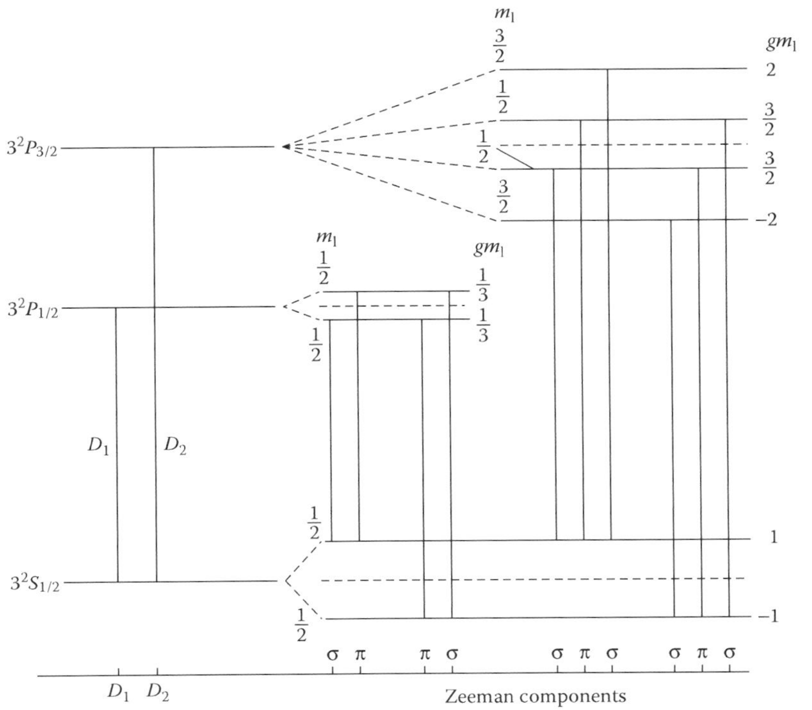

FIGURE 10.9 Splitting of the electronic D and P states in the anomalous Zeeman effect.

levels (compare with P level splitting into two new levels in Figure 10.8) and the s state into two new levels compared with none in the normal Zeeman effect.

The diagram of Figure 10.9 is kept relatively simple because the P state ($l = 1$) is split by the normal Zeeman effect into only two levels, which are each further split into two by the spin. If we had looked at a D state ($l = 2$), for example, there would have been a total of 10 levels in an applied field, five as a result of the normal Zeeman splitting ($\boldsymbol{m}_l = 2, 1, 0, -1, -2$), each of which are further split into two by the spin splitting $\left[\boldsymbol{m}_s = +(1/2)\,\text{or} - (1/2)\right]$.

10.2.9 Stern-Gerlach Experiment

How do we know that the electronic angular momentum is quantized?

In zero field, of course, all of the values of $\boldsymbol{m}_l$ represent degenerate energy levels, although the quantization is still present. In the presence of a field, the component of angular momentum along the field direction is determined by Larmor precession of the angular momentum, as shown in Figure 10.3. The total orbital angular momentum remains the same for each of these electrons but by precessing at an angle to the field direction, the component perpendicular to the field direction averages to zero. The quantization of the component of the angular momentum, or magnetic moment, along the direction of a magnetic field represented by the magnetic quantum numbers was first confirmed by the experiment of Stern and Gerlach [11].

Stern and Gerlach tested the space quantization of the angular momentum using atoms of silver in which $\boldsymbol{J} = \boldsymbol{S} = (1/2)$. The only contribution to the magnetic moment comes from the spin of a single electron, which according to theory should be quantized with value $+(1/2)$ or $-(1/2)$ along the field direction. The arrangement of the experiment is shown in Figure 10.10. Silver atoms leaving an oven at high speeds were collimated by slits and passed through an inhomogeneous magnetic field generated by the pole pieces shown. They were stopped by a photographic plate where their final positions were recorded. It was found that the locations of the atoms arriving on the photographic plate were not continuous but instead were in two lines corresponding to silver in the two allowed spin states of $\boldsymbol{m}_s = +(1/2)$ and $-(1/2)$.

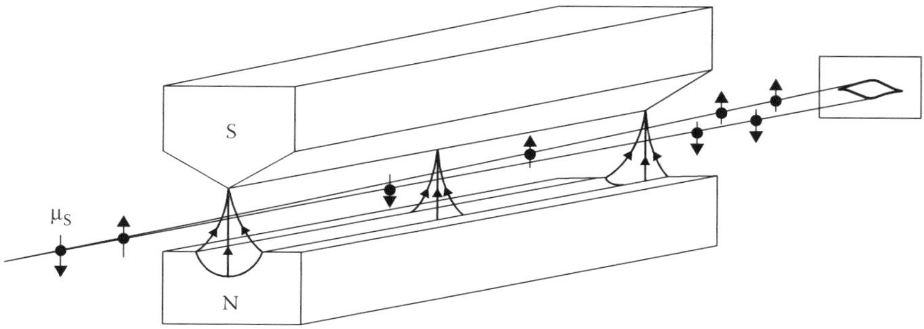

FIGURE 10.10 Experimental arrangement in the Stern-Gerlach experiment.

10.3 MAGNETIC PROPERTIES OF FREE ATOMS

What determines the magnetic moment of an atom?

The magnetic properties of an atom are determined principally by the magnetic moments of its electrons. We have shown in the previous chapter that the magnetic moments of these electrons can be calculated from the sum of orbital and spin angular momentum. The net magnetic moment of a filled electron shell is zero, so that only the unfilled shell needs to be considered. This simplifies the problem of calculation greatly.

The angular momentum of the atom can be found by summing vectorially the spin and orbital angular momenta of its electrons. This can be done in two ways: either the values of j for each electron can be found and the vector sum of these calculated to give $\boldsymbol{J}$. Alternatively, the orbital angular momentum $\boldsymbol{L}$ of all the electrons can be found by vector sum and the spin angular momentum $\boldsymbol{S}$ of all the electrons can be found. $\boldsymbol{L}$ and $\boldsymbol{S}$ can then be added vectorially to provide the $\boldsymbol{J}$ for the atom. These two methods do not give identical answers when the spin and orbit moments are coupled.

10.3.1 Magnetic Moment of a Closed Shell of Electrons

Does a closed shell of electrons have any net magnetic moment?

The total $\boldsymbol{J}$, orbital $\boldsymbol{L}$, and spin $\boldsymbol{S}$ angular momenta of a closed shell of electrons are always zero. This result enables us to simplify calculations of the net magnetic moment of an atom by considering only the contributions due to the partially filled shells.

To demonstrate that a closed shell of electrons has zero angular momentum, consider the Zn^{2+} ion that has 10 electrons in its 3d shell. For the 3d series $l = 2$, the electron configuration is as follows

m_l	2	1	0	−1	−2	2	1	0	−1	−2
m_s	1/2	1/2	1/2	1/2	1/2	−1/2	−1/2	−1/2	−1/2	−1/2
Occupancy	↑	↑	↑	↑	↑	↓	↓	↓	↓	↓

Consequently, $\boldsymbol{L} = \sum \boldsymbol{m}_l = 0$ and $\boldsymbol{S} = \sum \boldsymbol{m}_s = 0$.

This means that the Zn^{2+} ion has no net magnetic moment since the vector sum of the angular and spin contributions to the magnetic moment are both zero.

10.3.2 Atomic Magnetic Moment

How can we calculate the magnetic moment of an atom from knowledge of its electron structure?

We can find the magnetic moment of an atom from its total angular momentum. The vector model of the atom is a simple semiclassical model that allows us to calculate the total angular momentum $\boldsymbol{J}$ from the orbital and spin angular momenta of the electrons belonging to an atom. The model is semiclassical because, although the sum

$\boldsymbol{J} = \boldsymbol{L} + \boldsymbol{S}$ is made in the classical vector manner, the actual values of $\boldsymbol{L}$ and $\boldsymbol{S}$ are calculated from quantum mechanics.

We know that an electron has a total angular momentum j, which arises from its spin angular momentum s and its orbital angular momentum l. In the vector model of the atom, the total angular momentum of an electron is simply the vector sum of its spin and orbital angular momenta.

$$j = l + s \tag{10.37}$$

as indicated in Figure 10.11. Of course, the vector sum must always be a half integer. For any given value of l, the total angular momentum j can only have two possible values, $l \pm s$, depending on whether s is parallel or antiparallel to l. An exception is the case $l = 0$ when j must be 1/2.

For an atom with a single electron in an outer unfilled shell, and with all other shells filled, these values become the angular momenta of the atom. For a multielectron atom, that is, one with more than one electron in an unfilled shell, we use the terms $\boldsymbol{J}$, $\boldsymbol{L}$, and $\boldsymbol{S}$ to designate the total, orbital, and spin angular momenta of the atom respectively, where $\boldsymbol{L}$ is the vector sum of all the orbital components $\boldsymbol{m}_\mathrm{l}$, $\boldsymbol{S}$ is the vector sum of all spin components $\boldsymbol{m}_\mathrm{s}$, and $\boldsymbol{J}$ is the vector sum of total angular momenta of the electrons. For an atom with completely filled electron shells $\boldsymbol{J} = 0$, $\boldsymbol{L} = 0$, and $\boldsymbol{S} = 0$.

Corrections are made to the model as a result of quantum mechanics in order to find the magnitude of the vectors $|\mathbf{L}|$, $|\mathbf{S}|$, and $|\mathbf{J}|$ in terms of the quantum numbers, $\boldsymbol{L}$, $\boldsymbol{S}$, and $\boldsymbol{J}$. Fortunately, the correction is very simple and is the same as the correction needed for individual electrons as discussed in the previous chapter.

We should mention that there is a very small contribution to the total angular momentum of the atom from the nucleus due to nuclear spin. This contribution is about one thousandth of the spin of an electron. The nuclear magnetic moment is measured

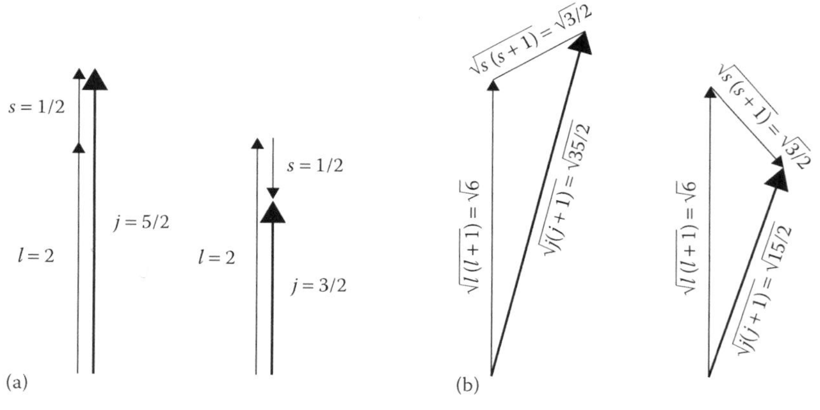

FIGURE 10.11 (a) Vector addition of the components of angular momentum $\boldsymbol{l}$ and $\boldsymbol{s}$ to form the total angular momentum $\boldsymbol{j}$ according to the semiclassical vector model of the atom and (b) vector addition of the components of angular momentum $\boldsymbol{l}$ and $\boldsymbol{s}$ to form the total angular momentum $\boldsymbol{j}$ with corrections due to quantum mechanics. Both cases are for a single electron.

in units of nuclear magnetons, denoted μ_N. The value of μ_N is 5.05×10^{-27} A m^2, compared with 9.27×10^{-24} A m^2 for the Bohr magneton.

10.3.3 Atomic Orbital Angular Momentum

How does the atomic orbital angular momentum depend on the electronic orbital angular momenta?

The atomic orbital angular momentum, denoted by the quantum number $\boldsymbol{L}$ is the vector sum of the orbital angular momenta of the electrons within the atom.

$$\boldsymbol{L} = \sum \mathbf{l}_l \tag{10.38}$$

An example of the vector summation for an atom with two electrons in its unfilled shell with quantum numbers $l = 1$ and $l = 2$ is shown in Figure 10.12. The magnitude of the orbital angular momentum vector $|\mathbf{L}|$ in terms of the orbital angular momentum quantum number $\boldsymbol{L}$ is

$$|\mathbf{L}| = \sqrt{L(L+1)} \tag{10.39}$$

which is identical in form to the relation between $|\mathbf{l}|$ and the orbital angular momentum quantum number l of a single electron.

10.3.4 Atomic Spin Angular Momentum

How does the atomic spin depend on the electron spins?

The spin angular momentum $\boldsymbol{S}$ for an atom can be found in a similar way using the vector sum of the spin angular momenta of the electrons.

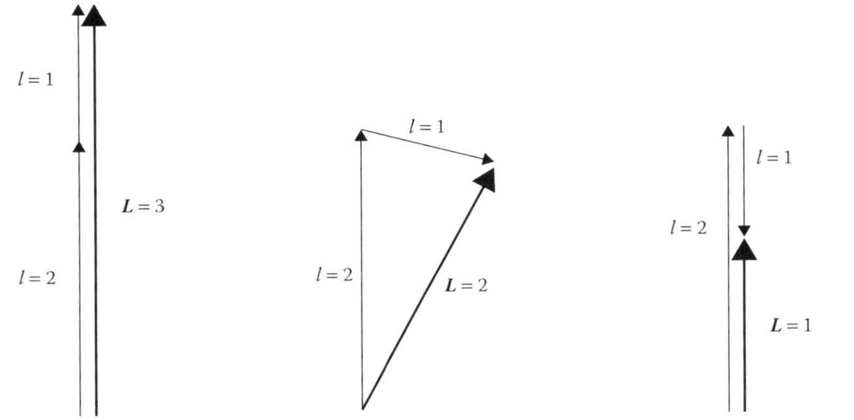

FIGURE 10.12 Vector addition of the orbital angular momentum in a two-electron system.

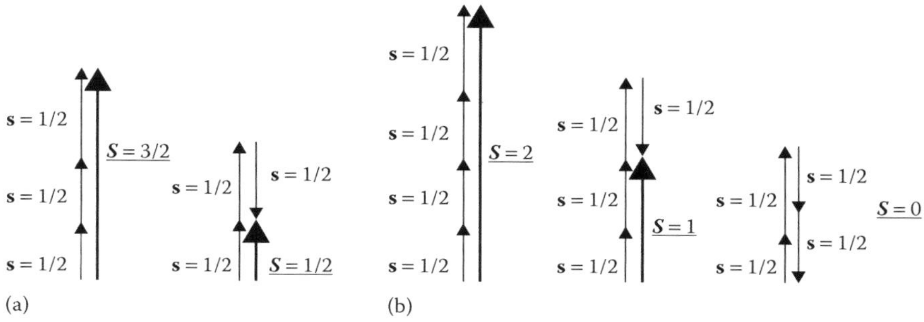

FIGURE 10.13 Vector addition of electron spins for: (a) a three-electron system and (b) a four-electron system.

$$S = \sum \mathbf{s}_i \tag{10.40}$$

The summation process is indicated in Figure 10.13a for an atom with three electrons in its unfilled shell and in Figure 10.13b for an atom with four electrons in its unfilled shell. The magnitude of the spin angular momentum vector $|\mathbf{S}|$ in terms of the spin quantum number S is according to quantum mechanics

$$|\mathbf{S}| = \sqrt{S(S+1)} \tag{10.41}$$

which is again identical in form to the relation for a single electron.

10.3.5 Hund's Rules: Occupancy of Available Electron States

How do electrons decide which values of **s** *and l to take?*

There is a set of rules which determine the occupancy of the available electronic states within an atom. These rules identify how the possible electronic states are filled and can be used therefore to calculate L, S, and J for an atom from the electron configuration in its unfilled shell.

The Hund rules [12] can be applied to electrons in a particular shell to determine the ground state of the atom. The three rules apply to the atomic spin S, the atomic orbital angular momentum L, and the atomic total angular momentum J, respectively. Electrons occupy available states so that

1. The maximum total atomic spin $S = \sum m_s$ is obtained without violating the Pauli exclusion principle.
2. The maximum value of total atomic orbital angular momentum $L = \sum m_l$ is obtained, while remaining consistent with the given value of S.
3. The total atomic angular momentum J is equal to $|L - S|$ when the shell is less than half full, and is equal to $|L + S|$ when the shell is more than half full. When the shell is exactly half full $L = 0$ so that $J = S$.

This means that the electrons will occupy states with all spins parallel within a shell as far as is possible. They will also start by occupying the state with the largest orbital angular momentum followed by the state with the next largest orbital angular momentum, and so on.

The next two examples show how Hund's rules are applied in specific cases to calculate $\boldsymbol{L}$, $\boldsymbol{S}$, and $\boldsymbol{J}$. The Sm^{3+} ion, which has five electrons in its 4f shell ($n = 4$, $l = 3$), must have the electrons arranged with the following configuration

$\boldsymbol{m}_\mathrm{l}$	3	2	1	0	−1	−2	−3
$\boldsymbol{m}_\mathrm{s}$	1/2	1/2	1/2	1/2	1/2	1/2	1/2
Occupancy	↑	↑	↑	↑	↑	↑	↑

Consequently, $\boldsymbol{L} = \Sigma \boldsymbol{m}_\mathrm{l} = 5$ and $\boldsymbol{S} = \Sigma \boldsymbol{m}_\mathrm{s} = 5/2$.

The shell is less than half full, so that $\boldsymbol{J} = \boldsymbol{L} - \boldsymbol{S} = 5/2$.

10.3.6 Total Atomic Angular Momentum

How is the total atomic angular momentum related to the spin and orbit components?

The coupling between the orbital angular momentum of the atom $\boldsymbol{L}$ and the spin angular momentum $\boldsymbol{S}$ gives the total atomic angular momentum $\boldsymbol{J}$. This summation can be obtained in two ways, by independently summing the orbital and spin moments of the unpaired electrons

$$\boldsymbol{J} = \sum \mathbf{l}_i + \sum \mathbf{s}_i \tag{10.42}$$

This form of coupling is known as *Russell-Saunders coupling* [13]. In this case, the **l** vectors of individual electrons are coupled together (**l–l**) and the **s** vectors are coupled together (**s–s**). These couplings are stronger than the coupling between the resultant **L** and **S** vectors.

Another method of obtaining $\boldsymbol{J}$ is by summing the total angular momenta of the electrons, which are obtained for each electron individually by summing **l** and **s** first

$$\boldsymbol{J} = \sum \mathbf{j}_i = \sum (\mathbf{l}_i + \mathbf{s}_i) \tag{10.43}$$

This is the so-called **j–j** coupling. This type of coupling occurs if there is a strong spin orbit coupling (**l–s**) for each electron. Then the coupling between **l** and **s** for a particular electron is stronger than the coupling between **l** and **l** for separate individual electrons.

In these calculations, the vectors that are strongly coupled together must always be added first. In practice, the most common form of coupling is the first kind. This Russell-Saunders (or **l–s**) coupling is the more commonly occurring form of coupling.

10.3.7 Russell-Saunders Coupling

How do the spin and orbital angular momenta combine to form the total angular momentum of an atom?

Russell-Saunders coupling, in which the spin-spin and orbit-orbit couplings are strongest, applies in most cases. It occurs in all of the light elements and situations in which the multiplet splitting is small compared with the energy difference of levels having the same electron structure but different values of **L**.

In this coupling, it is assumed that when several electrons are present in an unfilled shell, the orbital angular momenta are so strongly coupled with each other that states with a different total **L** have a different energy. Similarly, it is assumed that states with a different total **S** have a significant difference in energy. The resultant vectors **L** and **S** are then less strongly coupled with one another. In Russell-Saunders coupling, the orbital momentum vectors combine to form a total orbital angular momentum **L** while the spin momentum vectors combine independently to form a total spin momentum vector **S**. This means that the calculated **S** is not affected by the value of **L**. This leads to a total angular momentum **J** of the atom, which is simply the vector sum of the two noninteracting momenta **L** and **S**.

$$\mathbf{J} = \mathbf{L} + \mathbf{S} \tag{10.44}$$

as shown in Figure 10.14. There are certain restrictions on this however. **J** must be an integer if **S** is an integer, and **J** must be a half integer if **S** is a half integer. Russell-Saunders coupling applies to iron and the rare earth atoms.

10.3.8 *j–j* Coupling

Are there other ways of coupling to form the total angular momentum of an atom?

The alternative form of coupling, spin-orbit coupling, in which the orbital and spin angular momenta are dependent on each other, is called ***j–j*** *coupling*. This form of coupling is more applicable to heavy atoms. It assumes there is a strong spin-orbit (**l–s**) coupling for each electron and that this coupling is stronger than the **l–l** or **s-s** coupling between electrons. Since the strongly coupled vectors are summed first, these give a resultant **j** for each electron. The **j** vectors are then summed to obtain **J** for the whole atom.

10.3.9 Quenching of the Orbital Angular Momentum

Why do the magnetic moments of the 3d series elements imply that **L** = 0 *in all cases?*

In the 3d series of elements, it has been found experimentally that $\boldsymbol{J} = \mathbf{S}$ as shown in Table 10.1. This implies that $\mathbf{L} = 0$ throughout the 3d series, which is a surprising

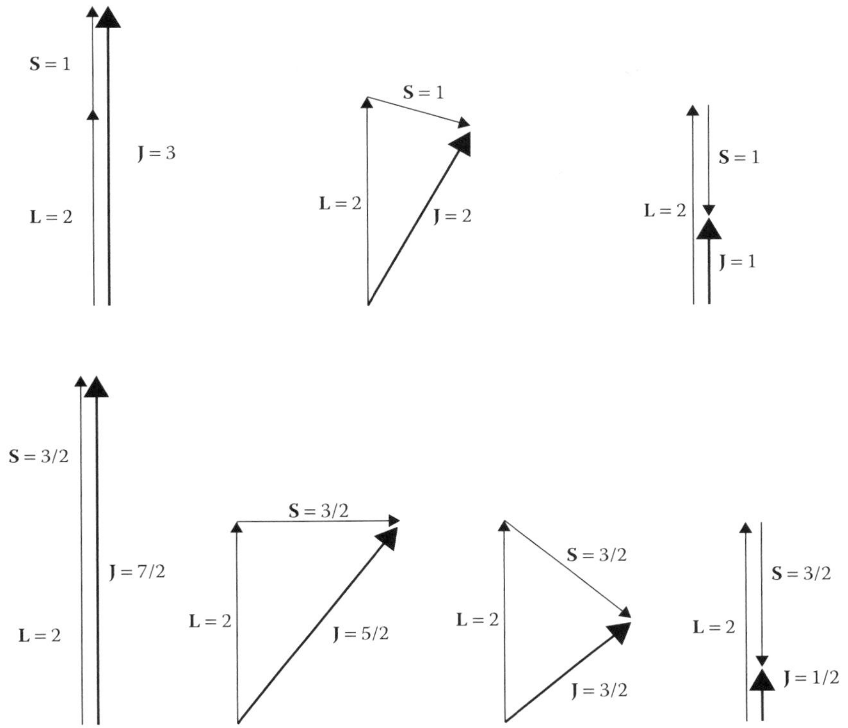

FIGURE 10.14 Vector addition of atomic orbital angular momentum **L** and atomic spin angular momentum **S** to give the total atomic angular momentum **J**.

result if we only consider the properties of electrons in isolated atoms. In these cases, we speak of the orbital angular momentum being *quenched*. The phenomenon has been discussed by Morrish [14] and Kittel [15]. Here, we note that under certain circumstances, the plane of the orbit can move about and this can average to zero over the whole atom. This results in a value of $\mathbf{L} = 0$. Notice from Table 10.2 that this does not occur in the 4f series, which is the other principal group of magnetic materials.

The Fe^{2+} ion has six electrons in its 3d shell. Since $l = 2$ for this series, the electron configuration is

m_l	2	1	0	−1	−2	2	1	0	−1	−2
m_s	1/2	1/2	1/2	1/2	1/2	−1/2	−1/2	−1/2	−1/2	−1/2
Occupancy	↑	↑	↑	↑	↑	↓				

Consequently $\mathbf{L} = \Sigma m_l = 2$ and $\mathbf{S} = \Sigma m_s = 2$.

This leads to $\mathbf{S} = 2$ and $\mathbf{L} = 2$. Since the orbital angular momentum is quenched in these elements $J = \mathbf{S}$ and therefore the total magnetic moment is

TABLE 10.1
Magnetic Moments of Isolated Ions of the 3d Transition Metal Series

Icon	Configuration	Calculated Moment $g\sqrt{[J(J+1)]}$	$g\sqrt{[S(S+1)]}$	Measured Moment
Ti^{3+}, V^{4+}	$3d^1$	1.55	1.73	1.8
Ti^{2+}, V^{3+}	$3d^2$	1.63	2.83	2.8
V^{2+}, Cr^{3+}	$3d^3$	0.77	3.87	3.8
Cr^{2+}, Mn^{3+}	$3d^4$	0	4.90	4.9
Mn^{2+}, Fe^{3+}	$3d^5$	5.92	5.92	5.9
Fe^{2+}	$3d^6$	6.70	6.70	5.4
Co^{2+}	$3d^7$	6.63	6.63	4.8
Ni^{2+}	$3d^8$	5.59	5.59	3.2
Cu^{2+}	$3d^9$	3.55	3.55	1.9

TABLE 10.2
Magnetic Moments of Isolated Ions of the 4f Transition Metal Series

Icon	Configuration	Calculated Moment $g\sqrt{[J(J+1)]}$	Measured Moment
Ce^{3+}	$4f^15s^2p^6$	2.54	2.4
Pr^{3+}	$4f^25s^2p^6$	3.58	3.5
Nd^{3+}	$4f^35s^2p^6$	3.62	3.5
Pm^{3+}	$4f^45s^2p^6$	2.68	–
Sm^{3+}	$4f^55s^2p^6$	0.84	1.5
Eu^{3+}	$4f^65s^2p^6$	0	3.4
Gd^{3+}	$4f^75s^2p^6$	7.94	8.0
Tb^{3+}	$4f^85s^2p^6$	9.72	9.5
Dy^{3+}	$4f^95s^2p^6$	10.63	10.6
Ho^{3+}	$4f^{10}5s^2p^6$	10.60	10.4
Er^{3+}	$4f^{11}5s^2p^6$	9.59	9.5
Tm^{3+}	$4f^{12}5s^2p^6$	7.57	7.3
Yb^{3+}	$4f^{13}5s^2p^6$	4.54	4.5

$$m = 2\mu_B\sqrt{S(S+1)} \tag{10.45}$$

which corresponds to 4.90 Bohr magnetons for the isolated Fe^{2+} atom. This agrees reasonably well with the observed value of 5.36 Bohr magnetons. The expected value without quenching of the orbital angular momentum is 6.7 Bohr magnetons, which is clearly in serious error.

We should note that these calculations apply only to the isolated paramagnetic (i.e., noninteracting) atoms or ions. When a large number of atoms are brought together in a solid, the electron energy levels are drastically altered and the magnetic moment per atom is in most cases significantly different from the values calculated above for the isolated atoms.

10.3.10 Electronic Behavior in Strong Magnetic Fields

Is the coupling between the spin and orbital angular momenta different in a strong magnetic field?

As the magnetic field is increased and the field splitting becomes greater than the multiple splitting, the anomalous Zeeman effect changes over to a normal Zeeman effect. The reason for this is that the precessional velocity of **J** about the field axis becomes greater than the precession of the **S** and **L** vectors about **J**. Therefore, this is described better as independent precessions of **S** and **L** about the field direction, that is the **L–S** coupling breaks down. We speak of **L** and **S** being decoupled by the magnetic field. This transition is known as the *Paschen-Back effect* [16] and only occurs in high magnetic fields.

REFERENCES

1. Bohr, N. *Phil. Magn.*, 26, 1, 1913.
2. Sommerfeld, A. *Annaln. Physik.*, 51(1), 125, 1916.
3. Larmor, J. *Phil. Magn.*, 44, 503, 1897.
4. Born, M. *Atomic Physics*, Blackie, Glasgow, Scotland, 1969.
5. Sherwin, C. W. *Introduction to Quantum Mechanics*, Holt, Rinehart & Winston, New York, 1959, pp. 151, 361.
6. Zeeman, P. *Phil. Magn.*, 43, 226, 1897.
7. Herzberg, G. *Atomic Spectra and Atomic Structure*, Prentice Hall, New York, 1937.
8. Weidner, R., and Sells, R. L. *Elementary Modern Physics*, 2nd edn, Allyn & Bacon, Boston, MA, 1968.
9. Lorentz, H. A. *Kon. Ak. van Wetenscheppen*, 6, 193, 1897; *Eel Electr.*, 14, 311, 1898.
10. Goudsmit, S., and Uhlenbeck, G. E. *Nature*, 117, 264, 1926.
11. Gerlach, W., and Stern, O. *Ann. Physik.*, 74, 673, 1924.
12. Hund, F. *Linienspektren und Periodische System der Elemente*, Springer, Berlin, Germany, 1927. (Trans. Line Spectra and Periodicity of the Elements.)
13. Russell, H. N., and Saunders, F. A. *Astrophys. J.*, 61, 38, 1925.
14. Morrish, A. H. *The Physical Principles of Magnetism*, Wiley, New York, 1965.
15. Kittel, C. *Introduction to Solid State Physics*, 6th edn, Wiley, New York, 1986.
16. Paschen, F., and Back, E. *Ann. Physik.*, 40, 960, 1913.

FURTHER READING

Chikazumi, S. *Physics of Magnetism*, Wiley, New York, 1964, Ch. 3.

Cullity, B. D. *Introduction to Magnetic Materials*, Addison-Wesley, Reading, MA, 1972, Ch. 3.

Hertzberg, G. *Atomic Spectra and Atomic Structure*, Prentice Hall, New York, 1937.

Semat, H. *Introduction to Atomic and Nuclear Physics*, Holt, Rinehart & Winston, New York, 1972, Chs. 8 and 9.

Sherwin, C. W. *Introduction to Quantum Mechanics*, Holt, Rinehart & Winston, New York, 1959.
Weidner, R., and Sells, R. L. *Elementary Modern Physics*, 2nd edn, Allyn & Bacon, Boston, MA, 1968.

EXERCISES

10.1 *Diamagnetic moment.* The diamagnetic effect of a magnetic field on an atom occurs in all materials, although in para- and ferromagnets, this is very small compared with other effects. Show that when a uniform magnetic induction $\boldsymbol{B}$ is applied normal to the plane of a circular electronic orbit, the induced change in angular velocity is $e\boldsymbol{B}/2m$ (where e is the electron charge and m is the electron mass). Assume that the change in angular velocity is small compared to the angular velocity at zero field.

10.2 *Relation between magnetic moment and angular momentum.* Derive an expression for the relationship between orbital angular momentum of an electron and the orbital magnetic moment using a classical approach. Then derive an expression relating the spin angular momentum of an electron to its magnetic moment using a similar approach. Comment on the results. Explain how it is possible to derive the correct relationship between angular momentum and magnetic moment.

10.3 *Tangential speed of electron—classical model.* Find the conditions under which the magnetic induction $\boldsymbol{B}$ generated by a nonmagnetic sphere of radius a carrying a uniform surface charge density ρ rotating about a diameter with angular velocity ω is the same as the magnetic induction generated by a magnetic sphere also of radius a with uniform magnetization $\boldsymbol{M}$. Then calculate the angular velocity and the tangential speed of an electron, which is needed to account for its known magnetic moment of 1 Bohr magneton. You may assume the classical electron radius. Comment on the result.

10.4 *Orbital and spin angular momentum of an electron.* (1) Explain the significance of the four quantum numbers and n, l, s, and $\boldsymbol{m}_l$. Describe how to calculate the orbital, spin, and total angular momentum of an electron from its quantum numbers using the *classical* vector model of the atom. Why do the values of l, s, and j differ from the classically expected results when calculated using quantum mechanics? (2) Using vector diagrams, determine the different values for the total orbital angular momentum of a two-electron system (i.e., atom) for which $l_1 = 3$ and $l_2 = 2$. What are the possible values for $\boldsymbol{L}$, for $\boldsymbol{S}$, and for $\boldsymbol{J}$ of the atom?

Find the values of $\boldsymbol{L}$, $\boldsymbol{S}$, and $\boldsymbol{J}$ for the Co^{2+} ion.

10.5 *The Zeeman effect.* Calculate the normal Zeeman effect separation in energy levels in cadmium when subjected to magnetic field strengths of 1.6×10^6, 0.4×10^6, and 0.064×10^6 A m^{-1} (equivalent to a free space magnetic induction of 2.0, 0.5, and 0.1 T). The wavelength of the main 6^1D_2–5^1P_1 transition is $\lambda = 643.8$ nm, which is a spectral line at the red end of the visible spectrum. Calculate the shift in wavelength.

Draw an energy level diagram showing the 6^1D_2 and 5^1i states before and after the application of a magnetic field and explain why the original singlet is split only into a triplet and not into a larger number of spectral lines.

10.6 *Determination of atomic angular momentum.* Using vector diagrams, determine the possible values of $\boldsymbol{J}$ for an atom with

a. $\boldsymbol{L} = 2, \boldsymbol{S} = 3$
b. $\boldsymbol{L} = 3, \boldsymbol{S} = 2$
c. $\boldsymbol{L} = 3, \boldsymbol{S} = 5/2$
d. $\boldsymbol{L} = 2, \boldsymbol{S} = 5/2$

What are the values of $\boldsymbol{L}$, $\boldsymbol{S}$, and $\boldsymbol{J}$ for the ground state of the paramagnetic carbon atom?

10.7 *Relationship between angular momentum and magnetic moment.* Using both the classical and quantum mechanical descriptions, explain why an electron has a magnetic moment and show how this magnetic moment is related to the orbital and spin angular momenta. Prove, by giving a specific example, that the spin and orbital components of magnetic moment of an electron can have different relations to their respective angular momenta.

10.8 *Magnetic moment of a classical rotating sphere.* Show that provided $\boldsymbol{M} = \sigma\omega r$, identical magnetic $\boldsymbol{B}$ fields are produced by: (1) a ferromagnetic sphere of radius r with uniform magnetization $\boldsymbol{M}$ throughout and (2) a nonferromagnetic sphere also of radius r, carrying a uniform surface density of charge σ and rotating about a diameter with angular velocity ω.

10.9 *Failure of the classical model of electron magnetic moment based on spinning charge.* Calculate the tangential velocity of a classical spherical particle with mass 9.1×10^{-31} kg, charge 1.6×10^{-19} C, radius 2.8×10^{-15} m that is needed to generate a magnetic moment of 1 Bohr magneton. (You may make an assumption about the distribution of charge and mass in the electron but state your assumption explicitly.) Comment on the result.

10.10 *Magnetic moment of an isolated atom in terms of orbital and spin quantum numbers.* Applying the constraints of the quantum mechanical description of electrons moments in atoms, calculate the expected magnetic moment a single isolated Mn^{2+} ion, assuming (1) that the orbital component is operative or assuming (2) that the orbital moment is *quenched*. How does this compare with the actual measured value of the moment on an Mn^{2+} ion? If there is a difference between your calculated value and the measured value, give some possible reasons for this.

11 Quantum Theory of Magnetism

In Chapter 10, we discussed the origin of the atomic magnetic moment. In this chapter, we look at how these magnetic moments interact to give cooperative magnetic phenomena such as ferromagnetism and paramagnetism in situations where there is a large number of atoms together in solids. We then consider theories of magnetism based on two mutually exclusive models: (1) the localized moment model and (2) the band or itinerant electron model. The localized model works well for the rare earth metals, while the itinerant electron model is more appropriate for the magnetic properties of the 3d transition metals and their alloys.

11.1 ELECTRON-ELECTRON INTERACTIONS

The exchange interaction is obviously crucial to an understanding of ordered magnetic states, but where does it come from?

We have seen that the behavior of the susceptibility of many ferromagnets above their Curie temperature follows a Curie-Weiss law and that this implies the existence of an internal field, which has been called the Weiss *molecular* field. The field originates from quantum mechanical interactions between the electrons. A discussion of these together with a derivation of the exchange interaction from considering overlap of the electron wave functions has been given in detail by Martin [1].

11.1.1 Wave Functions of a Two-Electron System

How do we form the wave function for a system consisting of two electrons?

The total wave function of two electrons on two atoms can be represented as $\Psi(r_1,r_2)$, where r_1 is the coordinate of the first electron and r_2 is the coordinate of the second electron. Then we may consider possible representations of this wave function using linear combinations of the separate wave functions of the individual electrons $\psi(r_1)$ and $\psi(r_2)$. There are four possibilities:

$$\begin{aligned}
\Psi(r_1,r_2) &= \psi_a(r_1)\psi_b(r_2)\\
\Psi(r_1,r_2) &= \psi_a(r_2)\psi_b(r_1)\\
\Psi(r_1,r_2) &= \psi_a(r_1)\psi_b(r_2)+\psi_a(r_2)\psi_b(r_1)\\
\Psi(r_1,r_2) &= \psi_a(r_1)\psi_b(r_2)-\psi_a(r_2)\psi_b(r_1)
\end{aligned} \tag{11.1}$$

where:

$\psi_a(r_1)$ represents the wave function of electron 1 on atom a
$\psi_b(r_2)$ represents the wave function of electron 2 on atom b and so on

We need a solution such that the *observed* properties $\Psi^*\Psi$ are unaltered by interchanging electrons, while the electrons remain distinct, that is, $\Psi(r_1,r_2)$ must be antisymmetric. This is an important point; if as a result of exchange of electrons the wave function is symmetric and $\Psi^*\Psi$ is identical, then the wave functions of the two electrons are identical and hence they have the same four quantum numbers, which is not allowed.

In the first two cases above, the interchange of electrons, r_1 for r_2, alters the value of $\Psi^*\Psi$ and hence alters the electron distribution at all points. Therefore, these cannot be solutions of the wave function. Another possibility is

$$\Psi(r_1,r_2) = \psi_a(r_1)\psi_b(r_2) + \psi_a(r_2)\psi_b(r_1) \tag{11.2}$$

In this case, the function remains symmetric after interchanging the electrons and $\Psi^*\Psi$ is also unaltered. However, according to the Pauli exclusion principle, the probability of finding both electrons in an identical state (i.e., with the same set of quantum numbers) is zero.

If the wave functions of the separate electrons are otherwise identical, the Pauli principle can only be satisfied among this group of candidate wave functions by

$$\Psi(r_1,r_2) = \psi_a(r_1)\psi_b(r_2) - \psi_a(r_2)\psi_b(r_1) \tag{11.3}$$

See, for example, Figure 11.1.

11.1.2 Heitler-London Approximation

What is the total energy of two interacting electrons on neighboring atoms according to quantum theory? Do we find a result that is just a modification of classical theory or something radically different?

The Heitler-London approximation [2] is merely a method of obtaining the orbital wave function of two electrons by assuming that it can be approximated by a linear combination of the atomic orbital wave functions of two electrons localized on the two atomic sites. This is analogous to the hydrogen molecule.

Consider the orbital functions and in particular the energy for such a system of two atoms with one electron each. Heitler and London calculated the energy of this system by evaluating the integral

$$E = \iint \Psi^*(r_1,r_2)\langle H\rangle\Psi(r_1,r_2)\,d\tau_1 d\tau_2 \tag{11.4}$$

and the Hamiltonian $\langle H\rangle$ must contain the separate Hamiltonians $\langle H_1\rangle$ and $\langle H_2\rangle$ for each electron on its atom and a separate interaction Hamiltonian $\langle H_{1,2}\rangle$

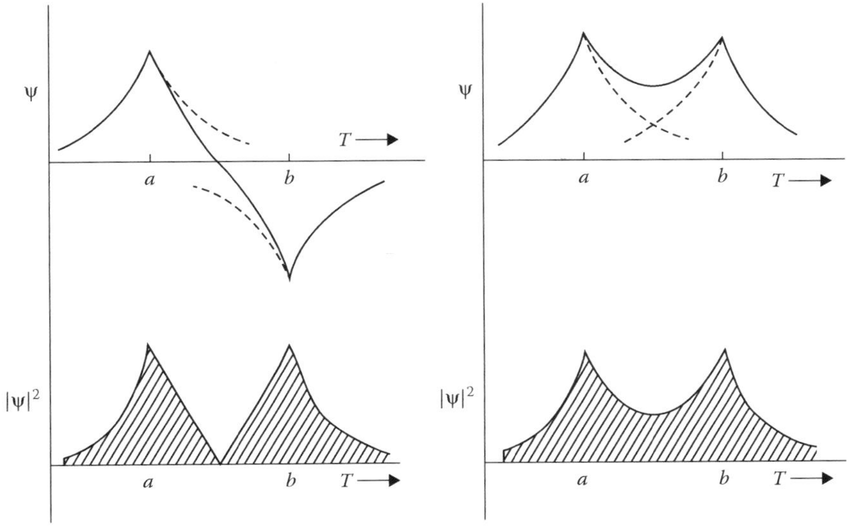

FIGURE 11.1 Wave function for a two-electron system using a linear combination of atomic orbitals as in the Heitler-London approximation. (After Heitler, W., and London, F., *Z. Phys.*, 44, 455, 1927.) ψ is antisymmetric, while $\psi^*\psi$ is symmetric. This means that a wave function of the type shown on the left is needed.

$$\langle H\rangle = \langle H_1\rangle + \langle H_2\rangle + \langle H_{1,2}\rangle \tag{11.5}$$

where:

$\langle H_1\rangle$ contains only the coordinate r_1
$\langle H_2\rangle$ contains only the coordinate r_2
$\langle H_{1,2}\rangle$ contains both

Therefore,

$$E_{\text{tot}} = \iint \Psi^*(r_1, r_2)\langle H_1 + H_2 + H_{1,2}\rangle \Psi(r_1, r_2)\,\mathrm{d}\tau_1\mathrm{d}\tau_2$$

and if the various terms in the Hamiltonian are separated, the operator H_1 leads to energy terms that only depend on electron a on atom 1, while the operator H_2 leads to energy terms that only depend on electron b on atom 2.

$$E_1 = \iint \Psi^*(r_1, r_2)\langle H_1\rangle \Psi(r_1, r_2)\,d\tau_1 d\tau_2 = E_a \iint \Psi^*(r_1)\Psi(r_1)\,d\tau_1 d\tau_2 \tag{11.6}$$

$$E_2 = \iint \Psi^*(r_1, r_2)\langle H_2\rangle \Psi(r_1, r_2)\,d\tau_1 d\tau_2 = E_b \iint \Psi^*(r_2)\Psi(r_2)\,d\tau_1 d\tau_2 \tag{11.7}$$

The $H_{1,2}$ operator includes the interaction terms. The remaining interaction energy term is

$$E_{1,2} = \iint \Psi^*(r_1, r_2)\langle H_{1,2}\rangle \Psi(r_1, r_2)\,d\tau_1 d\tau_2 = E_{\text{int}} \iint \Psi^*(r_1, r_2)\Psi(r_1, r_2)\,d\tau_1 d\tau_2 \tag{11.8}$$

The interaction energy has two components. The first component is a Coulomb interaction term between electrons that stay on their original atoms. This has a classical analog and introduces no new energy terms.

$$E_{\text{coulomb}} = \iint \left\{\psi_a^*(r_1)\psi_b^*(r_2)\right\}\langle H_{1,2}\rangle\left\{\psi_a(r_1)\psi_b(r_2)\right\} d\tau_1 d\tau_2 \tag{11.9}$$

The second component is an interaction energy caused by exchanging the electrons.

$$E_{\text{ex}} = \iint \left\{\psi_a^*(r_1)\psi_b^*(r_2)\right\}\langle H_{12}\rangle\left\{\psi_a(r_2)\psi_b(r_1)\right\} d\tau_1 d\tau_2 \tag{11.10}$$

This has no classical analog because if the electrons are exchanged the classical description gives no difference in energy. The interaction energy can be written as

$$E_{1,2} = E_{\text{coulomb}} + E_{\text{ex}} \tag{11.11}$$

This term E_{ex} is called the *exchange energy* because it arises from the exchange of the electrons. The symbol J is usually used to represent this energy. See Section 11.1.5 where it is italicized to distinguish it from $\boldsymbol{J}$ the atomic angular momentum.

11.1.3 Exchange Interaction

What is the cause of the extra energy term?

We can write this expression for the energy obtained from evaluating these integrals in the form

$$E_{\text{tot}} = E_1 + E_2 + E_{1,2} \tag{11.12}$$

In Equation 11.12, we have on the right-hand side, the energy of the electrons when they are on their separate atoms E_1 and E_2 without any interactions, and some additional energy, which results from their interactions. $E_{1,2}$ includes the coulomb electrostatic energy between the electrons and an energy, which arises from exchange of the electrons, as can be seen from the form of the wave function. We have already encountered this exchange energy in Chapter 6. The above description, however, only considers the total wave function, and so makes no explicit statement about spin ordering.

11.1.4 Wave Function Including Electron Spin

What is the form of the two-electron wave function if we include electron spin?

We have shown from the very simple quantum mechanical treatment above that there exists an exchange energy that has no classical analog. It is now necessary to show that the ordering of the electron spins arises from this. Suppose we write the spin wave functions as $\varphi_1(s)$ and $\varphi_2(s)$. Then the spin state for two electrons can be represented, using the same approximation as before, as a linear combination of the individual spin wave functions.

$$\begin{aligned}
\varphi(s_1,s_2) &= \varphi_a(s_1)\varphi_b(s_2) - \varphi_a(s_2)\varphi_b(s_1) \\
\varphi(s_1,s_2) &= \varphi_a(s_1)\varphi_b(s_2) + \varphi_a(s_2)\varphi_b(s_1) \\
\varphi(s_1,s_2) &= \varphi_a(s_1)\varphi_b(s_2) \\
\varphi(s_1,s_2) &= \varphi_a(s_2)\varphi_b(s_1)
\end{aligned} \tag{11.13}$$

The total wave function of the two-electron system with spins can then be represented as

$$\Psi_{tot} = \Psi(r_1,r_2)\varphi(s_1,s_2) \tag{11.14}$$

which as we already know must be antisymmetric with respect to interchange of the electrons. So, if $\Psi(r_1,r_2)$ is antisymmetric, then $\varphi(s_1,s_2)$ must be symmetric, and vice versa, in order to maintain the antisymmetric nature of the total electron wave function Ψ_{tot}.

Symmetric and antisymmetric φs correspond to parallel and antiparallel spins. Therefore, the exchange energy as derived above can be considered as really the interaction between the spins, since it is the spins that maintain the total wave function Ψ_{tot} as antisymmetric. In other words, the spin can be used to distinguish between two electrons, which are otherwise identical because the spins can be used to satisfy the Pauli principle when the other three quantum numbers of the two electrons are identical. For parallel spins, the exchange energy is then simply $[1/(1 \pm \alpha^2)](-J)$. This means that a positive J corresponds to parallel spin alignment, and hence to ferromagnetic ordering. A negative J leads to antiparallel alignment.

It is clear in the rather oversimplified case described above in Equations 11.13 and 11.14 that the constraints on the electrons spins will ensure that they will be antiparallel, but in principle, we can envisage parallel alignment in other situations.

11.1.5 Exchange Energy in Terms of Electron Spin

Can we separate the exchange energy and make it dependent only on spin?

Heisenberg took the Heitler-London approach and applied it to electron spins. From the above discussion, the exchange energy dictates a lower energy and hence ferromagnetic order when $J > 0$, and antiferromagnetic alignment when $J < 0$. The model has only considered two localized electrons on neighboring atoms and so has some obvious limitations when we wish to apply it to a solid. The exchange Hamiltonian can, however, now be written simply in terms of the spins on two electrons

$$\langle H \rangle = -2J_{ij}\mathbf{s}_i \cdot \mathbf{s}_j \tag{11.15}$$

which leads to the Heisenberg model of ferromagnetism [3], which was also proposed by Dirac [4].

A difficulty with the description of exchange on the basis of the *hydrogen molecule* in the Heitler-London treatment is that it can only produce negative exchange energies between the electrons because of the need to produce antiparallel alignment of the spins on the two electrons in identical orbitals. An alternative approach first suggested by Coles and discussed by Myers [5] used the oxygen molecule as a basis. This provides a description that allows a parallel alignment of spins of the unpaired outer electrons associated with each atom, and hence a net positive exchange interaction between the magnetic moments on the two atoms. This approach has some advantages as a conceptual basis for understanding the existence of positive exchange energies between electrons on neighboring atoms, and hence the net magnetic moment on the oxygen molecule, although the interaction between the electrons on neighboring molecules of oxygen is effectively nonexistent, and oxygen is therefore paramagnetic.

In the oxygen atom, the electron configuration is $1s^2 2s^2 2p^4$, with spins arranged as shown in Figure 11.2. The configuration of spins in the 2p state with three *spin-up* electrons and one *spin-down* electron is in accordance with Hund's rules and arises because the Coulomb interaction energy between the electrons is minimized if as many electrons as possible occupy different orbitals and are kept apart by having their spins aligned parallel. Since the 2p orbitals are all energy degenerate, the total energy is minimized under these conditions. This results in a spin moment

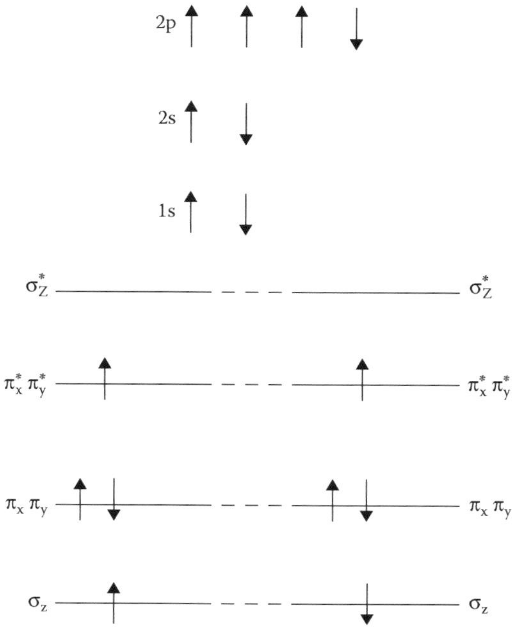

FIGURE 11.2 Alignment of electron spins in an oxygen atom.

per atom of $S = 1$, or, equivalently 2 Bohr magnetons. Thus, there appears to be an intra-atomic exchange coupling between the 2p electrons, which aligns their magnetic moments within the atom.

This approach can be extended to the molecular orbitals in the oxygen molecule O_2. Each atom shares its 2p electrons with the other. These eight electrons occupy molecular orbitals, which have different energies from the 2p orbitals in the isolated atoms. The energies of these states are shown in Figure 11.3. The *z*-axis is assumed to be the axis joining the atoms, whereas the *x*-*y* plane is normal to this axis. Therefore, by axial symmetry, the orbitals along the *x*- and *y*-directions are energy degenerate. These are conventionally denoted π_x and π_y and can hold two electrons each, one with *spin-up* and the other with *spin-down*. The orbital along the *z*-axis is conventionally denoted σ_z and can also hold two electrons. At higher energies are the *antibonding*" energy levels π_x^*, π_y^*, and σ_y^*, with relative energies as shown in Figure 11.3. The electrons occupy these available states beginning from the lowest energy. Therefore, the π_x, π_y, and σ_z molecular orbitals are completely filled with six electrons, with a net spin and orbital magnetic moment of zero. The remaining two electrons must occupy the lowest remaining energy states, which are the π^* orbitals. Since these orbitals are energy degenerate, the electrons occupy them in such a way that they minimize the Coulomb energy, as in the case of the oxygen atoms described above. This means that the electrons will occupy π_x^* and π_y^* orbitals with their spins aligned parallel. The ground state of the oxygen molecule therefore has a net spin moment of 2 Bohr magnetons. In fact, oxygen is paramagnetic as a result of this permanent magnetic moment per molecule.

In the 3d transition metals such as iron, cobalt, and nickel, the Fermi energy lies in a high density of states of such *antibonding* orbitals. From the foregoing discussion, it is therefore apparent why these electrons would likely be arranged with their

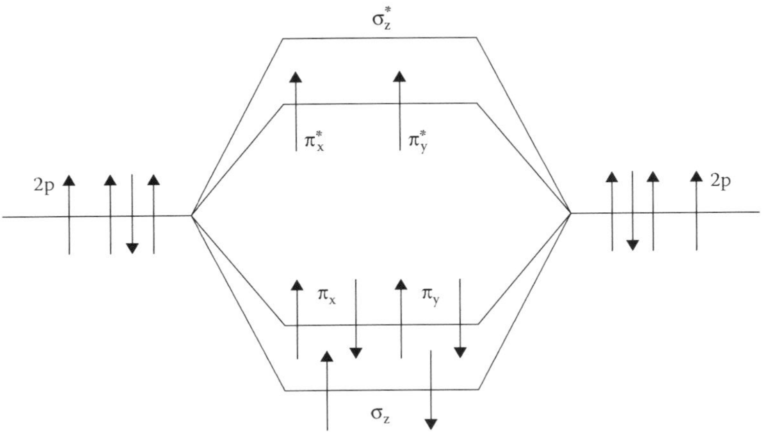

FIGURE 11.3 A schematic of the available molecular orbital energy states in O_2. Each atom contributes four electrons and six states. The π states are degenerate because of symmetry in the plane perpendicular to the axis of the atoms.

spin moments parallel. On the other hand, if the increase in energy needed to achieve this parallel alignment is greater than the reduction in exchange energy, which happens to be the case in most metals, then the parallel alignment of spins is not energetically favorable, and hence ferromagnetism does not arise.

11.1.6 Heisenberg Model of Ferromagnetism

How might the direct exchange energy between two neighboring electrons account for long-range magnetic order in ferromagnets and antiferromagnets?

The Heisenberg model of ferromagnetism, like the classical Weiss model [6], is another local moment theory, which considers the quantum mechanical exchange interaction between two neighboring electrons with overlapping wave functions. The idea of direct exchange between electrons on neighboring atoms first occurred in the Heitler-London treatment of electron orbitals. Heisenberg was the first to include the electron spins in the wave function and then apply the same Heitler-London approximation to obtain the total wave function of a two-electron system. The energy integral was evaluated but now including spin and this showed that the relative orientations of the spins of two interacting electrons can be changed only at the expense of changing the spatial distribution of the charge.

If two electron wave functions overlap, then the Pauli exclusion principle [7] applies to the region of overlap. Since no two electrons can have the same set of quantum numbers, when the orbital wave function is symmetrical, the spins must be antisymmetric and vice versa. This leads immediately to a correlation between the spins on the two electrons, which is all that is needed to cause a magnetically ordered state. The correlation can be expressed in the form of a magnetic field (although strictly it is not of magnetic origin but is rather electrostatic) or as an energy. The interaction energy is proportional to the dot product of the spins.

$$E_{\text{Heis}} = -2J\boldsymbol{s}_1 \cdot \boldsymbol{s}_2 \tag{11.16}$$

Here, $J > 0$ results in ferromagnetism and $J < 0$ results in antiferromagnetism.

When considering a solid, it is then necessary to sum the exchange over all the electrons, which can contribute to this energy, so that

$$\langle H \rangle = -2\sum\sum J_{ij}\boldsymbol{s}_i \cdot \boldsymbol{s}_j \tag{11.17}$$

In many cases, we are only interested in the nearest-neighbor interactions and this simplifies the Heisenberg Hamiltonian considerably.

$$\langle H \rangle = -2J\sum \sum_{\substack{\text{nearest}\\\text{neighbors}}} \boldsymbol{s}_i \cdot \boldsymbol{s}_j \tag{11.18}$$

where it is assumed that the exchange integral J is identical for all nearest-neighbor pairs. The exchange integral J is difficult to understand from a classical viewpoint as there is no classical analog of this quantum mechanical effect. However, a brief consideration of the Pauli principle tells us that two electrons with like spins cannot approach one another closely. This leads to an effective energy, which influences the alignment of the moments.

Empirical values of the exchange interaction J for various ferromagnetic metals have been calculated from specific heat measurements and from spin wave considerations. Some of these have been reported by Hofmann et al. [8], who indicated that for iron and nickel $J \approx 0.01 - 0.02$ eV and for gadolinium $J \approx 0.0002$ eV. The exchange energy for gadolinium is comparable to that of iron and nickel at about 10^{-21} J (0.01 eV), but the magnetic moment per atom is much larger in gadolinium than in iron or nickel and that implies a smaller value of J for gadolinium in Equation 11.18.

Wohlfarth [9] had encountered problems with first principles calculation of J, obtaining the wrong sign when using spherical wave functions. Later Stuart and Marshall [10] made detailed calculations of the direct Heisenberg exchange in iron. They obtained a value of 6.8×10^{-3} eV for two neighboring localized electron interactions, and on this calculation, a similar result would be obtained for both cobalt and nickel. This is about a third of the value obtained from experimental measurements. Watson and Freeman [11,12] refined the calculations of Stuart and Marshall to include additional terms neglected in the earlier work and with improved orbital wave functions. They confirmed that the magnitude of J is in serious error, being much smaller than is required to explain magnetic order in these 3d metals; but furthermore they found that J should be negative.

Finally, in the rare earth metals, the magnetic 4f electrons, which determine the magnetic properties, are highly localized, so that there is no significant overlap and hence no direct exchange mechanism of the Heisenberg type can occur, although these metals show a range of different magnetic order. Therefore, an additional exchange mechanism involving indirect exchange needs to be invoked.

It seems therefore that although the Heisenberg model is a useful concept, the interactions between electrons in real solids are probably not the simple direct Heisenberg exchange.

11.1.7 Interaction Energy in Terms of Electron Spin and Quantum Mechanical Exchange Energy

How can the interaction energy be expressed in terms of the exchange interaction between two electrons on neighboring atomic sites?

The exchange coefficient J is an alternative and more fundamental representation of our nearest-neighbor interaction ϑ, which was discussed in Chapters 6 and 7. Typically, J has a value of 10^{-21} Joules in ferromagnets such as iron, nickel, and cobalt. The spin on one electron is always $s = 1/2$, which corresponds to a magnetic moment of one Bohr magneton or 9.27×10^{-24} A m^2. Summing all the spin contributions from electrons in an atom gives the net spin S per atom. In iron, each atom has a net spin of $S = 1$, which gives 2 Bohr magnetons per atom [6].

In practice, the magnetic moments on the atoms are not always integral multiples of a Bohr magneton because there is a contribution to the magnetic moment from the electron orbital motion and indeed the electrons that contribute to the net magnetization in solids such as iron, cobalt, and nickel may not be localized at the atomic sites, which can also lead to nonintegral values of magnetic moment.

In iron, the moment corresponds to a net spin on the atom of $S = 1$. However, in nickel, this is quite seriously an error, since if we take $S = ½$, this suggests a magnetic moment per atom of $0.93\mu_B$, whereas the true value is $0.6\mu_B$. Therefore, the use of the quantum mechanical exchange interaction J, apart from making first principles calculations for very simple systems, is limited. Practical estimates of the value of J can be made using the Curie temperature, the low-temperature saturation magnetization or the specific heat [7,8].

The relation between the classical nearest-neighbor interaction ϑ and the quantum mechanical exchange coefficient J can be obtained from the respective expressions for the exchange energy

$$\mu_0 \vartheta \boldsymbol{m}^2 = J\boldsymbol{s}^2 \tag{11.19}$$

and where appropriate, using the mean-field model

$$\mu_0 \alpha N \boldsymbol{m}^2 = \mu_0 \alpha \boldsymbol{m} M \tag{11.20}$$

$$= nJS^2 \tag{11.21}$$

where:

n is the number of nearest-neighbor atoms
S is the net electron spin per atom
J is the quantum mechanical exchange coefficient

From this, it can be seen that there is a direct link between the phenomenological mean field $\boldsymbol{H}_e$, which is used to explain magnetic ordering on the classical continuum scale, and the quantum mechanical exchange coefficient J on the scale of a few atoms.

$$H_e = \alpha N \boldsymbol{m} = \frac{nJS^2}{\mu_0 \boldsymbol{m}} \tag{11.22}$$

11.1.8 Exchange Interactions between Electrons in Filled Shells

In order to determine the type of magnetic order, is it then necessary to find the exchange energy between every electron in one atom and every electron in a neighboring atom?

It is easy to show on the basis of the Heisenberg model that the exchange energy between electrons in a filled shell of one atom and electrons in a filled shell of a neighboring atom is zero.

$$E_{ex} = -2J\sum\sum s_i \cdot s_j = -2J\sum s_i \sum s_j \tag{11.23}$$

and in a filled shell $\sum s_i = 0$. The exchange energy is consequently zero. This is a useful result because in order to determine exchange energy, we only need to take into account interactions from electrons in partially filled shells, which simplifies the analysis considerably.

11.1.9 Bethe-Slater Curve

Can we make any simple and verifiable predictions of magnetic order in materials based on the quantum theory of exchange interactions?

The magnetic behavior of the 3d elements chromium, manganese, iron, cobalt, and nickel was of interest to the early investigators of the quantum theory of magnetism. Slater [13,14] published values of the interatomic distances r_{ab} and the radii of the incompletely filled d subshell r_d of some transition elements. It was found that the values of r_{ab}/r_d seemed to correlate with the sign of the exchange interaction. For large values of r_{ab}/r_d, the exchange was positive while for small values, it was negative.

Bethe [15] made some calculations of the exchange integral in order to obtain J as a function of interatomic spacing and radius of d orbitals. He found the exchange integral for the electrons on two one-electron atoms, which is given by

$$J = \int\int \psi_a^*(r_1)\psi_b^*(r_2)\left[\frac{1}{r_{ab}} - \frac{1}{r_{a2}} - \frac{1}{r_{b1}} + \frac{1}{r_{21}}\right]\psi_b(r_1)\psi_a(r_2)\,d\tau \tag{11.24}$$

where:

r_{ab} is the distance between the atomic cores
r_{21} is the distance between the two electrons
r_{a2} and r_{b1} are the distances between the electrons and the respective nuclei

Evaluation of this integral was made using the results of Slater and it was found that the exchange integral J becomes positive for small r_{12} and small r_{ab}. It also becomes positive for large r_{a2} and r_{b1}. The exchange integral can therefore be plotted against the ratio r_{ab}/r_d, where r_d is the radius of the 3d orbital. This gives the Bethe-Slater curve as shown in Figure 11.4, which correctly separates the ferromagnetic 3d elements such as iron, cobalt, and nickel from the antiferromagnetic 3d elements chromium and manganese.

This means that if two atoms of the same kind are brought closer together without altering the radius of their 3d shells, then r_{ab}/r_d will decrease. When r_{ab}/r_d is large, then J is small and positive. As the ratio is decreased, J at first increases and then after reaching a maximum decreases and finally becomes negative, indicating antiferromagnetic order at small values of r_{ab}. The exact nature of the exchange interaction is therefore dependent on the interatomic and interelectronic spacing.

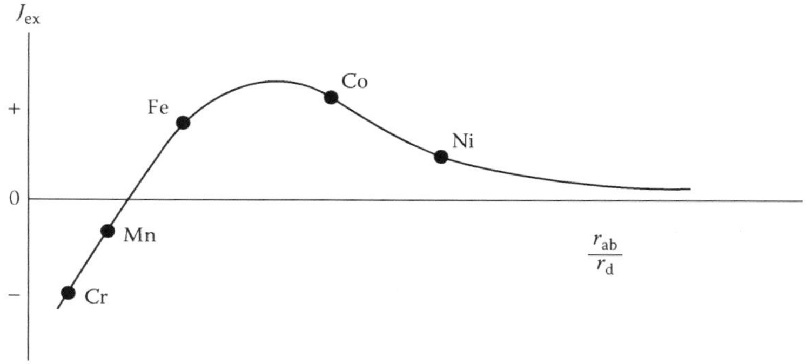

FIGURE 11.4 The Bethe-Slater curve representing the variation of the exchange integral J with interatomic spacing r_{ab} and radius of unfilled d shell r_d. The positions of various magnetic elements on this curve are indicated. The rare earth elements lie to the right of nickel on the curve.

Subsequently, Slater [16], Wohlfarth [9], and Stuart and Marshall [10] expressed dissatisfaction with the Bethe-Slater model after difficulties with the sign and magnitude of J. Finally, Herring [17] showed that the Heitler-London method of calculating the spin coupling is ultimately unreliable.

Therefore, although the whole approach of Heitler-London, Heisenberg, and Bethe still provides a useful conceptual framework for discussing the magnetic interactions of electrons, the method seems to be ultimately inadequate and we await a better description, which can give more accurate values of J from first principles.

11.1.10 Heusler Alloys

Can ferromagnetic alloys be made from elements that are not themselves ferromagnetic?

Perhaps one of the most striking examples of how the magnetic properties of materials are determined by their electron structures, rather than by the properties of the atoms themselves, is given by the Heusler alloys. These alloys are formed from elements that themselves are not ferromagnetic, such as manganese, copper, and aluminum. Yet when alloyed together, these elements can form ferromagnetic materials as a result of the electron configuration in the conduction band.

The Heusler alloys have the general composition A_2MnB, where A can be copper or palladium, and B can be any one of aluminum, gallium, arsenic, indium, tin, antimony, or bismuth. A large number of such alloys have been produced to date, but the two most widely studied are the first two alloys, Cu_2MnAl and Cu_2MnSn. These two alloys have saturation magnetizations comparable with that of nickel.

It is thought that the Heusler alloys owe their magnetic properties to manganese, which is usually antiferromagnetic. However, in the Heusler alloys, the separation between the manganese atoms is about 60% larger than in the pure metal. This observation of ferromagnetism at larger interatomic spacing is often cited as evidence in support of the Bethe-Slater criterion for ferromagnetism. Specifically, this empirical rule states that when the atomic separation of the atoms exceeds one and one half times the diameter of the 3d orbital, then the exchange integral becomes positive, resulting in ferromagnetic order.

11.2 LOCALIZED ELECTRON THEORY

Can the magnetic properties of a solid be described in terms of the properties of electrons localized at the atomic sites?

We now go on to look at the magnetic properties in solids rather than isolated atoms. The most natural extension of the above discussion of atomic magnetism is to consider that the magnetic moments of atoms in a solid are due to electrons localized at the ionic sites. This means that we can deal with the magnetic properties of solids as merely a perturbation of the magnetic properties of the individual atoms. We shall see how far this proves to be correct in the two main groups of magnetic elements, the 3d and 4f series.

11.2.1 Atomic Magnetic Moment Due to Localized Electrons

How is the atomic magnetic moment determined from the angular momentum of the atom?

We can apply the above ideas to calculate the magnetic moment of an atom if we assume that the electrons in the unfilled shell, which contribute to the magnetic moment, are all localized at the ionic sites. The magnetic moment of an atom is determined on this basis from the total angular momentum of the isolated atom $(h/2\pi)\boldsymbol{J}$, which is obtained as a vector sum of the orbital and spin angular momenta of electrons in unfilled shells.

$$\boldsymbol{m} = -g\mu_B \boldsymbol{J} \tag{11.25}$$

where:

μ_B is the Bohr magneton

g is the Lande splitting factor where

$$g = 1 + \frac{J(J+1) + S(S+1) - L(L+1)}{2J(J+1)} \tag{11.26}$$

The magnetic moment can equally be well expressed as

$$\boldsymbol{m} = \gamma\left(\frac{h}{2\pi}\right)\boldsymbol{J} \tag{11.27}$$

as discussed in Chapter 3, where γ is the gyromagnetic ratio and h is Planck's constant. We should note, however, that the values of total angular momentum $\boldsymbol{J}$ will not always be the same in the solid as in the free atoms discussed above. In the 3d series metals, there are significant differences, even allowing for the quenching of the orbital angular momentum, but in the 4f or lanthanide series, there is good agreement between the magnetic moments in the solid and the isolated atoms.

11.2.2 Quantum Theory of Paramagnetism

How do all these ideas combine to provide an explanation of paramagnetism, and how different is this theory from the classical models?

11.2.2.1 Single-Electron Atoms

By single-electron atoms, we mean atoms with only one unpaired electron contributing to the magnetic properties. A good example is nickel. If we consider the energy of a single magnetic moment $\boldsymbol{m}$ under the action of a magnetic field $\boldsymbol{H}$, the energy is given by

$$E = -\mu_0 \boldsymbol{m} \cdot \boldsymbol{H} \tag{11.28}$$

We have now an expression for the magnetic moment of an isolated, and hence paramagnetic, atom

$$\boldsymbol{m} = g\mu_B \boldsymbol{J} \tag{11.29}$$

$$E = -\mu_0 g \mu_B \boldsymbol{J} \cdot \boldsymbol{H} \tag{11.30}$$

If we consider only the spins on the electrons and ignore the orbital contribution, then $g = 2$ and $J = S = \boldsymbol{s} = 1/2$ for a single-electron spin. As the spin can only have the values $\boldsymbol{m}_s = \pm 1/2$, this means that, in the case of a single-electron atom, the electronic magnetic moment, as a result of quantization, can align only either parallel $(\boldsymbol{m}_s = 1/2)$ or antiparallel $(\boldsymbol{m}_s = -1/2)$ to the field direction.

An atom with one unpaired electron therefore has two possible energy states in the magnetic field. The estimation of the magnetization using Boltzmann statistics leads to

$$\boldsymbol{M} = N\boldsymbol{m} \tanh\left(\frac{\mu_0 \boldsymbol{m} H}{k_B T}\right) = NgJ\mu_B \tanh\left(\frac{\mu_0 gJ\ \mu_B H}{k_B T}\right) \tag{11.31}$$

where:

$\boldsymbol{M}$ is the bulk magnetization or magnetic moment per unit volume
N is the number of atoms per unit volume
$\boldsymbol{m}$ is the magnetic moment per atom

Solutions of this equation as a function of $\boldsymbol{H}/T$ are shown as a special case of the Brillouin function with $J = 1/2$ in Figure 11.5.

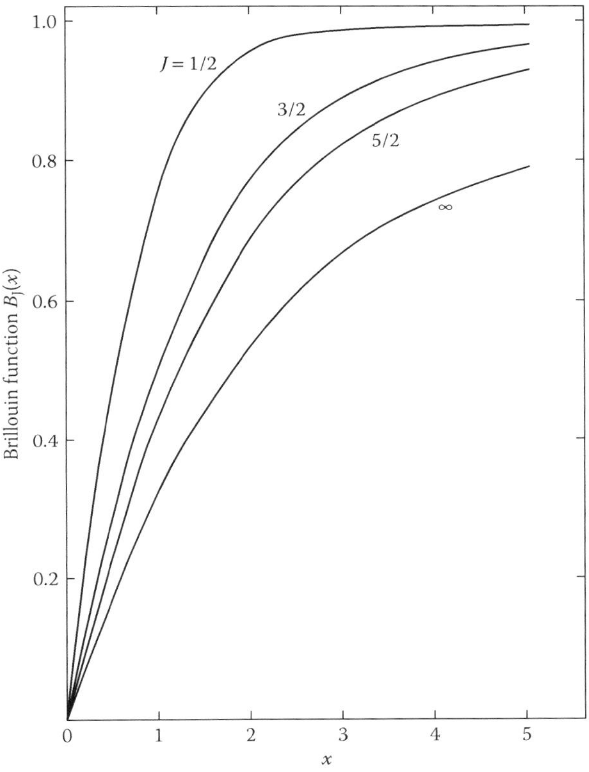

FIGURE 11.5 The value of the Brillouin function $B_J(x)$ for various values of J ($1/2, 3/2, 5/2$, and infinity) and as a function of x. In the quantum theory of paramagnetism, for example, $x = \mu_0 gJ\,\mu_B H/k_B T$. (After Brillouin, L., *J. de Phys. Radium*, 8, 74, 1927.)

For values of $\mu_0 \boldsymbol{m} H / k_B T << 1$, this leads to the approximation

$$\boldsymbol{M} = \frac{N\boldsymbol{m}^2 \mu_0 H}{k_B T} \tag{11.32}$$

and

$$\chi = \frac{N\boldsymbol{m}^2 \mu_0}{k_B T} \tag{11.33}$$

which is the Curie law.

11.2.2.2 Multielectron Atoms

By multielectron atoms, we mean simply atoms with more than one electron contributing to the magnetic properties. Nickel is, by this definition, a single-electron atom, while iron is a multielectron atom. In these more complex situations, there are $2J + 1$ energy levels and so the expression for the magnetization is

$$\boldsymbol{M} = NgJ\mu_B B_J\left(\frac{gJ\mu_B\mu_0\boldsymbol{H}}{k_BT}\right) \tag{11.34}$$

where $B_J(x)$ is the Brillouin function [18], which is defined as

$$B_J(x) = \left[\frac{(2J+1)}{2J}\right]\coth h\left[\frac{(2J+1)x}{2J}\right] - \left(\frac{1}{2J}\right)\coth h\left(\frac{x}{2J}\right) \tag{11.35}$$

Equations 11.34 and 11.35 for the single-electron atoms are then seen to be merely special or restricted cases of this general function.

The equation for the magnetization of the multiple electron atom is the quantum analog of the Langevin function [19] given in Chapter 9. In the limiting case when $\boldsymbol{J} \to \infty$; that is, when there are no quantum mechanical restrictions on the allowed directions and values of magnetic moment in the atom, we arrive at the classical Langevin expression for magnetization.

Solutions for the paramagnetic magnetization equation are shown for different values of S in Figure 11.6 together with experimental values for three paramagnetic salts as determined by Henry [20].

11.2.2.3 Curie Law

The Curie law of paramagnetic susceptibilities can be derived from the quantum theory of paramagnetism. For $\mu_0\boldsymbol{mH}/k_BT << 1$, the Brillouin function becomes

$$B_J\left(\frac{\mu_0\boldsymbol{mH}}{k_BT}\right) = \frac{\mu_0\boldsymbol{mH}}{3k_BT} \tag{11.36}$$

$$\boldsymbol{M} = \frac{N\mu_0\boldsymbol{m}^2\boldsymbol{H}}{3k_BT} \tag{11.37}$$

$$\chi = \frac{\boldsymbol{M}}{\boldsymbol{H}} = \frac{N\mu_0\boldsymbol{m}^2}{3k_BT} = \frac{N\mu_0 g^2\mu_B^2 J(J+1)}{3k_BT} = \frac{C}{T} \tag{11.38}$$

which is the Curie law, this time with

$$C = \frac{N\mu_0 g^2\mu_B^2 J(J+1)}{3k_B} \tag{11.39}$$

We see from this discussion that the quantum theory of paramagnetism provides qualitatively similar results to the classical theories; however, the classical Langevin function needs to be replaced by its quantum mechanical analog, the Brillouin function, which leads to different numerical results.

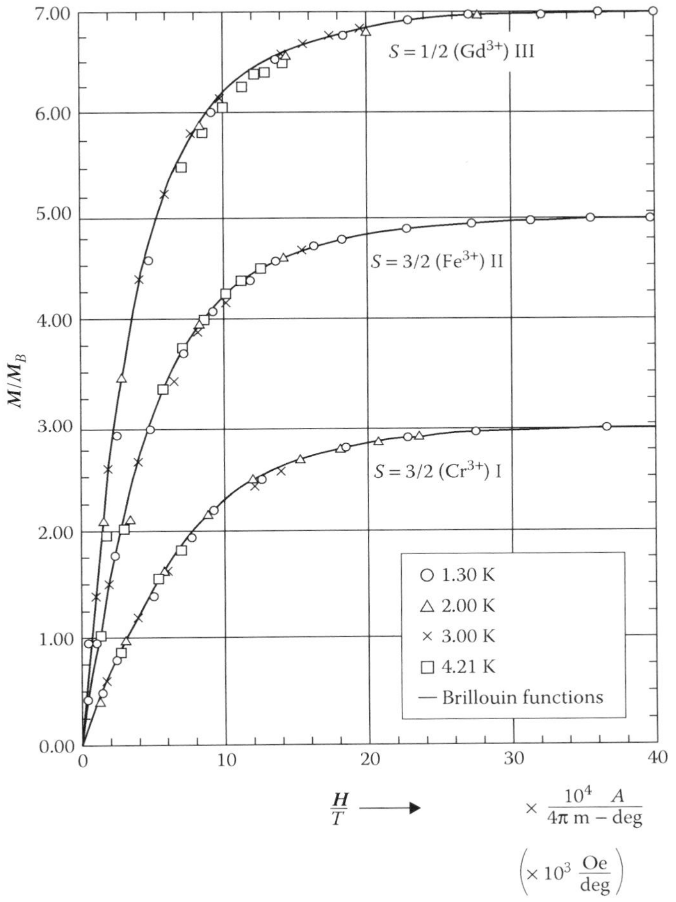

FIGURE 11.6 Comparison of theory and experiment for the magnetization curves of three paramagnetic salts containing Gd^{3+}, Fe^{3+}, and Cr^{3+} ions, respectively. These salts are potassium chromium alum, ferric ammonium alum, and gadolinium sulfate octahydrate. (After Henry, W.E., *Phys. Rev.*, 88, 559, 1952.)

11.2.3 Quantum Theory of Ferromagnetism

How can we extend the above theories to a quantum theory of ferromagnetism and how different is this from the classical models?

The quantum theory of ferromagnetism is derived from the quantum theory of paramagnetism in much the same way as the classical Weiss theory of ferromagnetism was derived from the Langevin theory of paramagnetism. A perturbation in the form of an interaction, or exchange coupling, is introduced into the quantum theory of paramagnetism, so that the electrons on neighboring atoms interact with one another. The energy of an electron in a magnetic field therefore becomes, as before

$$E = -\mu_0 \boldsymbol{m}(\boldsymbol{H} + \alpha \boldsymbol{M}) \tag{11.40}$$

where $\alpha\boldsymbol{M}$ represents a mean field form of the interaction of the electron magnetic moment with those other electrons close by.

11.2.3.1 Magnetization

In the case of an atom with one electron contributing to the net magnetic moment, this leads via statistical thermodynamics to

$$\boldsymbol{M} = NgJ\mu_B \tanh\left[\frac{\mu_0 gJ\mu_B(\boldsymbol{H} + \alpha\boldsymbol{M})}{k_B T}\right] \tag{11.41}$$

and for atoms with many electrons contributing to the net magnetic moment this leads to

$$\boldsymbol{M} = NgJ\mu_B B_J\left[\frac{gJ\mu_B\mu_0(\boldsymbol{H} + \alpha\boldsymbol{M})}{k_B T}\right] \tag{11.42}$$

where $B_J(x)$ is the Brillouin function.

11.2.3.2 Curie-Weiss Law

At higher temperatures, this system will be paramagnetic, so that $\boldsymbol{M}$ will be uniform throughout the material, and hence

$$\boldsymbol{M} = \frac{N\mu_0 g^2\mu_B^2 J(J+1)(\boldsymbol{H} + \alpha\boldsymbol{M})}{3k_B T} \tag{11.43}$$

$$\chi = \frac{\boldsymbol{M}}{\boldsymbol{H}} = \frac{N\mu_0 g^2\mu_B^2 J(J+1)/3k_B T}{\left[1 - \alpha N\mu_0 g^2\mu_B^2 J(J+1)\right]/3k_B T} \tag{11.44}$$

$$= \frac{N\mu_0 g^2\mu_B^2 J(J+1)}{3k_B T - \alpha N\mu_0 g^2\mu_B^2 J(J+1)} \tag{11.45}$$

which is the Curie-Weiss law, in which

$$C = N\mu_0 g^2\mu_B^2 J(J+1) \tag{11.46}$$

$$T_c = \alpha\frac{N\mu_0 g^2\mu_B^2 J(J+1)}{3k_B T} \tag{11.47}$$

We see therefore that the quantum theory of ferromagnetism also leads to qualitatively similar results to the classical theory.

In the ferromagnetic regime, the magnetization $\boldsymbol{M}$ will not be uniform throughout the solid. Since any individual magnetic moment is likely to be more affected by other moments nearby than by distant magnetic moments, the moments will couple to the spontaneous magnetization within the domain $\boldsymbol{M}_s$ rather than the bulk magnetization $\boldsymbol{M}$ when in the ferromagnetic state. Therefore, a nearest-neighbor model, with possible extension to next and higher-order neighbors, is more appropriate in the ferromagnetic regime.

11.2.4 Temperature Dependence of the Spontaneous Magnetization within a Domain

How does the magnetization within a domain vary with temperature according to the quantum theory of ferromagnetism?

The spontaneous magnetization within a domain $\boldsymbol{M}_s$ is determined by solving the ferromagnetic Brillouin function in the absence of an applied field.

$$\boldsymbol{M}_s = NgJ\mu_B B_J\left(\frac{\mu_0 gJ\mu_B \alpha \boldsymbol{M}_s}{k_B T}\right) \tag{11.48}$$

In the case of nickel, we can use the $S = 1/2$ solution so that

$$\boldsymbol{M}_s = NgJ\mu_B \tanh\left(\frac{\mu_0 gJ\mu_B \alpha \boldsymbol{M}_s}{k_B T}\right) \tag{11.49}$$

Solutions of this equation are shown in Figure 11.7. It is seen from this that the spontaneous magnetization is only weakly dependent on temperature until it reaches $0.75T_c$. Above that, it decreases rapidly toward zero at the Curie temperature.

11.2.5 Exchange Coupling in Magnetic Insulators

How are the magnetic moments coupled in magnetic insulators, in which the wave functions of the magnetic *electrons localized on the metallic atom sites do not overlap?*

Of course, not all magnetic materials are metallic, and so the question that arises is, how is the exchange mechanism maintained when the wave functions of the relevant electrons do not overlap? In these cases, it is clear that the simple direct exchange mechanism described above cannot be responsible for ordering of the electron moments. When the metallic ions in the material are separated by a nonmetallic ion,

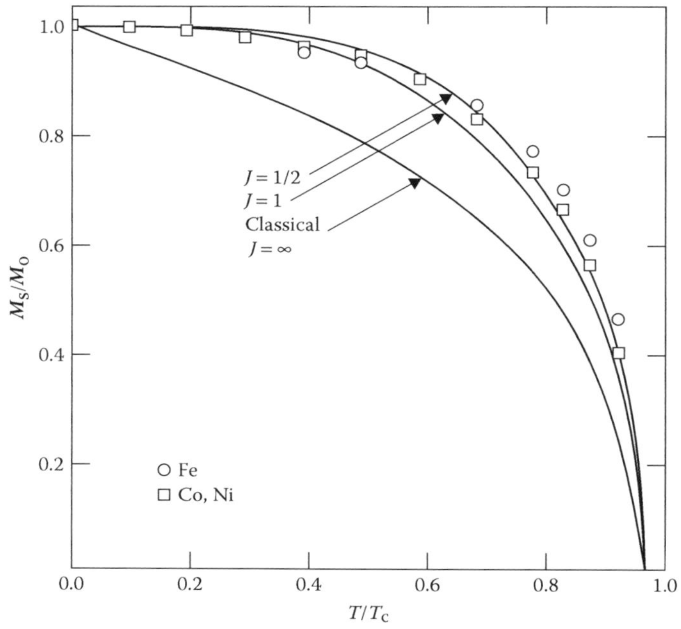

FIGURE 11.7 Temperature dependence of the spontaneous magnetization within a domain for iron, cobalt, and nickel, compared with calculations based on the ferromagnetic Brillouin function.

it is possible for the electrons on the metallic ions to be coupled together via their interactions with electrons on the nonmetallic ion. This form of exchange coupling is known as *superexchange*.

11.2.6 Critique of the Local Moment Model

How well does this model work and how realistic is it?

The local moment model works well in that it does provide relatively simple mathematical functions for the dependence of magnetization on magnetic field and temperature. It is quite realistic for the lanthanides with their closely bound 4f electrons, which determine the magnetic properties of the solid. In most other cases, the magnetic properties of paramagnets and the other main group of ferromagnets, the 3d series, the "magnetic" electrons are outer electrons, which are relatively free to move through the solid. This situation, in which the electrons that determine the magnetic properties of the material are now itinerant, violates the basic assumptions of the local moment theory. Furthermore, the magnetic moments per atom in iron, cobalt, and nickel in their solid form are not integral multiples of the electron spin, the Bohr magneton, as required under the Heitler-London-Heisenberg local moment theory. Therefore, the localized model, despite its successes, is not a realistic model for these cases.

11.3 ITINERANT ELECTRON THEORY

What happens if the magnetic *electrons are not localized at the atomic sites?*

If the magnetic electrons are in unfilled shells, then in a number of metals, these magnetic electrons are unlikely to be localized as described above. The unfilled shells are in most cases among the furthest removed from the nucleus, and it is these electrons that are most easily removed. This means that alternative models must be sought for these metals in order to provide a realistic theory.

11.3.1 Magnetism of Electrons in Energy Bands

What are the magnetic properties of electrons in energy bands?

So far, we have only considered the magnetic properties of solids in terms of localized magnetic moments, which behave as if they were attached to the atomic cores in the material. Thus, we have talked about atomic magnetic moments. However, we have come to realize that these magnetic moments are really due to the angular momentum of unpaired electrons in unfilled shells. With the exception of the lanthanides, the unpaired *magnetic* electrons are usually outer electrons and so are unlikely to be closely bound to the atoms. This is true of the 3d transition elements iron, nickel, and cobalt. We therefore need to find a theoretical description of magnetism due to itinerant electrons in these cases.

Metals such as the alkali metal series lithium, sodium, potassium, rubidium, and caesium all show temperature independent paramagnetism, for example, which cannot be explained by the local moment model. In this case, the paramagnetism is due to the outer electrons, which behave as a free electron gas.

11.3.2 Pauli Paramagnetism of Free Electrons

Can paramagnetism be described simply on the basis of changes in population of nearly free electrons in bands?

The Langevin theory of paramagnetism and its quantum mechanical analog work for dilute paramagnets such as the hydrated salts of transition metal ions. However, it is well known that the paramagnetic susceptibility of most metals is independent of temperature and hence does not follow the classical Langevin theory. In addition, the paramagnetism of most metals is considerably weaker than would be expected on the basis of the localized model.

The reason that the paramagnetic susceptibility is so much lower is that the electrons are in general not free to rotate into the field direction as required by the Langevin model. This is because, as a result of the Pauli exclusion principle [7], the electron states needed for reorientation are already occupied by other electrons. Only those electrons within an energy $k_B T$ of the Fermi level are able to change

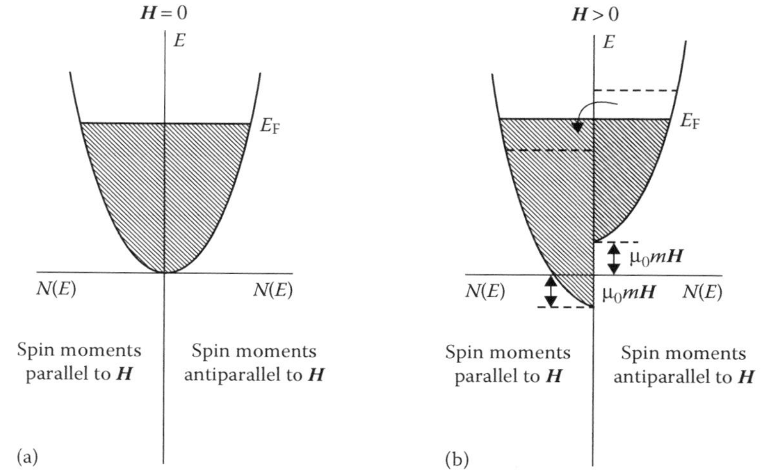

FIGURE 11.8 Pauli paramagnetism for nearly free electrons in a magnetic field $\boldsymbol{H}$. In (a), the field is zero so that both spin-up and spin-down sub-bands have identical energy. In (b), the field causes an energy difference between sub-bands with spins aligned parallel and antiparallel to the field.

orientation. We therefore need to consider an alternative model of paramagnetism due to the band electrons as conceived by Pauli [21].

Consider therefore the parabolic distribution of nearly free electrons as shown in Figure 11.8. Only the fraction T/T_f of the total number of electrons can contribute to the magnetization, where T is the thermodynamic temperature and T_f is the Fermi temperature, defined such that $E_f = k_B T_f$. Therefore,

$$\boldsymbol{M} = \left(\frac{N\mu_0\boldsymbol{m}^2\boldsymbol{H}}{3k_B T}\right)\left(\frac{T}{T_f}\right) = \frac{N\mu_0\boldsymbol{m}^2\boldsymbol{H}}{3k_B T_f} \tag{11.50}$$

The number density of electrons parallel to the field is

$$N_+ = \frac{1}{2}\int f(\varepsilon)D(\varepsilon + \mu_0\boldsymbol{mH})d\varepsilon \approx \frac{1}{2}\int f(\varepsilon)D(\varepsilon)d\varepsilon + \frac{1}{2}\mu_0\boldsymbol{mH}D(\varepsilon_f) \tag{11.51}$$

and the number density of those antiparallel to the field

$$N_- = \frac{1}{2}\int f(\varepsilon)D(\varepsilon - \mu_0\boldsymbol{mH})d\varepsilon \approx \frac{1}{2}\int f(\varepsilon)D(\varepsilon)d\varepsilon - \frac{1}{2}\mu_0\boldsymbol{mH}D(\varepsilon_f) \tag{11.52}$$

The magnetization is then given by

$$\boldsymbol{M} = \boldsymbol{m}(N_+ - N_-) = \boldsymbol{m}^2\mu_0\boldsymbol{H}D(\varepsilon_f) = \frac{3N\boldsymbol{m}^2\mu_0\boldsymbol{H}}{2k_B T_f} \tag{11.53}$$

Equation 11.53 gives the Pauli spin magnetization of the paramagnetic conduction band electrons. The susceptibility is then

$$\chi = \frac{3N\boldsymbol{m}^2\mu_0}{2k_B T_f} \tag{11.54}$$

which is temperature independent since T_f is a constant.

11.3.3 Band Theory of Ferromagnetism

How do electrons in bands behave when there is an exchange interaction, which causes alignment of the spins?

In metallic ferromagnets such as iron, cobalt, and nickel, the magnetic properties are due principally to the conduction electrons, and in these cases, an argument can be made in favor of *itinerant exchange* between these electrons. This form of exchange interaction, which has been examined in great detail by Herring [22], is conceptually different from the direct and superexchange theories, which have been developed for localized moments.

The band theory of ferromagnetism is a simple extension of the band theory of paramagnetism by the introduction of an exchange coupling between the electrons. The simplest case is to consider the electrons to be entirely free, that is, a parabolic energy distribution. This simplifying assumption does not alter the main conclusions of the theory. The band theory of ferromagnetism was first proposed by Stoner [23] and then independently by Slater [24].

Since magnetic moments can only arise from unpaired electrons, it is immediately clear that a completely filled energy band cannot contribute a magnetic moment since in such an energy band all electron spins will be paired giving $L = 0$ and $S = 0$. In a partially filled energy band, it is possible to have an imbalance of spins leading to a net magnetic moment per atom. This arises because the exchange energy removes the degeneracy of the spin-up and spin-down half bands as shown in Figure 11.9. The larger the exchange energy, the greater the difference in energy between these two half bands. Electrons fill up the band by occupying the lowest energy levels first. If the half bands are split as in Figure 11.9a, the electrons can begin to occupy the spin-down half band before the spin-up half band is full. This usually leads to a nonintegral number of magnetic moments per atom. A large exchange splitting can lead to a separation between the half bands as in Figure 11.9b. In this case, the spin-up band must be filled before electrons can enter the spin-down band. This leads to an integral number of magnetic moments per atom.

For example, consider the situation in Figure 11.10, where 10 electron states, provided by 10 atoms each donating one electron, exist in close proximity in an electronic band. A spin imbalance of 2 can be created by flipping one of the moments into the opposite direction. This then gives rise to a magnetic moment of $0.2\mu_B$ per atom. We now see the possibility of atomic magnetic moments that are nonintegral multiples of the Bohr magneton.

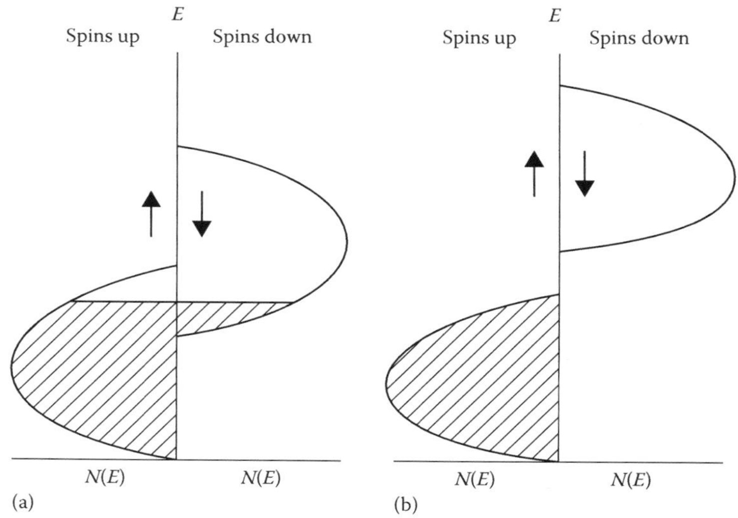

FIGURE 11.9 (a) A schematic band structure density of states, showing exchange splitting of the spin-up and spin-down half bands, with remaining overlap between the two half bands leading to a magnetic moment with nonintegral number of Bohr magnetons per atom and (b) a schematic band structure density of states with a larger exchange splitting leading to complete energy separation of the half bands and to a magnetic moment with an integral number of Bohr magnetons per atom.

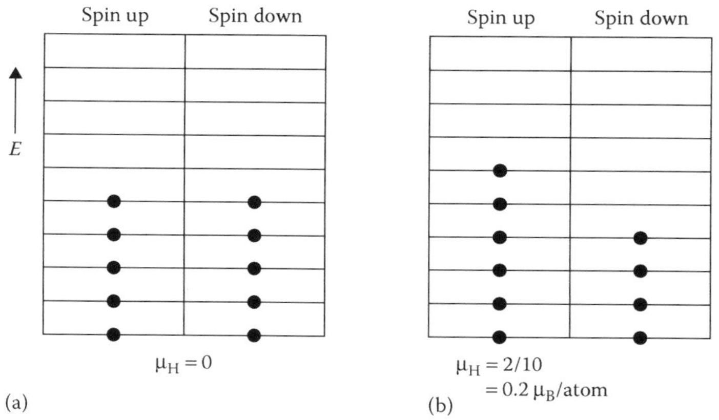

FIGURE 11.10 Diagram representing the occupation of allowed energy states in an electron band: (a) balanced numbers of spin-up and spin-down electrons and (b) unbalanced numbers of spin-up and spin-down electrons due to the exchange interaction, leading to a net magnetic moment per atom.

11.3.4 Magnetic Properties of 3d Band Electrons

Can the band theory provide a satisfactory description of the magnetic properties of the 3d metals?

In the transition metals such as iron, nickel, and cobalt, which are the three ferromagnetic elements with electronic structure for which the band theory should apply, the magnetic properties are due to the 3d band electrons. Of course, there is also a 4s electron band but this contains two paired electrons and so does not affect magnetic properties.

The 3d band can hold up to 10 electrons (5 up and 5 down), and it is here that we must concentrate attention in order to explain the observed properties. The exchange interaction is responsible for creating the imbalance in the spin-up and spin-down states. In the absence of the exchange energy, the spin imbalance would be an excited state, but this does not require too much energy in the 3d band because of the high density of states and therefore a positive exchange interaction can be sufficient to cause the alignment, resulting in a spin imbalance and a net magnetic moment per atom.

If we suppose that in these metals, the exchange interaction causes five of the 3d electrons to align *up* and the remainder *down*, we arrive at an equation, which approximates observed magnetic moments in these elements. Let n be the number of 3d + 4s electrons per atom, x be the number of 4s electrons per atom, and $n - x$ be the number of 3d electrons per atom.

$$\boldsymbol{m} = \left[5 - (n - x - 5)\right]\mu_B = \left[10 - (n - x)\right]\mu_B \tag{11.55}$$

We can approximate observed magnetic properties by the assumption that $x = 0.6$.

$$\boldsymbol{m} = (10.6 - n)\mu_B \tag{11.56}$$

which for nickel with $n = 10$, cobalt with $n = 9$, and iron $n = 8$ gives the following:

$$\text{Ni} : \boldsymbol{m} = 0.6\mu_B \tag{11.57}$$

$$\text{Co} : \boldsymbol{m} = 1.6\mu_B \tag{11.58}$$

$$\text{Fe} : \boldsymbol{m} = 2.6\mu_B \tag{11.59}$$

These results are close to the known values. In this way, the band theory can account for the nonintegral atomic magnetic moments in these metals, a result that is more difficult to justify on the localized moments model.

11.3.5 Slater-Pauling Curve

How well does the itinerant electron model describe the magnetic properties of 3d alloys?

The above argument can be used to explain the moment per atom of several of the 3d transition metals from manganese to copper. The Slater-Pauling curve [25,26] gives the magnetic moments of these 3d metals and their alloys from the premises of the itinerant electron theory. This is shown in Figure 11.11. Most alloys fall on a locus consisting of two straight lines beginning at chromium with 0 Bohr magnetons rising to 2.5 Bohr magnetons between iron and cobalt and dropping to 0 Bohr magnetons again between nickel and copper. The metals in this range, chromium, manganese, iron, cobalt, nickel, and copper, have total numbers of electrons ranging from 24 to 29, while the number of 3d electrons ranges from 5 in chromium to 10 in copper.

The interpretation of the results in Figure 11.11 is in terms of the rigid band model. It is considered that the alloy metals share a common 3d band to which both elements contribute electrons. The maximum magnetic moment occurs at a point between iron and cobalt. It appears from this model that, as expected, the 3d and 4s electrons are responsible for the magnetic properties of these metals and alloys, and that they are relatively free. Therefore, it is a reasonable assumption that they are shared between the ions in a common 3d band.

It has been suggested [26] that the 3d band is broken into two parts: the upper part capable of holding 4.8 electrons (2.4 up and 2.4 down) and the lower part capable of holding 5.2 electrons (2.6 up and 2.6 down). This means that as electrons are removed beginning with zinc, for example, the magnetic moment is increased by depletion

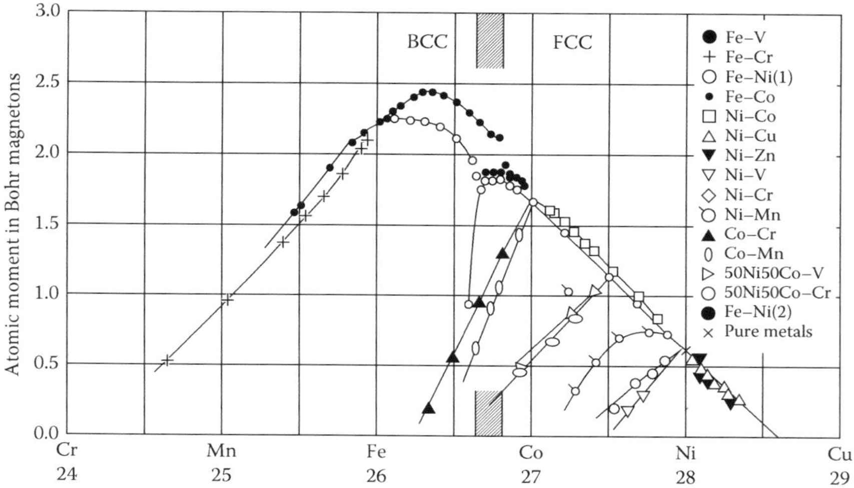

FIGURE 11.11 The Slater-Pauling curve that gives the net magnetic moment per atom as a function of the number of 3d electrons per atom. The various 3d elements are shown at the relevant points on the *x*-axis, but the curve is more general, being primarily of interest in explaining the magnetic moments of various intra-3d alloys.

of spin down electrons in the upper part of the band until a magnetic moment of 2.4 Bohr magnetons is reached between iron and cobalt. Then the moment begins to decrease again toward chromium because the removal of further electrons results in a reduction in spin-up electrons from the upper part of the 3d band.

11.3.6 Critique of the Itinerant Electron Model

What are the strengths and weaknesses of the itinerant electron model?

The itinerant electron model has had successes and these include the ability to explain nonintegral values of atomic magnetic moments and to predict some aspects of the magnetic behavior of the 3d series metals and alloys. Furthermore, the localized moment model, which is the main alternative, is certainly open to the objection that it is unrealistic in so far as the magnetic electrons of most atoms, with the exception of the lanthanides, are relatively free being in the outer shells.

The drawback of the itinerant electron theory is that it is extremely difficult to make fundamental calculations based on it. Unlike the local moment theory, which lends itself readily to simple models such as the Weiss [27] and Heisenberg [28] models of ferromagnetism, the itinerant electron theory does not provide any simple model from which first principles calculations can be made. Therefore, although the current opinion is that the itinerant theory is intrinsically closer to reality in most cases, interpretations of magnetic properties are still more often made on the basis of the localized moment model.

11.3.7 Correlation Effects among Conduction Electrons

Since 3d magnetic metals seem to have features of both itinerant and localized moment models, how can this be explained?

The local moment and the itinerant electron models have clearly identified limitations as described above. However, ferromagnetic metals, particularly the 3d transition metals, have features that are characteristic of both types of model. Given that the assumptions of the two models appear to be mutually exclusive, it is difficult at first sight to see how these features can be reconciled. As examples, we can consider the occurrence of spin-wave phenomena in these materials and the strong temperature dependence of the susceptibility. These are indicative or characteristic of a local moment model. However, the occurrence of magnetic moments with nonintegral numbers of Bohr magnetons per atom is characteristic of an itinerant electron model.

These apparent contradictions have been resolved by the model suggested by Hubbard [29]. In this treatment, the 3d electrons bands are considered to be narrow enough that with the electron charge density concentrated in the vicinity of the atomic sites, it becomes meaningful to describe an electron as being located on, or at least associated with, a particular atom. The correlations of the electron spins on the atomic sites can then be understood, because although the actual conduction electron associated with each atomic site will be continually changing, the atom will tend to attract electrons with the same spin as a result of Hund's rules and repel electrons with opposite spin. This means that the magnetic moment of a particular atomic site will tend to persist even though it is caused by conduction electrons, which are always changing as a result of their itinerant or bandlike character. In this way, it becomes apparent why the magnetic moments in these metals appear to have some characteristics normally attributable to localized electronic magnetic moments.

In real materials, the situation is itself usually somewhere in between the localized and itinerant models. The conditions of the Hubbard model can be relaxed to represent either localized or bandlike behavior in the appropriate limits. Thus strong correlation of the electron spins on a particular atom, but weak correlation with other conduction band electrons, tends to produce behavior characteristic of the local moment model. Weak correlations of electron spins on a particular atom together with itinerant exchange coupling with other conduction electrons tend to produce behavior characteristic of the band model of ferromagnetism.

Ultimately, this model is itself rather oversimplified, but it does at least contain a bare minimum of features that enable it to address electron characteristics ranging from itinerant to localized moment behavior. However, even with these simplifications, the model is too difficult for exact general solutions to be provided, although in special cases, solutions have been obtained, which provide some interesting insights into these mechanisms. Nevertheless, the Hubbard model continues to be used because it provides a simple method of treating electron correlations in the itinerant electron theory of ferromagnetism, as in the recent work of Edwards and von der Linden [30] and Samson [31].

11.3.8 Indirect Exchange

How can the electron spins be coupled on localized atomic sites if their wave functions do not overlap?

In the lanthanide metals, the 4f *magnetic* electrons are highly localized and so at first, it is difficult to see how the exchange interaction is propagated. The idea of an indirect exchange between the unpaired magnetic electrons localized on neighboring atoms via the conduction electrons was introduced at first for the 3d elements by Vonsovski [32] and Zener [33]. In this form of exchange, the polarization of the spins on the conduction electrons as a result of exchange with the unpaired bound electrons could allow ordering of the electron spins on neighboring atomic sites, and could also account for the nonintegral number of Bohr magnetons per atom (i.e., nonintegral number of net electron spins per atom) that are known to occur in these materials.

This concept of indirect exchange was subsequently investigated by Ruderman and Kittel [34], Kasuya [35], and Yosida [36] as a means of explaining the exchange coupling in the rare earth metals, in which the unpaired, highly localized, 4f electrons are ordered, and yet are so closely bound to the atomic sites that their wave functions do not overlap. This later theory of indirect exchange, which became known as the *RKKY* theory, relies also on the propagation of exchange coupling via the polarized conduction electrons, and can lead to ferromagnetic, antiferromagnetic, sinusoidal, and helical arrangements of the localized magnetic moments under different conditions. This model therefore takes into account the presence of both conduction and localized electronic magnetic moments in providing an explanation of magnetic ordering in these materials.

11.3.9 Giant Magnetoresistance in Multilayers

How can the large change in resistance with applied field in magnetic multilayers be explained?

In Section 5.4, we described magnetoresistance, including the special case of giant magnetoresistance in multilayers. This giant magnetoresistance occurs in some multilayers when the relative orientations of the magnetic moments are changed as shown in Figure 11.12. If the magnetic moments in the successive layers are parallel, the material can have relatively high conductivity, but when the moments are antiparallel, the material can have relatively high resistivity.

The band theory of magnetism as described above provides an explanation of this effect, as described by Fert et al. [37]. The change in resistance with change in relative orientation of the directions of magnetization in the successive magnetic layers can be attributed to the spin-dependent conduction of the electrons in these itinerant electron ferromagnets. The majority (e.g., spin up) and the minority (e.g., spin down) electrons have different densities of states and mobilities at any given energy due to the exchange splitting of the energy bands. Consequently, the conductivities of these two groups of electrons can be very different. Therefore, electrons with a particular energy and spin orientation, which pass from a layer where the

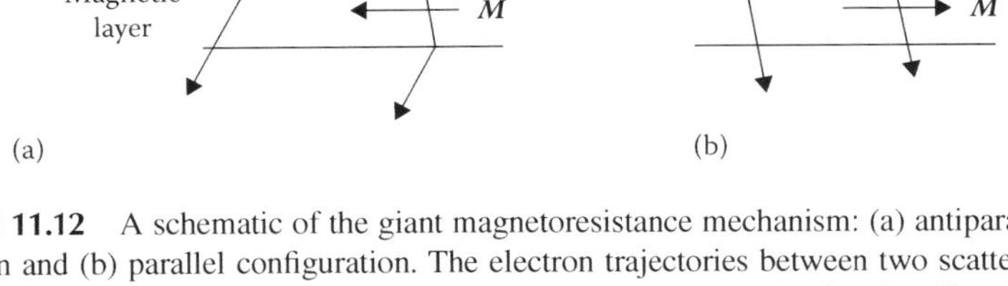

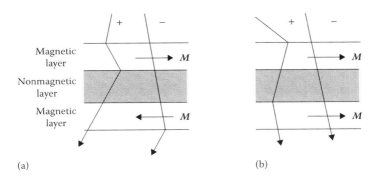

FIGURE 11.12 A schematic of the giant magnetoresistance mechanism: (a) antiparallel configuration and (b) parallel configuration. The electron trajectories between two scatterings are represented by straight lines and the scatterings by abrupt changes in direction (for simplicity, the figure is drawn with only interface scattering). The signs + and − are for electrons spins $s_z = +1/2$ and $-1/2$, respectively. The arrows represent the majority spin direction in the magnetic layers. In the ferromagnetic (F) configuration at the right, the spin—electrons are not scattered, which gives short-circuit effect and leads to a small resistivity. In the antiferromagnetic (AF) or opposite configuration at the left, each spin direction is scattered in every second magnetic layer, there is no short-circuit effect and the resistivity is higher. The schematic is for mean free path length much larger than the layer thickness. (After Fert, A. et al., *JMMM*, 140, 1, 1995.)

majority spins are, for example, pointing to the right, as shown in Figure 11.12, to a layer where the majority spins are pointing to the left, will experience a dramatic change in conductivity.

In the situation, where the electron mean free path is larger than the width of the layers, these electrons, which move across the layer boundaries, will find that in the antiparallel configuration shown in Figure 11.12a, they are strongly scattered in every second layer, whereas in the parallel configuration shown in Figure 11.12b, one group of electrons will be only weakly scattered in both magnetic layers, while the other group will be strongly scattered in both magnetic layers. This gives the so-called spin valve, because with the parallel configuration, there is one group of electrons that are only weakly scattered everywhere, a phenomenon that Fert calls the *short-circuit effect*, so that the resistivity is much lower in the parallel configuration than in the antiparallel configuration. A survey of the main ideas behind giant magnetoresistance in multilayers by White [38] provides a useful introduction to the subject.

REFERENCES

1. Martin, D. H. *Magnetism in Solids*, Illife, London, 1967, Ch. 5.
2. Heitler, W., and London, F. *Z. Phys.*, 44, 455, 1927.
3. Heisenberg, W. *Z. Phys.*, 49, 619, 1928.
4. Dirac, P. A. M. *Proc. Roy. Soc. A*, 123, 714, 1929.
5. Myers, H. P. *Introductory Solid State Physics*, Taylor & Francis, London, 1990.
6. Weiss, P. *J. de Phys.*, 6, 661, 1907.
7. Pauli, W. *Z. Phys.*, 31, 765, 1925.
8. Hofmann, J., Paskin, A., Tauer, K. J., and Weiss, R. J. *J. Phys. Chem. Sol.*, 1, 45, 1956.
9. Wohlfarth, E. P. *Nature*, 163, 57, 1949.
10. Stuart, R., and Marshall, W. *Phys. Rev.*, 120, 353, 1960.
11. Watson, R. E., and Freeman, A. J. *Phys. Rev.*, 123, 2027, 1961.
12. Watson, R. E., and Freeman, A. J. *J. Phys. Soc. Jap.*, 17 (B-l), 40, 1962.
13. Slater, J. C. *Phys. Rev.*, 35, 509, 1930.
14. Slater, J. C. *Phys. Rev.*, 36, 57, 1930.
15. Bethe, H. *Handb. d. Phys.*, 24, 595, 1933.
16. Slater, J. C. *Phys. Rev.*, 49, 537, 1936; 49, 931.
17. Herring, C. *Rev. Mod. Phys.*, 34, 631, 1962.
18. Brillouin, L. *J. de Phys. Radium*, 8, 74, 1927.
19. Langevin, P. *Ann. de Chem. et Phys.*, 5, 70, 1905.
20. Henry, W. E. *Phys. Rev.*, 88, 559, 1952.
21. Pauli, W. *Z. Phys.*, 41, 81, 1926.
22. Herring, C. in *Magnetism*, Vol. 2B (eds., G. T. Rado and H. Suhl), Academic Press, New York, 1966.
23. Stoner, E. C. *Phil. Magn.*, 15, 1080, 1933.
24. Slater, J. C. *Phys. Rev.*, 49, 537, 1936.
25. Slater, J. C. *J. Appl. Phys.*, 8, 385, 1937.
26. Pauling, L. *Phys. Rev.*, 54, 899, 1938.
27. Weiss, P. *J. de Phys.*, 6, 661, 1907.
28. Heisenberg, W. *Z. Phys.*, 49, 619, 1928.
29. Hubbard, J. *Proc. Roy. Soc. Lond. A*, 276, 238, 1963.

30. Edwards D. M., and von der Linden, W. *JMMM*, 104, 739, 1992.
31. Samson, J. H. *JMMM*, 140, 205, 1995.
32. Vonsovskii, S. V. *JETP*, 24, 419, 1953.
33. Zener, C. *Phys. Rev.*, 81, 440, 1951.
34. Ruderman, M. A., and Kittel, C. *Phys. Rev.*, 96, 99, 1954.
35. Kasuya, T. *Prog. Theor. Phys.*, 16, 45, 1956.
36. Yosida, K. *Phys. Rev.*, 106, 893, 1957.
37. Fert, A., Grunberg, P., Barthelemy, A. et al. *JMMM*, 140, 1, 1995.
38. White, R. L. *IEEE Trans. Magn.*, 28, 2842, 1992.

FURTHER READING

Chikazumi, S. *Physics of Magnetism*, Wiley, New York, 1964, Ch. 3.
Cullity, B. D. *Introduction to Magnetic Materials*, Addison-Wesley, Reading, MA, 1972, Ch. 3.
Herring, C. in *Magnetism*, Vol. 2B (eds., G. T. Rado and H. Suhl), Academic Press, New York, 1966.
Morrish, A. H. *The Physical Principles of Magnetism*, Wiley, New York, 1965.
http://en.wikipedia.org/wiki/Exchange_interaction.

EXERCISES

11.1 *The exchange interaction.* Explain the origin of the exchange interaction J. Does this have a classical analog? If not, how can it be interpreted classically? Since the exchange interaction is only invoked to prevent two electrons occupying the same energy levels with the same set of quantum numbers, it would seem that the exchange interaction should lead only to antiferromagnetism by ensuring that the *exchanged* electrons have antiparallel spins. Explain why the exchange interaction does not lead only to antiferromagnetism.

11.2 *Magnetic moment of dysprosium ions.* Dy^3+ has nine electrons in its 4f shell. What are the values of L, S, and J? Calculate the susceptibility of a salt containing 1 g mole of Dy^3+ at 4 K.

11.3 *Paramagnetism of $S = 1$ system.* Find the magnetization as a function of magnetic field and temperature of a system with $S = 1$, moment $\boldsymbol{m}$, and concentration N atoms per unit volume. Show that the limit $\mu_0 \boldsymbol{m} H \gg k_B T$ leads to $M \approx (2\mu_0 N \boldsymbol{m}^2 H)/3k_B T$.

11.4 *Moments on various ions.* Use the electron configurations to calculate the magnetic moments in the following ions: Nb^+, Gd^{3+}, Pt^{2+}, Fe^{2+}, Fe^{3+}.

11.5 *Saturation magnetization of ferrite.* The chemical formula for the ferrite Fe_3O_4 can be expressed as $Fe^{(2+)}O.Fe_2{}^{(3+)}O_3$, so that there are twice as many Fe^{3+} ions as Fe^{2+} ions. Calculate the saturation magnetization of Fe_3O_4 given a density of 5.18×10^3 kg m^{-3}. You are asked to redesign an inductor core using Mn^{2+} instead of Fe^{2+} in the ferrite, [MnO][Fe_2O_3]. If the dimensions of the unit cell are identical, calculate the saturation magnetization of the new inductor core material.

11.6 *Atomic moments from crystal data.* Calculate the magnetic moment per atom in iron, cobalt, and nickel, both in A m^2 and Bohr magnetons using the following information.

Element	Crystal Type	Lattice Parameters (nm)	Saturation Magnetization (A m^{-1})	Number of Nearest Neighbors (z)
Fe	Body-centered cubic	0.29	1.7×10^6	8
Co	Hexagonal close packed	$a = 0.26$ $c = 0.41$	1.4×10^6	12
Ni	Face-centered cubic	0.35	0.5×10^6	12

The Curie temperatures of these metals are: 1040, 1400, and 630 K, respectively. Calculate the mean-field constant and hence the strength of the internal mean (or exchange) field for each metal. From these, calculate for each metal the exchange energy $E_{ex,vol}$ in joules per cubic meter, the exchange energy E_{ex} per atom, and the exchange energy $E_{ex,nn}$ between two nearest-neighbor atoms in joules.

Section III

Magnetics

Technological Applications

12 Soft Magnetic Materials

The most widely used soft magnetic materials are electrical steels, which are produced in grain-oriented and nonoriented forms, soft ferrites and amorphous magnetic metals. Applications of soft ferromagnetic materials are almost exclusively associated with electrical circuits, in which the high permeability of the soft magnetic material is used to amplify the magnetic flux density $\boldsymbol{B}$ for a given magnetic field $\boldsymbol{H}$ generated by the electric currents. In this chapter, we will consider both ac and dc applications of soft magnetic materials. Each requires somewhat different material properties, although in general high permeability and low coercivity are needed in all cases.

12.1 PROPERTIES AND APPLICATIONS OF SOFT MAGNETS

Soft ferromagnetic materials find extensive applications as a result of their ability to enhance the magnetic flux density produced by an electric current. Consequently, the uses of soft materials are closely connected with electrical applications such as generation, transmission, and distribution of electric power, reception of radio signals, microwaves, inductors, relays, and electromagnets. The basic requirement to achieve good soft magnetic properties of high permeability and low coercivity is to have low anisotropy in the magnetic material, since this determines how easy or difficult it is to change the orientation of the magnetic moments when a magnetic field is applied. The available range of magnetic properties of soft magnetic materials is continually being expanded and has been described in a review by Herzer [1]. Among these, iron and iron-silicon alloys are the most widely used and they account for about 80% of the market in soft magnetic materials. Soft magnetic materials comprise about one-third of the total magnetic materials market.

12.1.1 Permeability

Permeability is the most important parameter for soft magnetic materials since it indicates how much magnetic induction is generated by the material in a given magnetic field. In general, the higher the permeability the better for these materials. Initial permeabilities of magnetic materials range from $\mu_r = 1{,}000{,}000$ in materials such as permalloy down to as low as $\mu_r = 1$ in some of the permanent magnet materials, as described in Section 4.2. It is well known that initial permeability and coercivity have in broad terms an inverse relationship, so that materials with high coercivity have a low initial permeability and vice versa.

12.1.2 Coercivity

Coercivity is the parameter that is used to distinguish hard and soft magnetic materials. Traditionally, a material with a coercivity of less than 1000 A m^{-1} is considered magnetically *soft*. A material with a coercivity of greater than 10,000 A m^{-1} is considered magnetically *hard*. Low coercivities are achieved in nickel alloys such as permalloy in which the coercivity can be as low as 0.4 A m^{-1} and in amorphous magnetic materials in which coercivity can be as low as 0.1 A m^{-1}.

12.1.3 Saturation Magnetization

The highest saturation magnetization available in bulk magnetic materials is 2.43 T, which is achieved in an iron-cobalt alloy containing 35% cobalt. The possible values of saturation magnetization in other materials then range downward continuously to effectively zero.

A comparison of the ranges of saturation magnetizations and coercivities of a variety of soft magnetic materials is shown in Figure 12.1.

12.1.4 Hysteresis Loss

The hysteresis loss is the area enclosed by the dc hysteresis loop on the ***B-H*** plane, as discussed in Section 5.1. It represents the energy expended during one cycle of the hysteresis loop. The hysteresis loss increases as the maximum magnetic field

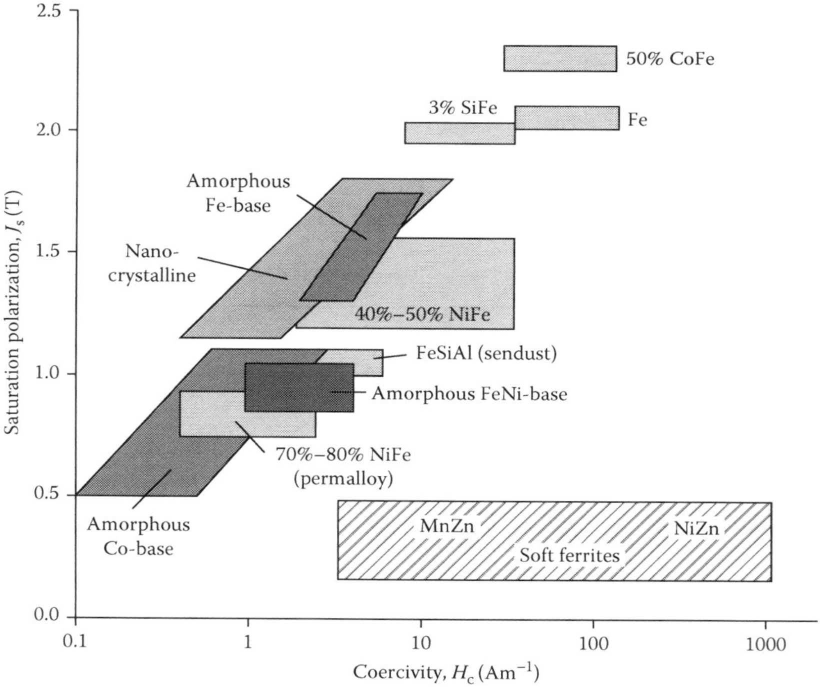

FIGURE 12.1 Comparison of the saturation magnetic polarizations ($= \mu_0 M_s$) and coercivities (H_c) of a variety of soft magnetic materials. (After Herzer, G., *Acta Mater.*, 61, 718, 2013.)

reached during the cycle increases. Clearly, for ac applications, in which energy dissipation should be minimized to improve energy conversion efficiency, the hysteresis loss should be as low as possible. This loss is closely related to the coercivity, so that processing of materials to reduce coercivity also reduces the hysteresis loss. However, in ac applications, the hysteresis loss is not the only dissipative or loss mechanism.

12.1.5 Conductivity and AC Electrical Losses

One of the most important parameters of a magnetic material for ac applications is its electrical loss. The electrical losses are shown for various materials in Table 12.1 at 50 Hz and at an amplitude of excitation of 1 T [2, p. 46].

The electrical losses depend on the frequency of excitation ν, the amplitude of magnetic induction $\boldsymbol{B}_{max}$, the hysteresis loss W_H, the sheet thickness t (due to the penetration depth of the ac magnetic field), and the eddy current dissipation W_{ec}. In addition, there is usually a discrepancy between the measured loss and the loss expected from the sum of hysteresis and eddy current losses and this is usually referred to as the *anomalous loss* W_a. The total electrical power loss W_{tot} is usually expressed as the sum of these components, that is,

$$W_{tot} = W_H + W_{ec} + W_a \tag{12.1}$$

where the hysteresis loss and eddy current loss are frequency dependent. The hysteresis loss increases linearly with frequency, while the eddy current loss increases with the frequency squared. This method of loss separation expressed in Equation 12.1 is a simplification and some authors hold the view that the separation of losses in this way is invalid [3].

The classical eddy current loss W_{ec} in watts per cubic meter depends on $(d\boldsymbol{B}/dt)^2$ and the cross section d^2 as discussed in Section 2.5.1. In the special case of a sinusoidal variation in the magnetic induction, $d\boldsymbol{B}/dt$, the classical eddy current loss takes a particularly simple form, and because of the widespread use of sinusoidal induction

TABLE 12.1
Total Electrical Losses of Various Soft Magnetic Materials*

Material	Total Losses W_{tot} (W kg^{-1})
Commercial iron	5–10
Si-Fe hot rolled	1–3
Si-Fe cold rolled, grain-oriented	0.3–0.6
50% Ni-Fe	0.2
65% Ni-Fe	0.06

[2, p. 46]
* With sheet thicknesses t between 0.2 and 0.5 mm at a frequency of 50 Hz, and an induction amplitude of 1 T

in power transmission and conversion, this form for the loss is frequently encountered in practice. From Section 2.5.1,

$$W_{ec} = \frac{d^2}{2\rho\beta}\left(\frac{dB}{dt}\right)^2 \tag{12.2}$$

where β is the shape factor, and if

$$\boldsymbol{B} = \boldsymbol{B}_{max}e^{i\omega t} \tag{12.3}$$

$$\frac{d\boldsymbol{B}}{dt} = j\omega\boldsymbol{B}_{max}e^{i\omega t} = j\omega\boldsymbol{B} \tag{12.4}$$

which has an average value of $\omega\boldsymbol{B}_{max}/\sqrt{2}$, so that

$$W_{ec} = \frac{B_{max}^2 d^2\omega^2}{4\rho\beta} \tag{12.5}$$

or

$$W_{ec} = \frac{\pi^2 B_{max}^2 d^2 v^2}{\rho\beta} \tag{12.6}$$

where:
- d is the cross-sectional dimension
- ρ is the bulk resistivity in Ω m
- β has different values for different geometries

For example, β = 6 for laminations, in which d represents the thickness t. For cylinders, β = 16 and d represents the diameter. For spheres, β = 20 and d represents the diameter. Stephenson [4] has shown that for low alloy, nonoriented electrical steel sheets Equation 12.6 and gives an eddy current power loss in watts per kilogram of

$$W_{ec} = \frac{1.644 B_{max}^2 t^2 v^2}{\rho D} \tag{12.7}$$

where D is the density in kilograms per cubic meter.

The anomalous loss at 60 Hz and an induction amplitude of 1.5–1.7 T was found to be dependent on t^2/ρ, so that

$$W_a = W_{ao} + \frac{kt^2}{\rho} \tag{12.8}$$

where W_{ao} and k are empirical coefficients. The anomalous power loss results primarily from domain-wall motion as the domain structure changes under the action of the applied field. This has been investigated in detail by Bertotti [5], who has provided a mathematical model for these processes from which it was concluded that the anomalous loss depends on the rate of change of magnetic induction according to the relation

$$W_a = \left(\frac{GwdH_o}{\rho} \right)^{1/2} \left(\frac{dB}{dt} \right)^{3/2} \qquad (12.9)$$

where:

G is a dimensionless constant, which is determined by the choice of the units system ($G = 0.1356$ for SI units)
w is the width
d is the thickness of a lamination
$\boldsymbol{H}_o$ is an internal *magnetic field*, which represents the strength and distribution of the internal potential encountered by the domain walls as they move within the material

If the material is completely homogeneous, then there are no fluctuations of the internal potential, and $\boldsymbol{H}_o = 0$, resulting in $W_a = 0$, so that the power losses will only be the sum of hysteresis and classical eddy current losses. All of these power loss mechanisms have been incorporated into an integrated theory of frequency-dependent hysteresis [6].

From the empirical Steinmetz law, the dc hysteresis loss is known to depend on induction amplitude $\boldsymbol{B}_{max}$ and frequency v according to the relation

$$W_H = \eta B^n_{max} v \qquad (12.10)$$

where:

η is a material constant
n is an exponent, which lies in the range of 1.6–2.0

The total losses can be reduced if the conductivity of the material is reduced. This is exploited in transformer materials such as silicon-iron, in which silicon is added principally to reduce the conductivity. Although it has an adverse effect on permeability, it does reduce the coercivity. Losses in Ni-Fe alloys are lower than in silicon-iron, and this is also used in ac applications such as inductance coils and transformers, but silicon-iron has a higher saturation magnetization.

12.1.6 Electromagnets and Relays

Among the most important dc uses that soft magnetic materials find are in electromagnets and relays. An electromagnet is any device in which a magnetic field $\boldsymbol{H}$ is generated by an electric current and the resulting flux density $\boldsymbol{B}$ is increased by the use of a high-permeability core. The simplest example is a solenoid carrying a dc current wound around a ferromagnetic core. In electromagnets, soft iron is still the most widely used material because it is relatively cheap and can produce high magnetic flux densities. It is often alloyed with a small amount of carbon (<0.02 wt%) without seriously impairing its magnetic properties for this application. Also, the alloy Fe-35% Co is used in electromagnet pole tips to increase saturation magnetization and hence increases the flux density $\boldsymbol{B}$ in the gap of the electromagnet.

A relay is a special form of dc electromagnet with a moving part, the armature, which operates a mechanical switch. This can be used, for example, to open and close electrical circuits and therefore is useful as a control device.

12.1.7 Transformers, Motors, and Generators

A transformer is a device that can transfer electrical energy from one electrical circuit to another although the two circuits are not connected electrically. This is achieved by the process of electromagnetic induction, whereby a magnetic flux links the two circuits through an inductance coil in each circuit. The two coils are usually connected by a high-permeability magnetic core. The material used for transformer cores is almost exclusively grain-oriented silicon-iron, although some small cores use nonoriented silicon-iron. High-frequency transformers use cobalt-iron, although this represents only a small volume of the total transformer market.

Generators are devices for converting mechanical energy into electrical energy. Motors are devices for converting electrical energy into mechanical energy. Both are constructed from high-permeability magnetic materials. The most common material used for these applications is nonoriented silicon-iron but many smaller motors use silicon-free, low-carbon steels.

12.2 MATERIALS FOR AC APPLICATIONS

We will look at uses in power generation and transmission since these are the most important applications in which soft ferromagnetic materials are employed. The desirable properties here are, as in most soft magnetic materials applications, high permeability and high saturation magnetization with low coercivity and low power loss. Soft iron and low carbon steels were once the materials of choice for transformers, motors, and generators but have been superseded by silicon-iron, both in its grain oriented form for transformers and in its nonoriented form for motors and generators. Amorphous magnetic materials are widely used for more specialized low power applications.

12.2.1 Iron-Silicon Alloys (Electrical Steels)

There are several ways in which the properties of iron can be improved in order to make it more suitable for electrical power conversion at low frequencies. First, the resistivity can be increased so that the eddy current losses are reduced. This is achieved by alloying of silicon with iron. The resulting material is known as *silicon iron*, *electrical steel*, or *silicon steel*. The variation of resistivity of silicon-iron with silicon content is shown in Figure 12.2. Iron containing 3% silicon has four times the resistivity of pure iron [7]. Over time, there have been substantial improvements in the core losses of silicon-iron. These improvements for 0.35 mm thick sheets are shown in Figure 12.3.

Silicon of course is a cheap material, which is an important consideration when so much electrical steel is needed. In addition, suitable mechanical properties are a consideration since silicon-iron becomes brittle when the silicon content is high.

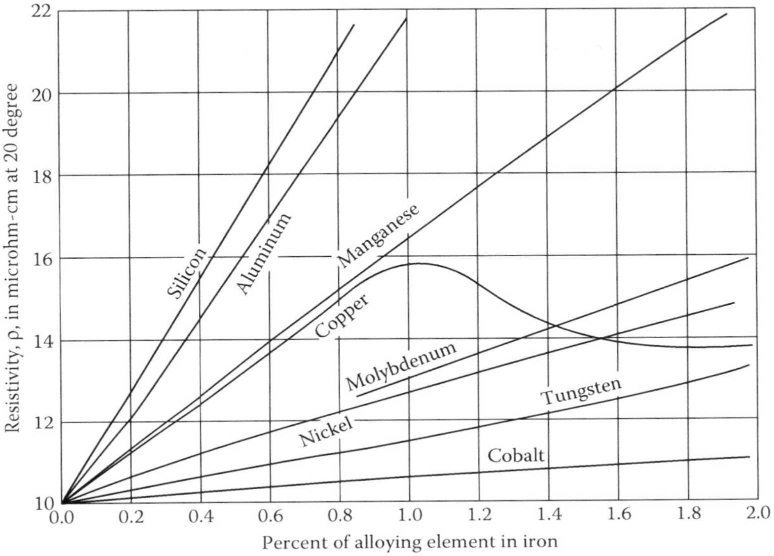

FIGURE 12.2 Variation of the electrical resistivity of iron with the addition of different alloying elements.

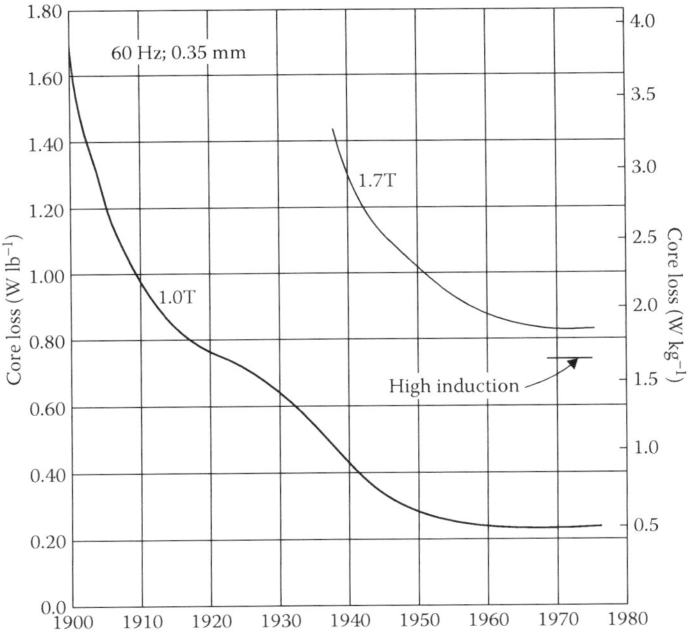

FIGURE 12.3 Reduction of core loss of 0.35 mm thick silicon-iron at 60 Hz from the year 1900 until 1975.

Normally, it is not possible to have all of these desirable properties in a single material, so it is necessary to decide which is the most crucial property in any given application. For high-power applications, silicon-iron is widely used. Nonoriented silicon-iron is the material of choice in motors and generators, while grain-oriented silicon-iron is used for transformers.

The magnetic properties of silicon-iron are dependent on the microstructure and texture [8]. These are altered by rolling mechanical treatment (*rolling*) and by heat treatment. Depending on the type of rolling and heat treatment silicon iron can be produced in a nonoriented form [9] or a grain-oriented form [10].

In electrical power transmission and distribution, the greatest demand is for transformer cores. In this application, silicon-iron is used to the exclusion of all others. In the power industry, the electrical voltage is almost always low-frequency ac at 50 or 60 Hz. This leads to an alternating flux in the cores of the motors, generators, or transformers and consequently to the generation of *eddy currents*. Eddy currents reduce the energy conversion efficiency of motors, generators, and transformers because some of the energy is dissipated as heat.

The addition of silicon to iron has two main beneficial effects on the properties of the iron. The conductivity is reduced as silicon is added and the magnetostriction is reduced as shown in Figure 12.4. For ac applications, this reduction of

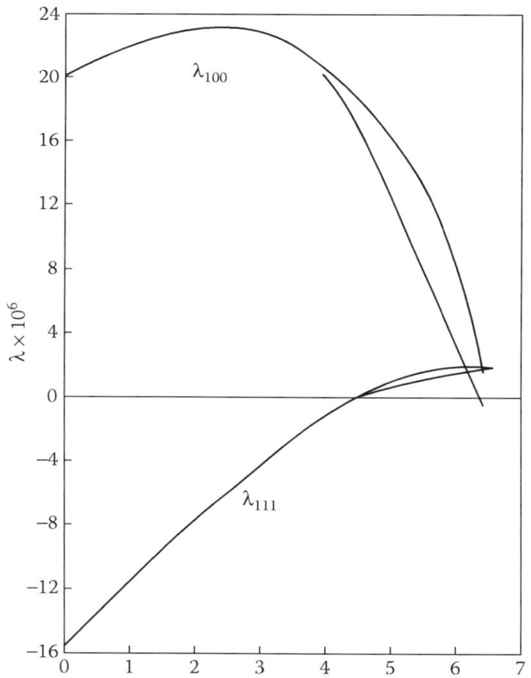

FIGURE 12.4 Dependence of the magnetostriction coefficients λ_{100} and λ_{111} of silicon-iron on silicon content. The lower curves in the range of 4%–7% Si are for slowly cooled samples.

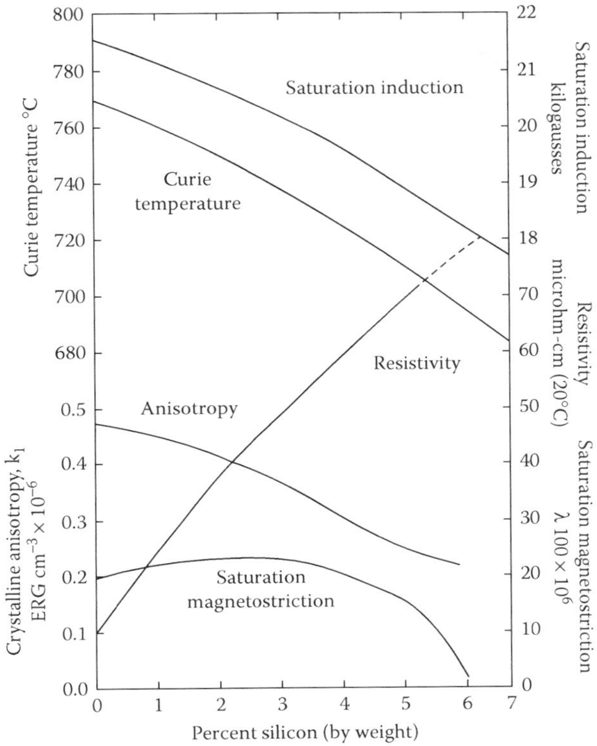

FIGURE 12.5 Various magnetic and electrical properties of silicon-iron as a function of silicon content. (Adapted from Littmann, M. F., *IEEE Trans. Mag.*, 7, 48, 1971.)

magnetostriction is an additional advantage since the cyclic stresses resulting from magnetostrictive strains at 50 or 60 Hz produce acoustic noise. Therefore, any reduction of magnetostriction is desirable, particularly if it arises as a result of modifying the material to suit other unrelated requirements. A third benefit caused by the addition of silicon is that it reduces the anisotropy of the alloy as shown in Figure 12.5, leading to an increase in permeability in the nonoriented silicon-iron.

It is also beneficial to laminate the cores in such a way that the laminations run parallel to the magnetic field direction. This does not interfere with the magnetic flux path but does reduce the eddy current losses, by only allowing the eddy currents to exist in a narrow layer of material. The dependence of core losses at 60 Hz on sheet thickness is shown in Figure 12.6. Furthermore, the coating of laminations with an insulating material also improves the eddy current losses by preventing current passing from one layer to the next. The thickness of the laminations for optimum performance is comparable with the skin depth at 50 or 60 Hz, which is typically 0.3–0.7 mm [11].

Since the magnetic flux only passes in one direction along the laminations in transformers, it is advantageous to ensure that the permeability is highest along this direction. Techniques have been devised for producing grain-oriented silicon-iron by

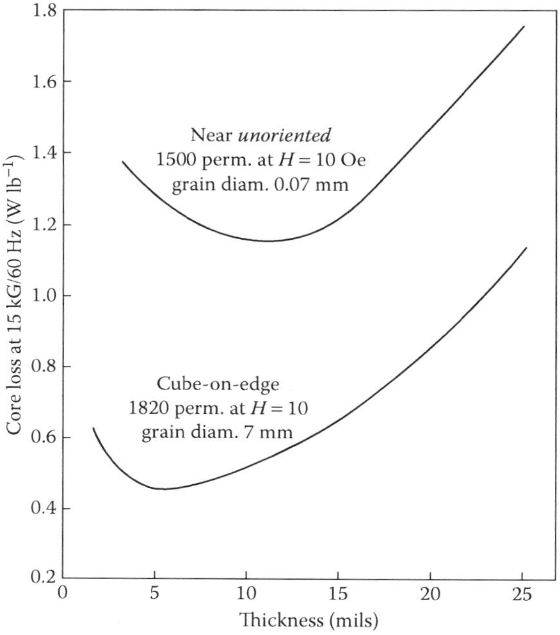

FIGURE 12.6 Dependence of core loss on sheet thickness in 3.15% silicon-iron.

hot and cold rolling and annealing, which result in the ⟨001⟩ direction lying along the length of the laminations. The ⟨001⟩ directions are the magnetically easy axes. The addition of silicon to iron increases the grain size so that the grain diameters in 3% silicon-iron can be as large as 10 mm. The core losses of 3% silicon-iron vary with grain size as shown in Figure 12.7, in which a minimum in core loss occurs at a grain diameter of between 0.5 and 1.0 mm.

There are nevertheless some disadvantages involved in the addition of silicon to iron. At higher silicon contents, the alloy becomes brittle and this imposes a practical limitation on the level of silicon that can be added without the material becoming too brittle to use. This limit is about 4% and most silicon-iron transformer materials have a composition of 3%–4% Si, although material with a silicon content of 6.5% with adequate mechanical properties for transformer applications has been developed. Another disadvantage resulting from the addition of silicon to iron is the reduction of saturation induction.

A summary of advances in silicon-iron materials, including the 6.5 wt% Si alloys, by Fiorillo [12] shows that the need to improve energy conversion efficiency in these materials has resulted in improvements in magnetic properties, driven principally by the enormous economic impact of such improvements. This has been achieved through a combination of better understanding of the energy loss mechanisms in the magnetization process, coupled with materials processing methods such as rolling and heat treatments, which can advantageously control grain size and domain structures [13,14]. Figure 12.8 shows how the coercivities of various soft magnetic materials vary with grain size.

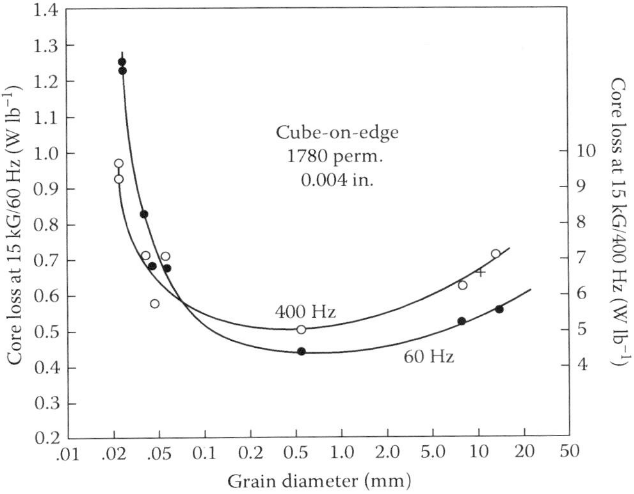

FIGURE 12.7 Dependence of core loss on grain size in 3.15% silicon-iron sheets.

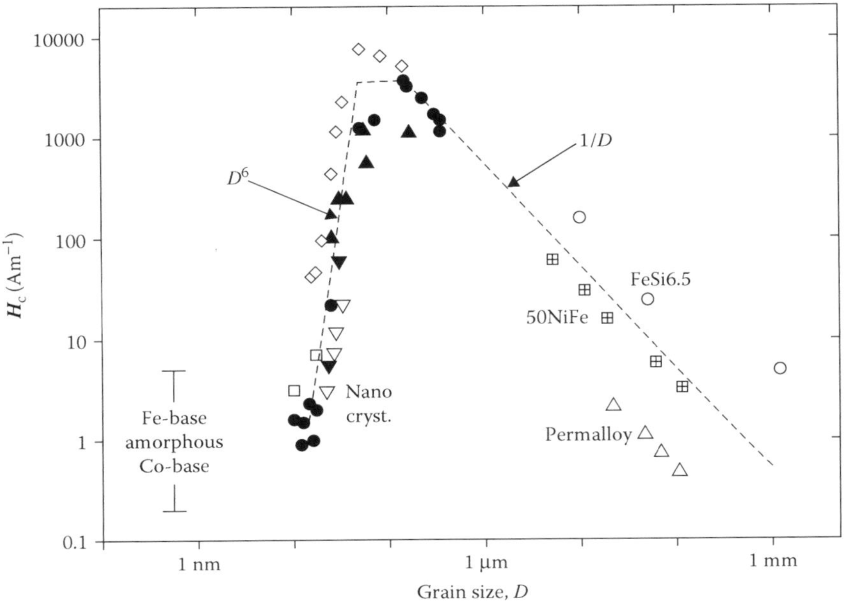

FIGURE 12.8 Variation of coercivity H_c with grain diameter D for various soft magnetic materials. (After Herzer, G., *Acta Mater.*, 61, 718, 2013.) Fe-Nb-Si-B (solid triangles), Fe-Cu-Nb-Si-B (solid circles), Fe-Cu-V-Si-B (solid nablas and open down triangles), Fe-Zr-B (open squares), Fe-Co-Zr (open diamonds), NiFe alloys (+ center squares and open triangles), and Fe-6.5 wt% Si (open circles).

12.2.2 Iron-Aluminum Alloys

The properties of aluminum-iron are very similar to those of silicon-iron but since aluminum is more expensive than silicon, these alloys are unlikely to replace silicon-iron in applications where they both compete. Furthermore, the presence of Al_2O_3 particles in iron-aluminum alloys causes rapid wear of punching dies, which is disadvantageous in the production process.

The magnetic properties of some aluminum-iron alloys are shown in Figures 12.9 and 12.10. Alloys of up to 17% Al are ferromagnetic but at higher aluminum contents the alloys become paramagnetic. Often, aluminum is used as an addition in silicon-iron because it promotes grain growth, which can lead to lower losses. Furthermore, the addition of aluminum produces higher resistivity with less danger from brittleness. Therefore, ternary alloys of iron, silicon, and aluminum are used in electrical steels for special applications.

12.2.3 Nickel-Iron Alloys (Permalloy)

These alloys are the most versatile of all soft magnetic materials for electromagnetic applications. Only the alloys with above 30% nickel are widely used because at lower nickel contents there is a lattice transformation, which occurs at different temperatures

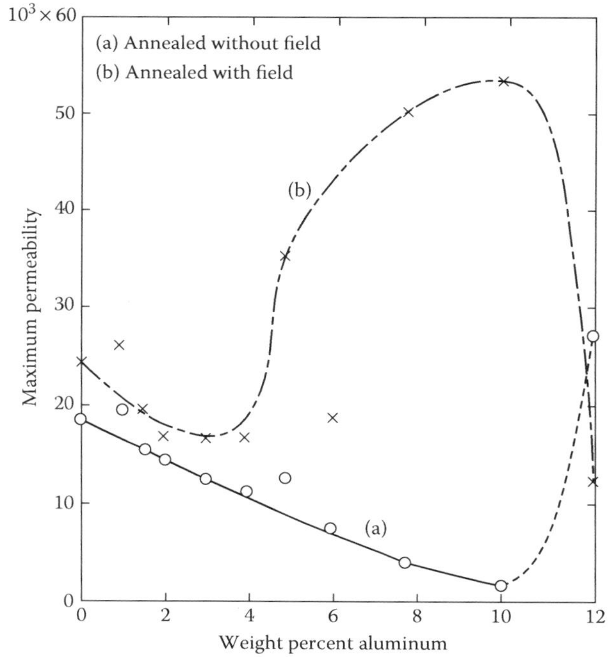

FIGURE 12.9 Maximum permeability of iron-aluminum alloys as a function of aluminum content after two different types of annealing. The field anneal was in a magnetic field of 136 A m^{-1}. Specimens were 0.35 mm thick laminations.

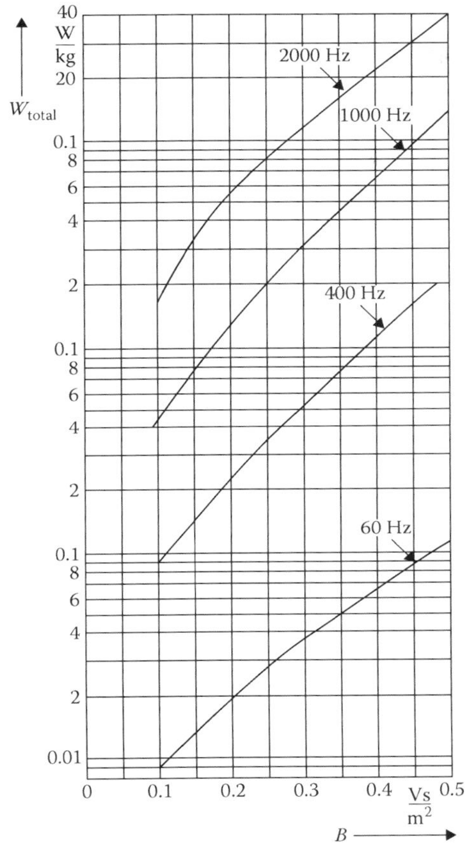

FIGURE 12.10 Total core loss as a function of peak magnetic induction for laminated 0.35 mm sheets of iron-16% aluminum.

depending on the actual chemical composition. This transformation exhibits temperature hysteresis and hence there is no well-defined Curie temperature under these conditions. As a result of this complication, the alloys in this range are not widely used.

Three groups of these alloys are commonly encountered [15]. These have nickel contents close to 80%, 50%, or in the range of 30%–40%. The permeability is highest for the alloys close to 80% Ni, as shown in Figure 12.11. The saturation magnetization is highest in the vicinity of 50% Ni, as shown in Figure 12.12. The electrical resistivity is highest in the 30% Ni range, as shown in Figure 12.13. These are the three magnetic properties, which are of most interest in soft magnetic material applications, and so the alloys used are usually close to one of these compositions depending on the specific application.

These alloys are used in inductance coils and transformers, particularly power supply transformers. They are used at audio frequencies as transformer cores and also for higher frequency applications. They can be made with low, or even zero,

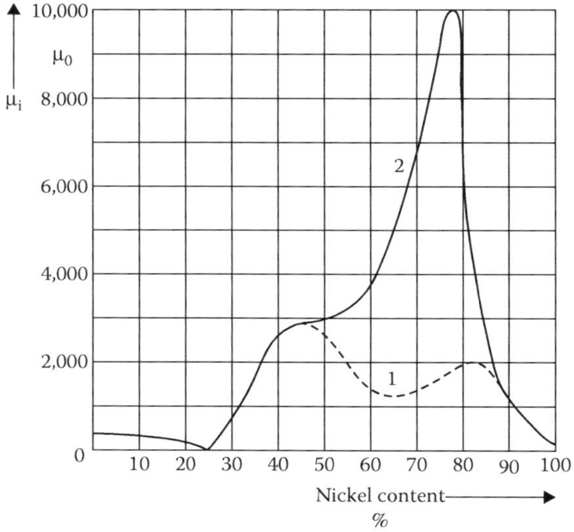

FIGURE 12.11 Initial permeability of iron-nickel alloys: 1 = slow cooled and 2 = normal permalloy treatment.

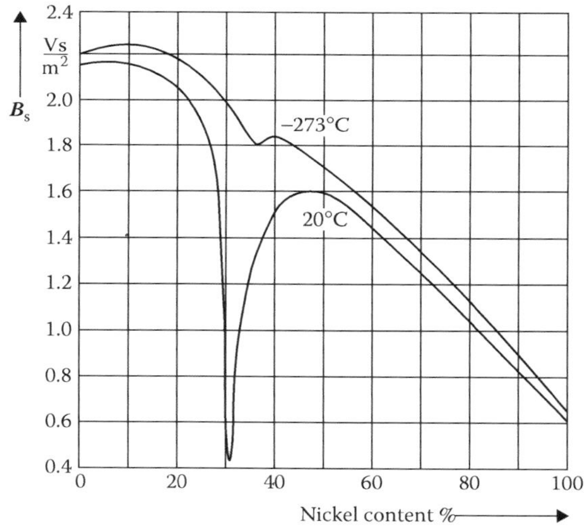

FIGURE 12.12 Saturation magnetic induction in iron–nickel alloys.

magnetostriction, as shown in Figure 12.14. Some of the high permeability alloys, mu-metal and supermalloy, have relative permeabilities of up to 3×10^5 and coercivities down to 0.4 A m^{-1}. They also have low anisotropy and this contributes to their high permeability in the polycrystalline form.

Permalloy and mu-metal, both Ni-Fe alloys with close to 80% Ni, are used in magnetic shielding because of their very high permeability. This application

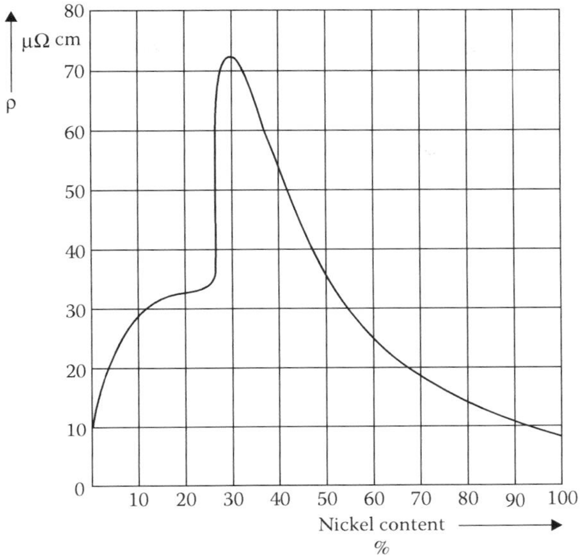

FIGURE 12.13 Electrical resistance in iron-nickel alloys.

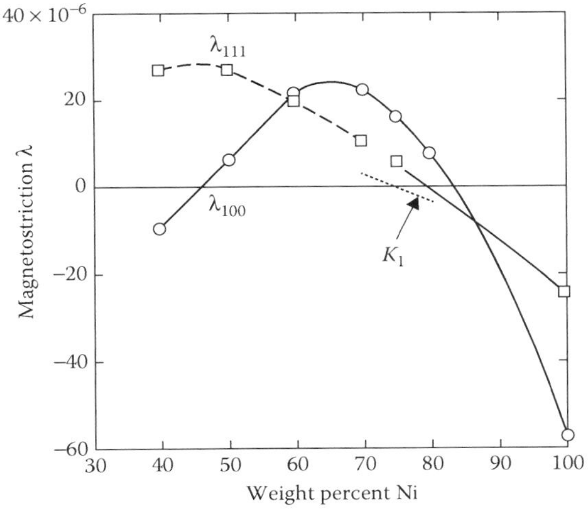

FIGURE 12.14 Dependence of magnetostriction coefficients λ_{100} and λ_{111} with nickel content in iron-nickel alloys.

is discussed in Section 12.4. Some of these alloys can reach an induction of 0.6 T in a field as low as 1.6 A m^{-1}, corresponding to a relative permeability of $\mu_r = 3 \times 10^5$. Alloy additions to the basic iron-nickel alloy combined with processing allows the magnetic properties of these alloys to be varied within wide limits. Cold working by rolling gives rise to high permeability perpendicular to

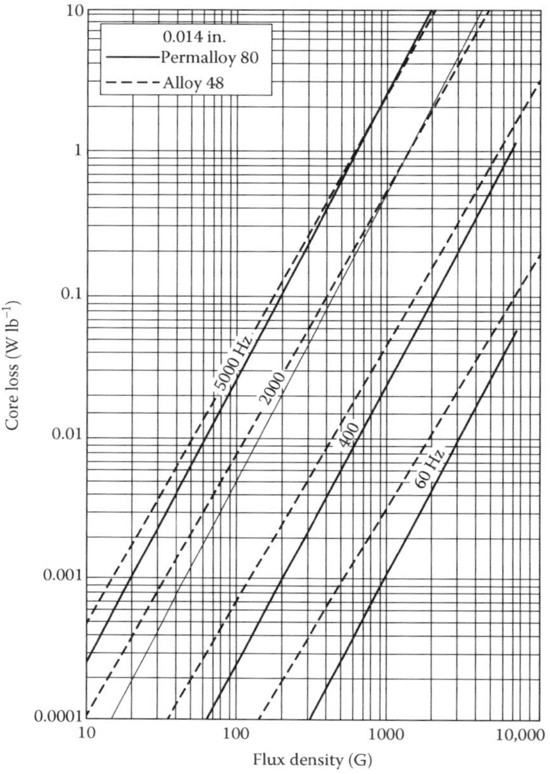

FIGURE 12.15 Core losses against peak magnetic induction and frequency for 0.35 mm thick laminations of two commercial nickel-iron alloys: permalloy 80 and alloy 48.

the field as in isoperm, a 50%–50% Fe-Ni alloy. Invar is a 64% Fe 36% Ni alloy with zero thermal expansion.

High-quality transformers are often made of permalloy. Relative permeabilities of up to 100,000 are attainable with coercivities in the range of 0.16–800 A m^{-1} (0.002–10 Oe) and these can be adjusted with precision by suitable processing of the material. The core losses of two commercial Ni-Fe alloys are shown in Figure 12.15.

This alloy system is also used in some magnetic memory devices and amplifiers. For high-frequency applications of up to 100 kHz, the alloy can be used in the form of powdered cores in which each particle is electrically insulated from others and therefore the bulk conductivity of the material is low.

12.2.4 Amorphous Magnetic Ribbons (Metallic Glasses)

The main interest in amorphous soft magnetic materials arises from the low coercivities, as shown in Figure 12.16. These are an order of magnitude smaller than silicon-iron while the permeabilities are about an order of magnitude

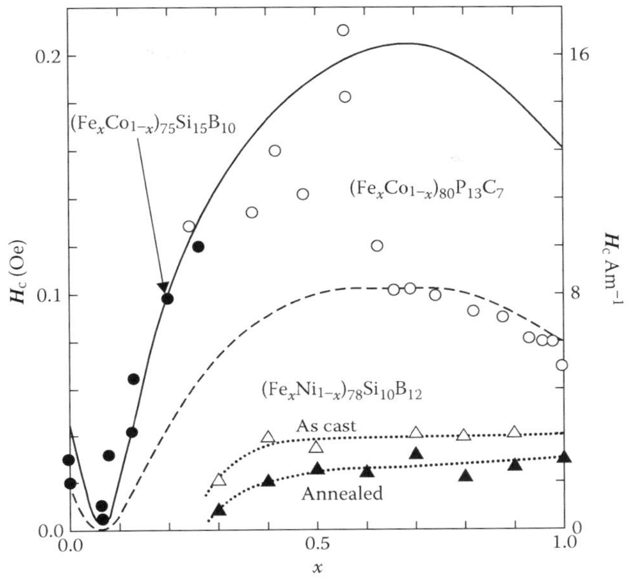

FIGURE 12.16 Variation of coercivity of a number of amorphous alloys with chemical composition.

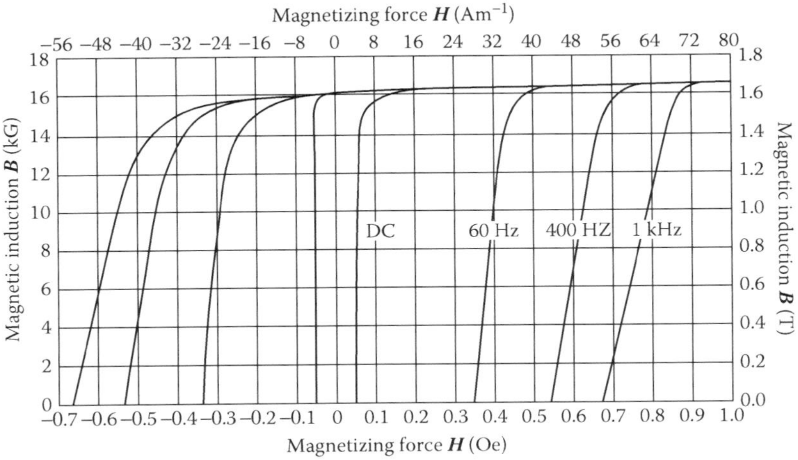

FIGURE 12.17 Upper two quadrants of the hysteresis loop of Metglas 2605CO at different frequencies.

greater. The hysteresis loops of the $Fe_{80}B_{20}$ material (Metglas 2605CO) at different frequencies are shown in Figure 12.17. Core losses are also very low, as indicated in Figures 12.18 and 12.19. Such properties are a distinct advantage for soft magnetic material application; however, certain disadvantages also emerged.

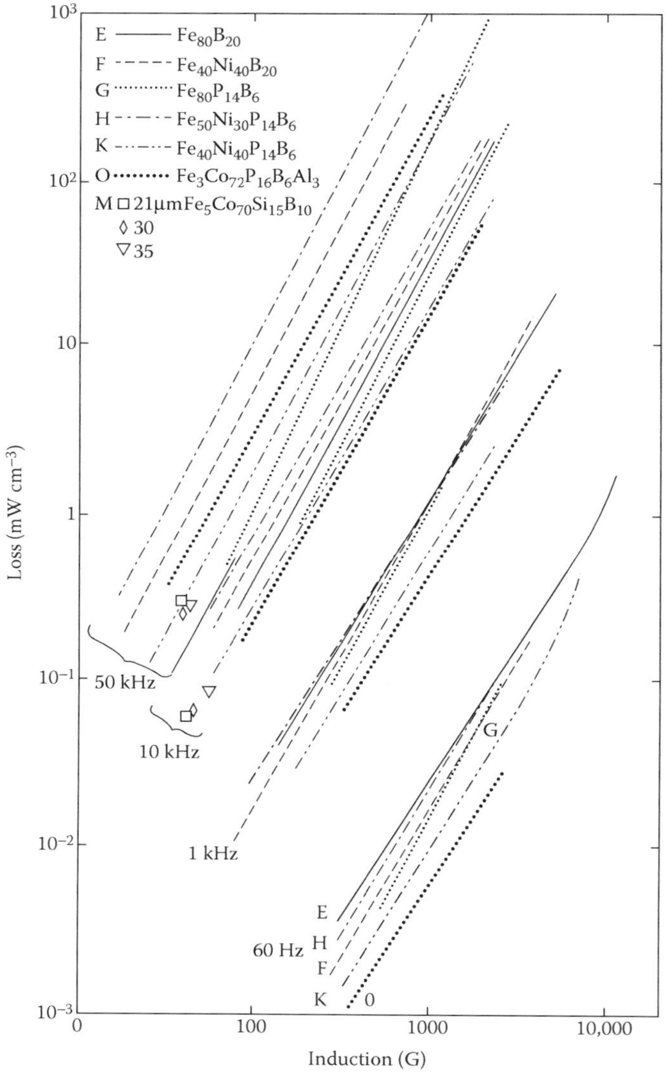

FIGURE 12.18 Dependence of core loss on magnetic induction and frequency for various amorphous alloys. All sample thicknesses were in the range of 25–50 μm.

These materials are produced by rapid cooling (quenching) from the melt of magnetic alloys consisting of iron, nickel, and/or cobalt together with one or more of the following elements: phosphorus, silicon, boron, and sometimes carbon [16]. The molten metal is sprayed in a continuous jet under high pressure on to a rapidly moving cold surface such as the surface of a rotating metal wheel (to produce ribbons)

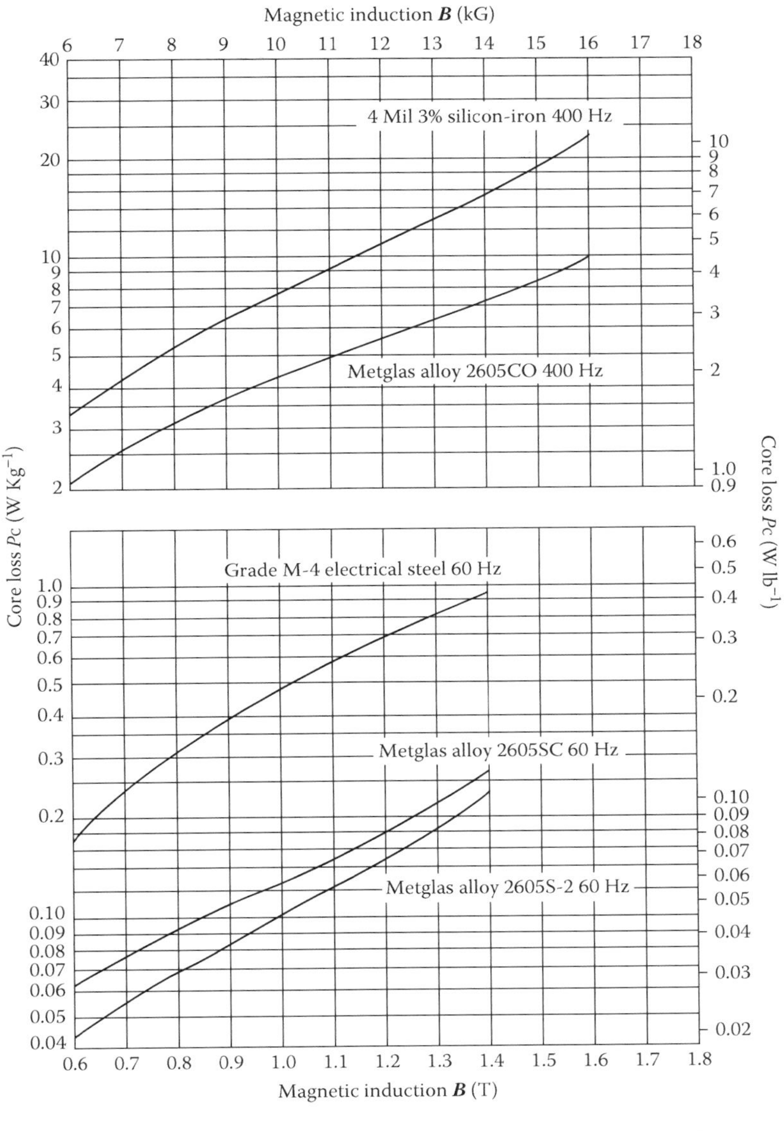

FIGURE 12.19 Comparison of core losses for different soft magnetic materials as a function of peak magnetic induction.

or a rotating water bath (to produce wires or fibers). As a result of the rapid cooling, the materials do not form a crystalline state but instead produce a solid with only short-range order but otherwise microstructurally disordered. These materials can be considered as a random packing of atoms. They are also known as *metallic glasses* because of this random structure.

The materials produced in this way have large internal strains, which contribute to coercivity and limit permeability. Strain relief annealing is therefore used to reduce coercivity and increase permeability, but this annealing must not be enough to cause crystallization, which would result in the emergence of magnetocrystalline anisotropy that would increase coercivity again and reduce permeability.

Even in the *as cast* condition these alloys have very soft magnetic properties but the annealed nanocrystalline form of the material can have even better properties for soft magnetic material applications. The internal residual stresses in amorphous magnetic alloys can be altered by annealing the material at intermediate temperatures in order to relieve the strains without leading to complete recrystallization, which results in a nanocrystalline form of the alloy. Amorphous alloys with widely differing magnetic hysteresis properties can be produced by annealing in the presence of an applied magnetic field.

One of the disadvantages of these materials is the low saturation magnetization of the alloys, as shown in Figure 12.20, which limits their use in heavy current engineering when compared, for example, with silicon-iron. Second, at higher flux densities, their core losses begin to increase rapidly. There is a better market for these

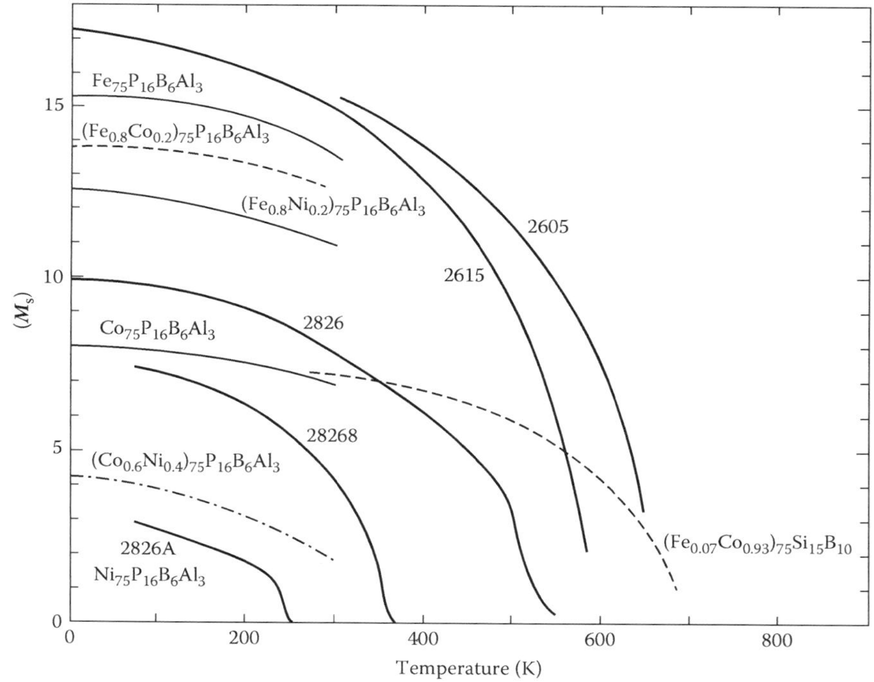

FIGURE 12.20 Temperature dependence of the saturation magnetic induction for different amorphous alloys. Allied chemical Metglas alloys are as follows: 2605 is $Fe_{80}B_{20}$; 2615 is $Fe_{80}P_{16}C_3B$; 2826 is $Fe_{40}Ni_{40}P_{14}B_6$; 2826A is $Fe_{32}Ni_{36}Cr_{14}P_{12}B_6$; and 2826B is $Fe_{29}Ni_{49}P_{14}B_6Si_2$.

TABLE 12.2
Magnetic Properties of Amorphous Alloys under DC Conditions

		As Cast			Annealed		
Alloy	**Shape**	H_c (A m^{-1})	M_r/M_s	μ_{max} (10^3)	H_c (A m^{-1})	M_r/M_s	μ_{max} (10^3)
Metglas 2605 $Fe_{80}B_{20}$	Toroid	6.4	0.51	100	3.2	0.77	300
Metglas 2826 $Fe_{40}Ni_{40}P_{14}B_6$	Toroid	4.8	0.45	58	1.6	0.71	275
Metglas 2826 $Fe_{29}Ni_{44}P_{14}B_6Si_2$	Toroid	4.6	0.54	46	0.88	0.70	310
$Fe_{4\cdot7}Co_{70.3}Si_{15}B_{10}$	Strip	1.04	0.36	190	0.48	0.63	700
$(Fe_{.8}Ni_{.2})_{78}Si_8B_{14}$	Strip	1.44	0.41	300	0.48	0.95	2000
Metglas 2615 $Fe_{80}P_{16}C_3B$	Toroid	4.96	0.4	96	4.0	0.42	130

alloys in low power, low current applications, and specialized small-device applications in which transformers are needed with only moderate flux densities, where the amorphous alloys can compete successfully with the nickel-iron alloys such as permalloy. These amorphous alloys are being produced in large quantities and have found uses in pulsed power transformers, in magnetic sensors, in magnetostrictive transducers, and in communication equipment.

The magnetic properties of a range of amorphous alloys is shown in Table 12.2.

12.2.5 Amorphous Magnetic Fibers

Amorphous magnetic fibers are similar in many respects to amorphous magnetic ribbons. Like the amorphous magnetic ribbons, these are produced by rapid solidification from the melt, but in this case the material is ejected in a jet from a nozzle and quenched in a stream of liquid, usually water, but in some cases, oil can be used for particularly reactive metals. The final product is a fiber or wire typically 50 μm in diameter, which has no crystalline structure. Narrower *nano-wires* can also be produced.

The compositions of amorphous magnetic fibers that have been most widely investigated are of the form $TM_x\ Si_y\ B_z$, where TM is one of the transition metals Fe, Co, or Ni. Typically $x = 0.7–0.8$, while y and z are each in the range of 0.1–0.2. In some cases, other transition metals are added such as Cr, which improves the corrosion resistance. Fibers with positive magnetostriction based on Fe can be produced as well as fibers with negative magnetostriction based on Co. Those containing Co:Fe in the ratio of 16:1 result in an alloy with nearly zero magnetostriction.

Much of the original work on these amorphous magnetic fibers was carried out by Ohnaka et al. [17] and by Mohri et al. [18]. The first application identified by Mohri for these materials was in *jitterless* pulse generator elements, in which the sharp magnetization changes that occur in these materials over a wide range of frequencies of applied field are ideal. The rapid magnetization changes are due to large Barkhausen effect jumps inside the material, which can be induced by field strengths as low as 10 A m^{-1} (0.125 Oe).

In most cases, the amorphous fibers that have been produced and studied are iron-based materials such as the $Fe_{81}Si_4B_{14}C_1$ composition. Other compositions based on the FeSiB glass-forming alloy include FeCoSiB, FeCrSiB, and FeNiSiB, as reported by Mohri et al. [19], who suggested a domain model for the as-cast fibers, which can explain the unusually large Barkhausen jumps in these materials. These large Barkhausen jumps have continued to be a source of great interest because of possible applications in switching devices.

Accordingly, the basic model of the amorphous magnetic fiber consists of an axially oriented *core* domain running along the length of the fiber, surrounded by outer or *shell* domains. The large Barkhausen jumps are caused by a sudden reorientation of the core domain. The orientation of the *shell* domains depends crucially on the magnetostriction λ_s of the material because the quenching process induces high levels of radial stress and hence stress-induced anisotropy in the radial direction. For materials with positive λ_s, this results in radially directed magnetization in the *shell* domains as shown in Figure 12.21(a), whereas for materials with negative λ_s, the *shell* domains orient circumferentially as shown in Figure 12.21(b).

The magnetic properties of these amorphous magnetic fibers are strongly affected by postproduction heat treatment, as demonstrated by Atkinson et al. [20], in which the FeSiB alloy fibers were studied. Annealing at 425°C relieves the casting stresses in the fibers and results in an alteration of the ratio of core domain volume to shell domain volume. This has dramatic effects on the coercivity, remanence, and field-induced magnetostriction. For example, the field-induced magnetostriction can be reduced by 50%, from 30×10^{-6} to 15×10^{-6}, by simply annealing the fibers for 600 s at 425°C. The coercivity was also reduced from 7.0 to 2.9 A m^{-1} and the remanence ratio $\boldsymbol{M}_r/\boldsymbol{M}_s$ was increased from 0.46 to 0.77.

Squire et al. [21] have reviewed the properties of these amorphous fibers and their applications, which provided a comprehensive summary of previous work.

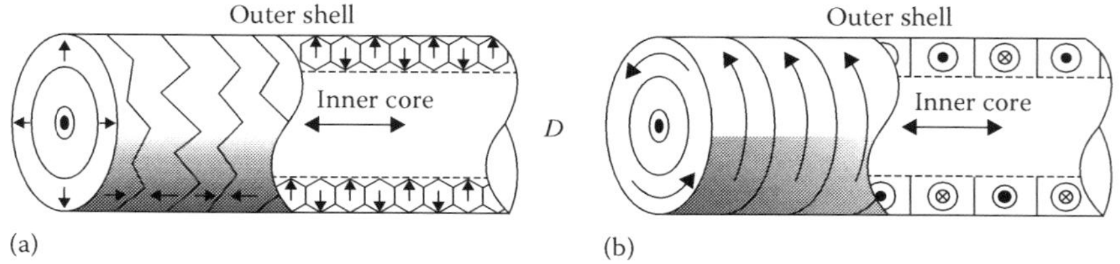

FIGURE 12.21 Domain configuration in the two main classes of amorphous fibers: (a) positive magnetostriction λ ($\lambda > 0$) with the core domain and radially directed sheath domains and (b) negative magnetostriction λ ($\lambda < 0$) with core domain and circumferentially directed sheath domains. (After Mohri, K. et al., *IEEE Trans. Magn.*, 26, 1789, 1990.)

12.2.6 Nanocrystalline Magnetic Materials

The range of available soft magnetic materials was significantly increased by the discovery of nanocrystalline magnetic materials by Yoshizawa et al. [22]. The exceptional properties of these materials, which have coercivities below 1 A m^{-1} (0.0125 Oe) and high relative permeabilities of typically 10^5 combined with relatively high saturation magnetization of 1.05×10^6 A m^{-1} (13 kG) and resistivities as high as 1.15×10^6 Ω m, make them suitable for applications in magnetic cores for ground fault circuit interrupters, high-frequency transformers, and chokes.

These materials are mostly based on iron alloys and have grain diameters of typically 10–15 nm. The most widely investigated alloy is $Fe_{73.5}Si_{13.5}B_9Nb_3Cu_1$, which is produced by rapid solidification and is then annealed above its crystallization temperature, but about 200° below its melting temperature to produce a nanocrystalline structure. The properties of these materials have been described in detail by Herzer [1,23].

12.2.7 Artificially Structured Magnetic Materials

Soft magnetic materials also include artificially structured materials in which the structure of heterogeneous materials is carefully controlled to produce the desired magnetic properties. The nanocrystalline or nanostructured materials described in Section 12.2.6 come into this category, as do the magnetic multilayers described in Section 11.3.9, which exhibit giant magnetoresistance when subjected to a magnetic field.

These are multiphase materials consisting of particulate and matrix materials with different magnetic and structural phases in which the eventual properties of the entire material can be controlled by processing to cause structural changes, which produce a material with the desired combination of anisotropy and Curie temperature. A typical heterogeneous magnetic material consists of nanocrystalline particles embedded in an amorphous magnetic matrix phase.

Yoshizawa et al. [22] found that the crystallization of Fe-Si glasses with the addition of small amounts of Cu or Nb formed an ultrafine grain structure of body centered cubic (bcc) Fe-Si with grain sizes of typically 10–15 nm embedded in an amorphous matrix of FeNbB. Figure 12.22 shows the structure in detail. These nanostructured alloys have soft magnetic properties that are comparable to permalloys and Co-based amorphous alloys, but with a significantly higher saturation induction of 1.2 T and above.

The magnetic properties of the particulate and matrix phases are essentially different and that allows control of the magnetic properties through structural changes brought about by annealing. Hernando et al. [24] described such a material in which iron-silicon-boron nanocrystals were embedded in an amorphous matrix of the same material as shown in Figure 12.23. This resulted in a material with a desirable low magnetic anisotropy, and hence low loss, because of the amorphous phase. However, the Curie temperature of the material, which is normally low for the amorphous alloys, could be controlled by the presence and size of the nanocrystalline particles.

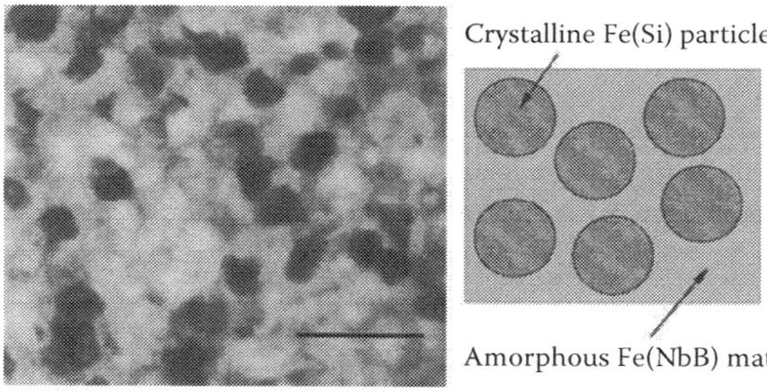

FIGURE 12.22 Transmission electron micrograph of the two-phase structure of nanocrystalline FeCuNbSiB (left) and idealized schematic representation of the two phase structure (right). (After Yoshizawa, Y. et al., *J. Appl. Phys.*, 64, 6044, 1988.)

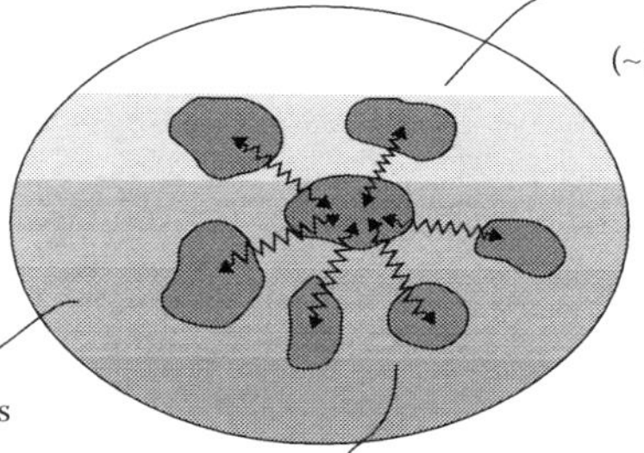

FIGURE 12.23 Artificially structured magnetic material consisting of nanocrystalline particles embedded in an amorphous matrix. (After Hernando, A. et al., *Phys. Rev. B*, 51, 3281, 1995.)

Curie temperatures in the range of 200°C–350°C were obtained by annealing at different temperatures and for different times [25].

12.2.8 Soft Ferrites

For high-frequency applications, the conductivity of metals limits their use and so we turn to magnetic insulators. These materials must of course exhibit the usual properties associated with soft ferromagnets: high permeability, low coercivity, and high saturation magnetization. Soft ferrites are widely used in these applications. Ferrites are ceramic magnetic solids, which first appeared commercially in 1945. They are ferrimagnetic rather than ferromagnetic, but on the bulk scale, they behave in much the same way as ferromagnets with the presence of domains, a saturation magnetization, a Curie temperature and hysteresis in their ***B-H*** characteristics.

There are two most commonly used soft ferrites. (1) Manganese zinc ferrite, with the formula $Mn_aZn_{(1-a)}Fe_2O_4$. Manganese zinc ferrites have higher permeability and saturation induction than nickel zinc ferrites and therefore are more suitable for lower frequency applications. (2) Nickel zinc ferrite, with the formula $Ni_aZn_{(1-a)}.Fe_2O_4$. Nickel zinc ferrites exhibit higher resistivity than manganese zinc ferrites, and are therefore more suitable for higher frequencies. The appropriate frequency ranges are discussed below.

The cubic or soft ferrites all have the general chemical formula $MOFe_2O_3$, where M is a transition metal such as nickel, iron, manganese, magnesium, and zinc. The most familiar of these is Fe_3O_4. The ferrite $CoO.Fe_2O_3$ although of the same general type is nevertheless a hard ferrite rather than a soft ferrite. The magnetic garnets were discovered by Bertraut and Forret [26]. Yttrium-iron garnet is the best-known example. Soft ferrites can be further classified into the *non-microwave ferrites* [27] for frequencies from audio up to 500 MHz and *microwave ferrites* for frequencies from 500 MHz to 500 GHz [28]. Microwave ferrites, such as yttrium-iron garnet, are used as waveguides for electromagnetic radiation and in phase shifters.

Soft ferrites are also used in frequency selective circuits in electronic equipment, for example, in phone signal transmitters and receivers. Manganese zinc ferrite, which is sold under the commercial name of Ferroxcube, is widely used for applications at frequencies of up to 1 MHz. Above that frequency range, nickel-zinc ferrites are preferred because they have lower conductivity. Another area where ferrites find wide application is in antennae for radio receivers. Almost all radio receivers using amplitude modulation of signals have ferrite rod antennae. Other applications include waveguides and wave shaping, for example, in pulse-compression systems.

The permeability of these soft ferrites does not change much with frequency up to a critical frequency but then decays rapidly with increasing frequency. The critical frequency of these materials varies between 1 and 100 MHz. The saturation magnetization of ferrites is typically 0.5 T, which is low compared with iron and cobalt alloys. For very high-frequency applications, beyond 100 MHz, there are other materials such as some of the hexagonal ferrites, which have properties that make them suitable for use at these frequencies. These materials have their magnetic moments confined by anisotropy to the hexagonal base plane.

Snelling has given a survey of the properties and applications of ferrites [29]. This included preparation, properties, basic ac magnetic theory and magnetic circuit design using ferrites.

12.3 MATERIALS FOR DC APPLICATIONS

In dc applications, the need for a low conductivity does not arise and so there are fewer constraints on the type of material suitable for particular applications. These applications generally require low coercivity and high permeability. High permeability is best achieved through high saturation magnetization and this means that alloys of iron and cobalt are widely used. Chin and Wernick [30] have reviewed soft magnetic materials for dc applications.

12.3.1 Iron and Low-Carbon Steels (Soft Iron)

Soft iron is used as a core material for dc electromagnets such as laboratory electromagnets for which it remains the best material. The prime concern is to obtain either high magnetic flux densities and or very uniform magnetic flux densities. Iron with low levels of impurities such as carbon (0.05%) and nitrogen has a coercivity of about 80 A m^{-1} (1 Oe) and a maximum relative permeability of the order of 10,000. By annealing in hydrogen, the impurities can be removed and this results in a reduction in coercivity to 4 A m^{-1} (0.05 Oe) and an increase in maximum relative permeability to about 100,000 as shown in Table 12.3. The highest relative permeability obtained for pure iron is 1.5×10^6; however, this material is too expensive for many applications.

In most applications, ultra-high-purity iron is unnecessary. A typical commercial soft iron for electromagnet applications will therefore contain about 0.02% C, 0.035% Mn, 0.025% S, 0.015% P, and 0.002% Si in the form of impurities. The magnetization curves of high-purity iron and a commercial soft iron after annealing in hydrogen to remove impurities are shown in Figure 12.24. For electromagnets, the principal question that arises is: What field is necessary to produce an induction of 1.0 or 1.5 T? For the commercial soft iron the values are typically 200 and 700 A m^{-1}, respectively.

Any form of mechanical deformation will result in a deterioration of the magnetic properties of soft iron for electromagnet applications. The internal stresses produced by cold working can be removed by annealing at temperatures between 725°C and 900°C, provided the material does not suffer oxidation during annealing, which would also result in impaired magnetic properties. The usual procedure is to anneal in a hydrogen atmosphere, which has the additional advantage of removing some of

TABLE 12.3
Magnetic Properties of Various High-Purity Forms of Iron

	Saturation Induction B_s (T)	Coercivity H_c (A m^{-1})	Relative Permeability At 80 A m^{-1}	Relative Permeability At 800 A m^{-1}	Maximum Relative Permeability μ_{max}
Cast magnetic ingot iron	2.15	68	3500	1500	–
Magnetic ingot iron (2 mm sheet)	2.15	89	1800	1575	–
Electromagnet iron (2 mm sheet)	2.15	81.6	2750	1575	–
Ingot iron (vacuum melted)	–	24.8	–	–	21,000
Electrolytic iron (annealed)	–	18.4	–	–	41,500
Electrolytic iron (vacuum melted and annealed)	–	7.2	–	–	61,000
Puron (H_2 treated)	2.16	4.0	–	–	100,000

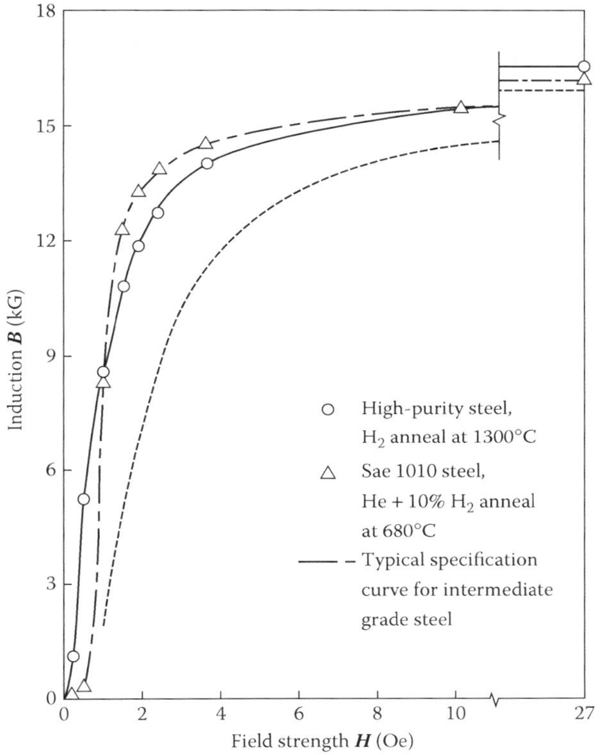

FIGURE 12.24 Initial magnetization curves for two high-purity steels, after annealing in a reducing atmosphere of hydrogen, and a typical intermediate-grade steel. (Data from Swisher, J. H. et al., *Trans. A.S.M.*, 62, 257, 1969; Swisher, J. H., and Fuchs, E. O. *J. Iron. Steel. Inst.*, August, 777, 1970.)

the impurities. The variation of coercivity with impurity nitrogen and carbon content is shown in Figures 12.25 and 12.26.

12.3.2 Iron-Nickel Alloys (Permalloy)

Iron and nickel form a number of commercially important alloys, most of which are in the *permalloy* range, with nickel contents above 35%. For dc applications, the iron-nickel alloys are very versatile since by suitable alloying, they can be produced with a wide range of properties. Among these, it is possible to produce for example an alloy with zero magnetostriction (19% Fe-81% Ni).

The nickel-iron alloys in general have very high permeabilities, as shown in Tables 12.4 and 12.5. The maximum permeability of the polycrystalline alloys occurs in those alloys for which the anisotropy and magnetostriction are small. The value of the anisotropy constant K_1 is zero at a nickel content of 78%. The addition of 5% copper to permalloy produces the alloy known as mu-metal, although commercial

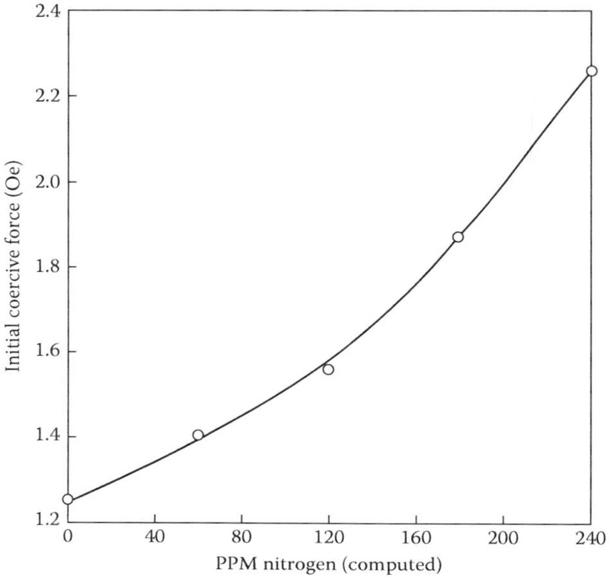

FIGURE 12.25 Dependence of coercivity on nitrogen content in high-purity iron. (Data from Swisher, J. H. et al., *Trans. A.S.M.*, 62, 257, 1969; Swisher, J. H., and Fuchs, E. O. *J. Iron. Steel. Inst.*, August, 777, 1970.)

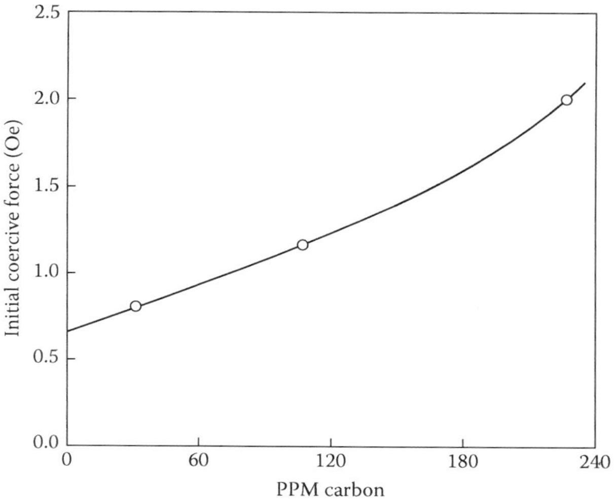

FIGURE 12.26 Dependence of coercivity on carbon content in high-purity iron. (Data from Swisher, J. H. et al., *Trans. A.S.M.*, 62, 257, 1969; Swisher, J. H., and Fuchs, E. O. *J. Iron. Steel. Inst.*, August, 777, 1970.)

TABLE 12.4
Selected Magnetic Properties of Materials Used for Relays

	Relative Permeability μ_{max}	Saturation Induction B_s (T)	Coercivity H_c (A m^{-1})	Remanence B_r (T)
Unalloyed Iron				
Auto machining iron	4,300–8,000	2.05	60–140	0.8
Open smelted iron	2,200–7,500	2.15	24–120	0.9
Vacuum smelted iron	–	2.15	16–40	–
Carbonyl iron	30,000	2.15	8–24	0.8
Carbonyl iron (critically stretched)	40,000	2.15	6–10	0.8
Silicon Steels				
Fe-1% Si	4,000–15,000	2.1	30–120	0.9–1.45
Fe-2.5% Si	4,000–12,000	2.0	12–120	0.8–1.2
Fe-4% Si	5,000–20,000	2.0	5–90	0.8–1.2
Nickel Steels				
Fe-36% Ni	6,000–14,000	1.3	8–24	0.8
Fe-50% Ni	15,000–60,000	1.55	5–14	0.8–1.2
Fe-78% Ni	5,000–300,000	0.7	1–8	0.5–0.75

TABLE 12.5
Selected Magnetic Properties of Different Soft Magnetic Materials

		Relative Permeability		Coercivity H_c (A m^{-1})	Saturation Induction B_s (T)
	Composition	μ_i	μ_{max}		
Iron	100% Fe	150	5,000	80	2.15
Silicon-iron (nonoriented)	96% Fe, 4% Si	500	7,000	40	1.97
Silicon-iron (grain-oriented)	97% Fe, 3% Si	1,500	40,000	8	2.0
78 Permalloy	78% Ni, 22% Fe	8,000	100,000	4	1.08
Hipernik	50% Ni, 50% Fe	4,000	70,000	4	1.60
Supermalloy	79% Ni, 16% Fe, 5% Mo	100,000	1,000,000	0.16	0.79
Mu-metal	77% Ni, 16% Fe, 5% Cu, 2% Cr	20,000	100,000	4	0.65
Permendur	50% Fe, 50% Co	800	5,000	160	2.45

(*Continued*)

TABLE 12.5 (Continued)
Selected Magnetic Properties of Different Soft Magnetic Materials

	Composition	Relative Permeability μ_i	Relative Permeability μ_{max}	Coercivity H_c (A m^{-1})	Saturation Induction B_s (T)
Hipereo	64% Fe 35% Co, 0.5% Cr	650	10,000	80	2.42
Supermendur	49% Fe 49% Co, 2% V		60,000	16	2.40

mu-metal also contains 2% Cr. Its magnetic properties are no better than permalloy but mu-metal is rather more ductile than permalloy and is therefore used in the form of thin sheets in magnetic shielding in order to prevent stray magnetic fields from affecting sensitive components.

The low coercivity of these alloys can be seen from Table 12.4 and this has made them ideal for relays (electromechanical actuators) with a short release time. Both permalloy and mu-metal are used for the cores. However, the low saturation magnetization of the permalloys with higher levels of nickel (e.g., Fe 20%-Ni 80%) has meant that they are not so widely used in relays.

Addition of cobalt to iron-nickel gives the so-called perminvar ternary alloys, which have constant permeability and almost zero hysteresis losses at fields of up to 200 A m^{-1}.

12.3.3 Iron-Cobalt Alloys (Permendur)

Cobalt is the only element that causes an increase in saturation magnetization, Figure 12.27, and Curie temperature when alloyed with iron. The cobalt-iron alloys

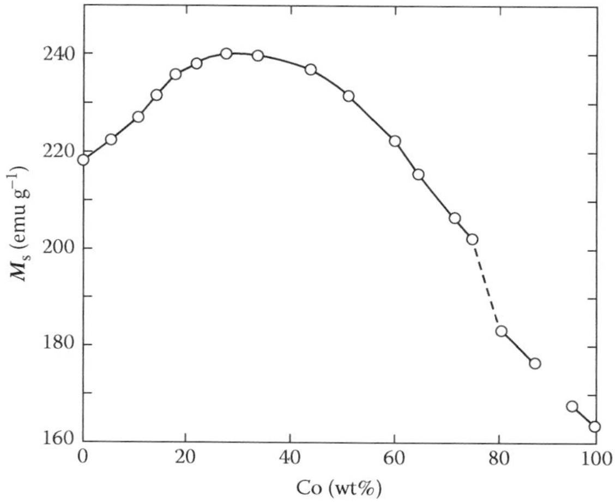

FIGURE 12.27 Variation of saturation magnetization with composition in iron-cobalt alloys. (Data from Weiss, P., and Forrer, R. *Annalen der Physik*, 12, 279, 1929.)

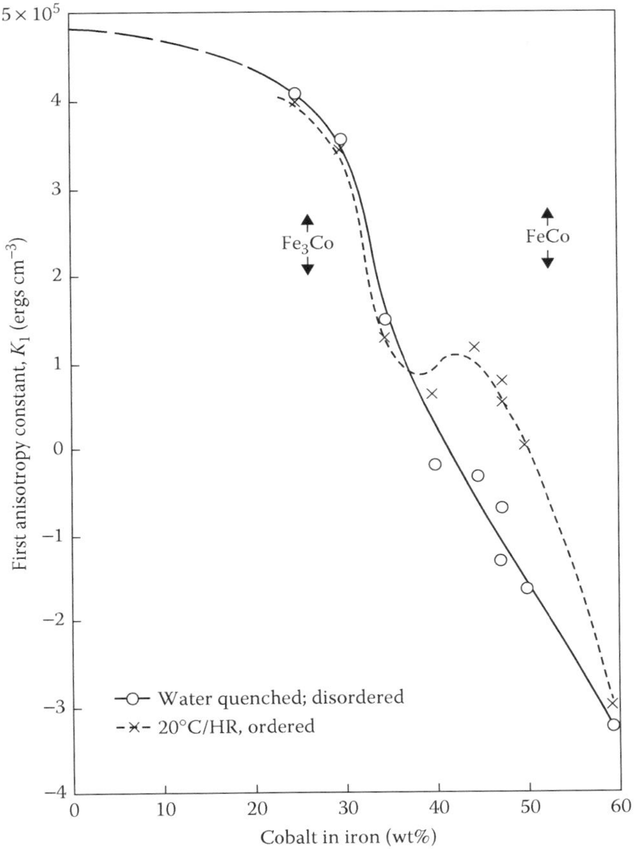

FIGURE 12.28 Variation of the first anisotropy constant with composition in iron-cobalt alloys. (Data from Hall, R. C. *Trans. Met. Soc. AIME*, 218, 268, 1960.)

are therefore of some significant interest for these reasons. These alloys have low anisotropy, as shown in Figure 12.28, and high permeability, as shown in Figure 12.29. They have found some applications in both ac and dc devices but the high cost of cobalt has been a limiting factor. Nickel and niobium are also used as alloying constituents in some of the iron-cobalt alloys.

The highest known saturation magnetization occurs in 65% iron-35% cobalt alloys in which it reaches 1.95 MA m^{-1}. These binary alloys are brittle, but improved mechanical properties can be obtained by alloying with vanadium. Permendur that has a composition of 49% Fe, 49% Co, and 2% V is known as vanadium permendur. This has a saturation magnetization that is close to the maximum 1.95 MA m^{-1} and also has a permeability, which remains constant with $\boldsymbol{H}$, over a wide range of field strengths. The high cost of cobalt has also limited the use of these alloys in relay armatures, where they would otherwise be the ideal material in view of their high

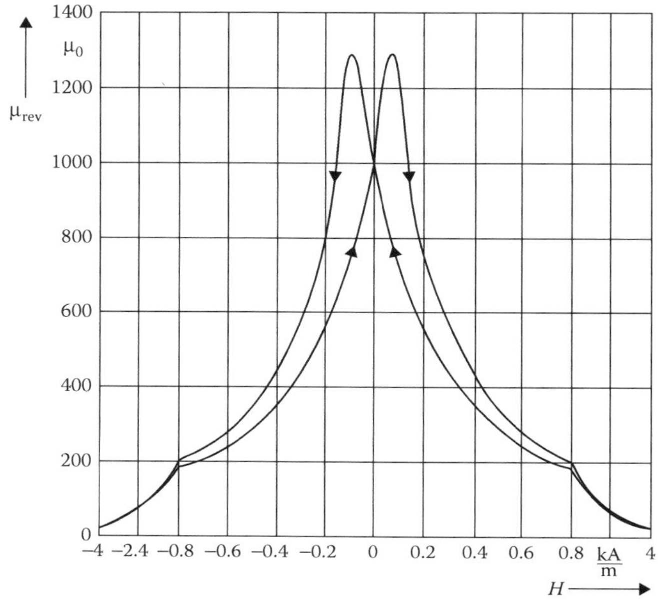

FIGURE 12.29 Reversible permeability of permendur as a function of field strength measured with a small superimposed cyclic field of amplitude 0.34 A m^{-1} at a frequency of 200 Hz.

saturation magnetization, which would generate a large attractive force to operate the moving parts of the relay.

The ternary alloys of iron-cobalt and vanadium are used in magnetic amplifiers and in some switching and memory-storage cores. They are used in diaphragms of high-quality phone receivers, where the high value of reversible permeability at high flux density is important, and as the pole pieces of servomotors in the aircraft industry, where high flux density is crucial. These alloys were used previously as magnetostrictive transducers; however, there are now far superior magnetostrictive transducer materials available.

12.4 MATERIALS FOR MAGNETIC SHIELDING

In some applications, it is necessary to provide a region that is as far as possible free from magnetic fields. This is achieved by enclosing the region within a magnetic shield, which is simply an enclosure made from high permeability material, which diverts the magnetic flux through the shield material and away from the region that is being shielded. The requirement may be to prevent either static or time-dependent magnetic fields from entering the region.

In practice, magnetic shielding is sometimes needed to protect sensitive equipment or devices, particularly measuring equipment, or to protect switching components from stray magnetic fields, which can alter their magnetic status. The latter is particularly important in data storage applications. The usual procedure is to surround the sensitive equipment or component with one or more layers of high permeability magnetic material. The most commonly used material for shielding purposes is mu-metal, which contains 73% Ni, 20% Fe, 5% Cu, and 2% Cr. For higher field strengths, other alloys in the composition range of 35% Ni-65% Fe are used because they have a higher saturation magnetization.

Magnetic shielding material is usually supplied in the form of sheets with thickness ranging from 0.2 to 1.0 mm [2], although in exceptional cases thicknesses of up to 2 mm are used. Thin foils which are flexible and so can be used to wrap sensitive components are also widely used. These are typically 0.05–0.1 mm thick, and therefore are less effective as shields than the thicker sheets of the same material simply because for a given flux density they divert less total flux. Therefore, these thin foils are used in situations where the shielding requirements are less stringent.

Probably the most widespread use of magnetic shielding is in magnetic read heads in magnetic recording systems, in which it is desirable to shield the read head from any stray fields emanating from other regions of the recording medium than the region currently being *read*. A discussion of magnetic read/write heads is given in Chapter 14. Large magnetic shields are used in medical diagnostic systems, such as magnetic resonance, where it is essential to control the magnetic fields precisely and to avoid interference from unexpected, time-dependent stray magnetic fields generated from nearby equipment.

12.4.1 Shielding Factor

The shielding effectiveness is measured by the shielding factor S. This is the ratio of the magnetic field at a point inside the shield $\boldsymbol{H}_i$ to the strength of the magnetic field at the same location in the absence of the shield $\boldsymbol{H}_a$ [31]:

$$S = \frac{\boldsymbol{H}_a}{\boldsymbol{H}_i} \tag{12.11}$$

The higher the value of the shielding factor S, the better the shielding and the lower the field inside the shield for a given external field. Equations for the shielding factor can only be given in the simplest geometrical situations. In shielding of steady fields, it is clear that the shielding factor should increase with the permeability μ_r of the shielding material and also with its thickness d. For a hollow sphere, the shielding factor is given approximately by

$$S = \frac{4}{3}\frac{\mu_r d}{D} \tag{12.12}$$

where D is the diameter of the sphere. For a long hollow cylinder perpendicular to an applied field

$$S = \frac{\mu_r d}{D} \tag{12.13}$$

where D is the diameter of the cylinder. Finally, for a cubic box with length of side a the shielding factor is

$$S = \frac{4}{5}\frac{\mu_r d}{a} \tag{12.14}$$

These shielding factors and their variation with permeability of the shield material and with the strength of the applied field have been discussed by Boll [2, p. 288].

12.4.2 Multiple Shields

In the case of multiple shields, which means in effect multiple layers of shielding material, the total shielding factor is in general a combination of all the individual shielding factors. This has been discussed by Freake and Thorp [32], who showed that by suitable geometrical arrangement of the shield layers, the total shielding factor can be skewed toward the product of the individual shielding factors, thereby leading to an increased overall shielding factor.

12.4.3 Shielding of Alternating Fields

In the majority of cases where shielding is employed, it is to reduce or eliminate alternating or time-dependent electromagnetic fields, which can cause noise by generating unwanted voltages in circuits in equipment. Similar shielding principles apply to both steady and time-dependent magnetic fields; however, there is an increase in the shielding factor of a given shield as the frequency of the applied field increases.

The main reason for this increase in shielding factor with frequency is that the penetration depth for electromagnetic fields in conductors δ decreases with frequency, as discussed in Section 2.5. The penetration depth δ is inversely proportional to the square root of the product of relative permeability, conductivity, and frequency of the applied field. The shielding factor for an alternating magnetic field can be calculated approximately from the shielding factor for a steady field using the following equation:

$$S(\omega) = S(0)K \tag{12.15}$$

where the correction factor K depends on the ratio of the thicknesses of the shield layer d to the penetration depth δ. The larger d/δ, the greater the correction factor K and therefore the greater the shielding factor.

The applications of magnetic materials in shielding are very diverse, and there is a wide range of different shielding materials in different thicknesses to meet a variety of needs. Comprehensive descriptions of magnetic shielding have been given by Hemming [33], who dealt with large-scale structural shielding, and Tsaliovich [34], who dealt with shielding for smaller scale electronics applications.

REFERENCES

1. Herzer, G. Modern soft magnets: Amorphous and nanocrystalline materials, *Acta Mater.*, 61, 718, 2013.
2. Boll, R. *Vacuumschmelze Handbook of Soft Magnetic Materials*, Heyden, London, 1978.
3. Moses, A. J. Energy efficient electrical steels: Magnetic performance prediction and optimization, *Scripta Mater.*, 67, 560, 2012.
4. Stephenson, E. T. *J. Appl. Phys.*, 57, 288, 4226, 1985.
5. Bertotti, G. *IEEE Trans. Magn.*, 24, 621, 1988.
6. Jiles, D. C. *J. Appl. Phys.*, 76, 5849, 1994.
7. Chen, C. W. *Magnetism and Metallurgy of Soft Magnetic Materials*, North Holland, Amsterdam, the Netherlands, 1977, p. 276.
8. Gomes, E., Schneider, J., Verbeken, K., Barros, J., and Houbaert, Y. Correlation between microstructure, texture, and magnetic induction in non-oriented electrical steels, *IEEE Trans. Magn.*, 46, 310, 2010.
9. Fischer, O., and Schneider, J. Influence of deformation process on the improvement of non-oriented electrical steel, *J. Magn. Magn. Mater.*, 254–255, 302, 2003.
10. Kubota, T., Fujikura, M., and Ushigami Y. Recent progress and future trends in grain oriented silicon steel, *J. Magn. Magn. Mater.*, 215, 69, 2000.
11. Cullity, B. D. *Introduction to Magnetic Materials*, Addison-Wesley, Reading, MA, 1972, p. 494.
12. Fiorillo, F. *JMMM*, 157, 428, 1996.
13. Verbeken, K., Schneider, J., Verstraete, J., Hermann, H., and Houbaert, Y. Effect of hot and cold rolling on grain size and texture in Fe-2.4wt%Si strips, *IEEE Trans. Magn.*, 44, 3820, 2008.
14. Gomes, E., Schneider, J., Verbeken, K., Hermann, H., and Houbaert, Y. Effect of hot and cold rolling on grain size and texture in Fe-Si strips with Si content larger than 2 wt%, *Mater. Sci. Forum*, 638, 3561, 2010.
15. Heck, C. *Magnetic Materials and Their Applications*, Crane and Russak, New York, 1974, p. 392.
16. Luborsky, F. E. Amorphous ferromagnets, in *Ferromagnetic Materials*, Vol. 1 (ed. E. P. Wohlfarth), North Holland, Amsterdam, the Netherlands, 1980.
17. Ohnaka, I., Fukusako, T., Ohmichi, T. et al. *Proceedings of the Fourth Conference on Rapidly Quenched Metals*, 1982, p. 31.
18. Mohri, K., Humphrey, F. B., Yamosaki, J., and Okamura, K. *IEEE Trans. Magn.*, 20, 1409, 1984.
19. Mohri, K., Humphrey, F. B., Kawashima, K. et al. *IEEE Trans. Magn.*, 26, 1789, 1990.
20. Atkinson, D., Squire, P. T., Gibbs, M. R. J., and Hogsdon, S. N. *J. Phys. D. (Appl. Phys)*, 27, 1354, 1994.
21. Squire, P. T., Atkinson, D., Gibbs, M. R. J., and Atalay, S. *JMMM*, 132, 10, 1994.
22. Yoshizawa, Y., Oguma, S., and Yamauchi, K. *J. Appl. Phys.*, 64, 6044, 1988.
23. Herzer, G. *JMMM*, 157, 133, 1996.
24. Hernando, A., Navarro, I., and Gorria, P. *Phys. Rev. B*, 51, 3281, 1995.

25. Hernando, A., and Kulik, T. *Phys. Rev. B*, 49, 7064, 1994.
26. Bertraut, F., and Forret, F. *Comptes Rend. Hebd. Seance. Acad. Sci.*, 242, 382, 1956.
27. Slick, P. I. Ferrites for non-microwave applications, in *Ferromagnetic Materials* (ed. E. P. Wohlfarth), North Holland, Amsterdam, the Netherlands, 1980.
28. Nicolas, J. Microwave ferrites, in *Ferromagnetic Materials*, Vol. 2 (ed. E. P. Wohlfarth), North Holland, Amsterdam, the Netherlands, 1980.
29. Snelling, E. C. *Soft Ferrites: Properties and Applications*, 2nd edn, Butterworths, London, 1988.
30. Chin, G. Y., and Wernick, J. H. Soft magnetic materials, in *Ferromagnetic Materials*, (ed. E. P. Wohlfarth), North Holland, Amsterdam, the Netherlands, 1980.
31. Mager, A. J. *IEEE Trans. Magn.*, 6, 67, 1971.
32. Freake, S. M., and Thorp, T. L. *Rev. Sci. Inst.*, 42, 1411, 1971.
33. Hemming, L. H. *Architectural Electromagnetic Shielding Handbook*, IEEE Press, Piscataway, NJ, 1992.
34. Tsaliovich, A. B. *Cable Shielding for Electromagnetic Compatibility*, Van Nostrand Reinhold, New York, 1995.
35. Swisher, J. H., English, A. T., and Stoffers, R. C. *Trans. A.S.M.*, 62, 257, 1969.
36. Swicher, J. H., and Fuchs, E. O. J. *Iron. Steel. Inst.*, August, 777, 1970.
37. Weiss, P., and Forrer, R. *Annalen der Physik*, 12, 279, 1929.
38. Hall, R. C. *Trans. Met. Soc. AIME*, 218, 268, 1960.

FURTHER READING

Bottauscio, O. (ed.). Selected papers from the 21st soft magnetic materials conference, Budapest, Hungary, *IEEE Trans. Magn.*, 50, 4, April 2014.

Coey, J. M. D. *Magnetism and Magnetic Materials*, Cambridge University Press, Cambridge, 2009.

Gutfleisch, O., Willard, M. A., Brück, E., Chen, C. H., Sankar, S. G., and Liu, J. P. Magnetic materials and devices for the 21st century: Stronger, lighter, and more energy efficient, *Adv. Mater.*, 23, 821, 2011.

Herzer, G. Modern soft magnets: Amorphous and nanocrystalline materials, *Acta Mater.*, 61, 718, 2013.

Inoue, A., and Hashimoto, K., *Amorphous and Nanocrystalline Materials: Preparation, Properties, and Applications*, Springer, Berlin, Germany, 2001.

Jenkins, K. W. Current status and future developments of grain oriented electrical steel, in *Proceedings of the 6th International Conference on Magnetism and Metallurgy*, Cardiff, June 17–19, 2014.

Jiles, D. C. Recent advances and future directions in magnetic materials, *Acta Mater.*, 51, 5907, 2003.

Littmann, M. F. *IEEE Trans. Mag.*, 7, 48, 1971.

Schneider, J., Stoecker, A., Franke, F., Schroeder, C., Li, G., and Hermann, H., Evolution of microstructure and texture along the processing route of ferritic non-oriented FeSi steels, in *Proceedings of the 6th International Conference on Magnetism and Metallurgy*, Cardiff, June 17–19, 2014.

MATERIALS

Soft Magnetic Materials. http://www.softmagneticalloy.com/soft_magnetic_materials.htm.
Grain Oriented Electrical Steels. http://www.cogent-power.com/grain-oriented/.

CONFERENCES

Soft Magnetic Materials Conference. http://en.wikipedia.org/wiki/Soft_Magnetic_Materials_Conference.
International Magnetics Conference. http://www.intermagconference.com/.
Magnetism and Magnetic Materials Conference. http://www.magnetism.org/.

13 Hard Magnetic Materials

In this chapter, we consider the properties of ferromagnets that make them useful as permanent magnets. A range of different permanent materials is available, many of which include rare earth metals, but there has been increasing interest in recent years to develop better rare-earth free permanent magnets. Improvements in properties such as increased coercivity, remanence, and maximum energy product continue to be made. The control of magnetic properties through alteration of the nanostructure is central to the development of improved permanent magnets as in hot deformed neodymium-iron-boron permanent magnets and two-phase *exchange spring* magnets. However, in addition to the materials properties, the shape of the permanent magnets is important in providing the desired flux density at a given location, so demagnetizing effects resulting from a combination of magnetization and shape need to be considered.

13.1 PROPERTIES AND APPLICATIONS OF HARD MAGNETS

A permanent magnet is a passive device used for generating a magnetic field. It does not need an electric current flowing in a coil or solenoid to maintain the field. The development of permanent magnets and the principles of the operation and uses have been described by Coey in a review [1]. The energy needed to maintain the magnetic field has been stored previously when the permanent magnet was *charged*—magnetized initially in a high field strength and then brought to remanence when the applied field was removed. In order to prevent demagnetization, a high coercivity is needed because a permanent magnet is only of use if it has a relatively high magnetization when removed from the applied magnetic field. The need for high remanence inevitably means a high saturation magnetization is needed. However, most of the other properties that were considered desirable in soft magnetic materials, such as high permeability and low conductivity, are less relevant in hard magnetic materials. The main categories of permanent magnet materials are hard ferrites, neodymium-iron-boron, samarium-cobalt, and alnico. The world market for permanent magnets was estimated at \$8 billion per year in 2007 and \$11 billion per year in 2010 [2]. This comprised 34% hard ferrite, 65% rare earth magnets, and 1% alnico as shown in Table 13.1. The growth rate of total permanent magnets is estimated to be 12% annually and the fastest growing segment of this market is for Nd-Fe-B-based magnets.

Permanent magnets are used for magnetic flux generation in a variety of situations in which it is difficult or impossible to provide electrical power, as in portable equipment, or where geometrical constraint or space restrictions dictate their use rather than electromagnets. The applications include large-scale power engineering, such as in permanent magnet motors and generators, low-power small-scale uses, for example in *voice coil* motors in hard disk drives, control devices for

TABLE 13.1
Permanent Magnet Market by Material (%)

Material	Percentage
Sintered ferrite	22
Bonded ferrite	11
Sintered NdFeB	32
Bonded NdFeB	9
Sintered SmCo	22
Bonded SmCo	2
Alnico	1
Other	1

Source: Gutfleisch, O. et al., *Adv. Mater.*, 23, 821, 2011.
Note: Total = 11 × 10^9 US dollars per year in 2010. Growth = 12% per year.

electron beams and moving-coil meters, and intermediate power-range applications such as microphones, loudspeakers, magnetic separators, frictionless bearings and magnetic levitation systems, and various forms of holding magnets such as door catches. The approximate market segments are motors (60%), loudspeakers (15%), communication (8%), electron tubes (7%), mechanical devices (4%), and other magnetic devices (6%). The application of permanent magnets in motors includes motors for hard disk drives in computers, which is the largest market segment for permanent magnets such as NdFeB. The range of available permanent magnet materials and their many device applications have been reviewed by Gutfleisch et al. [2].

When ferromagnetic materials are used as permanent magnets, they must operate in conditions where at best they are subject to their own demagnetizing field and at worst can be subjected to various demagnetizing effects of other magnetic materials or magnetic fields in their vicinity. It is therefore essential that they are not easily demagnetized. The permanent magnets need to be shaped accordingly and the shape of the magnets determines their demagnetizing field, which is important for achieving the required magnetic flux density at a given location [3].

13.1.1 Coercivity

Coercivity is used to distinguish between hard and soft magnetic materials. Hard magnetic materials are generally classified as having coercivities above 10 kA m^{-1} (125 Oe) but some permanent magnet materials have coercivities two orders of magnitude greater than this. For example, the intrinsic coercivity is typically 1.1 MA m^{-1} (14,000 Oe) in neodymium-iron-boron, 0.69 MA m^{-1} (8700 Oe) in samarium-cobalt, and 56 kA m^{-1} (700 Oe) in alnico [4]. All of the above values are at room temperature.

Since the permanent magnets operate without an applied field, their ability to resist demagnetization is an important property and consequently a high coercivity is desirable. Over the years, there has been continual progress in identifying new permanent magnet materials with higher coercivities [5]. Some permanent magnet materials have intrinsic coercivities in the range of 1.2×10^6 A m^{-1}.

Unlike soft magnetic materials, in which $\boldsymbol{B}$ the magnetic flux density is approximately equal to $\mu_0\boldsymbol{M}$, the magnetization in permanent magnets is not simply a linear function of the flux density $\boldsymbol{B}$ because $\boldsymbol{B} = \mu_o(\boldsymbol{H} + \boldsymbol{M})$ and the values of magnetic field $\boldsymbol{H}$ used in the permanent magnets are generally much larger than in soft magnetic materials. The result is that two coercivities can be specified. One is the *intrinsic* coercivity ${}_{m}\boldsymbol{H}_c$, the field at which the magnetization $\boldsymbol{M}$ in the material is zero, and the other is the coercivity ${}_{B}\boldsymbol{H}_c$, the field at which the magnetic flux density $\boldsymbol{B}$ in the material is zero. These quantities have quite different values in hard magnetic materials, and the greater the difference, the better the material is as a permanent magnet. It should be noted that ${}_{m}\boldsymbol{H}_c$ is always greater than ${}_{B}\boldsymbol{H}_c$.

13.1.2 Remanence

No matter what the coercivity of the permanent magnet is it will be of little use if the remanent magnetization is low. Therefore, a high remanence combined with a high coercivity is essential. The remanence $\boldsymbol{M}_R$ is the maximum residual magnetization, which can be obtained only in a closed-loop configuration, in which there is no demagnetizing field. Since, to be of any use, permanent magnets must be operated in an *open circuit* configuration, the residual magnetization at which the permanent magnet operates in open circuit will always be below the value of remanence $\boldsymbol{M}_R$. The remanence in neodymium-iron-boron is, for example, typically $\boldsymbol{M}_R = 1.05$ MA m^{-1} (1050 emu cm^{-3}), $\boldsymbol{B}_R = 1.3$ T (13 kG).

13.1.3 Saturation Magnetization

The remanence of course cannot exceed the saturation magnetization, and for this reason, the saturation magnetization of a permanent magnet should be large. While this condition is necessary, it is not sufficient since the *squareness ratio* or *remanence ratio* $\boldsymbol{M}_R/\boldsymbol{M}_s$ must also be as close to 1 as possible in order to ensure a large remanence. The saturation magnetization in neodymium-iron-boron can reach as high as 1.27 MA m^{-1} [6], in samarium-cobalt it is 0.768 MA m^{-1} [7], and in alnico alloys typically 0.87–0.95 MA m^{-1} [8]. A comparison of the intrinsic coercivities ${}_{m}\boldsymbol{H}_c$ and saturation magnetic polarizations $\mu_0\boldsymbol{M}_s$ for a variety of hard and soft magnetic materials is given in Figure 13.1.

13.1.4 Energy Product

One parameter that is often of interest to permanent magnet manufacturers and users is the maximum energy product, which is the maximum value of $|\boldsymbol{BH}|$ obtained in the second quadrant, as shown in Figure 13.2. This is closely related to the total hysteresis loss or area enclosed by the hysteresis loop.

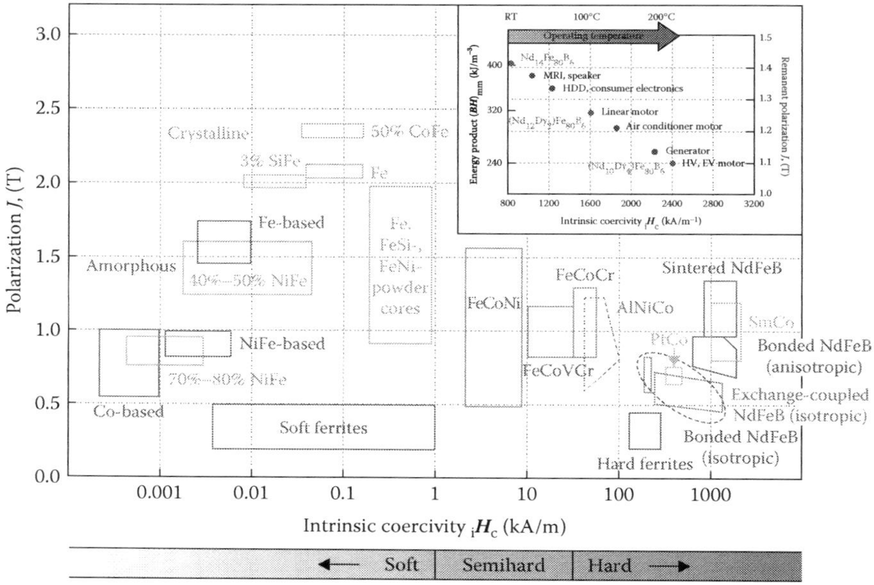

FIGURE 13.1 Comparison of the saturation magnetic polarizations ($\mu_o \boldsymbol{M}_s$) and intrinsic coercivities ($_m\boldsymbol{H}_c$) of a variety of magnetic materials. (After Gutfleisch, O. et al., *Adv. Mater.*, 23, 821, 2011.)

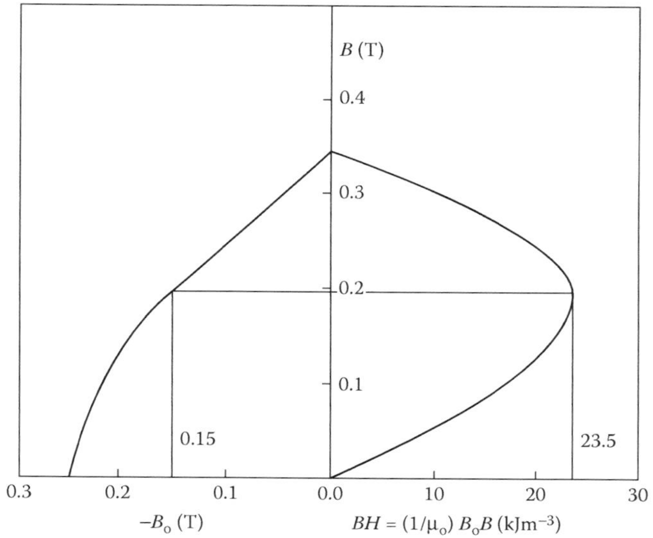

FIGURE 13.2 Second quadrant *demagnetization curve* of the ceramic magnet barium ferrite (left half of the diagram) and energy product as a function of induction (right half of the diagram).

The interest of researchers in the maximum energy product of permanent magnets deserves some explanation. It stems from the total energy associated with a permanent magnet and the work that it can do. The integral $\int \boldsymbol{B} \cdot \boldsymbol{H} dV$ over all space will necessarily be zero for a permanent magnet. This can then be split into two integrals, one representing the field inside the magnet and the other outside the magnet

$$\int_{\text{allspace}} \boldsymbol{B} \cdot \boldsymbol{H} dV = \int_{\substack{\text{inside}\\\text{magnet}}} \boldsymbol{B} \cdot \boldsymbol{H} dV + \int_{\substack{\text{outside}\\\text{magnet}}} \boldsymbol{B} \cdot \boldsymbol{H} dV = 0 \quad (13.1)$$

Therefore, since $\boldsymbol{B} = \mu_0 \boldsymbol{H}$ outside the magnet, the energy of the field generated outside the magnet is equal to the integral of $\boldsymbol{B} \cdot \boldsymbol{H}$ over the volume of the magnet.

$$\text{Useful energy of magnet} = \int_{\substack{\text{outside}\\\text{magnet}}} \mu_0 \boldsymbol{H}^2 dV = \int_{\substack{\text{inside}\\\text{magnet}}} \boldsymbol{B} \cdot \boldsymbol{H} dV \quad (13.2)$$

so the maximum energy product is simply a measure of the maximum amount of useful work that a permanent magnet is capable of doing outside the magnet.

There have been improvements in the maximum energy product of various permanent magnet materials over the years as shown in Figure 13.3 [9]. We can estimate the ultimate limits of the maximum energy product that may be achievable in the future. For a material with a remanence $\boldsymbol{M}_R$ and a very square hysteresis loop the coercivity ${}_B\boldsymbol{H}_c$ cannot exceed $\boldsymbol{M}_R$. Therefore, in this limiting case, the maximum energy product is $\mu_0 \boldsymbol{M}_R^2/4$. Clearly, remanence can never be greater than the

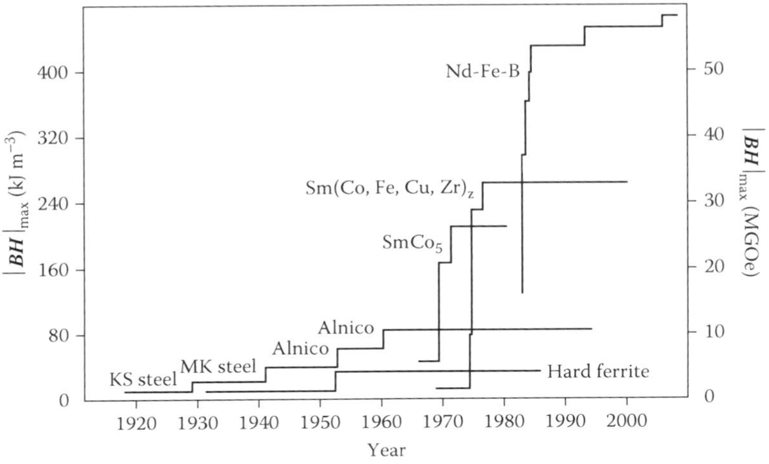

FIGURE 13.3 Improvements in maximum energy product $|\boldsymbol{BH}|_{max}$ in kJ m^{-3} in different permanent magnet materials during the years 1920 to 2010. (After Electron Energy Corporation, http://www.electronenergy.com/about/permanent-magnets.htm.)

saturation magnetization, and so the energy product cannot exceed $\mu_0 \boldsymbol{M}_s^2/4$ [10]. The largest known saturation is $\mathbf{M}_s = 1.95 \times 10^6$ A m^{-1} in 35% cobalt-65% iron alloys. Therefore, the largest possible maximum energy product among known magnetic materials is 1.19×10^6 J m^{-3} (150 MGOe).

In most cases in industry, the maximum energy product is still measured in non-SI units of megaGauss-Oersted (MGOe). The conversion factors between MGOe, J m^{-3}, and ergs cc^{-1} are as follows:

$$1\ \text{MGOe} = \left(10^6/4\pi\right)\text{ergs/cc} = 7.96\ \text{kJ m}^{-3} \tag{13.3}$$

A high-energy product results from high coercivity and remanence; however, it is not sufficient merely to choose the material with the highest energy product for any given application since the optimum operating conditions, which are affected by geometrical considerations and operating temperature, may dictate that other properties are more important than maximum energy product in determining performance. In order to understand this, it is necessary to consider the complete demagnetization curve of the permanent magnet.

13.1.5 Demagnetization Curve

The maximum energy product by itself gives insufficient information about the properties of a permanent magnet in order to determine its performance in a given situation. A more useful way of displaying the magnetic properties of a permanent magnet is to plot the portion of the hysteresis loop in the second quadrant of the $\boldsymbol{B}$-$\boldsymbol{H}$ curve from the remanence to the coercivity. This is known as the *demagnetization curve* and is the information that is needed in order to decide on the suitability of a permanent magnet for particular applications. Such a curve contains information about the maximum energy product, but also contains additional information for the designer if the permanent magnet cannot be operated at $|\boldsymbol{BH}|_{max}$, its optimum operating condition.

The operating conditions for a permanent magnet are determined by the demagnetizing factor in the form of the *load line* and the locus of the magnetic hysteresis loop in the second quadrant of the $\boldsymbol{B}$-$\boldsymbol{H}$ plane as shown in Figure 13.4. This region of the hysteresis loop is referred to as the *demagnetization curve* by the permanent magnet community. If the geometry is known, that is, the length of the air gap and the total length of magnet that can be used, the designer should select a material with the largest possible value of $\boldsymbol{BH}$ on the load line for that geometry. The load line is the locus of possible operating points as dictated by the demagnetizing factor for the particular geometry of specimen.

The strength of the demagnetizing field of a permanent magnet in an open-circuit configuration depends therefore on the shape of the permanent magnet. It becomes clear that the choice of material is dependent as much on the required shape for the permanent magnet and air gap as on the intrinsic material properties. This is discussed in the following section.

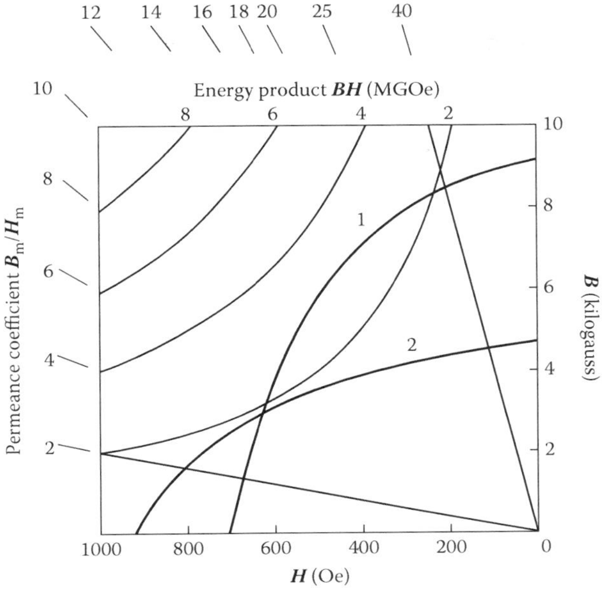

FIGURE 13.4 Demagnetization curves, load lines, and energy product values in the second quadrant of the magnetization curve. Two load lines are shown, one for a short specimen with $l{:}d = 1{:}1$, giving a permeance coefficient of 2 ($N_d = 0.23$), and the other for a longer specimen with $l{:}d = 7{:}1$, giving a permeance coefficient of 40($N_d = 0.244$).

13.1.6 Permanent Magnet Circuit Design

The operating point of a permanent magnet is the point of intersection of the load line with the demagnetization curve. The load line is a straight line that passes through the origin of the ***B***-***H*** plane. The slope of the load line is determined by the demagnetizing factor and hence by the shape of the magnet. Consider, for example, a permanent magnet with length to diameter ratio $l{:}d$, which has a demagnetizing factor N_d. Under these circumstances, it is fairly easy to show [8] that the demagnetizing field $\boldsymbol{H}_d$ is related to the induction $\boldsymbol{B}$ by

$$\boldsymbol{H}_d = -\left[\frac{N_d}{\mu_0(1-N_d)}\right]\boldsymbol{B} \tag{13.4}$$

Hence a load line $\boldsymbol{B} = -(\mu_0(1 - N_d)/N_d)\boldsymbol{H}$ passing through the origin on the ***B***-***H*** plane gives the locus of possible operating conditions for a magnet with the given geometry. The term $(1 - N_d)/N_d$ is called the *permeance coefficient*. It is one of the parameters often supplied to the designer because it contains information on the performance of the permanent magnet with a given geometry and therefore is important in determining the suitability of a permanent magnet for a particular application.

The field in the gap of a permanent magnet is given by [8]

$$\boldsymbol{H}_{\mathrm{g}} = \sqrt{\left(\frac{\boldsymbol{H}_{\mathrm{m}}\boldsymbol{B}_{\mathrm{m}}V_{\mathrm{m}}}{\mu_0 V_{\mathrm{g}}}\right)} \tag{13.5}$$

where:

$\boldsymbol{H}_{\mathrm{m}}$ is the field inside the magnet
$\boldsymbol{B}_{\mathrm{m}}$ is the flux density inside the magnet
V_{m} is the volume of the magnet
V_{g} is the volume of the gap

The important result that emerges here is that the maximum field $\boldsymbol{H}_{\mathrm{g}}$ is obtained by operating the permanent magnet at maximum energy product $|\boldsymbol{BH}|_{\mathrm{max}}$.

Combining these two results, it is clear that the most appropriate material for a given application is the one with the largest value of $\boldsymbol{BH}$ along the load line dictated by geometrical constraints of the given application. Conversely, given a particular material for an application, it should be shaped such that its load line passes through, or as close as possible to $|\boldsymbol{BH}|_{\mathrm{max}}$ in order to optimize its performance.

For example, if a short permanent magnet is required, the demagnetizing field may be large so that the operating conditions are far from the $|\boldsymbol{BH}|_{\mathrm{max}}$ point. Under these circumstances, the material with the largest coercivity may be the best. If a long permanent magnet is used, then the demagnetizing field may be quite low so that the magnet operates closer to the remanence. Under this condition, the material with the largest remanence may be the best. These two conditions are depicted by the two load lines shown in Figure 13.4.

The design of a permanent magnet assembly for producing a given magnetic field in a given volume is not unique, and in principle, many possible configurations can produce identical fields. However, Jensen and Abele [11] have established an algorithm for identifying the arrangement of permanent magnets, which will generate a given field most efficiently. In this case, *most efficiently* means maximizing the magnetic field energy in the volume of interest for a given amount of magnetic energy stored in the permanent magnet assembly. This takes into account both the volume of the permanent magnet assembly and its stored energy density.

In practice, numerical methods, such as three-dimensional finite-element methods, are used to determine the fields generated by a given assembly of magnets. These methods have been exploited to great effect in the work of Leupold, Potenziani, and Tilak [12], who have designed a variety of permanent magnet solenoid structures, with high magnetic flux densities. Moskowitz [13] has given a comprehensive guide to permanent magnet design.

13.1.7 Stoner-Wohlfarth Model of Rotational Hysteresis

The Stoner-Wohlfarth model [14] describes the magnetization curves of an aggregation of single-domain particles with uniaxial anisotropy. The uniaxial anisotropy can arise either as a result of particle shape anisotropy or from the magnetocrystalline

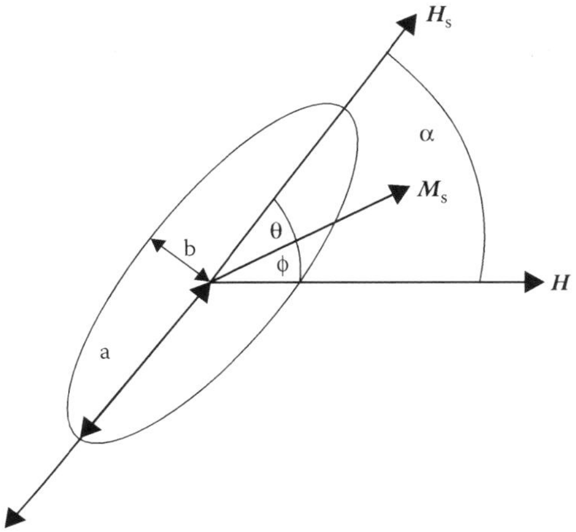

FIGURE 13.5 Shape anisotropy of an ellipsoidal single-domain particle assumed to have neither crystal nor stress anisotropy. The particle has higher demagnetizing factor N_d along the short axis than along the long axis. This leads to shape anisotropy.

anisotropy. If this anisotropy energy E_{an} is represented as a single coefficient expression, then

$$E_{an} = K \sin^2 \theta \tag{13.6}$$

When the magnetization is oriented at an angle θ to the easy direction, as shown in Figure 13.5, this will give rise to a torque of

$$\tau_{an} = \frac{-dE_{an}}{d\theta} = -2K \sin\theta \cos\theta \tag{13.7}$$

The torque produced by a field $\boldsymbol{H}$ τ_H will be dependent on the angle ϕ between the magnetization and the field direction as shown in Figure 13.5,

$$\tau_H = \mu_0 \boldsymbol{M}_s \times \boldsymbol{H} = \mu_0 \boldsymbol{H}\boldsymbol{M}_s \sin\phi \tag{13.8}$$

and when the torque produced by the field $\boldsymbol{H}$ equals the torque due to the anisotropy, we will have equilibrium

$$\tau_H + \tau_{an} = 0 \tag{13.9}$$

$$\mu_0 \boldsymbol{H}\boldsymbol{M}_s \sin\phi - 2K \sin\theta \cos\theta = 0 \tag{13.10}$$

The field strength $\boldsymbol{H}_s$ needed to saturate the magnetization in a polycrystalline specimen is therefore the field needed to overcome the anisotropy and rotate the magnetic moments from the easy axes at 90° to the field direction into the field direction. This saturation field strength is easily shown from Equation 13.10 to be

$$H_s = \frac{2K}{\mu_0 M_s} \tag{13.11}$$

If $\boldsymbol{H}$ is perpendicular to the anisotropy field, this gives completely reversible changes in magnetization as shown in Figure 13.6. If $\boldsymbol{H}$ is antiparallel to the anisotropy field, there arises an irreversible switching of magnetization as soon as $\boldsymbol{H}$ exceeds $2K/\mu_0 \boldsymbol{M}_s$, as shown in Figure 13.7. If $\boldsymbol{H}$ is at some arbitrary angle θ to the anisotropy field, then the behavior is partly reversible and partly irreversible as shown in Figure 13.8. In these cases, whenever θ is greater than 45°, it is found that τ_H

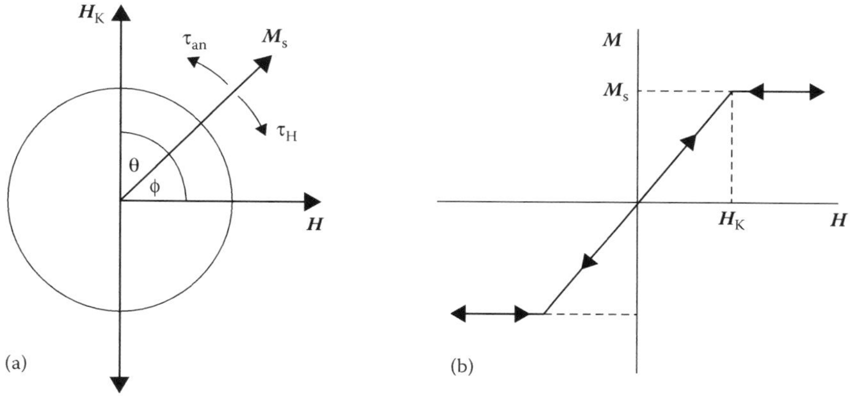

FIGURE 13.6 (a) A spherical single-domain particle with anisotropy field $\boldsymbol{H}_K$ and with the magnetic field $\boldsymbol{H}$ perpendicular to the easy axis and (b) magnetization curve obtained for the situation depicted in (a).

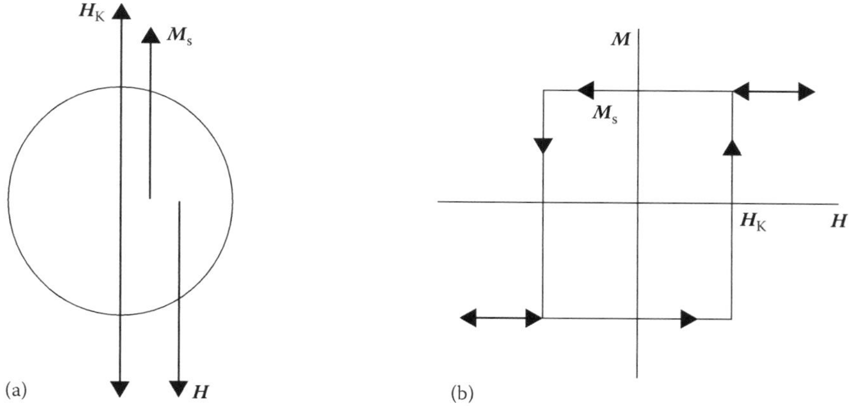

FIGURE 13.7 (a) A spherical single-domain particle with anisotropy field $\boldsymbol{H}_K$ and with the magnetic field $\boldsymbol{H}$ parallel to the easy axis and (b) magnetization curve obtained for the situation depicted in (a).

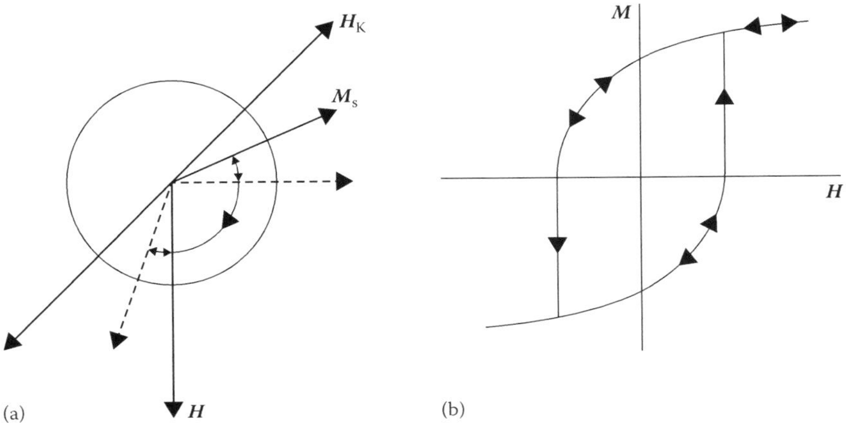

FIGURE 13.8 (a) A spherical single-domain particle with anisotropy field $\boldsymbol{H}_K$ and with the magnetic field $\boldsymbol{H}$ at an arbitrary angle to the easy axis and (b) magnetization curve obtained for the situation depicted in (a) with a coercivity of $\boldsymbol{H}_c = K/\mu_0\boldsymbol{M}_s$.

increases with θ while τ_{an} decreases and a discontinuity occurs in the magnetization at a critical field $\boldsymbol{H}_c$, where

$$\boldsymbol{H}_c = \frac{K}{\mu_0 \boldsymbol{M}_s} \tag{13.12}$$

When various different possible combinations of domain direction are present, this forms a domain distribution and we obtain the kind of curves shown in Figure 13.9, in which the critical field $\boldsymbol{H}_c$ for switching of a particular domain depends on the orientation of the applied field relative to the easy axis that the domain's magnetic moment lies along. Stoner and Wohlfarth considered a random assembly of such single domain particles, each with uniaxial anisotropy. From their calculations the composite hysteresis loop of Figure 13.10 was determined.

The Stoner-Wohlfarth model has been used by permanent magnet producers to indicate ways in which improved properties can be obtained, essentially by increasing the anisotropy. Despite the wide use of this theory, there are questions about its general validity for most real permanent magnet materials. One of the most serious weaknesses of the original theory is that it takes no account of interactions between the single-domain particles. In addition, real permanent magnet materials are not usually arrays of isolated single-domain particles, although polymer-bonded permanent magnets, which consist of permanent magnet particles contained in a nonmagnetic matrix, come closer to the conditions assumed in the Stoner-Wohlfarth model.

Therefore, there has been a shift in emphasis toward explaining properties of these materials in terms of domain-wall motion, which is a more valid physical model. The problem is that domain-wall motion is much more difficult to describe theoretically than domain rotation. Chikazumi [15] has indicated that domain-wall

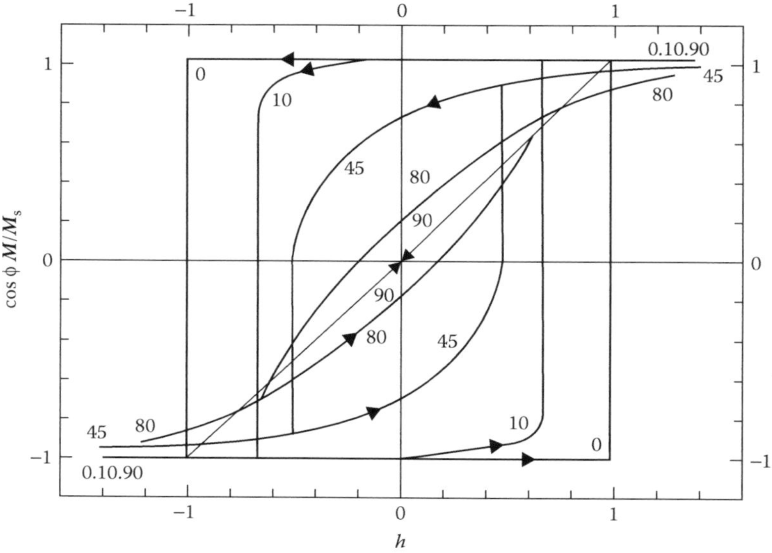

FIGURE 13.9 Magnetization curves obtained on the Stoner-Wohlfarth model for various angles between the direction of the magnetic field and the easy axis, showing critical switching fields for angles of 45° and above.

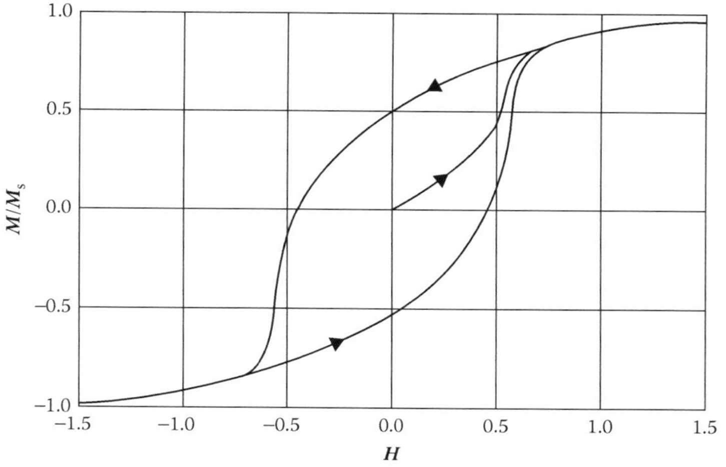

FIGURE 13.10 Composite hysteresis loop obtained from summing the elementary magnetization curves for a particular distribution of easy axes (i.e., magnetic texture) with respect to the field direction. In this case, the distribution of easy axes is random.

mechanisms are probably the dominant factor in permanent magnet materials such as samarium-cobalt and neodymium-iron-boron. Some authors have argued that the high coercivity is due to difficulty in nucleating domains after the material has been magnetized to near saturation [16]. Others suggest that the high coercivity is due principally to domain-wall pinning [17]. Evidence seems to favor the strong pinning mechanism [18].

Higher anisotropy therefore necessarily leads to improved properties for permanent magnet materials by increasing the coercivity [19]. However, this result is not unique to the Stoner-Wohlfarth model; it is far more general and applies just as well to domain-wall motion models. The essential feature of the Stoner-Wohlfarth model that has proved so useful in permanent magnets, namely the need for a high uniaxial anisotropy, is more generally applicable than the model itself. The values of the anisotropy field $\boldsymbol{H}_K$ and the anisotropy coefficient K for some widely used permanent magnets are shown in Table 13.2 and their magnetic properties are shown in Table 13.3.

TABLE 13.2
Anisotropy Field (H_K) and Anisotropy Coefficient (K) for Various Permanent Magnet Intermetallic Compounds

Materials	H_K (T)	K (MJ m^{-3})
$Nd_2Fe_{14}B$	6.1	4.9
$SmCo_5$	32	17
Sm_2Co_{17}	5.4	3.3
$Sm_2Fe_{17}N_3$	11.2	8.6
FePt	9.2	6.6

Source: Poudyal, N., and Liu, J. P., *J. Phys. D: Appl. Phys.*, 46, 043001, 2013.

TABLE 13.3
Comparison of Selected Magnetic Properties of Some Rare Earth Transition Metal Permanent Magnet Materials

Materials	M_s (MA m^{-1})	T_c (°C)	B_r (T)	${}_mH_c$ (kA m^{-1})	${}_BH_c$ (kA m^{-1})	BH_{max} (kJ m^{-3})
$SmCo_5$	0.91	747	1.1	1500	750	160
Sm_2Co_{17}	1.0	917	1.2	900	850	240
$Nd_2Fe_{14}B$	1.27	312	1.6	1200	750	320
NdFeB/αFe (exchange enhanced)	1.27	312	1.2	280	200	80–160
Sm_2Fe_{17}	0.85	115	—	880	640	—
$Sm_2Fe_{17}N_3$	1.22	476	1.5	2400	750	160

The idea of producing a very fine microstructure for permanent magnets is also well established. This has little to do with the anisotropy (shape or otherwise), but a lot to do with the creation of a large number of impediments to domain wall motion, both through the particle or grain boundaries and through the presence of localized residual strains, which pin domain walls and lead to higher coercivity.

13.1.8 Applications

There are many different applications of permanent magnets [20,21] and, as we have seen, considerations of geometry lead to many different materials requirements. Therefore, there remains a need for a wide range of permanent magnet materials to meet the needs for many different applications. Permanent magnet electric motors, in which electrical energy is converted into mechanical energy, and electric generators, in which mechanical energy is converted into electrical energy, are important applications of permanent magnets. The size of electric motors and generators can be reduced greatly by use of stronger permanent magnet materials and this is often an important consideration, which outweighs the additional cost of the high-performance magnets. Another example of this is the need for compact magnets for voice coil motors in computer data storage applications. Neodymium-iron-boron permanent magnets are used in magnetic resonance imaging systems [22]. This latter application requires a very high field homogeneity, typically 5 parts per million over a volume 0.6 m in diameter [23]. Previously, such fields had to be produced using superconducting magnet technology. As a result of the development of improved permanent magnets smaller magnetic resonance imaging systems have become possible.

13.1.9 Stability of Permanent Magnets

It is important to know under what conditions a permanent magnet will perform to its design specifications. Two types of problems may arise: (1) temporary effects such as reduction of remanent magnetization due to operating at temperatures beyond those for which the material was designed and (2) permanent deterioration of the magnetic properties caused by exposure to very high fields (field demagnetization) or high temperatures (thermal demagnetization) or by alteration of microstructure caused by exposure to elevated temperatures over a prolonged period of time (aging).

The temporary or reversible changes in the magnetic properties with temperature include the reduction of spontaneous magnetization within the domains as the temperature is raised and the reduction in coercivity. These effects become more significant the closer the temperature is to the Curie temperature as shown in Figure 13.11. The permanent changes that occur as a result of exposure to elevated temperatures are caused by acceleration of the aging process. Many permanent magnet materials exist in a metastable metallurgical state so that a phase transformation does occur, but at room temperature this proceeds very slowly. At higher temperature, the transformation proceeds more rapidly. Other factors such as mechanical treatment, mechanical damage, corrosion, and radiation effects can alter the properties of permanent magnets. These have been discussed in detail by McCaig [24].

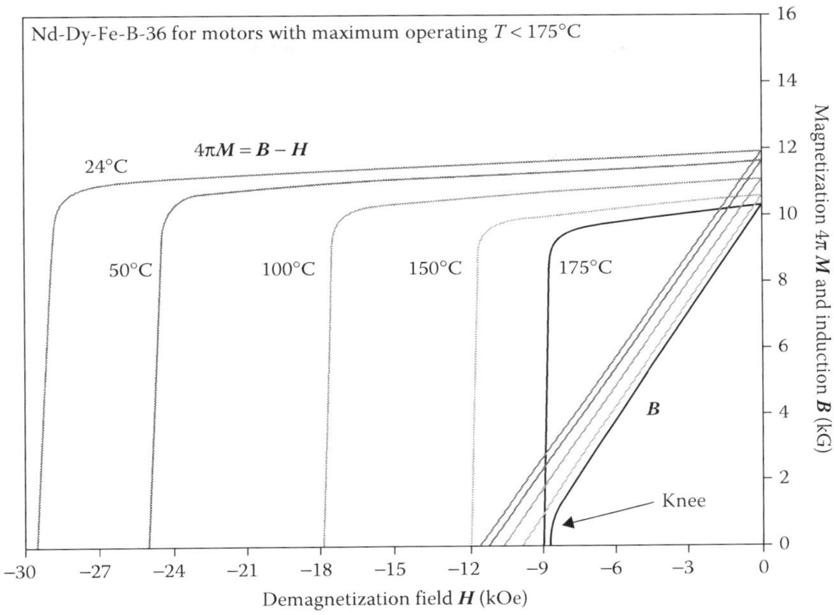

FIGURE 13.11 Demagnetization curves at various temperatures for Nd-Dy-Fe-B magnets. The values of $|BH|_{max}$ at 25°C are shown in parentheses. (After Chen, C.H., *Engineering Magnetic Materials and Their Applications Course MAT-512*, University of Dayton, Ohio, 2006–2010.)

13.2 PERMANENT MAGNET MATERIALS

In this section, we look at the various different materials that are used as permanent magnets. Materials that were considered *hard* magnetic materials in the past are in many instances not recognized as hard materials today because of the great improvement of magnetic properties such as coercivity and maximum energy product, which have taken place. Hadjipanayis et al. [25] have reviewed progress in rare earth permanent magnet research, focusing specifically on isotropic neodymium-iron-boron, anisotropic neodymium-iron-boron and magnets with a SmCo hard phase.

It is also important to note that it is not just the composition of the permanent magnets that determines their magnetic properties. The micro/nanostructure and the domain structure in permanent magnets are closely related, so that changes in microstructure also strongly affect the bulk magnetic properties [26]. Therefore, in fabricating permanent magnets, it is a combination of the chemical elements present together with the nano/microstructure produced by thermal and mechanical processing that ultimately determines the magnetic properties. Becker, Luborsky, and Martin [27] gave a summary of the properties of various permanent magnet materials up to the discovery of samarium cobalt permanent magnets and a review by Buschow [28] focused on the rare earth-iron and rare earth-cobalt permanent magnet materials.

13.2.1 Magnetite

The material Fe_3O_4, which is a naturally occurring oxide of iron, was the first *permanent magnet* material to be recognized and was known historically as lodestone. Today, given the advances in properties of magnetic materials, it is no longer considered to be a permanent magnet.

13.2.2 Permanent Magnet Steels

The addition of carbon to iron has long been known to increase coercivity and hysteresis loss. The first commercially produced permanent magnets were high-carbon steels containing about 1% carbon. These were also mechanically hard, while the low-carbon steels and iron were mechanically soft. Hence the classification *hard* and *soft*, which, for those involved in magnetism, later came to be a measure of coercivity rather than of mechanical properties.

Later permanent magnet steels were made with the addition of tungsten and chromium, which improved the coercivity compared with the carbon steels. Later still came the cobalt steels. In these materials, the improved magnetic properties arose from the presence of second-phase particles, which impeded the motion of domain walls, thereby leading to higher coercivity and maximum energy product. These permanent magnet steels have coercivities of up to 20 kA m^{-1} and maximum energy product of up to 7 kJ m^{-3}. The magnetic properties of some chromium and cobalt steels are shown in Figure 13.12 [29].

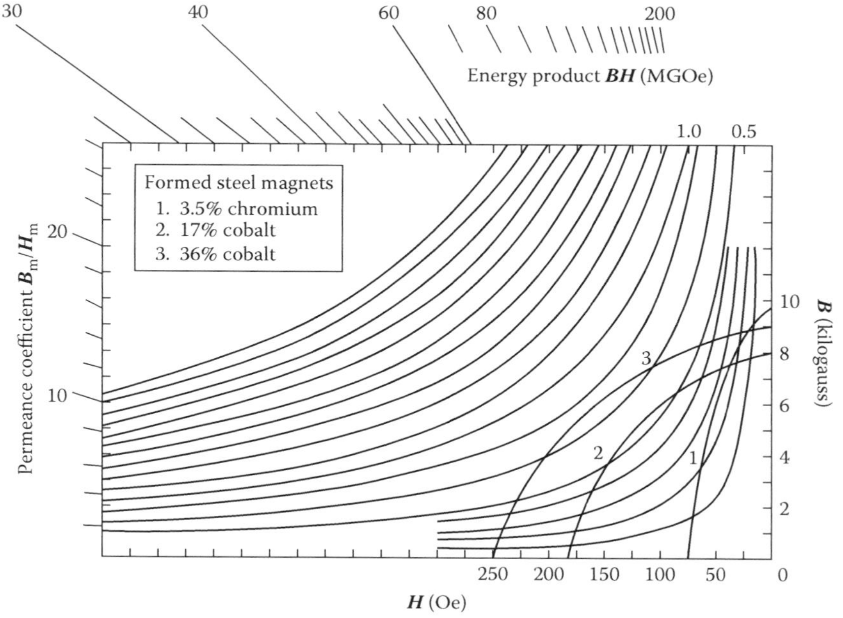

FIGURE 13.12 Demagnetization curves of three permanent magnet steels.

13.2.3 Alnico Alloys

The discovery of the important magnetic properties of alnico alloys was quite fortuitous during the development of a new type of steel for other purposes. The alnico alloys consist mainly of iron, cobalt, nickel, and aluminum, with small amounts of other metals such as copper [30]. These constituents form a finely intermixed two-phase alloy consisting of a strongly magnetic α_1 phase (Fe-Co) and a very weakly magnetic α_2 phase (Ni-Al), which provides pinning sites to restrain the motion of the magnetic domain walls.

The magnetic properties of these alloys were superior to other materials available at the time. These properties arose as a result of the formation of long rod-shaped grains of iron-cobalt, which gave rise to shape anisotropy. The permanent magnet properties of the alloy are improved by suitable heat treatment involving quenching followed by tempering at 700°C. They are also improved by annealing in a magnetic field. This raises the coercivity and maximum energy product. One disadvantage of these alloys is that they are very hard and brittle and therefore can only be shaped by casting or by pressing and sintering of metal powder.

Alnico alloys have remanences in the range of 50–130 kA m^{-1} with maximum energy products of 50–75 kJ m^{-3}. Typical magnetic properties of some alnico alloys are shown in Figures 13.13 through 13.16. These alloys represent a mature technology and no significant improvements in their magnetic properties have occurred in recent years, although efforts are now being made to identify high-performance permanent magnets that do not include rare earth metals and alnico provides a case study of how the magnetic properties can be improved for permanent magnet applications by manipulation of the microstructure.

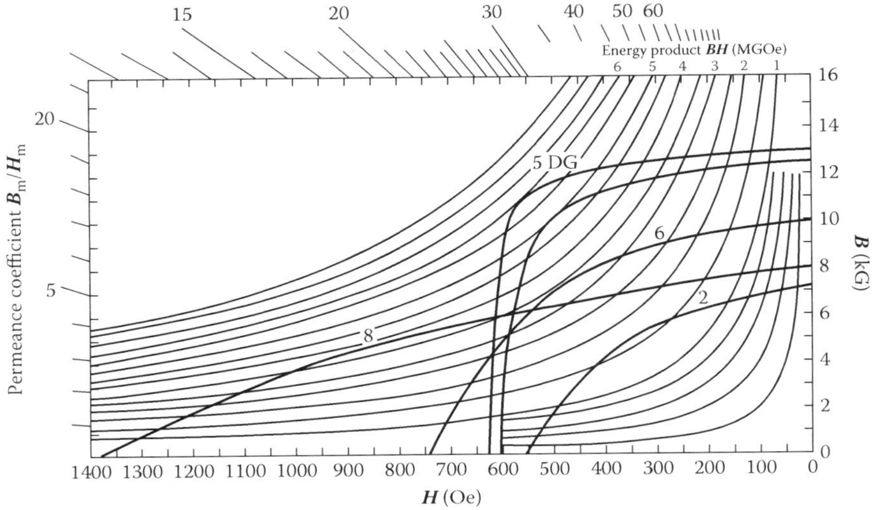

FIGURE 13.13 Demagnetization curves of various forms of alnico. (After Parker, R.J., and Studders, R.J., *Permanent Magnets and Their Applications*, Wiley, New York, 1962.)

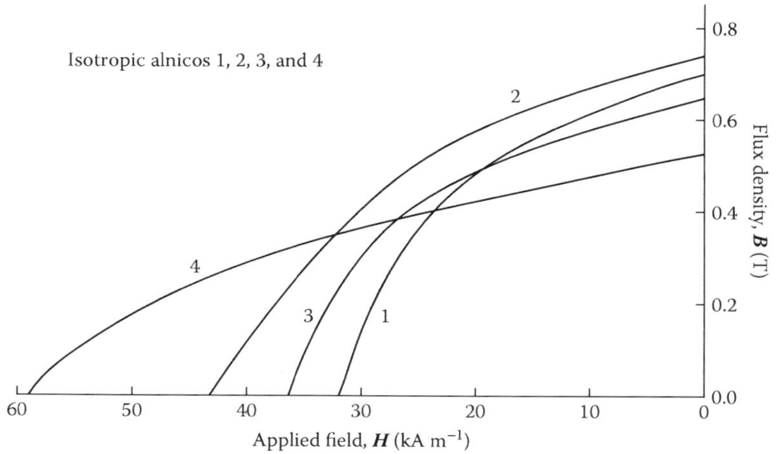

FIGURE 13.14 Demagnetization curves of the isotropic forms of alnico: alnicos 1, 2, 3, and 4.

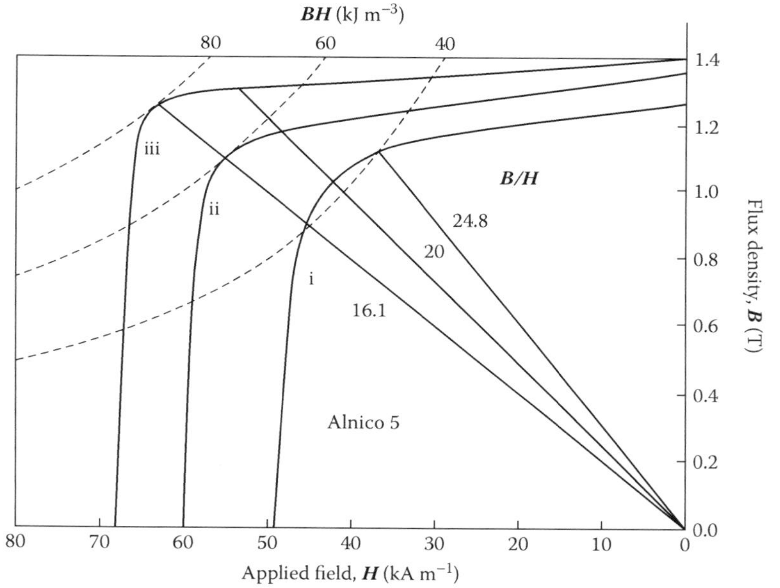

FIGURE 13.15 Demagnetization curves of oriented forms of alnico 5: i, equiaxed; ii, grain oriented; and iii, single crystal.

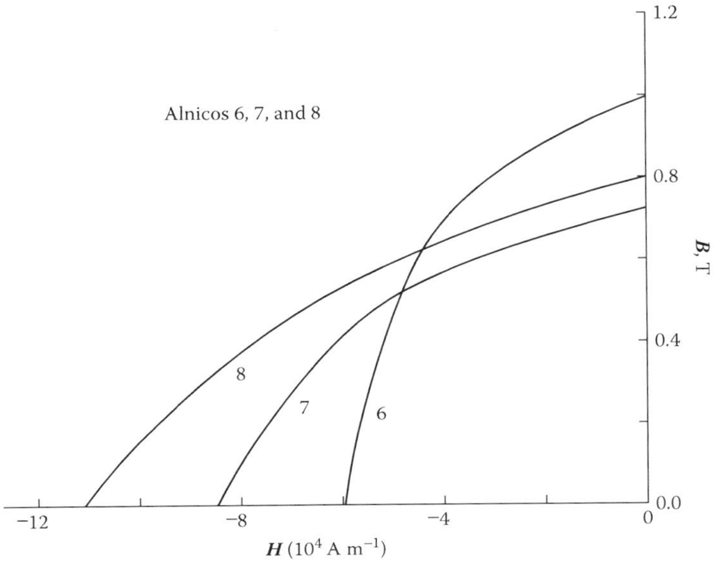

FIGURE 13.16 Demagnetization curves of the anisotropic forms of alnico: alnicos 6, 7, and 8.

13.2.4 Hard Ferrites

These materials, also known as *ceramic magnets*, were developed as a result of the Stoner-Wohlfarth theory, which indicated that the coercivity of a system of single-domain particles was proportional to the anisotropy. The theory thus provided a direction for the permanent magnet industry by indicating the types of materials that should prove to be good permanent magnets. This led the permanent magnet manufacturers to try to develop highly anisotropic materials in the form of aggregations of single-domain particles. The cause of the anisotropy could be either crystalline or particle/grain shape effects. However, it has been found that the coercivities of real materials have always been much smaller than the theoretical predictions of the Stoner-Wohlfarth model because mechanisms other than coherent domain rotation are almost always available and these can take place at lower field strengths.

The most commonly used hard ferrites are: (1) strontium ferrite, $SrFe_{12}O_{19}$ ($SrO{\cdot}6Fe_2O_3$), used in microwave devices, recording media, magneto-optic media, telecommunication, and electronic industry; (2) barium ferrite, $BaFe_{12}O_{19}$ ($BaO{\cdot}6Fe_2O_3$), used for permanent magnet applications where the need is for stability to moisture and corrosion-resistance; they are used in loudspeakers and as a magnetic recording medium, such as on magnetic stripe cards; and (3) cobalt ferrite, $CoFe_2O_4$ ($CoO{\cdot}Fe_2O_3$), which is used in some media for magnetic recording and in stress sensors.

The hard hexagonal ferrites barium or strontium ferrite ($BaO{\cdot}6Fe_20_3$ or $SrO{\cdot}6Fe_20_3$) are relatively cheap to produce and remain commercially important

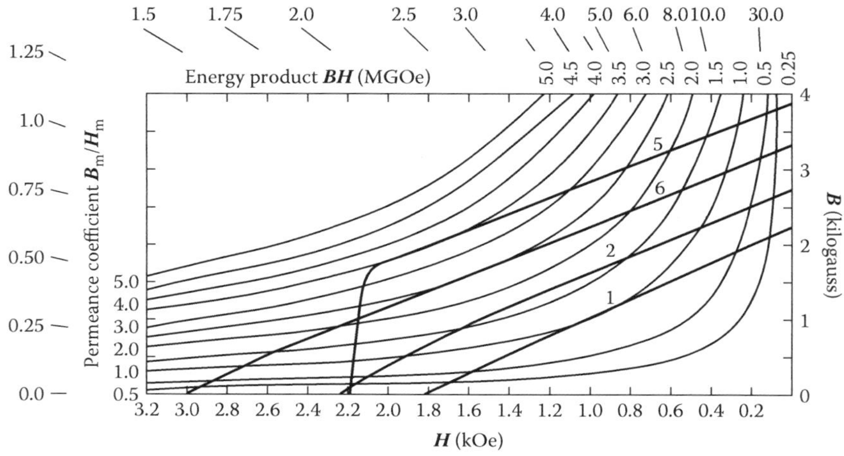

FIGURE 13.17 Demagnetization curves for: 1, isotropic barium ferrite; 2, 5, and 6, anisotropic barium ferrite permanent magnets.

among the permanent magnet materials. Their coercivities are larger than the alnicos, for example, being typically 150–250 kA m^{-1} (2–3 kOe) and remanences are 200–300 kA m^{-1} (200–300 emu cm^{-3}), but their saturation magnetizations are relatively low and their maximum energy product is also low, being typically 20 kJ m^{-3}. The magnetic properties of some barium ferrite magnets are shown in Figure 13.17. Reviews of the properties of the hard ferrites have been given by Stablein [31] and by Kojima [32]. The ferrites are often used to produce *plastic magnets* by embedding the ferrite in a flexible plastic matrix.

13.2.5 Platinum-Cobalt

This permanent magnet material was developed in the late 1950s. Although its magnetic properties were an improvement over other materials that were available at the time, its cost made it impractical for any application except the smallest magnets. Other cheaper magnets with superior properties are now available so this permanent magnet material is no longer in practical use. It has coercivities of typically 400 kA m^{-1} and a maximum energy product of typically 80 kJ m^{-3}.

13.2.6 Samarium-Cobalt

Samarium-cobalt permanent magnet material was developed as a result of a concerted research effort to identify new permanent magnet materials based on alloys of the rare earths with the 3d transition series ferromagnets iron, cobalt, and nickel. The motivation for the work was to take advantage of the high anisotropies of some of the rare earth metals in combination with the high Curie temperatures of the 3d transition metals iron, cobalt, and nickel, in order to make permanent magnets with high coercivity, which could operate at ambient temperatures and above. It was found that the cobalt-rare earth alloys had higher anisotropies than the nickel-rare earth or iron-rare earth alloys. Furthermore, the alloys with the light rare earths generally had higher saturation magnetizations.

The first of these alloys to be developed was $SmCo_5$ [33], which has a saturation magnetization of 800 kA m^{-1}, a coercivity of typically $_B\boldsymbol{H}_c$ = 760 kA m^{-1} ($_m\boldsymbol{H}_c$ = 1.5 MA m^{-1}), a maximum energy product of 150–200 kJ m^{-3}, a remanence of $\boldsymbol{B}_R$ = 0.9 T, and a Curie temperature of 720°C. This was followed by Sm_2Co_{17} [34,35], which has a saturation magnetization of 1 MA m^{-1}, a coercivity of typically $_B\boldsymbol{H}_c$ = 500 kA m^{-1} ($_m\boldsymbol{H}_c$ = 700 kA m^{-1}), a maximum energy product of 240–260 kJ m^{-3}, a remanence of $\boldsymbol{B}_R$ = 1 T, and a Curie temperature of T_c = 820°C. Note that most of these magnetic properties are structure sensitive, so the values given above should only be considered as typical results within a range of possible values.

The demagnetization curves for some typical samples of these materials are shown in Figure 13.18. Composite magnets consisting of samarium cobalt, both 1:5 and 2:17, blended with micron-sized cobalt powder have been reported by Gabay et al. [36]. The magnets were hot deformed at 800°C and the specific deformation conditions markedly influenced the microstructure. It was found that these composites had permanent magnet properties that were superior to samarium cobalt magnets made from a single alloy. On the whole, the greater chemical stability and

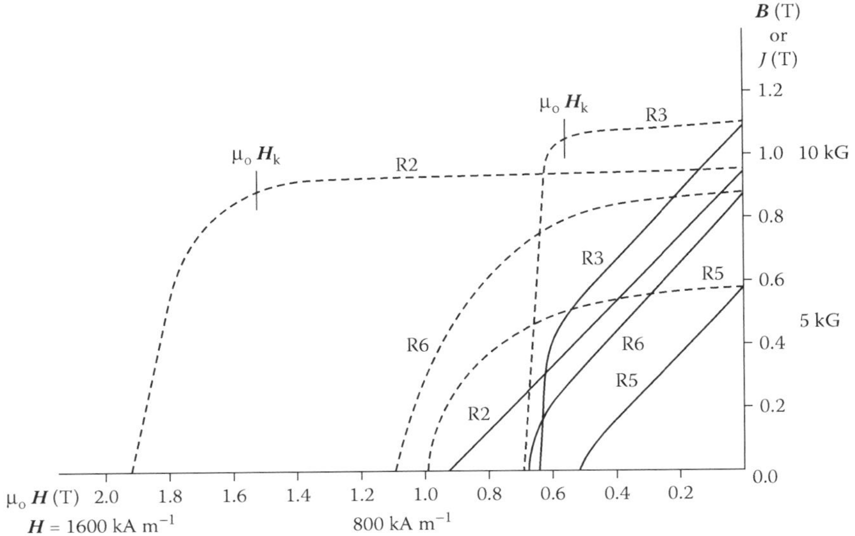

FIGURE 13.18 Second quadrant demagnetization curves for four specimens of samarium cobalt permanent magnet material. The broken lines are curves of $\boldsymbol{M}$ against $\boldsymbol{H}$, the full lines are curves of $\boldsymbol{B}$ against $\boldsymbol{H}$. R2 is the sintered $SmCo_5$, R3 is the sintered Sm_2Co_{17}, R5 is the bonded $SmCo_5$, and R6 is the bonded Sm_2Co_{17}. (After McCaig, M., and Clegg, A.G., *Permanent Magnets in Theory and Practice*, 2nd edn, Pentech Press, London, 1987.)

higher Curie temperature of Sm_2Co_{17} make it preferable to $SmCo_5$ for many applications, although the higher coercivity of $SmCo_5$ ensures that it maintains a position as material of choice for high coercivity applications at temperatures beyond which NdFeB is no longer viable.

As explained in Section 13.1.6, the exact choice of a permanent magnet material for a particular design application depends on many factors, including temperature of operation, shape, and demagnetizing factor, as determined by geometrical constraints and economic considerations.

13.2.7 Neodymium-Iron-Boron

This material was discovered largely because of problems with the supply of cobalt for making permanent magnets. There was a need for a new permanent magnet material to replace samarium-cobalt, even though the properties of the samarium-cobalt were adequate for the applications for which the new material was to be used. There had been some attempts to develop neodymium-iron materials, which were known to have large coercivities, but the properties of these alloys were not sufficiently reproducible. From this research, the addition of a small amount of boron was found to improve the properties dramatically.

The main neodymium-iron-boron alloy developed contained the $Nd_2Fe_{14}B$ phase [37,38], which has greater coercivity and energy product than samarium-cobalt. It is the presence of this very hard magnetic phase that leads to improved magnetic properties. Demagnetization curves of specimens of the material are shown in Figure 13.19. The magnetic properties of the neodymium-iron-boron material are sensitive to metallurgical processing. Two principal methods of production have been devised. The material is produced either by powdering and sintering, as in samarium-cobalt, or by rapidly quenching from the melt, like the process used to produce amorphous soft magnetic materials.

In the powder sintering method developed by Sagawa et al. [37], the constituents are induction melted in an alumina crucible under an inert atmosphere of argon, for example, to prevent oxidation. The alloy is then milled into a powder with particles of diameter 3 μm. The particles are aligned in a 800 kA m^{-1} field, compacted under 200 MPa pressure and then sintered under an argon atmosphere at temperatures in the range of 1050°C–1150°C. This process is followed by a postsintering anneal.

The rapid quenching process, known as *magnequench* was developed by Croat et al. [38]. This is similar to the process used for producing soft magnetic amorphous ribbons as described in Section 12.2.4. Constituents are melted together and then *melt spun* in an argon atmosphere by ejecting the molten alloy through a hole in the quartz crucible onto a rapidly rotating substrate, where the metal cools rapidly to form amorphous ribbons, which are then broken up into flakes. This gives a fine-grained microstructure of the $Nd_2Fe_{14}B$ phase. Particle sizes are in the range of 20–80 nm. The ribbon flakes then need to undergo one of two further processes to produce bulk permanent magnets: either they are bonded with epoxy to form isotropic *bonded magnets*, with intermediate maximum energy product typically of 72 kJ m^{-3}, or the ribbon fragments are vacuum hot pressed and vacuum die upset to form aligned anisotropic magnets, with high maximum energy products typically of 320 kJ m^{-3}. Zhang et al. [39]

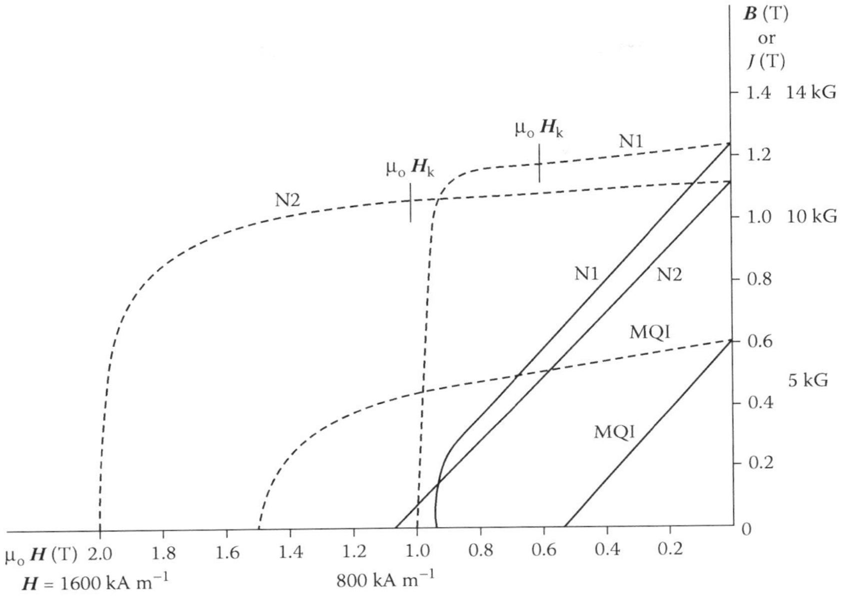

FIGURE 13.19 Second quadrant demagnetization curves for four specimens of neodymium-iron-boron permanent magnet material. The broken lines are curves of ***M*** against ***H***, the full lines are curves of ***B*** against ***H***. N1 is the sintered NdFeB, N2 is the high coercivity NdFeB, MQ1 is the bonded *magnequench* NdFeB. (After McCaig, M., and Clegg, A.G., *Permanent Magnets in Theory and Practice*, 2nd edn, Pentech Press, London, 1987.)

have shown that the addition of about 4% niobium to the $Nd_2Fe_{14}B$ alloy in the melt improves the glass forming ability of the alloy when it is rapidly quenched, so that bulk magnetic glass magnets may be formed with diameters up to 4 mm.

The principal advantage of these alloys compared with samarium-cobalt is that the alloy constituents, iron and neodymium, are cheaper than samarium and cobalt. One disadvantage of the alloy is its rather low Curie point of around 312°C. This means that the magnetic properties are rather more sensitive to temperature than are those of samarium-cobalt and are not suitable for some higher-temperature applications. Typical coercivities of these permanent magnet alloys are in the range of 1100 kA m^{-1} with maximum energy product of 300–350 kJ m^{-3}.

Sagawa et al. [40] and Herbst [41] have reviewed the development of neodymium-iron-boron permanent magnet materials. The high saturation, $\boldsymbol{M}_s = 1.3 \times 10^6$ A m^{-1}, $\boldsymbol{B}_s = 1.6$ T, and the high anisotropy of 4.6×10^6 J m^{-3}, which is equivalent to a field of 5.8×10^6 A m^{-1} (73 kOe) make it preferable to samarium-cobalt for many applications. However, the rather low Curie temperature of this material decreases with increased iron-to-rare earth ratio in these alloys. This is problematic because of the need to raise the Curie temperature of permanent magnet alloys such as neodymium-iron-boron in order to extend the useful temperature range of operation. Sometimes, cobalt is substituted for iron in neodymium-iron-boron permanent magnets to raise the Curie temperature.

The magnetization reversal mechanisms have been studied both experimentally and by computer simulations, and this has been useful for studying in a controlled way the relationship between structure and magnetic properties of these materials. Systematic studies of the nucleation field for domain reversal have been carried out by Kronmuller et al. [42,43]. Fidler and Schrefl [44] have examined the properties of nanocrystalline composites of NdFeB (see Section 13.2.8) both experimentally and by computer modeling. The materials showed an enhanced remanence due to improved exchange coupling, but this also resulted in a reduced coercivity.

In fine-grained materials with particle sizes of 10–30 nm, the remanence can be increased by 40% as a result of strong intergranular exchange coupling. By fabricating such composite magnetic materials consisting of permanent magnet particles within a magnetically soft matrix phase, such as α-iron, which occupies a volume fraction of up to 20%, the remanence was found to increase without a significant reduction in coercivity. In large-grained materials, it was found that dipolar magnetostatic interactions reduced the coercivity by up to 20% compared with noninteracting grains, and exchange coupling between the grains reduced the coercivity by 30%–40% of the *ideal* (i.e., noninteracting) coercivity. It is clear therefore that control of the nanostructure of the neodymium-iron-boron material in this way can have effects, which may be considered beneficial (e.g., increased remanence) or deleterious (e.g., reduced coercivity) depending on the specific application for which the material is being designed.

In general, neodymium-iron-boron has rather poor temperature stability of its magnetic properties and poor corrosion resistance. The former is caused by its relatively low Curie temperature. Kim and Camp [45] made significant progress on the problem of corrosion resistance. The addition of small amounts of copper and cobalt to the main $Nd_2Fe_{14}B$ composition was found to improve both coercivity and corrosion resistance without significantly reducing the remanence. For example, 0.03 wt% copper increased the coercivity from 1.12 MA m^{-1} (14 kOe) to 1.36 MA m^{-1} (17 kOe), and 1.2 wt% cobalt together with 0.15 wt% copper reduced the corrosion weight loss by more than three orders of magnitude under test conditions in an autoclave at 115°C and 70–100 kPa (10–15 psi) pressure of steam for 96 h. It was also noted that the presence of secondary nonmagnetic intergranular phases significantly improved the corrosion resistance.

Another method for producing the neodymium-iron-boron material for bonded magnets is the *hydrogenation-decomposition-desorption-recombination* (HDDR) process developed by Takeshita and Morimoto [46]. This process transforms the large crystalline grains of the cast ingot into very fine grains of diameter typically 0.3 μm. This size is close to that of single-domain $Nd_2Fe_{14}B$ particles. The isotropic bonded neodymium-iron-boron magnets produced in this way have enhanced coercivities of ${}_{m}\boldsymbol{H}_c = 1 \times 10^6$ A m^{-1} (12.5 kOe) and ${}_{B}\boldsymbol{H}_c = 0.56 \times 10^6$ A m^{-1} (7 kOe), with maximum energy products of 100 kJ m^{-3} (12.5 MGOe). This, however, is still significantly lower than the maximum energy products of the anisotropic bonded magnets, which have values of around 160 kJ m^{-3} (20 MGOe). Kaneko and Ishigaka [47] reported a high value of maximum energy product of 430 kJ m^{-3} (54.2 MGOe) in neodymium-iron-boron.

Hioki et al. [48] made an extensive study of the magnetic properties of hot deformed NdFeB permanent magnets and how these properties depended on the chemical composition and deformation temperature. It was shown that by optimizing the chemical composition and hot deformation conditions, Dy-free magnets with a coercivity of 1.6 MA m^{-1}, which is 23% higher than that of conventional magnets, were produced. Grain size and grain boundary phase composition were found to be the most critical microstructural factors in determining the magnetic properties of these permanent magnets. Therefore, in order to produce a high performance hot deformed magnet, these microstructural features need to be carefully controlled.

The relationship between the condition of the grain boundary phase and the magnetic properties for Dy-free hot deformed magnets were investigated by Liu et al. [49]. A thick Nd-rich boundary phase was found to give the highest coercivity and it was shown that there was a clear correlation between coercivity and concentration of Nd in the grain boundary phase in which coercivity increased as the Nd content of the grain boundary phase increased. These results were supported by micromagnetic simulation calculations.

Major applications of NdFeB magnets include the compact disk lens actuator, which focuses the laser spot on the disk, and voice coil motors for hard disk drives. Since the available space for these motors is minimal, the small, lightweight NdFeB magnets are ideally suited for this application. Other applications for the high-performance NdFeB magnets include magnets with highly homogeneous fields for medical magnetic resonance imaging applications, actuators in automobiles and automatic iris control mechanisms in video recorders.

13.2.8 Nanostructured Permanent Magnets

One of the disadvantages of NdFeB is its rather low remanent magnetization in the isotropic form of the material. This is about $\boldsymbol{M}_r = 0.64 \times 10^6$ A m^{-1} (0.8 T) compared with 1.0×10^6 A m^{-1} (1.2 T) in the anisotropic or oriented form of the material. The resultant energy product is 100 kJ m^{-3} (12.5 MGOe) compared with 320–400 kJ m^{-3} (40–50 MGOe) in some of the oriented material. Therefore, in a theoretical study, Kneller and Hawig [50] proposed an idea for improving both the saturation and remanent magnetizations in permanent magnet materials using a two-phase material with ultrafine (i.e., nanostructured) grain structure consisting of a magnetically hard phase embedded in a high saturation magnetically soft matrix. Such a material should allow good exchange coupling between the two phases, so that the high magnetization matrix phase helps to maintain a high magnetization in the hard phase. This idea became known as the *exchange spring* magnet.

A structure of fine and regularly dispersed particles with grain size typically 10 nm should lead to enhanced $\boldsymbol{M}_s$ and $\boldsymbol{M}_R$. An interesting prediction was that permanent magnet nanocomposites could be formed in this way consisting of 90% soft phase (such as α-Fe) and only 10% hard phase (such as $Nd_2Fe_{14}B$). From an economic viewpoint, this was very attractive because α-Fe is much less costly than $Nd_2Fe_{14}B$, $SmCo_5$, or Sm_2Co_{17}, and therefore, this raised the prospect of high-performance, low-cost permanent magnets. It was found that both samarium-cobalt (Figure 13.20) [51] and neodymium-iron-boron (Figure 13.21) [52,53] exhibited exchange coupled remanence enhancement if fabricated in the form of a suitably fine nanostructured composite.

Hadjipanayis et al. produced isotropic remanence-enhanced magnets based on these ideas, consisting of a fine-grained composite of $Nd_2Fe_{14}B$ in a matrix of α-Fe, which comprised 50–70 wt% of the material. The *squareness ratio* or *remanence*

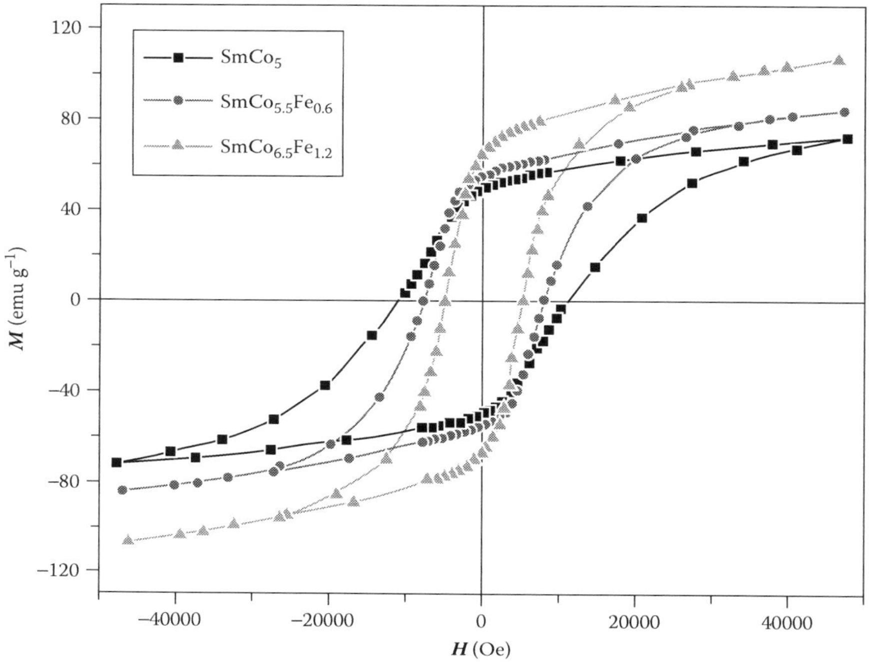

FIGURE 13.20 Room-temperature magnetic hysteresis loops of nanocrystalline $SmCo_5$ and $SmCo_5$/Fe nanocomposite permanent magnets, showing remanence enhancement in the exchange coupled nanocomposite. (After Chaubey, G.S. et al., *J. Alloy Compd.* 509, 2132, 2011.)

ratio $\boldsymbol{M_R/M_s}$ in these materials improved from 0.5 to 0.78, although the coercivity was much reduced, 320 kA m^{-1} (4 kOe) compared with 1.2×10^6 A m^{-1} (15 kOe) in the normal neodymium-iron-boron material as shown in Figure 13.21. Remanence-enhanced magnets subsequently received much attention and the idea has been extended to other materials.

Davies [54] reported that remanence enhancements from 0.8 to 1.2 T have been achieved in NdFeB nanocomposites with grain sizes below 40 nm as shown in Figure 13.22. The exchange coupling between the hard and soft phases becomes more important as the ratio of interfacial area to volume of the grains increases. While the intrinsic coercivity of these materials is lower than in NdFeB, the presence of the α-Fe on an ultrafine scale apparently does not result in serious deterioration of the maximum energy product, which remains at about 160 kJ m^{-3} (20 MGOe), a level comparable with NdFeB.

Modeling by Schrefl [55], incorporating the exchange coupling between the hard and soft phases, has shown excellent agreement with experimental results, indicating that the proposed mechanism is the reason for the enhancement in the properties. These calculations provided some of the most important modeling work for establishing a direct relationship between magnetic properties and the structure of the material.

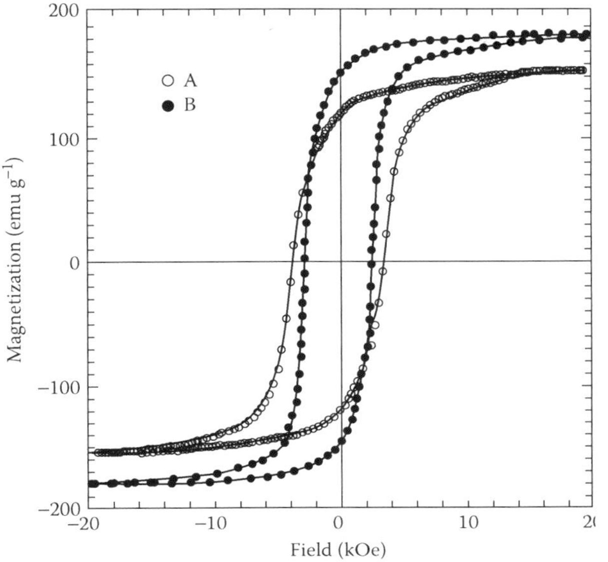

FIGURE 13.21 Room temperature hysteresis loops of nanocomposite exchange coupled neodymium-iron-boron permanent magnets, showing remanence enhancement in the exchange coupled material. (After Withanawasam, L. et al., *J. Appl. Phys.*, 75, 6646, 1994.) Sample A contained 50 wt% α-Fe, and was optimally annealed to obtain $\boldsymbol{B}_r$ = 110 emu g^{-1} (0.83 MA m^{-1}), while B contained 70 wt% α-Fe and had $\boldsymbol{B}_r$ = 145 emu g^{-1} (1.1 MA m^{-1}).

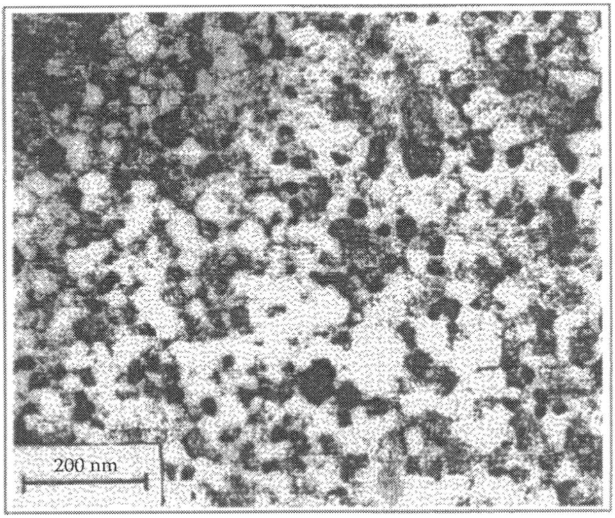

FIGURE 13.22 Transmission electron micrograph of nanocrystalline Nd_2Fei_4B material produced by melt spinning. (Reproduced by permission of Davies, H.A., *J. Magn. Magn. Mater.*, 157, 11, 1996.)

13.2.9 Samarium-Iron-Nitride

The rare-earth transition metal compounds of general composition R_2T_{17}, where R is a rare earth and T is a 3d transition metal, are chemically more stable than other rare earth transition metal compounds. This property is useful for producing materials with good corrosion resistance, which has been one of the main drawbacks of the NdFeB materials. The *2:17* compounds also have in general the highest saturation magnetization values among the families of rare earth transition metal compounds. This makes them at first sight good candidates from which hard magnets with high remanent magnetizations could be developed.

However, despite the high iron content, these compounds have low Curie temperatures, which lie in the range from −33°C to 207°C. For example, Sm_2Fe_{17} has a Curie temperature of only 115°C, and the highest Curie temperature in the series, T_c = 207°C, occurs in Gd_2Fe_{17}. Furthermore, the R_2Fe_{17} compounds exhibit easy plane anisotropy in either their tetragonal or their hexagonal form, which is undesirable for permanent magnets. Uniaxial anisotropy is much more beneficial for permanent magnets, in either the tetragonal or the hexagonal structure, because it makes it more difficult for the magnetic moments to change direction under the action of a magnetic field. The uniaxial form of anisotropy is therefore more conducive to permanent magnet properties such as high coercivity and remanence.

The addition of nitrogen to Sm_2Fe_{17} has dramatic and beneficial effects on its magnetic properties. These effects were first reported by Coey and Sun [56] and Otani et al. [57], and resulted in an increase in Curie temperature to 477°C in $Sm_2Fe_{17}N_{2\cdot3}$. This increase in T_c was attributed to a 280% increase in the Fe-Fe atom exchange interaction together with a slight decrease of the Fe-Sm exchange interaction. The nitrogen atoms enter the Sm_2Fe_{17} lattice interstitially resulting in a 6%–7% increase in the volume of the unit cell. The Fe-Fe interatomic spacings in Sm_2Fe_{17} are small and in accordance with the Bethe-Slater curve of Figure 11.4; this results in a low exchange energy, but the straining of the lattice caused by the nitrogen interstitials increases the exchange energy by almost a factor of 3, and this leads to the increased Curie temperature as a result of the relation described in Section 9.2.5.

Nitrogenation of other R_2T_{17} compounds has also been investigated, but only in $Sm_2Fe_{17}N_x$ did the resulting compound exhibit the necessary uniaxial anisotropy. Subsequently, efforts were made to improve the magnetic properties of this family of compounds by the addition of other interstitial elements to Sm_2Fe_{17}. A whole range of possible interstitial elements was investigated. Altounian et al. [58] showed that carbon also has a beneficial effect on the magnetic properties that is comparable to that of nitrogen. However, the enhancements were not quite so dramatic as on the addition of nitrogen. For example, the Curie temperature of $Sm_2Fe_{17}C_2$ was 410°C compared with the 477°C of $Sm_2Fe_{17}N_{2.3}$. It was found that the addition of both elements together to form the carbonitrided compound $Sm_2Fe_{17}C_yN_x$ ($x = 2$ and $y = 1$) gave improved thermal stability of the compound by reducing the outgassing of nitrogen.

Muller et al. [59] investigated a range of interstitially modified Sm-Fe compounds produced by the HDDR process. These materials had coercivity of 2.8×10^6 A m^{-1} (35 kOe) compared with 1.76×10^6 A m^{-1} (22 kOe) in coarse grained $Sm_2Fe_{17}N_3$.

The increase in Curie temperature from 115°C to 475°C and saturation magnetization from 0.85×10^6 A m^{-1} (1.05 T) to 1.23×10^6 A m^{-1} (1.55 T), and the emergence of uniaxial anisotropy on addition of the interstitial elements was confirmed. The anisotropy in $Sm_2Fe_{17}N_3$ was 17×10^6 A m^{-1} (22 T) compared with only 7×10^6 A m^{-1} (9 T) in $Nd_2Fe_{14}B$. Ding, McCormick and Street [60] have reported large remanence enhancements in SmFeN using a nanocomposite with an α-Fe matrix, for example.

In view of this, the $Sm_2Fe_{17}N_3$ compound has received attention from researchers as a possible new permanent magnet material. The result of nitrogenation gives a substantially higher Curie temperature, a higher saturation magnetization, remanence, and coercivity. It is therefore a candidate material for development of new high-performance permanent magnets for high-temperature applications.

13.2.10 Comparison of Various Permanent Magnet Materials

The various permanent magnet materials such as hard ferrites, alnico, samarium-cobalt, and neodymium-iron-boron provide a range of magnetic properties, which can be utilized for different applications. The demagnetization curves of samples of these materials are shown in Figure 13.23. In making this comparison, it is recognized that apart from the saturation magnetization, all properties listed are

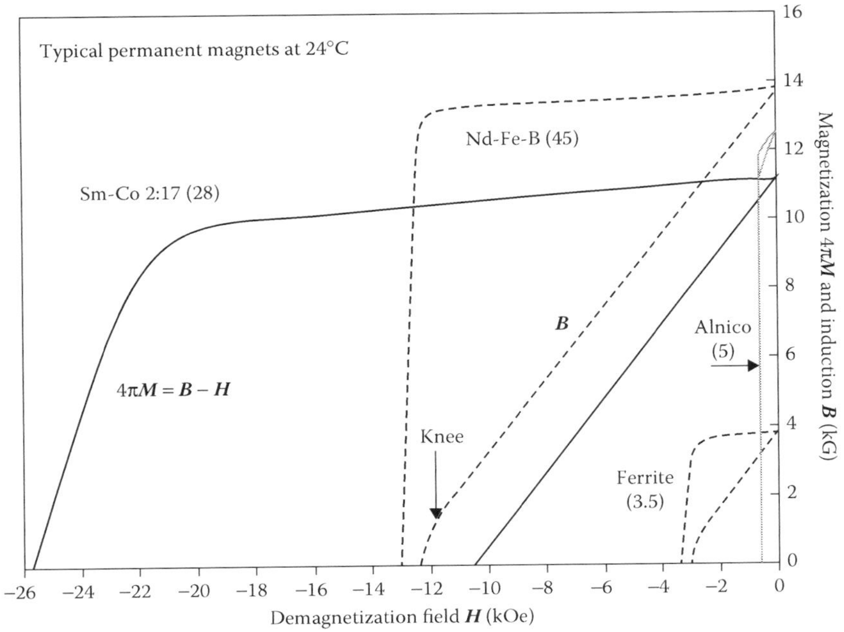

FIGURE 13.23 Demagnetization curves in the second quadrant for the four main classes of permanent magnet materials—neodymium iron boron, samarium cobalt, alnico, and ferrite. (After Chen, C.H., *Engineering Magnetic Materials and Their Applications Course MAT-512*, University of Dayton, Ohio, 2006–2010.)

structure sensitive and so can vary depending on the thermo-mechanical treatment of a material and its microstructure. Table 13.4 shows the magnetic properties of various permanent magnet materials. These values should therefore be considered as representative or typical values rather than exact materials constants.

The choice of a particular material in a given situation depends critically on the operational conditions, including the temperature, the demagnetization curve at the given temperature, and geometry expressed as the load line or the demagnetizing factor. These are shown in Figure 13.24 for a particular neodymium-dysprosium-iron-boron alloy.

The domain-wall surface energies, domain-wall widths, and single-domain particle diameters of the three most widely used rare-earth transition metal alloy permanent magnets are given in Table 13.5, which is taken from the review by Sagawa et al. [37]. These may be compared with the values for iron, cobalt, and nickel given in Table 7.1, from which it is seen that the permanent magnets have much higher domain-wall surface energies and correspondingly thinner domain walls than the soft magnetic materials.

Comprehensive treatment of permanent magnets and their applications can be found in the revision of McCaig's book by Clegg [61] and in Parker's book [62]. Moskowitz [13] gave another survey of permanent magnets in which a large amount of technical data can be found.

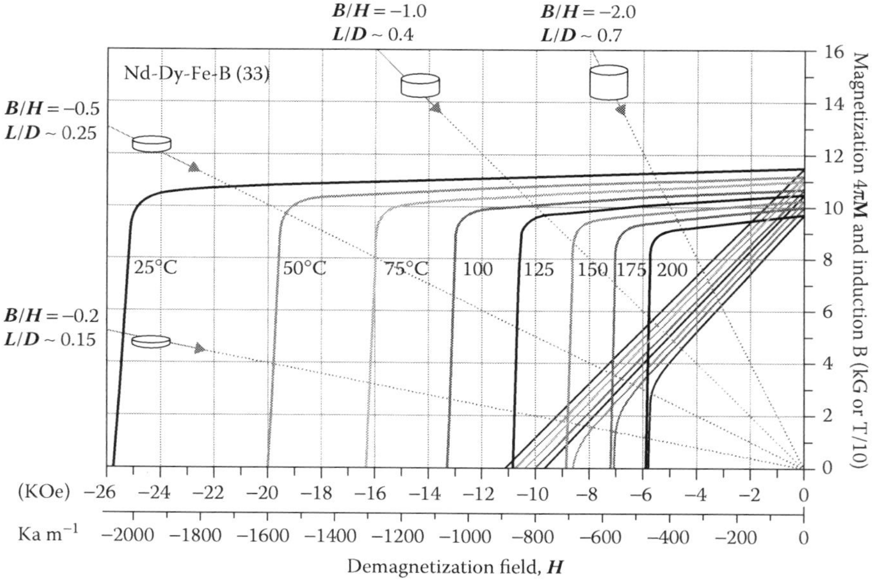

FIGURE 13.24 Demagnetization curves at 25°C to 200°C for an Nd-Dy-Fe-B magnet and the load lines with corresponding dimensions for cylindrical magnets. (After Chen, C.H., *Engineering Magnetic Materials and Their Applications Course MAT-512*, University of Dayton, Ohio, 2006–2010.)

TABLE 13.4
Important Magnetic Properties of Selected Permanent Magnetic Materials

Material	Composition	Remanence (T)	Coercivity (kA m^{-1})	$(BH)_{max}$ (kJ m^{-3})
Steel	99% Fe, 1%C	0.9	4	1.59
36Co Steel	36%Co, 5.75%Cr, 3.75%W, 0.8%C	0.96	18.25	7.42
Alnico 2	63%Fe 26%Ni, 12%Al, 3%Cu	0.7	52	13.5
Alnico 5	50%Fe, 24%Co, 15%Ni, 8%Al, 3%Cu	1.2	57.6	40
Alnico DG	50%Fe, 24%Co, 15%Ni, 8%Al, 3%Cu	1.31	56	52
Ba Ferrite	$BaO \cdot 6Fe_20_3$	0.395	192	28
PtCo	77%Pt, 23%Co	0.645	344	76
Remalloy	71%Fe, 17%Mo, 12%Co	1.0	18.4	9
Vicalloy	52%Co, 35%Fe, 13%V	1.0	36	24
Samarium-cobalt	$SmCo_5$	0.9	696	160
Neodymium-iron-boron	$Nd_2Fe_{14}B$	1.3	1120	320

TABLE 13.5
Magnetic Properties of Domains and Domain Walls in Permanent Magnet Materials Samarium-Cobalt and Neodymium-Iron-Boron

	Domain-Wall Surface Energy (mJ m^{-2})	Domain-Wall Width (nm)	Single-Domain Particle Diameter (μm)
$Nd_2Fe_{14}B$	30	5.2	0.26
$SmCo_5$	85	5.1	1.6
Sm_2Co_{17}	43	10.0	0.66

REFERENCES

1. Coey, J. M. D. Hard magnetic materials: A perspective, *IEEE Trans. Magn.*, 47, 1, 2011.
2. Gutfleisch, O., Willard, M. A., Brück, E., Chen, C. H., Sankar, S. G., and Liu, J. P. Magnetic materials and devices for the 21st century: Stronger, lighter, and more energy efficient, *Adv. Mater.*, 23, 821, 2011.
3. Martinek, G., Wyss, U., Maybury, D., and Constantinides, S. Optimizing magnetic effects through shaped field magnets, in *Proceedings of the 6th International Conference on Magnetism and Metallurgy*, Cardiff, June 17–19, 2014.
4. Hummel, R. *Electronic Properties of Materials*, Springer-Verlag, New York, 1985, p. 255.
5. Moskowitz, L. R. *Permanent Magnet Design and Application Handbook*, Robert Krieger, Malabar, FL, 1976.

6. Hilscher, G., Grossinger, R., Heisz, S., Sassik, H., and Wiesinger, G. *J. Magn. Magn. Mater.*, 54–57, 577, 1986.
7. Nesbitt, E. A., and Wernick, J. H. *Rare Earth Permanent Magnets*, Academic Press, New York, 1973, p. 80.
8. Cullity, B. D. *Introduction to Magnetic Materials*, Addison-Wesley, Reading, MA, 1972, pp. 560, 561, and 574.
9. Zijlstra, H. Permanent magnets; theory, in *Ferromagnetic Materials*, Vol. 3 (ed. E. P. Wohlfarth), North Holland, Amsterdam, the Netherlands, 1982.
10. Graham, C. D. *Conference on Properties and Applications of Magnetic Materials.* Illinois Institute of Technology, Chicago, IL, 1987.
11. Jensen, J. H., and Abele, M. G. *J. Appl. Phys.*, 79, 1157, 1996.
12. Leupold, H. A., Potenziani, E., and Tilak, A. S. *IEEE Trans. Magn.* 29, 2905, 1993.
13. Moskowitz, L. R. *Permanent Magnet Design and Application Handbook*, Robert Krieger, Malabar, FL, 1995.
14. Stoner, E. C., and Wohlfarth, E. P. *Phil. Trans. Roy. Soc. A*, 240, 599, 1948.
15. Chikazumi, S. *J. Magn. Magn. Mater.*, 54–57, 1551, 1986.
16. Durst, K. D., and Kronmuller, H. *J. Magn. Magn. Mater.*, 68, 63, 1987.
17. Pinkerton, F. E., and Van Wingerden, D. J. *J. Appl. Phys.*, 60, 3685, 1986.
18. Hadjipanayis, G. C. *J. Appl. Phys.*, 63, 3310, 1988.
19. Becker, J. J. *IEEE Trans. Magn.*, 4, 239, 1968.
20. Bradley, F. N. *Materials for Magnetic Functions*, Hayden, New York, 1971.
21. McCaig, M. *IEEE. Trans. Magn.*, 4, 221, 1968.
22. Molfino, P., Repetto, M., Bixio, A., Del Mut, G., and Marabotto, R. *IEEE Trans. Magn.*, 24, 994, 1988.
23. Bobrov, E. S., and Punchard, W. F. B. *IEEE Trans. Magn.*, 24, 553, 1988.
24. McCaig, M. *Permanent Magnets in Theory and Practice*, Wiley, New York, 1977.
25. Hadjipanayis, G. C., Liu, J., Gabay, A., and Marinescu M. Current status of rare-earth permanent magnet research in the USA, *J. Iron Steel Res. Int.*, 13 (Suppl. 1), 12–22, 2006.
26. Hioki, K., Morita, T., and Hattori, A., in *Proceedings of the 22nd International workshop on Rare Earth Permanent Magnets and their Applications*, Nagasaki, Japan, September 2–5, 2012, p. 89.
27. Becker, J. J., Luborsky, F. E., and Martin, D. L. *IEEE Trans. Magn.*, 4, 84, 1968.
28. Buschow, K. H. J. *Mater. Sci. Rep.*, 1, 1, 1986.
29. Parker, R. J., and Studders, R. J. *Permanent Magnets and Their Applications*, Wiley, New York, 1962.
30. McCurrie, R. A. The structure and properties of alnico permanent magnet alloys, in *Ferromagnetic Materials*, Vol. 3 (ed. E. P. Wohlfarth), North Holland, Amsterdam, the Netherlands, 1982.
31. Stablein, H. Hard ferrites, in *Ferromagnetic Materials*, Vol. 3 (ed. E. P. Wohlfarth), North Holland, Amsterdam, the Netherlands, 1982.
32. Kojima, H. Fundamental properties of hexagonal ferrites with magnetoplumbite structure, in *Ferromagnetic Materials*, Vol. 3 (ed. E. P. Wohlfarth), North Holland, Amsterdam, the Netherlands, 1982.
33. Strnat, K., Hoffer, G., Oison, J., Ostertag, W., and Becker, J. J. *J. Appl. Phys.*, 38, 1001, 1967.
34. Nesbitt, E. A., and Wernick, J. H. *Rare Earth Permanent Magnets*, Academic Press, New York, 1973.
35. Tawara, Y., and Strnat, K. *IEEE Trans. Magn.*, 12, 954, 1976.
36. Gabay, A. M., Hadjipanayis, G. C., Marinescu, M., and Liu, J. P., Hot-deformed Sm–Co/Co composite magnets fabricated from powder blends, *J. Appl. Phys.*, 107, 09A704, 2010.
37. Sagawa, M., Fujimura, S., Togawa, N., Yamamoto, H., and Matsuura, Y. *J. Appl. Phys.*, 55, 2083, 1984.

38. Croat, J. J., Herbst, J. F., Lee, R. W., and Pinkerton, F. E. *J. Appl. Phys.*, 55, 2078, 1984.
39. Zhang, J., Lim, K. Y., Feng, Y. P., and Li, Y. Fe-Nd-B based hard magnets from bulk amorphous precursor, *Scripta Mater.*, 56, 943, 2007.
40. Sagawa, M., Hirosawa, S., Yamamoto, H., Fujimura, S., and Matsuura, Y. *Jap. J. Appl. Phys.*, 26, 785, 1987.
41. Herbst, J. *Rev. Mod. Phys.*, 63, 819, 1991.
42. Kronmuller, H. *Phys. St. Sol. B*, 144, 385, 1987.
43. Kronmuller, H., Durst, K. D., and Sagawa, M. *J. Magn. Magn. Mater.*, 74, 291, 1988.
44. Fidler, J., and Schrefl, T. *J. Appl. Phys.*, 79, 5029, 1996.
45. Kim, A. S., and Camp, F. E. *Appl. Phys.*, 79, 5035, 1996.
46. Takeshita, T., and Morimoto, K. *J. Appl. Phys.*, 79, 5040, 1996.
47. Kaneko, Y., and Ishigaki, N. *J. Mater. Eng. Perform.*, 3, 228, 1994.
48. Hioki, K., Hattori, A., and Iriyama, T. Development of Dy-Free hot-deformed Nd-Fe-B magnets by optimizing chemical composition and microstructure, *J. Magn. Soc. Jap.*, 38, 79, 2014.
49. Liu, J., Sepehri-Amin, H., Ohkubo, T., Hioki, K., Hattori, A., Schrefl, T., and Hono, K. Effect of Nd content on the microstructure and coercivity of hot-deformed Nd–Fe–B permanent magnets, *Acta Mater.*, 61, 5387, 2013.
50. Kneller, E. F., and Hawig, R. *IEEE Trans. Magn.*, 21, 3588, 1991.
51. Chaubey, G. S., Poudyal, N., Liu, Y. Z., Rong, C. B., and Liu, J. P. *J. Alloy Compd.* 509, 2132, 2011.
52. Withanawasam, L., Hadjipanayis, G. C., and Krause, R. F. *J. Appl. Phys.*, 75, 6646, 1994.
53. Hadjipanayis, G. C., Withanawasam, L., and Krause, R. F. *IEEE Trans. Magn.*, 31, 3596, 1995.
54. Davies, H. A. *J. Magn. Magn. Mater.*, 157, 11, 1996.
55. Schrefl, T. *Phys. Rev. B,* 49, 6100, 1994.
56. Coey, J. M. D., and Sun, H. *J. Magn. Magn. Mater.*, 87, 251, 1990.
57. Otani, Y., Hurley, D. F. P., Sun, H., and Coey, J. M. D. *J. Appl. Phys.*, 69, 5584, 1991.
58. Altounian, Z., Chen, X., Liao, L. X. et al. *J. Appl. Phys.*, 73, 6017, 1993.
59. Muller, K. H., Cao, L., Dempsey, N. M., and Wendhausen, P. A. P. *J. Appt. Phys.*, 79, 5045, 1996.
60. Ding, J., McCormick, P. G., and Street, R. *J. Magn. Magn. Mater.*, 124, 1, 1993.
61. McCaig, M., and Clegg, A. G. *Permanent Magnets in Theory and Practice*, 2nd edn, Pentech Press, London, 1987.
62. Parker, R. J. *Advances in Permanent Magnets*, Wiley, New York, 1990.

FURTHER READING

Campbell, P. *Permanent Magnet Materials and Their Application*, Cambridge University Press, Cambridge, 1994.

Coey, J. M. D. *Rare Earth Iron Permanent Magnets*, Clarendon Press, Oxford, 1996.

Coey, J. M. D. *Magnetism and Magnetic Materials*, Cambridge University Press, Cambridge, 2009.

Gutfleisch, O., Willard, M. A., Brück, E., Chen, C. H., Sankar, S. G., and Liu, J. P. Magnetic materials and devices for the 21st century: Stronger, lighter, and more energy efficient, *Adv. Mater.*, 23, 821, 2011.

Jones, N. The pull of stronger magnets, *Nature*, 472, 22, 2011.

Jiles, D. C. Recent advances and future directions in magnetic materials, *Acta Mater.* 51, 5907, 2003.

Moskowitz, L. R. *Permanent Magnet Design and Application Handbook*, Robert Krieger, Malabar, FL, 1995.

Parker, R. J. *Advances in Permanent Magnets*, Wiley, New York, 1990.
Poudyal, N., and Liu, J. P. Advances in nanostructured permanent magnets research, *J. Phys. D (Appl. Phys.)*, 46, 043001, 2013.

MATERIALS

http://en.wikipedia.org/wiki/Magnet.
http://hyperphysics.phy-astr.gsu.edu/hbase/solids/magperm.html.
http://www.smma.org/pdf/permanent-magnet-materials.pdf.

CONFERENCES

International Magnetics Conference. http://www.intermagconference.com/.
Magnetism and Magnetic Materials Conference. http://www.magnetism.org/.

14 Magnetic Recording

In this chapter, we look at the various magnetic methods available for recording of information, including data, images, and sound. The most important of the recording media today are magnetic disks, which provide the main method for storing digital information on computers, and magnetic tapes, which are widely used for both audio and video recording and for long-term data storage and backup. Magnetic recording can be conveniently separated into two groups of related technologies: media and recording heads. *Media* consists of all technologies concerned with the production and use of magnetic disks and tapes for storing information, while *heads* cover all technologies concerned with the processes of writing or reading information to or from media.

14.1 HISTORY OF MAGNETIC RECORDING

Analog magnetic recording of the human voice was first demonstrated by Poulsen [1], a Danish engineer, more than one hundred years ago. In his device, acoustic signals were recorded on a ferromagnetic wire using an electromagnet connected to a microphone. However, the reproduction was very poor due to the absence of an amplifier. Later with the development of amplifiers, the signals from the magnetic medium could be re-created more strongly and the sound reproduction was more easily audible. However, there was also distortion and a low signal-to-noise ratio, due to the nonlinear nature of the recording process, which meant that the quality of the sound was still not good. The ac biasing method of recording [2] resulted in much better signal-to-noise ratios because the recorded magnetization could be made proportional to the original signal level. However, this was not fully exploited for another 20 years [3].

Magnetic tape was invented simultaneously in the United States, using a paper tape coated with dried ferrimagnetic liquid, and in Germany, using a tape containing iron powder. Oxide tapes were developed for the commercial market by 3M Corporation and both audio and video recorders became available soon thereafter. Of course, to make this of any practical use, it must be possible to store large amounts of data in as small a space as possible. Magnetic tape was first used to record digital computer data in 1951 on the *UNIVAC*1. In the recording industry, there has been a continual drive to increase the data storage density. These storage densities are conventionally measured in *bits per square inch*.

Although the storage densities on magnetic tapes are lower than on magnetic hard disks, magnetic tape has historically offered enough advantage in cost over hard disk storage to continue to make it viable, particularly for long-term data backup where the data can be stored at a different location. It also provides a more secure form of data storage because the data on a magnetic tape are not immediately available online.

IBM developed the first magnetic hard disk drive the *RAMAC* in 1956 [4]. This original hard disk drive had a data storage density of 2000 bits in^{-2} (or 3 bits mm^{-2}). The increase in data storage densities since then has been continual, so the storage densities of 1 Tbit in^{-2}, reached by Seagate in 2012, represent a storage density increase of a factor of 500,000,000 since the *RAMAC*. Similarly, the data transfer rate has increased over the same time period from 70 kbit s^{-1} to 1 Gbit s^{-1}. The progress in data storage densities on hard disk drives since 1990 is shown in Figure 14.1. A comparison of this progress with that of flash memory over the same period is shown in Figure 14.2.

Since these early developments in magnetic recording, there has been tremendous growth in demand for digital magnetic recording for the storage of computer data, together with the consumer demand for audio/visual recording, particularly storage of music and videos together with sound tracks. These form the most important commercial areas of the magnetic recording industry today. Magnetic storage densities have increased since 1956 at an average annual rate of 40%. From the mid-1990s until 2005, this rate even increased to 60% annual growth but now the growth rate has reduced to about 10% per year. The present generation of commercial hard disk drives have storage densities of typically 10^{12} bit in^{-2} (1.5×10^9 bit mm^{-2}). External hard disk drives with storage capacities of 5 TB are commercially available.

A trend in the market is the miniaturization of hard disk drives (HDDs) and also hard disk drives with the same size (*form factor*) but with greater storage capacity.

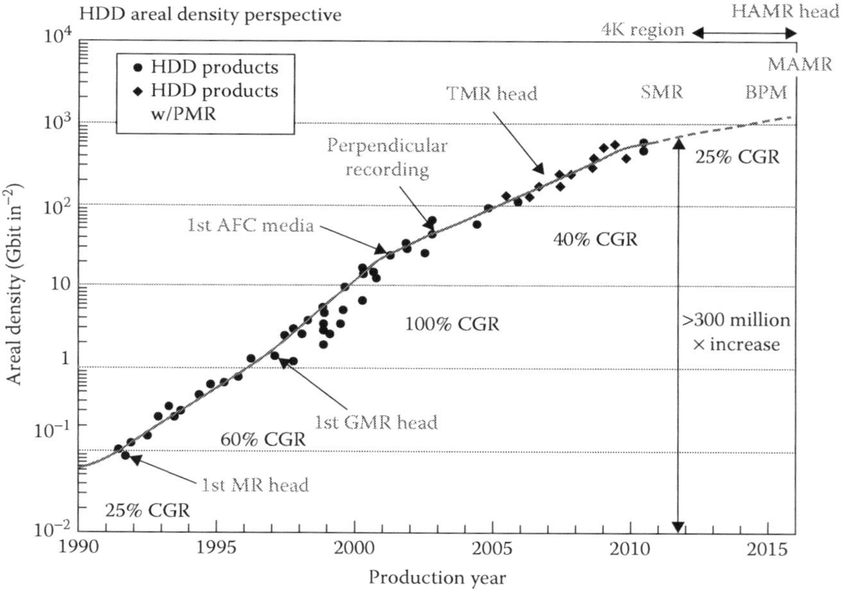

FIGURE 14.1 Progress in magnetic recording storage densities from 1990 to 2015. (After Coughlin, T., and Grochowski, E., *Years of Destiny: HDD Capital Spending and Technology Developments from 2012–2016*, IEEE Santa Clara Valley Magnetics Society, 2012, June 19. ewh.ieee.org/r6/scr/mag/.)

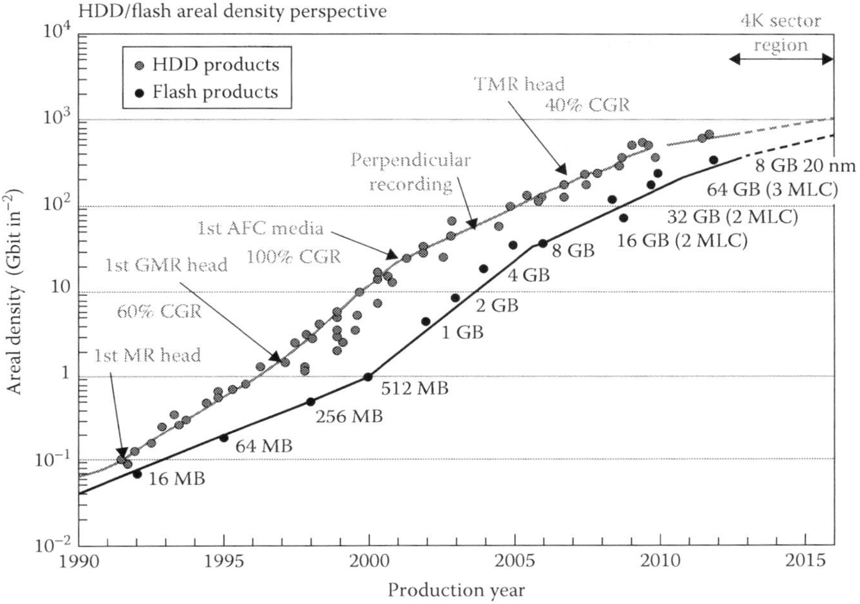

FIGURE 14.2 Improvements in storage densities for magnetic recording and flash memory technologies from 1990 to 2015. (After Coughlin, T., and Grochowski, E., *Years of Destiny: HDD Capital Spending and Technology Developments from 2012–2016*, IEEE Santa Clara Valley Magnetics Society, 2012, June 19. http://ewh.ieee.org/r6/scv/mag/MtgSum/Meeting2012_06_Presentation.pdf.)

The increase in area storage density has led to a reduction of form factors from 3.5 to 2.5 in. For instance, by increasing the area density the storage capacity of a nominal 500 GB hard disk drive has been increased to 640 GB while retaining the original form factor of 3.5 in. Similarly, the capacity of a nominal 320 GB hard disk drive has been increased to 375 GB while retaining the form factor of 2.5 in.

Disk drives can be categorized into disk drives for use in desktop computers, laptop computers, external disk drives, and enterprise hard disk drives for multiuser web or Internet-based computers. Desktop hard disk drives typically store between 60 GB and 4 TB and rotate at 5,400–10,000 rpm, with a media transfer rate of typically 1 Gbit s^{-1} or higher. In 2014, the highest capacity desktop hard disk drives stored 8 TB [5].

Mobile or laptop hard disk drives have smaller platters so that generally they have lower capacity than the desktop hard disk drives and they are slower. They rotate at 4200–7200 rpm, with 5400 rpm being typical. Drives of 7200 rpm are more expensive and have smaller capacities, while 4200 rpm models usually have higher storage capacities.

There are also *enterprise* hard disk drives that are used in the enterprise segment of the economy and are not intended to be used in laptops or desktops and are usually 2.5 in drives. They are used in multiuser computers running software for commercial transaction processing databases, web, or internet infrastructure (e-mail, web servers, and e-commerce) and scientific computing software. They commonly operate continuously (*24/7*) in demanding environments. Capacity is not the primary goal, and as a result, the drives are often offered in capacities that are relatively low in relation to their cost. The fastest enterprise hard drives can achieve maximum data transfer speeds above 1.6 Gbit s^{-1} and a sustained transfer rate up to 1 Gbit s^{-1} [6]. The drives run at 15,000 rpm and use smaller platters to offset power requirements.

External hard disk drives are usually connected to computers via USB ports and have slower data transfer rates when compared to internally mounted hard drives connected through serial advanced technology attachments (SATA). External HDDs are usually available in 2.5 and 3.5 in sizes. The 2.5 in variants are called *portable external drives*, while the 3.5 in variants are referred to as *desktop external drives*. Portable drives are packaged in smaller and lighter enclosures than the desktop drives; additionally, portable drives use power provided by the USB connection, while desktop drives often require external transformers to provide power. As of 2014, capacities of external hard disk drives generally range from 160 GB to 6 TB [7,8].

In 2014, hard disk drives accounted for a market of over $30 billion per year. Worldwide revenues for disk storage were $32 billion in 2013, down about 3% from 2012. This corresponds to shipments of 552 million disk drives in 2013 compared to 578 million in 2012 and 622 million in 2011 [9].

14.1.1 Magnetic Tapes

Magnetic recording tapes predated magnetic recording disks and are widely used for audio and video recording. In tape recording, a tape drive uses motors to wind the magnetic tape from one reel to another, passing tape heads, which can read, write, or erase information on the tape as it passes by the head. In magnetic tape recording, the contact between the tape and the read-and-write heads is a crucial factor determining performance. Contact between the recording heads and the recording medium is acceptable because of the relatively low number of head/tape passes expected (compared, for example, with a hard disk drive) so that wear of the tape is not a problem. The read/write heads are even contoured to improve contact with the tape.

Magnetic tape for recording sound originally used a ferric oxide (Fe_2O_3) powder coating on paper. This invention was made in Germany by Pfleumer and was further developed by the German company AEG, which manufactured the recording machines and BASF, which manufactured the magnetic tape. Schuller at AEG invented and patented the ring-shaped magnetic head that was another important invention in the development of magnetic tape recording. Schuller's ring focused a strong magnetic field on a small area of tape without needing to touch the surface. Other types of recording tape were also being developed, including steel tape invented by Stille in Germany. Matthias of BASF had expertise in producing high-frequency coils filled with carbonyl iron powder while another company I.G. Farben produced plastics, films, and a variety of coated materials. Combining these, Matthias developed a two-layer magnetic tape, bonding a top layer of carbonyl iron powder on a plastic base layer of cellulose acetate, similar to the kind of layered film that was being used in photography.

Video signals are similar to audio signals, although an important difference is that video signals need more bandwidth than audio signals. Audio tape recorders were therefore unable to adequately store and reproduce a video signal. Early video systems involved moving the tape across a fixed tape head at high speeds but these were not very successful until Ampex made the breakthrough of using a spinning recording head and normal tape speeds to achieve a very high head-to-tape speed that could record and reproduce the necessary high bandwidth signals.

Tape remains as a viable alternative to disk in some situations due to its lower cost per bit. Magnetic tape has the benefit of a comparatively long duration during which the medium can retain the stored data. The highest capacity tape media are generally on the same order as the largest available disk drives (about 5 TB in 2011). This is a significant advantage when dealing with large amounts of data particularly when there is less need for rapid access to the data and also when security is an issue. Though the data storage density of tape in bits per square inch is much lower than for disk drives, the available surface area on a tape is far greater and this can compensate for the lower data storage density. From 2002 onward, there was great interest in increasing the data capacity of magnetic tape. Manufacturers of modern data tape claim that magnetic tape is capable of 15–30 years of reliable archival data storage.

Magnetic tapes are produced in two main forms known as particulate and metal evaporated (ME) tapes. The advantage of particulate magnetic tapes is that they can be produced in a wide variety of widths at high coating speeds and at low cost. The disadvantage is that the magnetic particles occupy less than half of the tape volume. The inevitable trend therefore is toward the use of smaller particles with higher packing densities.

Particulate magnetic recording tapes consist of a coating of magnetic material, usually a cobalt-modified variant of gamma ferric oxide. The magnetic coating thickness is a few micrometers on a flexible, nonmagnetic substrate, which is usually polyethylene tetraphthalate. The thickness of the substrate is typically less than 10 μm. Thinner 5 μm tapes are used for long-playing tapes, although this is more expensive. The magnetic coating consists of the magnetic particles, a binder to contain and disperse the particles, lubricants to ease the motion of the tape, and abrasives such as particles of Al_2O_3 or CrO_2 to reduce wear of the tape. The lubricants in particulate tape are almost always *internal*, which means that they are included as an integral part of the coating.

The magnetic particle sizes vary depending on the choice of the magnetic material, but are acicular (elongated), and are typically 0.25 μm in length and 0.05 μm in width. These are single-domain particles, which can be easily magnetized parallel to their long axes. The typical coercivities of the magnetic particles on the tape are of the order of 175–200 kA m^{-1} (2200–2500 Oe).

Magnetic tapes are magnetically anisotropic. Conventionally, the elongated single-domain particles were aligned in the plane of the tape as shown in Figure 14.3, although perpendicular alignment of the particles as shown in Figure 14.4 gives scope for increased data storage capacity. In order to align the single-domain particles, the tapes are placed in a magnetic field oriented either parallel or perpendicular to the plane of the tape. The field is applied before evaporation of the solvent, which carries

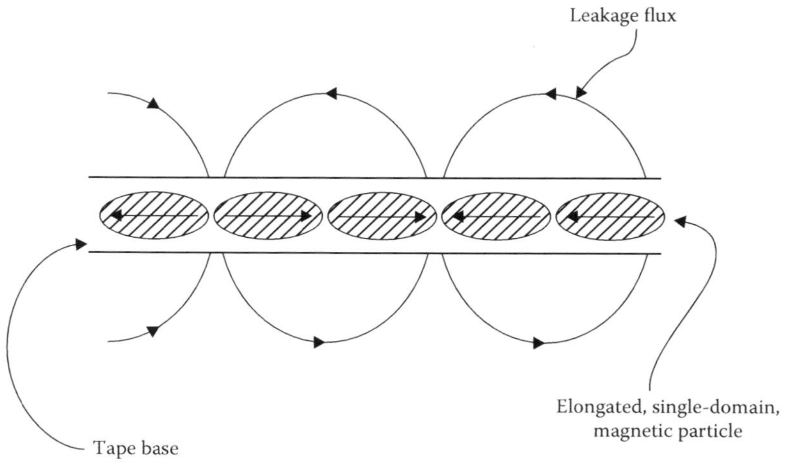

FIGURE 14.3 An arrangement of elongated, single-domain magnetic particles in conventional *longitudinal* magnetic recording tapes.

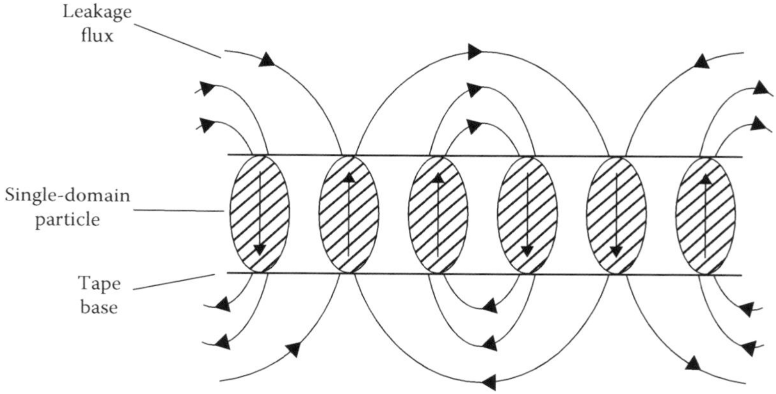

FIGURE 14.4 An arrangement of elongated, single-domain magnetic particles in *perpendicular* magnetic recording tapes.

the magnetic particles, leaving the dry binder, which holds the magnetic particles in place. The tape is then heated to completely dry the coating and is rolled or squeezed to densify the coating.

Cobalt-modified gamma iron oxide is used in preference to the simpler and older gamma iron oxide in magnetic tapes because of its higher coercivity. About 2–3 wt% cobalt is impregnated at the surface of the γ-Fe_2O_3 particles, and this results principally in an increase of coercivity from about 30 to about 60 kA m^{-1}. There is also a slight increase in saturation magnetization and a reduction in the temperature sensitivity of the coercivity. It is important that the cobalt is not absorbed into the bulk of the material as this leads to a deterioration of magnetic properties; in particular, the coercivity then becomes more sensitive to temperature, which is disadvantageous. The cobalt surface-modified gamma iron oxide has been widely used in video tapes, audio tapes, and floppy disks.

The thin film ME tapes have generally better performance characteristics than particulate tapes, but are more costly to produce. Advanced flexible media therefore consist almost exclusively of metal particle-based systems. Metallic particle tapes were first made available commercially for audio tapes by 3M Corporation. Later Sony used this form of tape in its video cassettes. The advantages of this form of tape include the high saturation magnetizations that can be achieved, up to 1.7×10^6 A m^{-1} in iron tapes. Coercivities are above 200 kA m^{-1} and particle sizes are below 0.05 μm [10]. Metal evaporated tapes, or simply *ME* tapes, were also introduced for audio cassettes by Matsushita in Japan, and then for video cassettes. Later ME tapes for digital video cassettes became available. In 2014, Sony and IBM announced that they had been able to reach a data storage density of 148 Gbit in^{-2}, with magnetic tape media developed using a new vacuum thin-film forming technology [11].

14.1.2 Magnetic Disks

Hard disk drives were developed originally to provide data storage for large *main frame* computers. Now hard disks are supplied as standard items on small personal computers such as laptops and desktops. In 2014, a typical personal computer came supplied with a hard disk of capacity between 1 and 2 TB ($1–2 \times 10^{12}$ B) of memory.

The principles of recording on magnetic disks are similar to those of recording on magnetic tapes. Magnetic recording disks come in two main types: floppy disks and hard disks. The materials used as a magnetic recording medium on floppy disks are also broadly similar to those used on particulate tapes. Floppy disks are made in the same way as tapes. The floppies were widely used in microcomputers but now have given way to miniature hard disk drives. Hard disks dominate the disk drive market at present. As of 2014, the primary competing technology for secondary storage is flash memory in the form of solid-state drives.

Magnetic hard disks consist of rotating disks (platters) coated with magnetic material [12]. They are produced by forming several layers of material deposited on a nonmagnetic disk substrate, including a nonmagnetic under layer, two or three magnetic layers separated by one or two *exchange layers* of ruthenium, an overcoat, and a top layer of lubricant as shown in Figure 14.5. The substrates, known as *platters*, are made from a nonmagnetic material, usually aluminum alloy, glass, or ceramic, and are coated with layers of magnetic material, typically 10–20 nm in depth, with an outer layer of carbon for protection [13,14]. The topmost magnetic layer, which consists usually of a cobalt-based alloy, is evaporated onto the surface to form a layer of 20 nm thick. Addition of platinum and tantalum has been found to enhance the anisotropy of the cobalt layer.

An overcoat is deposited to provide a protective layer of 15–20 nm thick. This is usually a carbon-based material, but sometimes zirconia- or tin-based materials are used. The disk lubricants are added to lower friction and prevent wear of the disk. These lubricants assist during *in-contact*, slow speed, sliding of the read/write

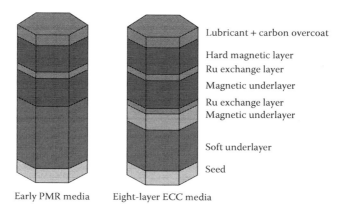

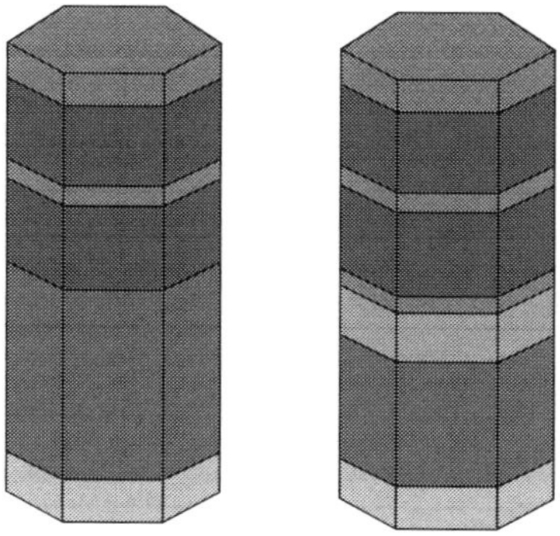

FIGURE 14.5 Different functional layers of hard disk perpendicular recording media with early perpendicular magnetic recording (*PMR*) media on the left and the more complicated eight-layer exchange-coupled composite (*ECC*) media on the right. (After Coughlin, T., and Grochowski, E., *Years of Destiny: HDD Capital Spending and Technology Developments from 2012–2016*, IEEE Santa Clara Valley Magnetics Society, 2012, June 19.)

assembly, which usually rests on the disk when stationary. As the disk begins to rotate, several tens of revolutions are needed before the air bearing begins to operate fully and lift the read/write head above the surface of the disk. The lubricant remains on the surface (unlike in floppy disks and particulate tape where they are embedded in the material) and is therefore known as *topical* lubricant. This usually consists of perfluoropolyether, a long-chain fluorocarbon compound, which is applied in monolayer thicknesses.

A major advantage of disks over tapes is that access time is much shorter on disks. This is mainly because the reading heads can be moved quickly to the right sector of the disk, whereas in tape recording, it is necessary to rewind the tape to find the data. In disk recording, the access time can also be improved by rotating the disk at a higher angular velocity. A typical 7200 rpm desktop HDD has a sustained *disk-to-buffer* data transfer rate in excess of 1 Gbit s^{-1} [15].

In hard disks, the read/write head is not in contact with the disk except when the disk is stationary and briefly as the disk comes up to full rotational speed. In order to optimize the performance of hard disks while ensuring that there is no direct contact between head and disk during the read/write process, an air bearing is used. The air flow is caused by the relative velocity between the disk and head and this maintains a small gap. The typical head to disk separation in today's hard disk drives is 5 nm [16]. In this way, the read/write head can be maintained close to, but not actually in direct contact with, the rotating disk. When this arrangement fails, as it does occasionally, we encounter the so-called disk head crash, which usually results in some damage in the form of lost data. The correlation between the head-disk separation, or flying heights, and storage density is shown in Figure 14.6. The head-disk separation can be reduced to about 2 nm before the situation becomes effectively *contact* rather than *noncontact* recording.

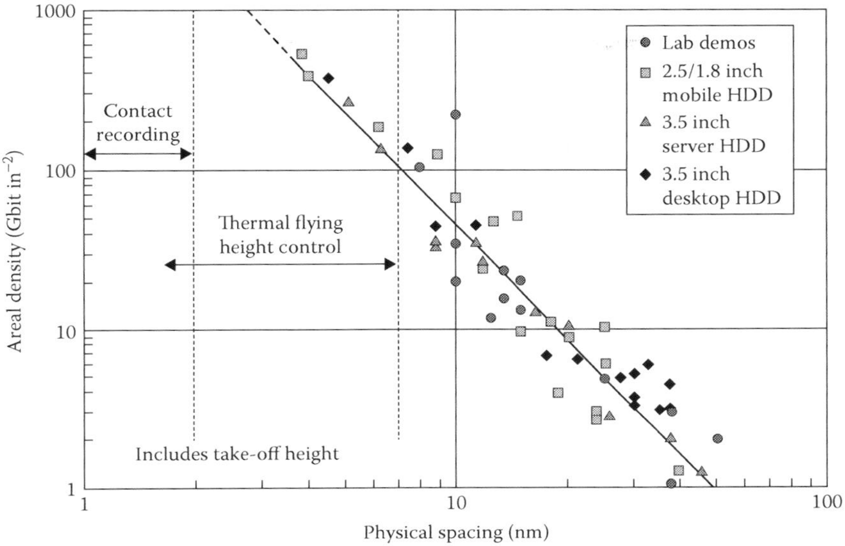

FIGURE 14.6 Relationship between head-media clearance (flying height) and areal storage densities. (After Coughlin, T., and Grochowski, E., *Years of Destiny: HDD Capital Spending and Technology Developments from 2012–2016*, IEEE Santa Clara Valley Magnetics Society, IEEE, Piscataway, New Jersey 2012, June 19.)

New magnetic storage technologies are being developed to enable continued progress in increasing data storage densities on hard disk drives. These include shingled magnetic recording, which was introduced in 2013 as the first step to reaching a 20TB HDD by 2020 [17]. Other HDD technologies that continue to be under development include *heat-assisted magnetic recording* [18], *microwave-assisted magnetic recording* [19], *two-dimensional magnetic recording* [20], *bit-patterned recording* [21], and *current perpendicular to plane giant magneto-resistance* heads [22,23].

14.1.3 Various Types of Recording Devices

At one time, the most common form of magnetic recorders were audio recorders. These have traditionally been analog recording devices, which use ac bias recording; that is, they make use of the linearity of the anhysteretic remanent magnetization curve to limit distortion of the reproduced signal. When signals are recorded in analog form, such as in audio recording, it is advantageous to have the recorded signal be proportional to the amplitude of the input signal. The recorded signal is the remanent magnetization on a region of the magnetic recording medium, while the input signal voltage is converted to an applied field in the recording head gap. Because the initial part of the magnetization curve is nonlinear, the remanent magnetization on the tape would be a nonlinear function of the applied field if the dc initial magnetization curve of the recording medium were used. However, by ac biasing, in which a sinusoidal field of decaying amplitude is superimposed on the dc bias field, the nonlinearity can be overcome. This produces the anhysteretic remanent magnetization curve. The variation of remanence with applied field is linear at low fields, and this produces a more desirable linear recording characteristic, which also improves the signal-to-noise ratio.

By this ac bias recording method, it became possible to make the magnetization imprinted on the recording medium proportional to the amplitude of the signal. In audio recording, particularly music, any distortion of the signal is undesirable. Therefore, the reading and writing processes take place relatively slowly. The typical tape velocity in professional audio recording is 0.4 m s^{-1} while on commercial audio cassettes it is less than 50 mm s^{-1}.

Digital recording of music has become standard and as a result little analog recording occurs now. In digital recording, it is only necessary to distinguish between *0* and *1* so these devices can function with much lower signal-to-noise ratios than was acceptable in analog recording. Furthermore, since in digital recording the actual level of the signal is not really crucial, providing that a *0* and *1* can be distinguished, the reading and writing process is very fast. However, even though the signal-to-noise ratio can be relatively small in digital recording the tolerable error rate is also very low.

14.1.4 Magneto-Optic Recording

Magneto-optic recording is a technology that is not currently in commercial use. However, technologies for recording data come and go, and so it is appropriate to mention it here briefly in this survey of magnetic recording.

Magneto-optic recording makes use of the Faraday and Kerr effects in which the direction of polarization of light is rotated in the presence of a magnetic field. In this way, two oppositely magnetized regions on a magnetic recording medium can be distinguished. The advantage of magneto-optic disks is that the storage density could be 1000 times greater than for floppy disks while access time for magneto-optic disks are typically 40 ms, which is about 10 times faster than for floppy disks. The data transfer rate in a magneto-optic drive is 5–10 Mbit s^{-1} [24].

The recording of information in magneto-optic recording depends on thermomagnetic magnetization in which an intense light source such as a focused laser beam is used to heat a small region of a thin film of ferrimagnetic material above its *compensation point* or Curie temperature and then it is allowed to cool again. Some of the original magneto-optic systems used a 780 nm wavelength laser light. If the recording medium is exposed to a reverse magnetic field throughout the heating process (i.e., is operating in the second quadrant of its magnetic hysteresis loop), then we know from earlier discussion of the anhysteretic magnetization that the optimum magnetic energy state will correspond to magnetization in the opposite direction. As it cools through the Curie temperature, the magnetization obtained in the region exposed to the laser beam will be the anhysteretic magnetization under the prevailing field, which will be in the third quadrant. This means that the regions that have been exposed to the laser beam will be magnetized in the opposite direction, as shown in Figure 14.7.

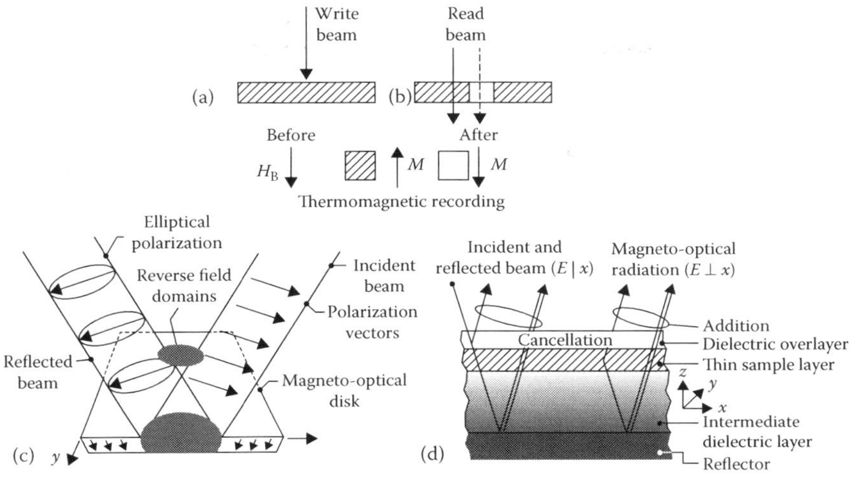

FIGURE 14.7 Magneto-optic reading and writing processes.

The reading of magnetic information on the medium depends on the magneto-optic Kerr effect. A polarized laser beam, with lower intensity (lower number of watts per square millimeter) than that used for writing is reflected from the surface of the magnetic recording medium, as shown in Figure 14.8. The laser light beam is then passed through a polarized analyzer before being detected. The presence or absence of the reverse domains can then represent either *0* or *1*. The film can later be wiped clean by saturating the magnetization in the original direction. It should be noted that for purposes of detection this technique works best in perpendicularly magnetized media. Signal-to-noise ratios are comparable with conventional magnetic disk recording. Magneto-optic disks typically have a 50 nm thick magnetic coating on a transparent 3 mm thick plastic substrate and use a 3 mW laser with a spot size of 2 × 5 μm to read the information on the disk by the Faraday and Kerr effects.

Two types of magnetic materials have been used in magneto-optic disks. The most widely used material is a rare earth-iron-cobalt film of general composition $R_x(FeCo)_{1-x}$, where R is usually terbium plus gadolinium, although sometimes dysprosium is used. The film is produced as an amorphous layer 50 nm thick. The rare earth-iron-cobalt films have high perpendicular anisotropy ($K_u > 10^5$ J m^{-3}), a coercivity of typically 240 kA m^{-1} (3 kOe), a high squareness ratio $\boldsymbol{M}_r/\boldsymbol{M}_s$, and a Curie temperature of 250°C–300°C depending on the specific composition. The advantages of using an amorphous film include (1) high uniformity of properties due to lack of crystallinity, (2) low-cost of deposition by sputtering, and (3) fine control of the magnetic properties, such as saturation magnetization and coercivity, by selection of the chemical composition.

The main alternative material for magneto-optic disks is a cobalt/platinum, which is deposited in a multilayer. These multilayers consist of 10–30 pairs of layers, each pair consisting of 0.3–0.5 nm of cobalt and 0.8–1.2 nm of platinum. The high

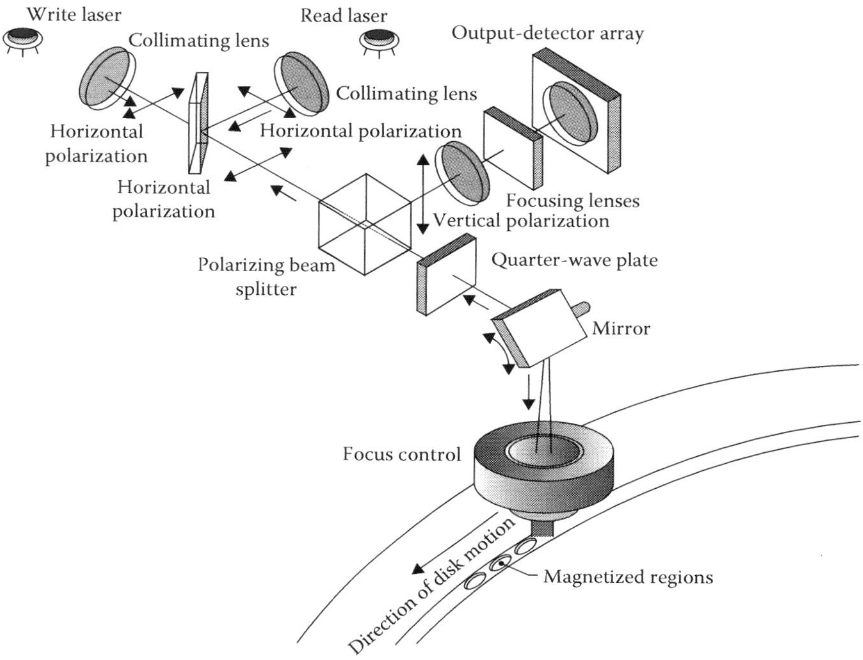

FIGURE 14.8 Components of a magneto-optic disk recording system.

uniaxial anisotropy of the multilayers ($K_u > 10^5$ J m^{-3} in these films also) ensures that the domain magnetizations align perpendicular to the plane of the film. The coercivities of these films are above 80 kA m^{-1} (1 kOe), but significantly they have high Kerr rotations at short wavelengths. One disadvantage of the Co/Pt multilayers is the relatively high Curie temperature of about 400°C, which means that the disks have to be heated to a higher temperature than $R_x(FeCo)_{1-x}$ in order to change the direction of magnetization.

However, a major problem for magneto-optic recording has been that the magneto-optic drive competes not with floppy disks, but instead directly with hard disk drives and other forms of data storage such as CDs and DVDs. The access times for magneto-optic drives are not competitive with access times for hard disks, which now are typically a few milliseconds. Magneto-optic recording using blue laser light at 429 nm achieved a recording density of 2.5 Gbit in^{-2} (3.9 Mbit mm^{-2}) and a data rate of a few megabytes per second [25]. In 2004, Sony demonstrated a 1 GB capacity optical MiniDisc known as *Hi-MD*. Hi-MD recorders were able to double the capacity of regular MiniDiscs with special formatting. However, as with other removable storage media, cheap CD/DVD drives and flash memory began to replace magneto-optic disk drives. Also, magneto-optic disks while highly reliable have slow writing time and therefore magnetic tape drives, such as the linear tape open format, surpassed magneto-optic media for high-capacity backup storage. Magneto-optic drives survived longer in some countries, such as Japan, because of the success of the Sony MiniDisc, but they have largely become obsolete in other countries.

14.2 MAGNETIC RECORDING MEDIA

The hysteresis of magnetization versus magnetic field in ferromagnetic and ferrimagnetic media is utilized in magnetic recording. Without hysteresis, the magnetic state of the material in zero field would be independent of the field direction to which it had last been exposed. However, in hysteretic systems, the remanent magnetization that has been written to a localized region of the medium acts as a memory of the last field to which that region of the medium has been exposed, both in magnitude and in direction. Therefore, data, usually in digital form for computers and related devices, or sometimes in analog form, as in sound recording, can be stored in the form of magnetic *imprints* on the magnetic medium.

The information stored on disks or tapes should be capable of being stored *permanently* and so disk drive manufacturers design disk drives to retain the data for a period of at least 10 years. The information stored on disks must be able to be retrieved with a minimum of distortion; that is, it must not be unexpectedly erased or significantly altered by exposure to unanticipated extraneous magnetic fields. Nor should the information be altered by the reading process since it is usually necessary to reread data many times without loss of information. Furthermore, it should be written and read with minimal power requirements.

The magnetic recording media must have high saturation magnetization to give as large a magnetic signal as possible during the reading process. The coercivity must be sufficient to prevent erasure, but small enough to allow the material to be reused for recording. Coercivities in the range of 24 to 240 kA m^{-1} (300–3000 Oe) are common for magnetic recording tapes [26]. For disk drives, in order to continue to increase storage densities, the coercivity of the media will need to increase. Therefore, disk drive media coercivities are in the range of 200–250 kA m^{-1} (2.5–3.1 kOe) [27,28].

14.2.1 Materials for Magnetic Recording Media

14.2.1.1 Gamma Ferric Oxide

For magnetic recording tapes and floppy disks, gamma ferric oxide (γ-Fe_2O_3) has been widely used as the magnetic recording medium. Gamma ferric oxide is not a commonly occurring form of Fe_2O_3 but is produced by oxidation of specially prepared Fe_3O_4. The particle size used is typically a few tenths of a micrometer with a length-to-diameter ratio of anything from 10:1 to 3:1. The shape anisotropy of the particles of course also determines their magnetization characteristics such as coercivity. The coercivity is in the range of 20–30 kA m^{-1} (250–375 Oe) [29]. Saturation magnetization of the gamma ferric oxide is 350 kA m^{-1} (350 emu cc^{-1}) while the Curie temperature is about 600°C, which is sufficiently high to avoid undue temperature dependence of the properties of the medium while operating under normal conditions at room temperature. Other properties are shown in Table 14.1.

TABLE 14.1
Magnetic Properties of Materials Used in Particulate Magnetic Recording Media

	γFe_2O_3	$Co/\gamma Fe_2O_3$	CrO_2	Fe	$BaO{\cdot}6Fe_2O_3$
M_s (kA m^{-1})	350	370	480	1700	370
T_c (°C)	600	600	128	770	320
H_c (kA m^{-1})	25–30	30–70	35–75	100–200	50–200
σ_s (A m^2 kg^{-1})	75	78	95	220	70
ρ (kg m^{-3})	4900	4900	4900	7800	5300
Particle size (μm)	0.5 × 0.1	0.5 × 0.1	0.4 × 0.05	0.15 × 0.05	0.15 × 0.05

14.2.1.2 Cobalt Surface-Modified Gamma Ferric Oxide

This is a variant of gamma ferric oxide. Cobalt surface-modified gamma ferric oxide [30] is preferred over gamma ferric oxide as a magnetic recording medium because it has a higher coercivity. Cobalt is added to the ferric oxide at the last stage of processing before it is coated onto the substrate. The cobalt accumulates preferentially on the surface of the gamma iron oxide particles to a depth of about 3 nm. The addition of cobalt increases the anisotropy of the material leading to higher coercivity for amounts of cobalt of up to 2% adsorbed on the surface. Above this amount of cobalt, the coercivity remains stable, but the saturation magnetization begins to decrease [31]. Most video tapes use cobalt surface-modified ferric oxide, which has a coercivity of 48 kA m^{-1} (600 Oe).

14.2.1.3 Chromium Dioxide

Chromium dioxide was also once popular as a recording medium for audio recording, before the cobalt-doped surface modification process was invented. It provides a magnetic recording medium with a higher coercivity of 40–80 kA m^{-1} than gamma ferric oxide. Chromium dioxide can usefully be employed with a rather smaller particle size of 0.4 by 0.05 nm and hence higher recording densities are possible than for gamma ferric oxide. Its saturation magnetization is slightly higher than gamma ferric oxide at 480 kA m^{-1} (480 emu cc^{-1}), but it has a rather low Curie temperature of 128°C, which makes its performance more temperature sensitive, a factor that is a distinct disadvantage. It is also more expensive than gamma ferric oxide, which reduces its commercial attraction. It has been largely replaced by the cobalt-doped gamma ferric oxide as a high-performance recording material for tapes and floppy disks.

14.2.1.4 Powdered Iron

Powdered iron has also been used as a recording medium. It has higher saturation magnetization than the oxide particulate media described above and so can be used in thinner coatings. The coercivity of these fine particles is typically 120 kA m^{-1}. The production of the iron particle tapes is a modification of the production process for iron oxide tapes, in which in the final part of the process, the oxide is reduced to metallic iron under a hydrogen atmosphere at 300°C. However, these tapes also need a surface coating of tin to prevent sintering whereby the particles coalesce and are no longer single domains. Typical magnetic properties are saturation magnetization of 1700 kA m^{-1} (1700 emu cc^{-1}) and a coercivity of 120 kA m^{-1} (1450 Oe).

14.2.1.5 Metallic Films

Metallic films are now the principal magnetic medium for high-performance, high storage density, hard disk drives. The main requirement in hard disk drive media is to obtain an adequate coercivity because ultimately this determines the bit sizes that can practically be used, and hence the recording density. The magnetic layer itself is usually cobalt with other chemical additives such as platinum, chromium, phosphorus, nickel, and tantalum. These additives are used to increase the anisotropy and hence the coercivity of the cobalt film. The magnetic layer is typically 50 nm thick and consists of an assembly of almost noninteracting particles. The absence of intergranular coupling helps to increase the coercivity of the film because each particle can then behave as a noninteracting single domain. Typically, the coercivities of these thin film media are up to $\boldsymbol{H}_c = 80$ kA m^{-1} (1 kOe), with typical remanences of $\boldsymbol{M}_R = 0.8 \times 10^6$ A m^{-1} (corresponding to $\boldsymbol{B}_R = 1$ T).

Cobalt-chromium thin film media have been preferred over pure cobalt because it was found that the chromium accumulates in paramagnetic intergranular regions, thereby reducing the exchange coupling between the grains. This increases coercivity and reduces noise fluctuations in the film. Tantalum has also been found to reduce noise fluctuations by reducing grain sizes in the magnetic film [32].

Metallic films are also used in some recording tapes because of their high saturation magnetization and remanence. They can be used in the form of thin coatings since the leakage fields, which are used in the reading process, are proportional to the remanent magnetization on the tape. The higher saturation magnetization therefore ensures that these leakage fields are rather larger than for similar thin films of other materials. The pickup voltage in the read head, which is proportional to the magnetic field from the tape, is therefore larger. Thinner recording media also allow higher recording densities since the rate of change of field with distance $d\boldsymbol{H}/dx$ along the tape can be made larger.

The magnetic layer is usually an evaporated film of cobalt-nickel alloy, which is formed as slanting columns in a porous matrix of cobalt and nickel oxides. These have remanences of typically $\boldsymbol{M}_r = 0.3 \times 10^6$ A m^{-1} (which corresponds to $\boldsymbol{B}_r = 0.37$ T) and coercivities of $\boldsymbol{H}_c = 80$ kA m^{-1} (1 kOe). The disadvantage of the metallic thin film tapes is that they do not wear well and so their lifetimes are relatively short. They need surface coating of lubricants (topical lubricants), which are usually fluorocarbons similar to those used on hard disk drives and corrosion inhibitor. However, while hard disk drive units are hermetically sealed to retain lubricant and keep out contaminants, the metallic tapes must necessarily be exposed and therefore retention of the lubricant is a problem.

14.2.1.6 Hexagonal Ferrites

Hexagonal ferrites have much higher coercivities than any of the above magnetic recording media materials and are used for more specialized permanent data storage applications such as credit cards and other cards with magnetic stripes to record data that does not need to be changed and where it is imperative that there is little chance of demagnetization by unanticipated exposure to low and moderate external magnetic fields and where there is little need for rerecording.

The main materials of interest are barium ferrite and strontium ferrite. Barium ferrite $BaO.6Fe_20_3$ has a coercivity of up to 480 kA m^{-1} in its pure form, with saturation magnetization of 370 kA m^{-1} (4.6 kOe). Barium ferrite particles of diameters 20 nm can be produced in films on substrates by controlling the particle growth conditions. These small particle sizes allow the possibility of storage densities of up to 1500×10^9 bits in^{-2} and high signal-to-noise ratios. The particles grow with their easy axis normal to the plane of the film and therefore barium ferrite is suitable for perpendicular recording media, as discussed below.

14.2.1.7 Perpendicular Recording Media

Perpendicular recording media in which the magnetic domains are oriented with magnetizations normal to the plane of the medium were pioneered in Japan [33]. These media offer higher recording density than conventional in-plane or *longitudinal* media but for a long time suffered from other problems such as the need for a very small head-to-medium distance and noise problems in the reading process, which prevented them becoming viable. The material that has been most widely used for this form of medium is a sputtered cobalt-chromium-platinum film, typically 20 nm thick and containing greater than 14% Cr in the form of columns normal to the surface of the substrate. In addition, due to the nature of the growth of these films, the magnetic moments remain perpendicular to the plane of the film, unlike the longitudinal recording materials in which the magnetic moments lie in the plane of the material. When the moments are perpendicular to the plane, the transitions between neighboring bits become much sharper leading to a higher recording density. Another material that has been tried is oriented barium ferrite, which is produced on a plastic substrate by a method similar to that used to produce particulate tapes.

14.3 RECORDING HEADS AND THE RECORDING PROCESS

The recording process involves the mechanism by which a magnetic imprint is written on the magnetic medium and the mechanism by which this imprint is read from the medium and the original information, whether an audio signal or some digital data, is re-created. The writing process is the means of transferring electrical impulses in a coil wound on an electromagnet (the write head) into magnetic patterns on the storage medium. The reading process is the inverse of this mechanism.

The writing or imprinting of magnetic signals or data on the hard disks is performed by inductive write heads. Magnetic recording heads are either inductive write heads, as shown in Figure 14.9, or magnetoresistive read heads, as shown in Figure 14.10. The reading of data from the medium is performed by magnetoresistive read heads. Therefore, the commercial disk drives employ different technologies for the reading and writing functions, unlike earlier disk drives, which used inductive heads to perform both operations.

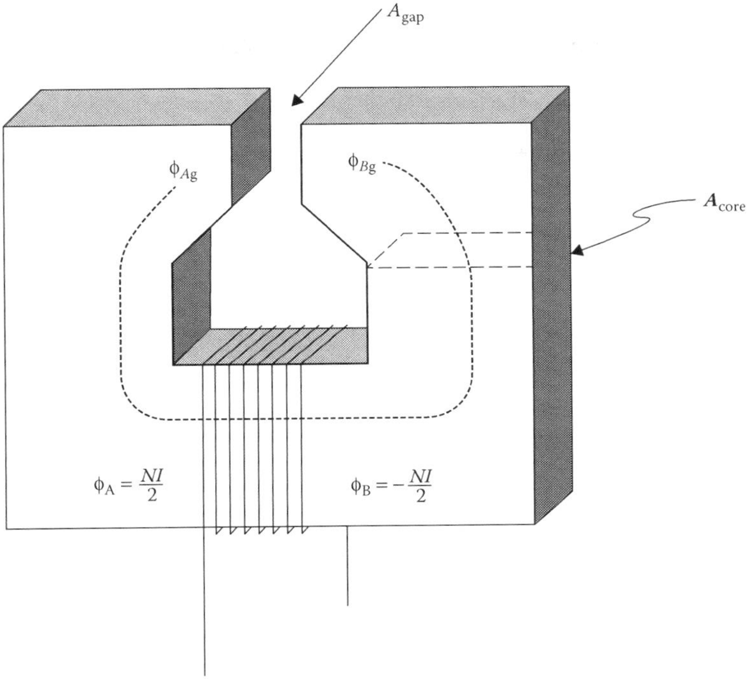

FIGURE 14.9 A schematic of an inductive write head.

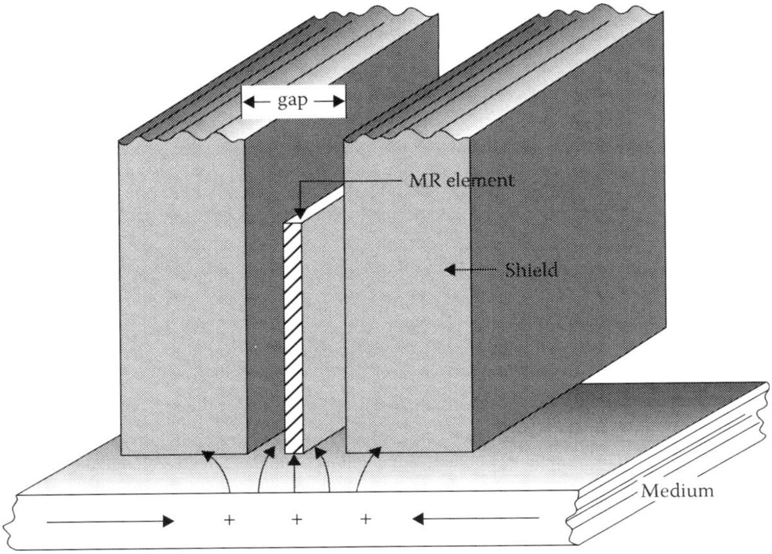

FIGURE 14.10 A schematic of a magnetoresistive read head.

Magnetoresistive read heads confer significant advantages over the earlier inductive read heads because of their low power requirements and high sensitivity. The magnetoresistive read heads that are currently used in hard disk drives employ giant magnetoresistive multilayers. These exhibit changes in resistance $\Delta R/R$ in the range of 80% or more [34].

The read process is quite well understood since it requires no knowledge of the magnetization characteristics such as the magnetic hysteresis loop of the medium. Only the remanent magnetization of the medium determines the response. However, the writing process, which involves the effect of an applied field on the magnetization of a magnetic medium, is more difficult to model. This is because it is difficult to accurately model the dependence of magnetization of the medium on the magnetic field even when the field is completely uniform, and in these cases, it is even more complicated because the field is not uniform and is changing with time.

14.3.1 Inductive Write Heads

Inductive recording heads that are used for writing data on disks and tapes consist of a *C* shaped high permeability core with a gap known as the *head gap*, of width of the order of 0.1 μm. The magnetic core is wound with a flux coil in order to generate the necessary magnetic flux in the gap when a current is passed through the coil. The field in the gap is called the *fringing field*. In magnetic tape recording, the heads are in direct contact with the tape, as shown in Figure 14.11, and usually there are separate heads for reading, writing, and erasing.

In hard disk recording, there is only one inductive write head, which rides on an air bearing above the surface of the disk. The write head material must have high saturation magnetization in order to leave a large imprint (high magnetization) on the recording medium but it must also have low remanence to ensure that there is no writing when the current in the coil is reduced to zero. Furthermore, a low coercivity

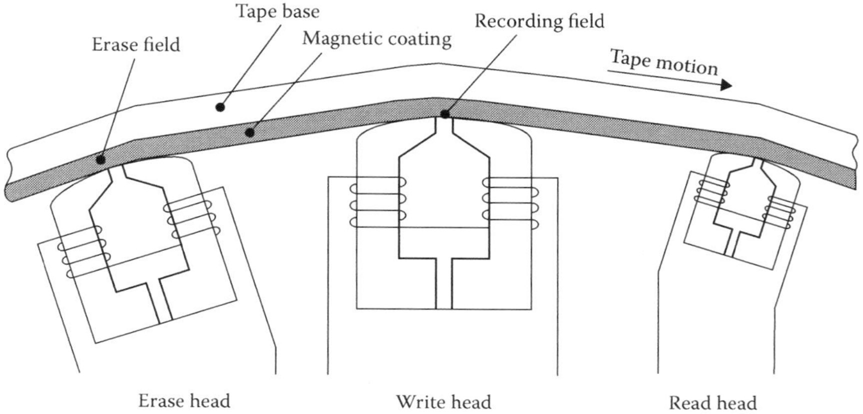

FIGURE 14.11 Magnetic erase, write, and read heads in a magnetic tape recording system.

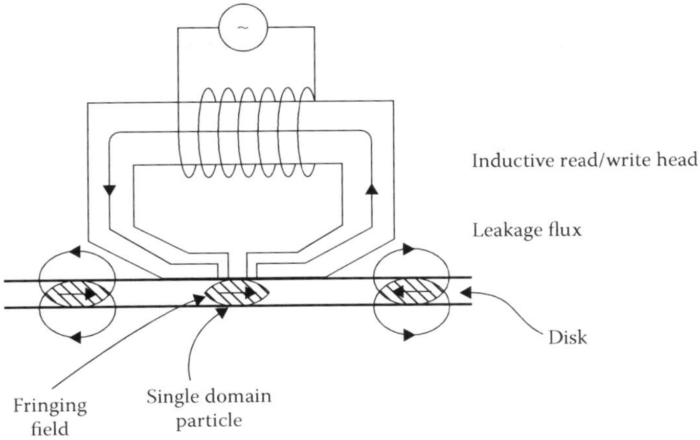

FIGURE 14.12 A schematic of an inductive read/write head above a magnetic disk.

is desirable. Recording heads are constructed of magnetically soft material; these include soft ferrites, Al-Fe, Al-Fe-Si, permalloy, and amorphous cobalt-zirconium.

In writing mode, the magnetic tape or disk passes the head where the fringing field causes a realignment of the magnetization within the recording medium as shown in Figure 14.12. The magnetization is then a record of the strength of the field in the gap of the recording head at the time that the disk or tape passed it.

The magnetic field in the gap of the write head, which is the region of main interest in the writing process, is shown in Figure 14.13. The field strength and direction at different locations in the write head gap can be determined by the finite-element techniques described in Chapter 1. With the older conventional, *in-plane* recording media, the component of the fringing field that was parallel to the tape surface determined the information stored on the recording medium. However, with perpendicular recording, which is now used in all commercial hard disk drives, the component of the fringing field that is normal to the plane of the recording medium determines the information stored on the medium.

14.3.2 Writing Process

We now consider the process of magnetizing the recording media. Specifically, how the local magnetization in the recording medium responds to an applied field in the head gap. In general, this is a rather difficult problem because of the nonlinear response of the magnetic recording medium to a magnetic field, and because the field that is generated in the recording medium is nonuniform. However, some empirically based models can be used to good effect. An obvious problem that arises for calculating the magnetic response of the recording medium to the field as the medium passes close to the head is that the magnetic fields at different depths in the magnetic medium are different, as shown in Figure 14.13. Second, as the recording medium passes the gap the strength of the fringing field, it experiences changes with time.

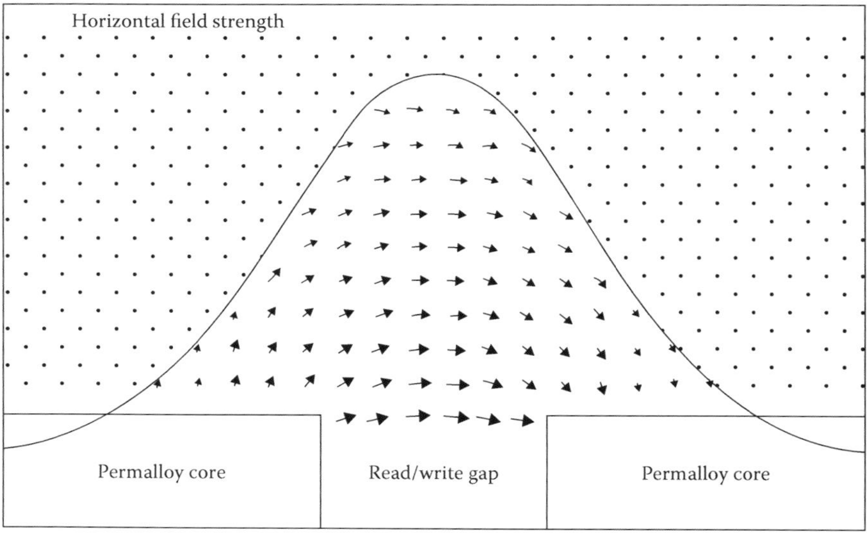

FIGURE 14.13 Variation of the magnetic field strength in both direction and magnitude close to a permalloy inductive recording head used for writing. (Adapted from Jiles, D. *Introduction to Magnetism and Magnetic Materials*, 2nd edn, Nelson-Thornes, Cheltenham, UK, Chapman & Hall, 1998.)

To give an example of the complexity of the writing process, suppose that a given region of the recording medium passing the head begins at the positive remanence point on its hysteresis curve, as shown in Figure 14.14. Then as it passes the gap with a negative field, the material passes down the second quadrant of the hysteresis loop to its coercive point $-\boldsymbol{H}_c$, for example. As it passes the head, the magnetic field that the recording medium experiences from the head gap decreases to zero and the material magnetization passes along a recoil minor hysteresis loop to $\boldsymbol{H} = 0$, ending with a small positive remanence.

This means that even when the magnetization has been reduced to zero at the coercive point $-\boldsymbol{H}_c$, the magnetization $\boldsymbol{M}$ will increase again as the field is reduced to zero to give a positive remanent magnetization. Therefore, to reach a demagnetized state $\boldsymbol{M} = 0$, the field experienced by the medium must reach a point $-\boldsymbol{H}_{cr}$, referred to as the *remanent coercivity*, which is beyond $-\boldsymbol{H}_c$. A square hysteresis loop is desirable in which the recoil minor loops are very flat giving $(d\boldsymbol{M}/d\boldsymbol{H}) \approx 0$.

During the writing process, the time-varying current in the write head coil changes with time, thereby altering the field in the gap. This causes localized changes in the magnetization in the recording medium as it passes the write head at a constant speed. It has proved difficult to determine the resultant magnetization of the disk or tape in two dimensions, and therefore theoretical models have only limited usefulness in the predictions of the resultant magnetization in the recording medium.

When a region of the recording medium has moved away from the write head the magnetization stabilizes. Models for the switch in direction of magnetization

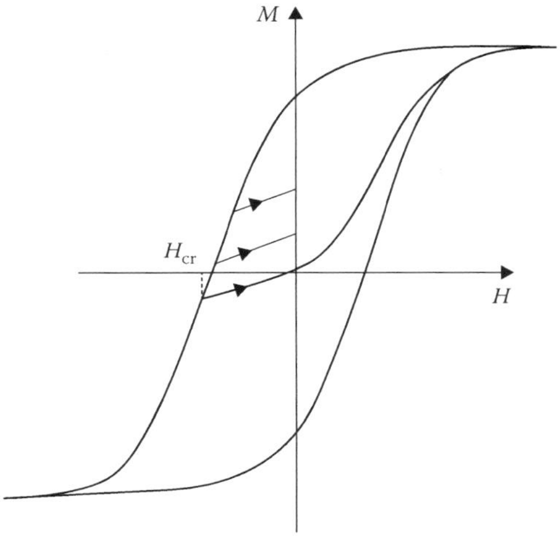

FIGURE 14.14 Recoil minor loops during the writing process for different regions of a magnetic recording medium as it passes a recording head.

M as a function of position in the recording medium usually make use of simplified approximations to the magnetization characteristics of the medium. For example, magnetization can be approximated as a single-valued function of the magnetic field. One such model is the Williams-Comstock [35] model. This is a one-dimensional model that employs the arctangent function to describe the change in magnetization from one domain (the *up* domain) to the next (the *down* domain) at different positions x in the recording medium as shown in Equation 14.1.

$$\boldsymbol{M}(x) = \left(\frac{2\boldsymbol{M}_R}{\pi}\right)\arctan\left(\frac{x}{a}\right) \tag{14.1}$$

where:

a is an adjustable parameter, which is determined by the rate of change of magnetization with distance

$\boldsymbol{M}_R$ is the remanent magnetization

This allows the magnetization $\boldsymbol{M}(x)$ to have the same magnitude but opposite signs +/− $\boldsymbol{M}_R$ at large values of x (meaning in the two domain at large distances from the transition region), while allowing the magnetization to be zero at $x = 0$, the transition region between the domains.

14.3.3 Writing Head Efficiency

Write head efficiency is defined as the ratio of magnetomotive force across the head gap to the magnetomotive force supplied by the energizing coil. This is determined from consideration of the magnetic circuit formed by the magnetic core and the air gap of the head.

In the air gap,

$$\boldsymbol{B}_g = \mu_0 \boldsymbol{H}_g \tag{14.2}$$

In the core,

$$\boldsymbol{B}_c = \mu_r \mu_0 \boldsymbol{H}_c \tag{14.3}$$

where:

μ_r is the relative permeability of the core
$\boldsymbol{H}_c$ is the magnetic field in the core

The magnetomotive force of the driving coil Ni is also the same as the magnetomotive force across the whole magnetic circuit. If l_g is the length of the gap and l_c is the length of the ferromagnetic core and assuming uniform magnetic fields $\boldsymbol{H}_g$ and $\boldsymbol{H}_c$ in the gap and the core, then by Ampere's law

$$Ni = \boldsymbol{H}_c l_c + \boldsymbol{H}_g l_g \tag{14.4}$$

The efficiency of the core is the ratio of magnetomotive force across the gap $\boldsymbol{H}_g l_g$ to the magnetomotive force supplied from the coil given in Equation 14.4. The efficiency is then

$$\eta = \frac{\boldsymbol{H}_g l_g}{\left(\boldsymbol{H}_c l_c + \boldsymbol{H}_g l_g\right)} = \frac{\boldsymbol{H}_g l_g}{Ni} \tag{14.5}$$

The efficiency can also be expressed in terms of the reluctances of the magnetic paths R_c in the core and R_g in the gap.

$$\eta = \frac{R_g}{R_c + R_g} = \frac{\left(l_g / A_g\right)}{\left[\left(l_c/\mu_r A_c\right) + \left(l_g / A_g\right)\right]} \tag{14.6}$$

where:

A_c is the cross-sectional area of the core
A_g is the cross-sectional area of the gap

To increase the field in the gap $\boldsymbol{H}_g$ for a given magnetomotive force Ni, it is necessary to have a larger permeability μ_r in the core and a larger ratio of cross-sectional areas A_c/A_g.

It can be seen therefore that the geometry (length and area of the gap and length and area of the core), together with the magnetic properties of the core material in terms of its relative permeability μ_r, determines the field in the gap, as indicated from the discussion of air gaps in magnetic circuits in Chapter 2. The fringing field in the vicinity of the gap $\boldsymbol{H}_g$ can be calculated from this using Equation 14.7 and consequently, since the magnetic field from the head needs to be able to change the magnetization of the recording medium, the maximum coercivity of the recording medium that can be used with the head can be estimated.

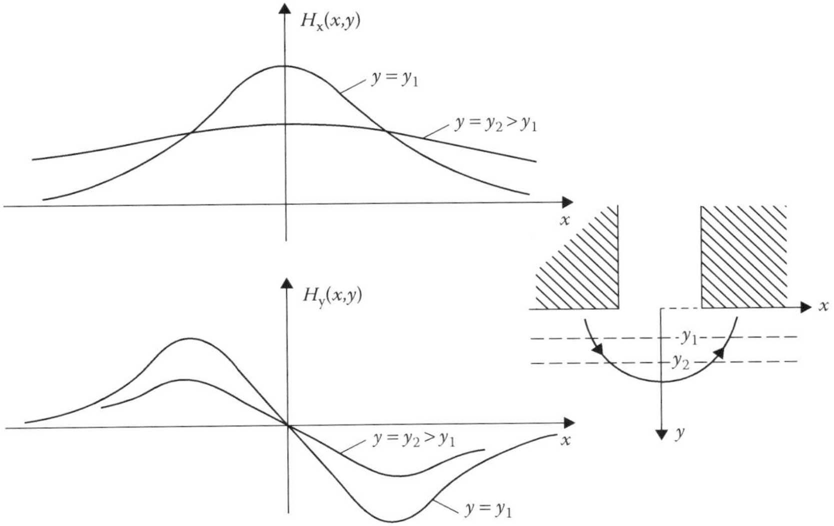

FIGURE 14.15 Vertical and horizontal components of the magnetic field in the vicinity of a simplified model of a read/write head known as a *Karlqvist head*.

$$\boldsymbol{H}_g = \frac{\eta Ni}{l_g} = \frac{(l_g/A_g)}{\left[(l_c/\mu_r A_c)+(l_g/A_g)\right]}\left(\frac{Ni}{l_g}\right) \tag{14.7}$$

Karlqvist heads [36] are an idealization that was often used for calculating the field in the vicinity of the head gap. In the Karlqvist approximation, the head gap is small compared with the length through the head between pole tips, such as in the head depicted in Figure 14.15. In the past, calculations based on the Karlqvist approximation provided a relatively simple geometrical situation for determining the magnetic field in the gap analytically. With the emergence of three-dimensional, finite-element codes available now for calculating the fields in the gap, the Karlqvist approximation is less important.

14.3.4 Magnetoresistive Read Heads

Magnetic recording technology made rapid progress in storage densities on the introduction of the magnetoresistive read head, which replaced the inductive read and led to the production of a new generation of high-sensitivity, high-spatial-resolution read heads [37]. The first commercially available magnetoresistive read heads employed anisotropic magnetoresistance, which has been described in Sections 3.3.2 and 5.4. The sensor material of choice was permalloy. Later, more sensitive magnetoresistive read heads based on giant magnetoresistance multilayers, as described in Section 11.3.9, were introduced and these are used in commercial disk drives now.

In read mode, the passage of the medium past the read head causes a variation in flux density in the read head, which changes the resistance of the read head.

The magnetoresistive head detects the leakage flux from the locally magnetized regions of the recording medium by means of a two-point resistance measurement. Spatial resolution is determined by the distance between the electrical contacts in the perpendicular or *cross-track* direction, which is usually around 1 μm, and by the shield separation along the track direction, which is usually 0.1 μm. To obtain an antisymmetric response that is approximately linear with strength of the leakage field, the magnetoresistive material also needs to be biased with a constant applied magnetic field. This is supplied using an adjacent magnetized film. The voltage generated by the magnetoresistive element also depends on the current, and so it is desirable to operate at high current to maximize sensitivity. Current densities of 2×10^{-3} A m^{-2} are typically used. The operating current density is of course limited by the power dissipation in the head.

14.3.5 Reading Process

The reading process in magnetic recording is relatively well understood. The recording medium (disk or tape) passes below the read head and causes a fluctuation in the flux density in the magnetic core of the read head. The fringing fields from the medium can be calculated using simple models. As the disk or tape passes near to the read head, the stray field associated with the magnetic imprint on the medium enters the reading head. This field passes through the magnetoresistive sensor element and changes its resistance. Therefore, the voltage in the read head will be dependent on the fringing flux density emanating from the recording medium.

14.3.6 Recording Density

The recording density in a medium depends on the magnetic properties of the medium and the characteristics of the write head. The recording density is determined by the product of bits per inch (BPI) and the number of tracks per inch (TPI). Progress since 1990 in increasing both BPI and TPI is shown in Figure 14.16. Projections of data storage densities toward 1 Tbit in^{-2} were made many years before such densities were actually achieved [38]. In 2014, the state-of-the-art hard disk drives had TPI values of 500,000 and BPI values of 2 Mbits in^{-1}, thereby reaching data storage densities of 1 Tbit in^{-2}. Predictions have been made of data storage densities of 10 Tbit in^{-2} using shingled writing and two-dimensional read back and signal processing [39].

The maximum attainable BPI can be measured by the parameter a, known as the *transition length*, which is the minimum distance on the recording medium that is needed to completely reverse the magnetization from saturation remanence in one direction to saturation remanence in the other. The transition length, in which a signal can be made to change, is dependent on $d\boldsymbol{M}/dx$ in the recording medium. This can be expressed as the product $(d\boldsymbol{M}/d\boldsymbol{H})(d\boldsymbol{H}/dx)$, where $(d\boldsymbol{M}/d\boldsymbol{H})$ is a property of the medium, specifically the slope of the hysteresis curve, while $(d\boldsymbol{H}/dx)$ is a property of the writing head, specifically how large is the rate change of field with distance. If we make the approximation that the slope of the hysteresis loop is constant then

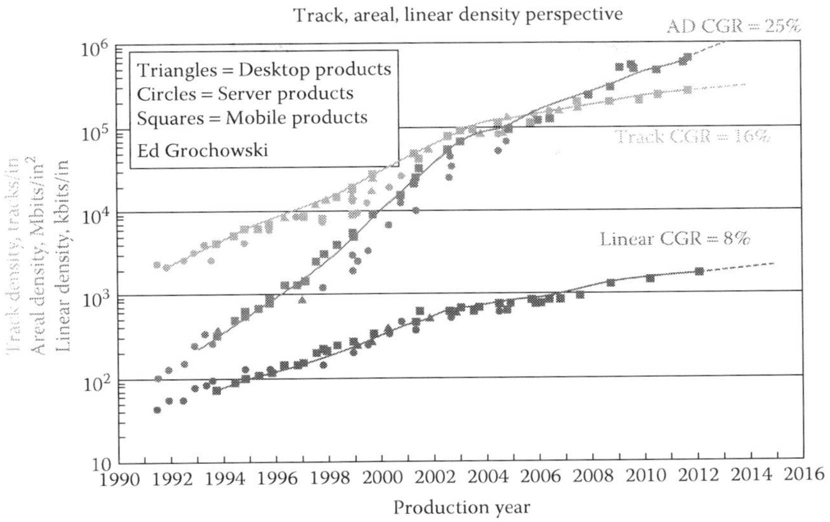

FIGURE 14.16 Progress in track density, linear bit density, and areal data storage density on hard disk drives from 1990 to 2016. (After Coughlin, T. and Grochowski, E., *Years of Destiny: HDD Capital Spending and Technology Developments from 2012–2016*.)

for a fixed field gradient in the head gap, we have the following expression for the transition length a

$$a = \frac{(2\boldsymbol{M}_{\mathrm{R}}/\pi)}{(d\boldsymbol{M}/dx)_{\max}} = \frac{(2\boldsymbol{M}_{\mathrm{R}}/\pi)}{(d\boldsymbol{M}/d\boldsymbol{H})(d\boldsymbol{H}/dx)_{\max}} \tag{14.8}$$

This can be verified from the Williams-Comstock equation (Equation 14.1) above. The transition length a is therefore smaller for larger field gradient $d\boldsymbol{H}/dx$ and for larger $d\boldsymbol{M}/d\boldsymbol{H}$ on the major hysteresis loop, that is, for a square hysteresis loop. Of course, there are other factors that have not been taken into account in this simple analysis, such as the demagnetizing field in the recording medium and the spatial variation in the transition region. However, these do not alter the basic conclusion about the desirability of a high-field gradient $(d\boldsymbol{H}/dx)_{\max}$ in the gap and of square hysteresis loop materials meaning high $(d\boldsymbol{M}/d\boldsymbol{H})_{\max}$. Notice that while large $d\boldsymbol{M}/d\boldsymbol{H}$ on the major hysteresis loop is desirable, this should ideally coincide with a small $d\boldsymbol{M}/d\boldsymbol{H}$ on the recoil minor loops.

14.4 MODELING THE MAGNETIC RECORDING PROCESS

14.4.1 Preisach Model

The magnetic properties of the recording medium, specifically the hysteresis loop, can be represented using a model for magnetization as a function of field that was devised in the 1930s by Preisach [40]. This model does not give much physical insight into the magnetic properties of materials and does not have predictive capabilities.

The model is in essence a curve-fitting procedure involving an array of up/down switching elements known as *hysterons*, each with a pair of switching fields, one in the up direction and one in the down direction. These switching fields may be the same magnitude in both directions (giving a symmetric hysteron) or they can be of different magnitude in the two directions (giving an asymmetric hysteron).

The essential idea of the Preisach model is that the measured magnetic hysteresis loop of a material is due to a summation of more elementary hysteresis loops of domains with differing switching fields. Within the confines of the model, these domains can only have two states, with magnetization parallel or antiparallel to a given direction. The model relies on a density function called the *Preisach function*, which is defined on a plane described by the positive and negative switching fields H_+ and H_-. This Preisach function is used to determine how many domains switch their orientation from + to –, or vice versa, as the applied magnetic field is swept between extreme values.

Once the hysteresis curves have been measured, the model can be used to give reasonable mathematical representations of hysteresis curves by calculating the Preisach function of the relative number density of hysterons with different switching fields. The Preisach model has been found useful for modeling the magnetic properties of digital recording media, which are deliberately designed to have two states, *spin-up* and *spin-down*, as, for example, in elongated single-domain particles, and is used in the recording industry [41].

The model works fairly well for weak interactions between domains such as occur in these recording media, which are usually aggregates of single-domain particles, and because the magnetic moments within the elongated single-domain particles can only have magnetic moments along one axis leading to a magnetization either parallel or antiparallel to the long axis of the particles. The model has been extended to take into account magnetic interactions between the single-domain particles and has also been extended beyond one dimension in the vector Preisach model. A comprehensive treatment of the Preisach model can be found in Mayergoyz's book on mathematical models of hysteresis [42].

14.4.2 Stoner-Wohlfarth Model

The Stoner-Wohlfarth theory [43], which was discussed in the previous chapter, has relevance to particulate recording media, perhaps more than to hard ferromagnetic alloys such as permanent magnets. In particulate recording media, the isolated single-domain particles are deliberately created on the recording medium (disks or tapes) and so these are well-suited for the application of the Stoner-Wohlfarth theory. The model has therefore found appropriate applications in determining the magnetization characteristics of particulate recording media [44].

REFERENCES

1. Poulsen, V. Danish patent No. 2653, 1899; US Patent No. 789, 336, 1905.
2. Carlson, W. L., and Carpenter, G. W. (1921) referenced in White, R. M. *Introduction to Magnetic Recording*, IEEE, Piscataway, 1984, p. 1.
3. Holmes, L. C., and Clark, D. L. *Electronics*, 18, 126, 1945.
4. http://en.wikipedia.org/wiki/IBM_305_RAMAC.

5. *8TB HDD Now Shipping....* 2014. Retrieved October 13, 2014, http://www.hgst.com/press-room/press-releases/HGST-unveils-intelligent-dynamic-storage-solutions-to-transform-the-data-center.
6. *Seagate Cheetah 15K.5 Data Sheet.* http://www.seagate.com/docs/pdf/marketing/po_cheetah_15k_5.pdf.
7. Graham, D. (2010, December 10). Pocket drive battle: 10 high speed external hard disks rated–PC & Tech Authority, Pcauthority.com.au. Retrieved April 26, 2012.
8. External hard drives: Buying guide for external hard drives, including portable and desktop devices, Helpwithpcs.com. Retrieved April 26, 2012.
9. *IDC Report: Worldwide Disk Storage Systems Quarterly Tracker*, 2014, June 19; *Any Future for Worldwide Storage Industry?* IDC, Press release. StorageNewsletter.com. Retrieved July 14, 2014; *Worldwide Global Disk Storage Systems Revenue 2012 = $32,564 Million; 2013 = $31,688 Million*, http://en.wikipedia.org/wiki/Hard_disk_drive.
10. Onodera, S., Kondo, H., and Kawana, T. *MRS Bull.*, 21 (9), 35, 1996.
11. *Sony Develops Magnetic Tape Technology with the World's Highest Areal Recording Density of 148 Gb/in² Able to Record (185 TB)*2 Approximately 74 Times More Data Than Conventional Magnetic Tape Media*, http://www.sony.net/SonyInfo/News/Press/201404/14-044E/index.html.
12. Arpaci-Dusseau, R. H., and Arpaci-Dusseau, A. C. *Operating Systems: Three Easy Pieces* [Chapter: Hard Disk Drives]. Arpaci-Dusseau Books, 2014.
13. http://en.wikipedia.org/wiki/Hard_disk_drive#cite_note-headcrash-24.
14. http://www.data-master.com/HeadCrash-explain-hard-disk-drive-fail_Q18.html.
15. http://en.wikipedia.org/wiki/Hard_disk_drive.
16. http://en.wikipedia.org/wiki/Flying_height.
17. *Seagate Delivers on Technology Milestone: First to Ship Hard Drives Using Next-Generation Shingled Magnetic Recording*, Seagate Technology, New York. September 9, 2013, Press release, seagate.com. Retrieved July 5, 2014; Shingled Magnetic Technology Is the First Step to Reaching a 20 Terabyte Hard Drive by 2020.
18. *"Report: TDK Technology" More Than Doubles "Capacity of HDDs."* Retrieved October 4, 2011.
19. Mallary, M. et al. Head and media challenges for 3 Tb/in² microwave-assisted magnetic recording, *IEEE Trans. Magn.*, 50 (7), 3001008, July 2014.
20. Wood, R. Shingled magnetic recording and two-dimensional magnetic recording (PDF), ewh.ieee.org, October 19, 2010. Hitachi GST. Retrieved August 4, 2014.
21. Will Toshiba's bit-patterned drives change the HDD landscape? *PC Magazine*, August 19, 2010. Retrieved August 21, 2010.
22. Coughlin, T., and Grochowski, E. *Years of Destiny: HDD Capital Spending and Technology Developments from 2012–2016*, IEEE Santa Clara Valley Magnetics Society, Piscataway, 2012, June 19. ewh.ieee.org/r6/scr/mag/.
23. Bai, Z., Cai, Y., Shen, L., Han, G., and Feng, Y. *All-Heuslergiant-Magnetoresistance Junctions with Matched Energy Bands and Fermisurfaces*, Cornell University Library, arXiv.org ≥ cond-mat ≥ arXiv:1301.6106, 2013.
24. Bloomberg, D. S. et al., in *Magnetic Recording Technology* (eds. C. D. Mee and E. D. Daniel), McGraw-Hill, New York, 1996.
25. Suzuki, T. *MRS Bull.*, 21 (9), 42, 1996.
26. *Fuji—Magnetic Tape Coercivities*, http://www.fujifilmusa.com/shared/bin/Degauss_Data_Tape.pdf.
27. Kryder, M. H. *MRS Bull.*, 21 (9), 17, 1996.
28. *Coercivity of Disk Drive M...*, http://en.wikipedia.org/wiki/Coercivity.
29. Mallinson, J. C. *The Foundations of Magnetic Recording*, Academic Press, San Diego, CA, 1987, pp. 31, 34, and 104.

30. Chakrabarti, S., Mandal, S. K., and Chaudhuri, S. Cobalt doped γ-Fe2O3 nanoparticles: Synthesis and magnetic properties, *Nanotechnology*, 16, 506–511, 2005.
31. Mee, C. D., and Daniel, E. D. *Magnetic Recording Technology*, McGraw-Hill, New York, 1996, p. 341.
32. Doerner, M. F., and White, R. L. *MRS Bull.*, 21 (9), 28, 1996.
33. Ouchi, K., and Iwasaki, S. *IEEE Trans. Magn.*, 23, 180, 1987.
34. Jiles, D. *Introduction to the Electronic Properties of Materials*, 2nd edn, Chapman & Hall, 2001, p. 19.
35. Williams, M. L., and Comstock, R. L. *AIP Conf. Proc.* 5, 758, 1971.
36. Karlqvist, O. *Trans. Roy. Inst. Techn. (Stockholm)*, 86, 1954.
37. Brug, J. A., Anthony, T. C., and Nickel, J. H. *MRS Bull.*, 21 (9), 23, 1996.
38. Wood, R. The feasibility of magnetic recording at 1 Terabit per square inch, *IEEE Trans. Magn.*, 36, 36, 2000.
39. Wood, R., Williams, M., Kavcic, A., and Miles, J. The feasibility of magnetic recording at 10 Terabits per square inch on conventional media, *IEEE Trans. Magn.*, 45, 917, 2009.
40. Preisach, F. *Zeit. Phys.*, 94, 277, 1935.
41. Vajda, F., and Della Torre, E., Relationship between model parameters and recording industry parameters, *IEEE Trans. Magn.*, 33, 1588, 1997.
42. Mayergoyz, I. D. *Mathematical Models of Hysteresis*, Springer-Verlag, New York, 1991.
43. Stoner, E. C., and Wohlfarth, E. P. *Phil. Trans. Roy. Soc. A*, 240, 599, 1948.
44. Chantrell, R. W., O'Grady, K., Bradbury, A., Charles, S. W., and Hopkins, N. *IEEE Trans. Magn.*, 23, 204, 1987.

FURTHER READING

Bertram, H. N. *Theory of Magnetic Recording*, Cambridge University Press, Cambridge, 1994.

Hard Disk Drives, http://en.wikipedia.org/wiki/Hard_disk_drive.

Mallinson, J. C. *The Foundations of Magnetic Recording*, 2nd edn, Academic Press, San Diego, CA, 1993.

Mee, C. D., and Daniel, E. D. *Magnetic Recording Technology*, McGraw-Hill, New York, 1996.

Wood, R. The feasibility of magnetic recording at 1 Terabit per square inch, *IEEE Trans. Magn.*, 36, 36, 2000.

Wood, R., Williams, M., Kavcic, A., and Miles, J. The feasibility of magnetic recording at 10 Terabits per square inch on conventional media, *IEEE Trans. Magn.*, 45, 917, 2009.

15 Magnetic Evaluation of Materials

In this chapter, we look at applications of magnetic measurements to materials characterization and to nondestructive evaluation (NDE) of materials. Magnetic methods can be used to solve two main classes of problem: (1) detection of defects and (2) evaluation of properties. The subject of magnetic NDE has received little attention in the past because the complexity of the magnetic responses of materials, for example, the hysteresis and the Barkhausen effect, has made it difficult to interpret results. Many of the magnetic techniques are directed toward the evaluation of steels because of the widespread use of steel as a constructional material in load-bearing applications. Magnetic imaging methods such as magnetic force microscopy and various other scanning techniques using magnetometers such as superconducting quantum interferences devices (SQUIDs), magnetoresistive sensors, and Barkhausen effect sensors are widely used.

15.1 METHODS FOR EVALUATION OF MATERIALS PROPERTIES

A number of magnetic nondestructive testing (NDT) techniques have appeared over the years. The subject of NDE as part of a broader strategy of materials monitoring is assuming a vital role as more industries have become aware of the potential benefits of quality control and assurance in the production of materials and components. Plant life extension is another application whereby the cost effectiveness of only retiring defective components (*retirement for cause*), and the possibility of avoiding potentially costly and catastrophic failures leads to important economic and safety benefits. This can be achieved by monitoring the condition of structures for defects and/or the presence of high levels of stress. Often, the approach is to look for changes in the detected signal from a used and/or damaged component compared with that of a new or undamaged part. These differences can provide a warning of impending failure.

While various measurement techniques may be used on magnetic materials, the magnetic methods have some advantages because they utilize the inherent ferromagnetic properties, which are sensitive to stress and microstructure. They can be used for evaluation of a wide range of material properties, residual strain, and cracks. In general, the changes in magnetic properties that are observed are easily measurable and do not need high-resolution electronics or signal processing for their use. Nevertheless, the magnetic methods have not yet been fully exploited when compared, for example, with ultrasound. This is because other techniques can be applied to a variety of other materials and previously there was more incentive for their development. Now, however, as the limitations of other techniques become apparent, as, for example, in the important area of detection and prediction of failure as occurs in fatigue or creep damage, attention has focused on the capabilities of magnetic methods.

15.1.1 Magnetic Hysteresis

All ferromagnetic materials exhibit hysteresis in the variation of flux density $\boldsymbol{B}$ with magnetic field $\boldsymbol{H}$ as shown in Figure 15.1. The hysteresis properties such as permeability, coercivity, remanence, and hysteresis loss are known to be sensitive to factors such as stress, strain, grain size, heat treatment, and the presence of precipitates of a second phase, such as iron carbide in steels. The measurement of hysteresis yields a number of independent materials parameters, each of which changes to some degree with stress, strain, and microstructure. Since it is understood that several independent parameters are needed in general to separate effects of stress state from microstructure, hysteresis measurement would seem to be ideally suited for the determination of the properties of steels because several independent parameters can be obtained from one measurement.

Despite this economy of means, there are certain difficulties that need to be overcome with the magnetic hysteresis method. First, the problem of demagnetizing effects due to finite geometries needs to be addressed, since results that may appear to be due to changes in sample properties can be caused by geometrical effects. Second, it has proved difficult to adequately model hysteresis in ferromagnets, so that it has been difficult to interpret changes in the hysteresis characteristics in terms of fundamental changes in sample properties. Of course, this has also been true of other magnetic methods, such as the Barkhausen effect, and that has not prevented

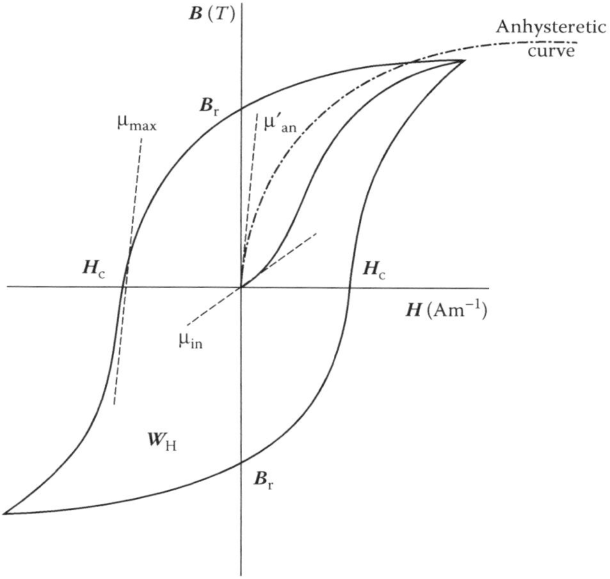

FIGURE 15.1 Typical hysteresis loop of a ferromagnetic material.

empirical use of these methods for NDE. Nevertheless, both hysteresis [1] and the Barkhausen effect [2] are much better understood than a generation ago and this has greatly assisted interpretation of NDE results obtained with these two methods.

Evaluation of the condition of steel components has been one area where NDE using hysteresis measurements has had great success. Numerous applications have been reported in the literature. Mikheev and coworkers made many investigations detecting the quality of heat treatment of steels from magnetic parameters [3,4], particularly the evaluation of hardness of various steels. In most cases, the magnetic properties were determined using a coercimeter and correlations made between chemical composition, microstructure, heat treatment and hardness, and the principal magnetic parameter of interest, the coercivity.

Kuznetsov et al. [5] also looked at the effects of heat treatments such as quenching, hardening, and tempering on the magnetic properties of steels. They have devised a method for determining the depth of case hardening. Zatsepin et al. [6] investigated the variation of coercivity with heat treatment and mechanical properties, while Rodigin and Syrochkin were able to use the effect of stress on coercivity to check mechanical hardness, thereby using hysteresis parameters as an accept–reject criterion for steel components [7].

The effects of stress on the magnetic hysteresis properties of steels are of interest because they have applications to NDE of large-scale structures. Typically, the hysteresis *signature* is changed by stress as shown by Langman [8] in Figure 15.2. Vekser et al. [9] showed that it was possible to measure the stress in rail steel from the permeability. The dependence of magnetic properties on static and dynamic stresses was the subject of a study by Novikov and Fateev [10]. Similar work was performed by Pravdin, who drew distinctions between the effects of static and dynamic stresses. The results revealed that dynamic loading changed the flux density $\boldsymbol{B}$ by different amounts depending on the applied field $\boldsymbol{H}$.

The effects of elastic stress on hysteresis have been reported by a number of investigators, including Atherton and Jiles [11] and Burkhardt and Kwun [12]. The interpretation of results is difficult without a working mathematical model of hysteresis, since none of the direct hysteresis parameters such as coercivity, remanence, and hysteresis loss is uniquely related to a single physical property such as stress. It is clear that any model, which has a hope of being used for interpretation of NDE results such as effects of stress, should have as few parameters as is reasonably possible. The stress or strain dependence of these parameters can then be determined. Some success has been achieved with the model of Jiles and Atherton [13]. Changes in the magnetic parameters with stress have been determined empirically by Szpunar and Szpunar [14] and from first principles by Sablik et al. [15], and these have been used to model the magnetic hysteresis properties under stress.

Theoretical work to provide a conceptual framework for these magnetic NDE techniques is essential for interpretation of results. This has been achieved in only limited cases. One example is the development of the theory of the magnetomechanical effect, which is used to interpret the effects of applied stress on the magnetization and magnetic properties of materials.

The magnetomechanical effect can be conveniently divided into two separate but related phenomena: the behavior of a material under constant stress and varying field

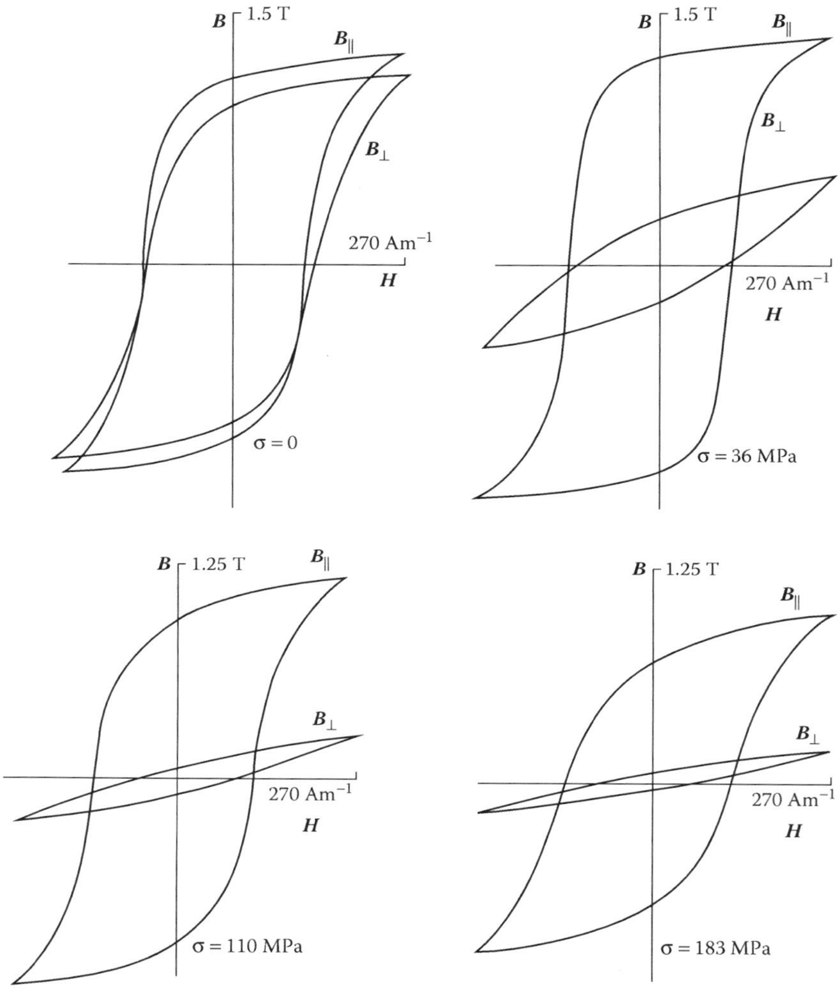

FIGURE 15.2 Changes in magnetic hysteresis *signatures* of mild steel with stress. (After Langman, R., *IEEE Trans. Magn.*, 21, 1314, 1985.)

(the *isostress* case) and the behavior under constant field and varying stress (the *isofield* case). In the isostress case, where a magnetic material is subjected to a uniaxial applied stress, the effect of the stress can be treated in some respects like an effective magnetic field, which changes the anisotropy of the material. This *stress equivalent magnetic field* $\boldsymbol{H}_\sigma$ was shown [15] to be

$$\boldsymbol{H}_\sigma = \frac{3}{2}\frac{\sigma}{\mu_0}\left(\frac{\partial \lambda}{\partial \boldsymbol{M}}\right)_T \tag{15.1}$$

and from this the correction to permeability and hysteresis curves resulting from stress can be determined. Based on a mean field model for the anhysteretic magnetization, Garikepati et al. [16] were able to combine that with Equation 15.1 to derive an equation for the dependence of the reciprocal of the maximum differential susceptibility χ'_{max} on stress

$$\frac{1}{\left[\chi'_{max}(0)\right]_{H=0}} - \frac{1}{\left[\chi'_{max}(\sigma)\right]_{H=0}} = \frac{3b\sigma}{\mu_0} \tag{15.2}$$

where b is a coefficient of proportionality linking magnetostriction λ to magnetization $\boldsymbol{M}$, according to the approximate relation $\lambda = b\boldsymbol{M}^2$. Equation 15.2 was verified experimentally, as shown in Figures 15.3 and 15.4.

Later the model (Equation 15.1) was extended to cover the case of a uniaxial stress at an arbitrary direction to the applied magnetic field by Kaminski, Jiles, and Sablik [17], who developed an expression for the angular dependence of the effective field due to an applied uniaxial stress. This was given by

$$\boldsymbol{H}_\sigma(\theta) = \frac{3}{2}\frac{\sigma}{\mu_0}\left(\cos^2\theta - \nu\sin^2\theta\right)\left(\frac{\partial\lambda}{\partial\boldsymbol{M}}\right)_T \tag{15.3}$$

where:

σ is the stress
θ is the angle between the stress axis and the direction of $\boldsymbol{H}_\sigma$
ν is Poisson's ratio

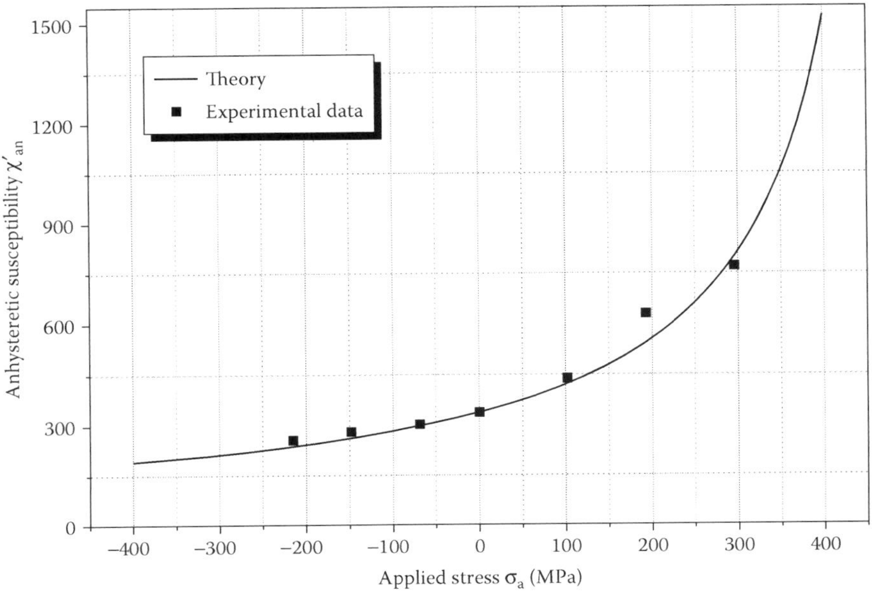

FIGURE 15.3 Anhysteretic susceptibility at the origin as a function of stress. (After Garikepati, P. et al., *IEEE Trans. Magn.*, 24, 2922, 1988.)

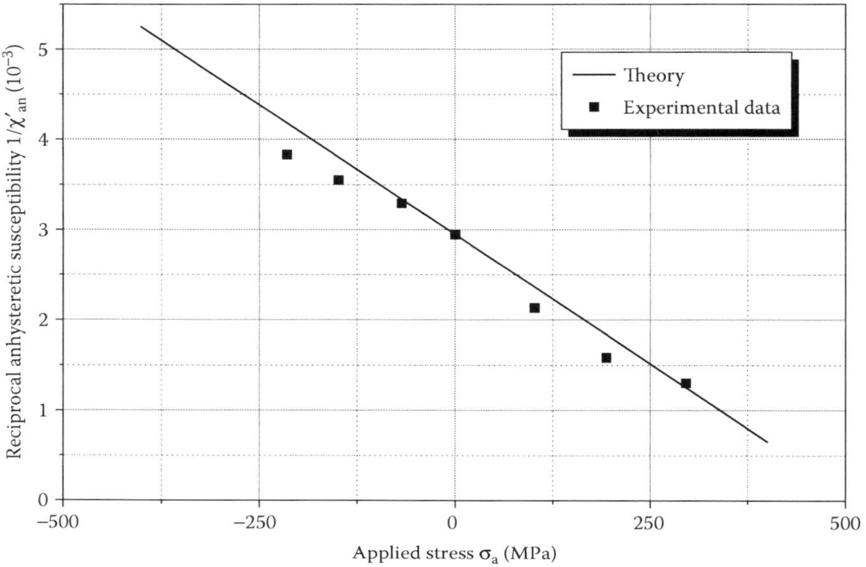

FIGURE 15.4 Reciprocal anhysteretic susceptibility at the origin as a function of applied stress. (After Garikepati, P. et al., *IEEE Trans. Magn.*, 24, 2922, 1988.)

In the isofield case, the situation is more complicated, but Jiles [18] has shown that there are basic principles behind the change in magnetization with stress at constant field, $(\partial \boldsymbol{M}/\partial \sigma)_H$, which can be used to derive the equation governing the mechanism. In this case, the guiding principle is the law of approach of the magnetization toward the anhysteretic when the applied stress is changed. For the reversible component of magnetization, the differential equation is

$$\frac{d\boldsymbol{M}_{\text{rev}}}{dW} = c\left(\frac{d\boldsymbol{M}_{\text{an}}}{dW} - \frac{d\boldsymbol{M}_{\text{irr}}}{dW}\right) \tag{15.4}$$

where:

W is the elastic energy resulting from the applied stress σ, $W = \sigma^2/2E$, that is, the energy that causes the domain walls to break away from their pinning sites

E is the Young's modulus

For the irreversible component of magnetization,

$$\frac{d\boldsymbol{M}_{\text{irr}}}{dW} = \frac{1}{\xi}\left(\boldsymbol{M}_{\text{an}} - \boldsymbol{M}_{\text{irr}}\right) \tag{15.5}$$

where ξ is simply a decay coefficient, which relates the derivative of the irreversible magnetization with respect to elastic energy to the displacement of the irreversible magnetization from the anhysteretic. Here, ξ has dimensions of energy per unit volume.

When these two equations are combined, an expression for the derivative of the total magnetization with respect to elastic energy and hence applied stress can be obtained:

$$\frac{d\boldsymbol{M}}{dW} = \frac{1}{\xi}\left(\boldsymbol{M}_{\text{an}} - \boldsymbol{M}\right) + c\frac{d\boldsymbol{M}_{\text{an}}}{dW} \tag{15.6}$$

Equation 15.6 conveniently expresses the underlying law of the magnetomechanical effect in its simplest form—the *law of approach to the anhysteretic*. The equation can be used to calculate the rate of change of magnetization with stress under a variety of situations. It should be noted that $\boldsymbol{M}_{\text{an}}$ is itself stress dependent, and this stress dependence can be determined by incorporating the addition of an effective field term caused by stress $\boldsymbol{H}_\sigma$ into the equation for the anhysteretic. Therefore, when the magnetization approaches the anhysteretic under the action of an applied stress, it is approaching a moving target.

Changes in magnetic properties with stress cycling have been used to predict fatigue life of specimens by Shah and Bose [19]. It was found that the hardness of the specimens, which can be inferred from coercivity measurements, began to change long before any crack appeared.

15.1.2 Magnetic Barkhausen Effect

The magnetic Barkhausen effect (MBE) [20] consists of discontinuous changes in flux density, known as *Barkhausen jumps*, as shown in Figure 15.5. These are caused mostly by sudden irreversible motion of magnetic domain walls when they break away from pinning sites as a result of changes in the applied magnetic field $\boldsymbol{H}$. The frequency range of Barkhausen emissions for NDE purposes is usually 10–500 kHz, which means that detected emissions originate only from the surface layer. The Barkhausen pulse height distribution, which is the number of events plotted against pulse height, as shown in Figure 15.6, is dependent on the number density and nature

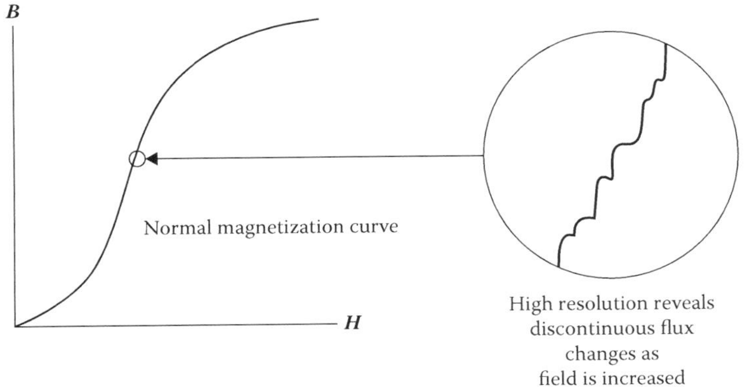

FIGURE 15.5 Discontinuous changes in the flux density $\boldsymbol{B}$ as the magnetic field $\boldsymbol{H}$ is changed.

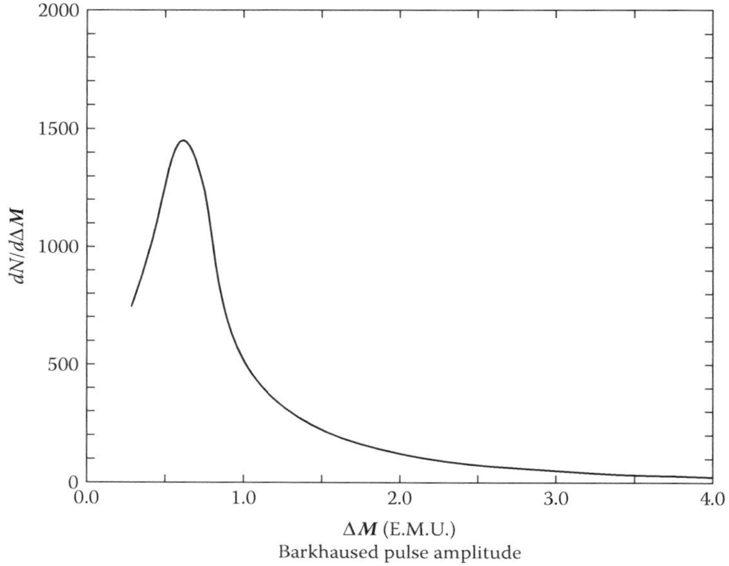

FIGURE 15.6 Magnetic Barkhausen pulse height distribution showing the number of Barkhausen events per unit change in magnetization plotted against the amplitude of the Barkhausen pulses.

of pinning sites within the material. These may be grain boundaries, dislocations, cavities, cracks, or precipitates of a second phase material with different magnetic properties from the matrix material, such as cementite (iron carbide, Fe_3C) in steel.

Most Barkhausen activity occurs close to the coercivity $\boldsymbol{H}_c$, where the slope $d\boldsymbol{M}/d\boldsymbol{H}$ of the magnetization curve reaches a maximum. Therefore, for a fixed rate of change of field $d\boldsymbol{H}/dt$, the rate of change of magnetization with time $[d\boldsymbol{M}/dt = (d\boldsymbol{M}/d\boldsymbol{H})/(d\boldsymbol{H}/dt)]$ reaches a maximum at the coercivity, resulting in more Barkhausen events in a given time interval. Double peaks in the count rate can occur, particularly if the material is composed of two phases each with a different coercivity [21]. Also, the location and size of the peaks can shift as a result of changes in the defect distribution, as shown in Figure 15.7, where results are for two specimens of the same material but with different defect distributions.

Early work showed distinct variations in Barkhausen signal amplitude with applied and residual stress. As stress increased in tension, the peak Barkhausen amplitude in steel was found to increase, while as stress increased in compression, it was found to decrease. Subsequently investigations by Tiitto [22] reported the effects of elastic and plastic strain on the MBE in silicon-iron and the microstructural dependence of MBE in steels. He was able to show that MBE could be used to nondestructively determine grain size in steels.

The effect of tensile cyclic stress loading on MBE signals during fatiguing of mild steel was the subject of an investigation by Karjalainen et al. [23]. They found that residual strains in unloaded specimens could be identified from the root-mean-square of the MBE signals. But changes occurring under cyclic loading (fatigue cycling)

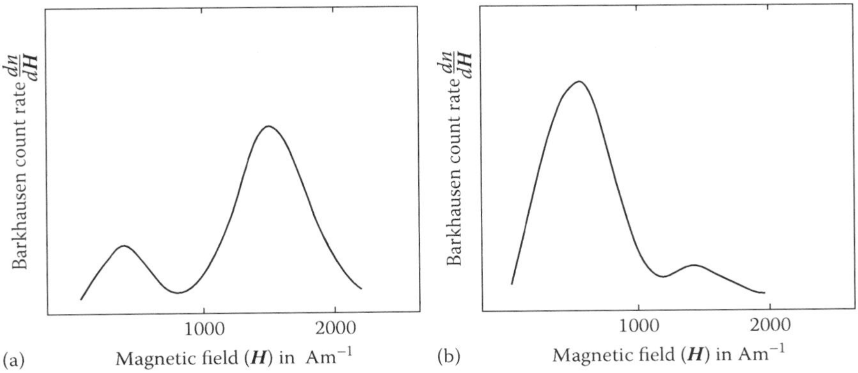

FIGURE 15.7 (a and b) Barkhausen count rate as a function of magnetic field ***H*** for specimens with two different defect distributions.

were very complex, so that the Barkhausen signals could not simply be related to the applied load. However, subsequent investigations by Ruuskanen and Kettunen [24] were able to demonstrate that the median Barkhausen pulse amplitude could be used to assess whether the applied stress amplitude was above or below the fatigue limit.

Lomaev [25] reviewed the literature relating to NDE applications of MBE. He identified five mechanisms by which MBE is caused: (1) discontinuous, irreversible domain-wall motion; (2) discontinuous rotation within a domain; (3) appearance and disappearance of Neel peaks; (4) inversion of magnetization in single-domain particles; and (5) displacement of Bloch or Néel lines in two 180° walls with oppositely directed magnetizations. The first of these mechanisms has been studied most intensively and is often incorrectly quoted as the sole mechanism for generation of MBE. It is interesting to note that, contrary to this, in the early years of MBE, the effect was erroneously attributed largely to the second mechanism, irreversible domain rotation.

Barkhausen emissions that are detected on the surface of a test sample originate only from the surface layer. The Barkhausen effect can therefore be used to probe surface plastic deformation of steel components using measurements at different frequencies that therefore represent properties averaged over different depths. This technique of detecting Barkhausen signals over controlled ranges of frequency can be used for evaluation of different types of surface condition such as case hardening or surface decarburization. A combination of MBE at different frequencies and hysteresis measurements was used by Mayos [26] for the determination of surface decarburization in steels. By this method, different depths of the material were inspected to investigate changes in magnetic properties.

Theiner et al. [27,28] used MBE in conjunction with incremental permeability and ultrasonic measurements for the evaluation of stress. As they noted, ferromagnetic NDE methods are sensitive to both mechanical stress and microstructure of the material. Therefore, in order to determine stress, it is necessary to use two or three independent measurement parameters. They found that Barkhausen effect, incremental permeability, X-ray diffraction, and hardness measurements were successful in estimating residual stress. Changes in the density of dislocations affected the detected MBE signals.

They also found that MBE could be used to distinguish between different microstructures, which cannot be distinguished on the basis of optical microscopy.

Theoretical work by Jiles et al. developed equations for the Barkhausen effect to describe both the stochastic and deterministic components [29,30] and a model for the stress dependence of the differential susceptibility at different angles to the stress [31]. When combining these resulted in a relation (Equation 15.7), in which the reciprocal of the Barkhausen voltage amplitude was linearly proportional to stress [32], then according to the following equation,

$$\frac{1}{V_{\mathrm{MBE}}(\sigma)} = \frac{1}{V_{\mathrm{MBE}}(0)} - \frac{3b'\sigma}{\mu_o} \tag{15.7}$$

where:

$V_{\mathrm{MBE}}(0)$ is the peak Barkhausen voltage under zero stress

$V_{\mathrm{MBE}}(\sigma)$ is the peak Barkhausen voltage under stress σ along the applied field direction, as shown in Figures 15.8 and 15.9

b' is an adjustable empirical model parameter that derives from the earlier work [16] (see Equation 15.1) and links Barkhausen voltage to differential susceptibility, magnetostriction, and magnetization

μ_o is the permeability of free space

This relation has been verified experimentally as shown in Figures 15.10 and 15.11.

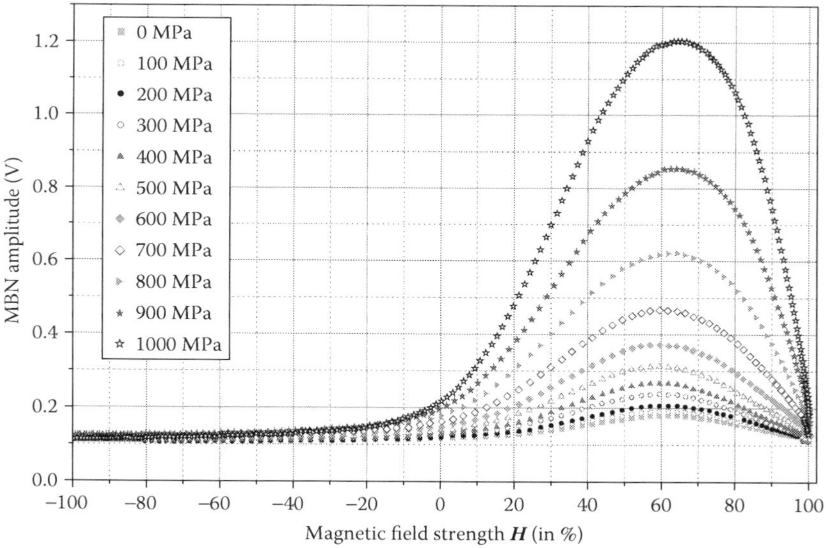

FIGURE 15.8 Barkhausen emission amplitude as a function of magnetic field for a specimen of carburized SAE 9310 steel under different levels of tensile applied stress. (Reproduced with permission from Mierczak, L., Jiles, D. C., and Fantoni, G. *IEEE Trans. Magn.*, 47, 459, 2011. Copyright 2011 IEEE.) The field location of the peak at 62% of maximum field is at the coercive field and this is largely independent of stress while in the elastic stress regime.

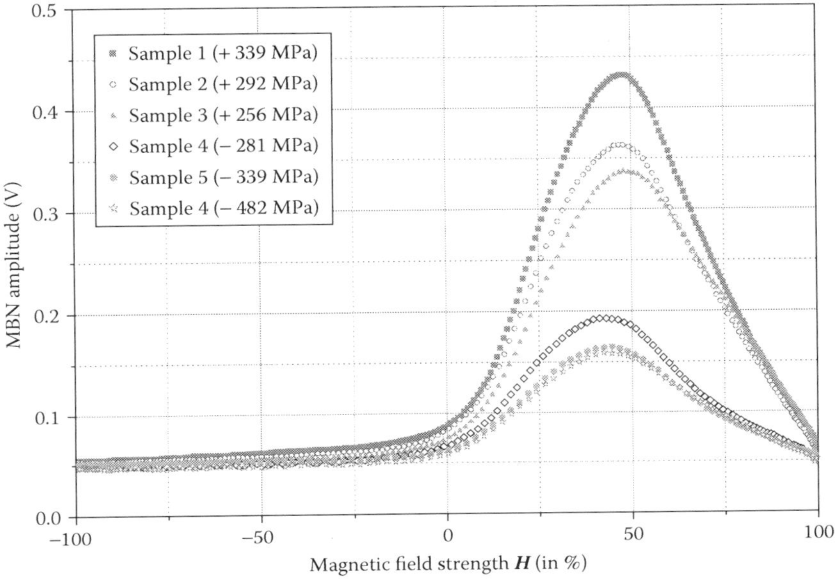

FIGURE 15.9 Barkhausen emission amplitude as a function of magnetic field for a specimen of carburized SAE 9310 steel under different levels of compressive applied stress. (Reproduced with permission from Mierczak, L., Jiles, D. C., and Fantoni, G. *IEEE Trans. Magn.*, 47, 459, 2011. Copyright 2011 IEEE.) The field location of the peak at 48% of maximum field is at the coercive field and this is largely independent of stress while in the elastic stress regime.

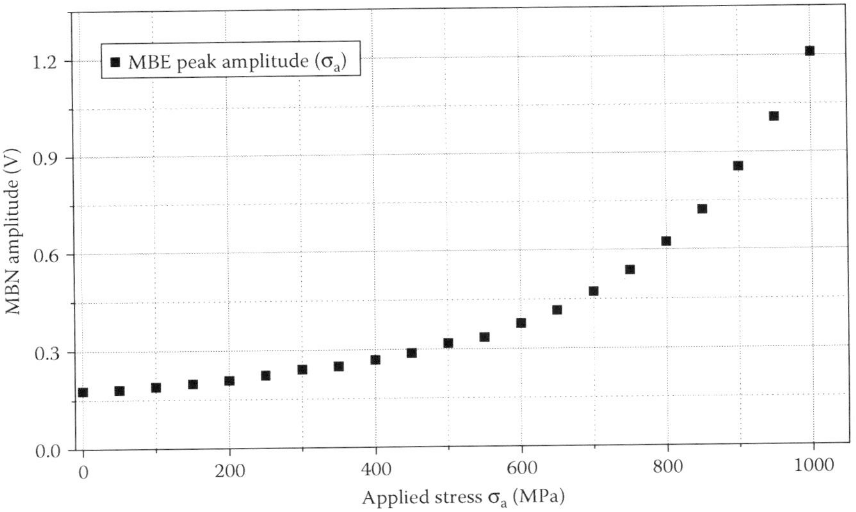

FIGURE 15.10 Magnetic Barkhausen effect (MBE) peak amplitude for carburized SAE 9310 specimen as a function of applied stress. (Reproduced with permission from Mierczak, L., Jiles, D. C., and Fantoni, G. *IEEE Trans. Magn.*, 47, 459, 2011. Copyright 2011 IEEE.)

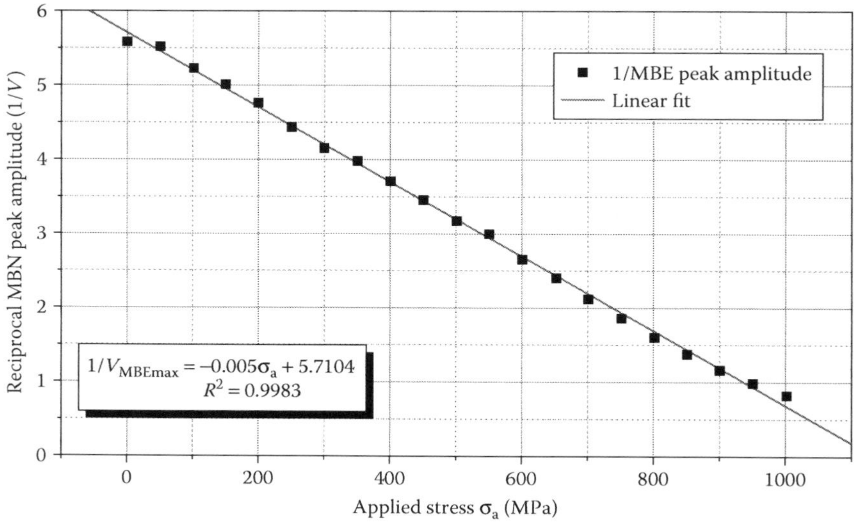

FIGURE 15.11 Reciprocal of MBE peak amplitude for carburized SAE 9310 specimen as a function of applied stress. (Reproduced with permission from Mierczak, L., Jiles, D. C., and Fantoni, G. *IEEE Trans. Magn.*, 47, 459, 2011. Copyright 2011 IEEE.)

Matzkanin, Beissner, and Teller have provided a description of practical NDE applications of the Barkhausen effect in the monograph *The Barkhausen Effect and Its Applications to NDE* [33].

15.1.3 Magnetoacoustic Emission

Magnetoacoustic emission (MAE) is an effect that is very closely related to MBE. It is sometimes referred to as *acoustic Barkhausen*. The effect is caused by microscopic changes in strain due to magnetostriction when discontinuous irreversible domain-wall motion of non-180° domain-walls or non-180° discontinuous domain rotation occurs. The 180° domain-wall motion and 180° domain rotation do not contribute to MAE because neither of these two processes causes any localized change in strain. MAE therefore occurs in ferromagnetic materials with non-180° domain processes when subjected to a time-dependent field as shown in Figure 15.12.

The acoustic emissions may be detected by a piezoelectric transducer bonded onto the test part. Some non-180° domain processes must occur to generate MAE, but the amplitude of MAE also depends on the spontaneous magnetostriction. The amplitude of MAE is zero if $\lambda_0 = 0$ and increases with λ_0. The amplitude of emissions is also a function of the frequency and amplitude of the driving field. In materials with nonzero spontaneous magnetostriction, MAE changes with applied stress, since stress alters the magnetocrystalline anisotropy through the magnetostriction, as shown in Equation 15.3. This results in a change in the relative numbers of 180°

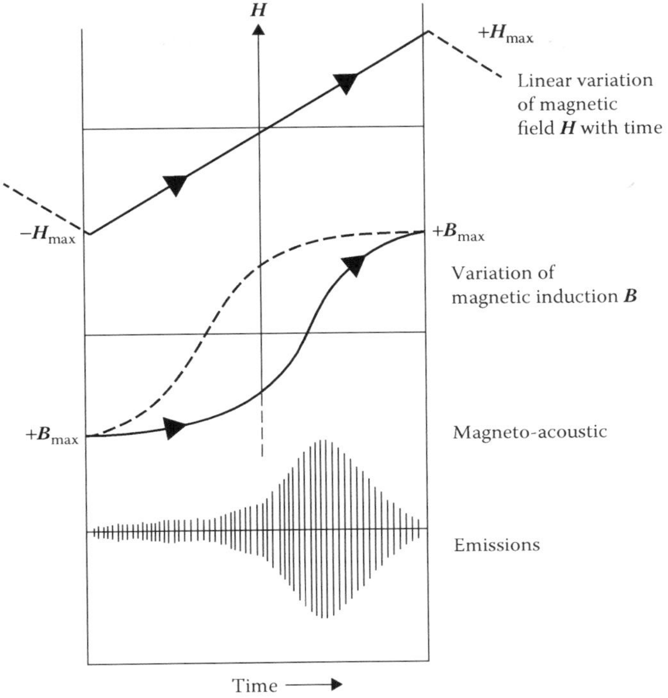

FIGURE 15.12 A schematic showing magnetic field H with time, variation in flux density over the same period, and the emergence of magneto-acoustic emission pulses as flux density changes.

and non-180° domain walls. Since 180° domain walls do not contribute to MAE, the amplitude of emissions and the total number of emissions changes with stress.

Despite its close affinity with the MBE, MAE has a much shorter history. It was first reported by Lord [34] during magnetization of nickel. Its significance for NDE was realized by Kusanagi, Kimura, and Sasaki [35], who were first to demonstrate the effect of stress on MAE. Shortly afterward, Ono and Shibata [36] reported MAE results on a number of steels. Their results indicated that the method could be used to determine the amount of prior cold work and differences in heat treatment.

Burkhardt et al. [37] have also investigated the dependence of MAE on the mechanical and thermal treatment of steels. They found that MAE was sensitive to the amount of plastic deformation. Theiner and Willems [38] used MAE in conjunction with other independent measurements such as incremental permeability, MBE, and magnetostriction. Their results showed that the MAE amplitude decreased with the mechanical hardness of steels but increased with tempering. This is consistent with the change in amplitude of Barkhausen emission, which also decreases with increasing hardness. The reason for this is that harder steels have more impediments to domain-wall motion and consequently there is less domain-wall motion in harder materials.

Edwards and Palmer [39] showed that MAE signals are affected not only by stress and frequency of field but also by factors such as sample shape. Ranjan, Jiles, and Rastogi [40] used MAE and MBE for the determination of grain size in decarburized steels. They used two types of measurement: the MAE peak height and the total number of emissions, both of which were found to increase with grain size. This is due to the fact that in larger grain materials, there are fewer impediments to domain-wall motion and therefore both Barkhausen and MAE emissions increase with grain size.

A related phenomenon has been reported by Higgins and Carpenter [41]. Acoustic and magnetic Barkhausen emissions due to domain-wall motion were observed in ferromagnetic materials when the applied stress was changed without changing the magnetic field. This can be understood in terms of the domain-wall pinning model. These emissions were also observed under dynamic stress by Jiles and Atherton during investigations of magnetomechanical effects [42]. This phenomenon of magnetomechanical emissions has received little attention, but would appear to have significant implications for the detection of dynamic stresses in magnetic materials.

15.1.4 Residual Field and Remanent Magnetization

This technique is closely related to the flux leakage method (see Section 15.2.3). However, while the flux leakage method is used to detect flaws from anomalies in magnetic flux, the residual field method is usually aimed at detecting changes in intrinsic properties such as strain, microstructure, and heat treatment from variations in the magnetic field close to the surface of a ferromagnetic structure or component as shown in Figure 15.13.

The magnetometers used with the technique are often fluxgates (also known as *ferroprobes*), but Hall probes and induction coils are also used and in cases where the coercivity needs to be measured, a coercimeter is used. A coercimeter magnetizes the specimens to saturation in one direction, using an electromagnet, and then determines $\boldsymbol{H}_c$ by reversing the field until the flux density in the specimen is

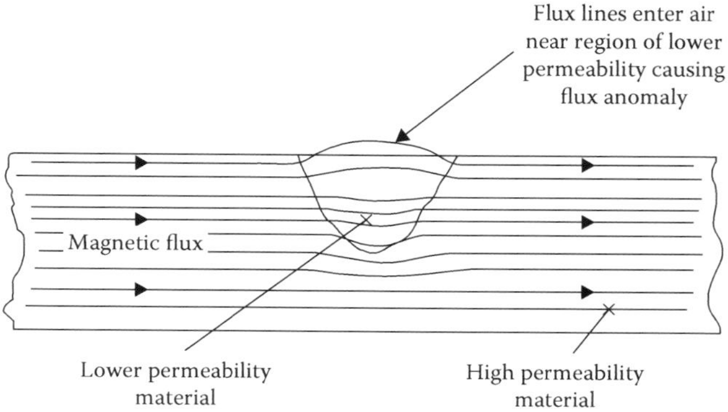

FIGURE 15.13 Flux leakage into the air caused by a region of lower permeability within a piece of steel.

reduced to zero. Konovalov, Golovko, and Roitman [43] determined the mechanical properties of steel pipes from a combination of coercivity and remanence obtained from measurements of field intensity close to the surface of pipes.

Suzuki, Komura, and Takahashi [44] used measurements of remanent magnetization for detection of stresses in pressure vessels. A detailed discussion of the variation of magnetic field and flux density with stress has been given by Langman, who detected and measured stress levels in steel plates by a novel method of rotating the magnetic field H and noted differences caused by stress-induced anisotropy [45]. A number of techniques for sorting steel components on the basis of magnetic field measurements, usually from the determination of coercivity, have been reported by Mikheev et al. [46].

15.2 METHODS FOR DETECTION OF FLAWS AND OTHER INHOMOGENEITIES

This section reviews magnetic detection and characterization of defects or flaws in magnetic materials. The objective is to provide a broad survey of the existing techniques, giving a summary of earlier work but without going into fine details, which have been covered in most cases by more specialized reviews of the particular techniques.

Systematic development of testing techniques based on perturbations of the magnetic flux in iron and steel due to the presence of defects began with the chance discovery by Hoke that iron filings accumulated close to defects in hard steels while in the process of being ground. The technique of magnetic particle inspection (MPI) based on this discovery was then developed by DeForest and Doane, who formed the Magnaflux company to exploit this method.

As the subject of flaw detection became more quantitative, related methods were developed in which the leakage field in the vicinity of the flaw was measured with a magnetometer. Once the field strengths of the leakage fields were being measured on a routine basis, it became desirable to relate these to flaw size and shape, and therefore, there arose the need for modeling the leakage fields resulting from different crack geometries. This is now achieved routinely by numerical techniques such as finite-element modeling.

15.2.1 MAGNETIC PARTICLE INSPECTION

The technique of MPI was the first magnetic NDE method in widespread use. It was discovered accidentally by Hoke in 1918, but it was left to DeForest to develop the method further for practical use. DeForest's work involved devising methods of generating a magnetic field of sufficient strength in any direction in a specimen. This was solved using electrical contact prods with heavy-duty cables being used to pass large currents through test specimens in desired directions. Furthermore, it became apparent that better results were obtained using magnetic powders with uniform properties such as particle shape, size, and saturation magnetization in order to obtain more reliable results. Particle sizes with diameters ranging from 0.3 up to 300 μm are normally used.

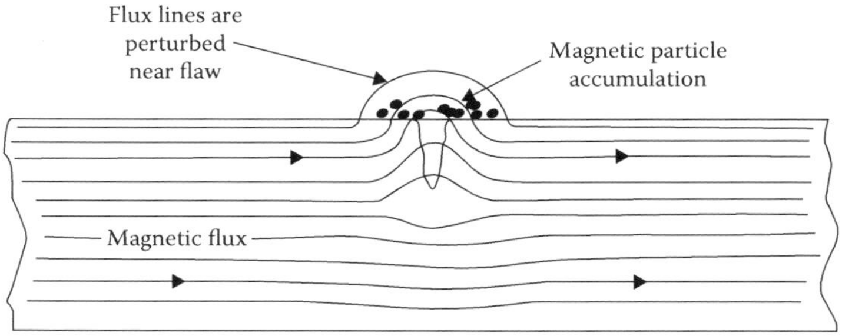

FIGURE 15.14 A schematic of leakage flux in the vicinity of a surface-breaking flaw with accumulation of magnetic particles. In the presence of narrow cracks, the particles can form a *bridge* across the crack. In the presence of wide cracks, the particles accumulate on both sides of the crack.

The MPI method is very simple in principle. It depends on the leakage of magnetic flux at the surface of a ferromagnetic material in the vicinity of surface-breaking, or near-surface flaws as shown in Figure 15.14. There are five main methods for generating a magnetic field in a material. Three of these methods depend on the generation of a magnetic field in the test specimen without necessarily inducing a current. These are the yoke method, the encircling coil method, and the internal conductor method (also known as the *threading bar* method) shown in Figure 15.15. Two other methods depend exclusively on the generation of high current densities in the test specimen, and these are the *prod* method and the current induction method shown in Figure 15.16. In each case, the best indication is given when the magnetic field is perpendicular to the largest dimension of the flaw or crack.

MPI is a very reliable method for finding surface flaws and gives a direct indication of the location and length of the flaw. There is little or no limitation on the size or shape of component being tested, although more care is needed in the application of the method to complex geometries. Nevertheless, the method does have some distinct limitations. It can of course only be used on ferromagnetic materials, and in addition, the magnetic field must lie at a large angle to the direction of the flaw to give the best indication. An angle of 90° gives optimum performance, but good indications can be obtained with angles as low as 30°. Flaws can be overlooked if the angle is smaller. Finally, although the length of the flaw is easily found, the depth of the flaw is difficult to ascertain.

Various enhancements have been made to the original dry powder method. These include the use of water-borne and oil-borne suspensions or *magnetic inks*, known as the *wet method*, and especially fluorescent powders or suspensions, which give a clearer indication of small flaws when viewed under ultraviolet light.

Another related method that has found use in detection of flaws in structural components is the *replica* or magnetographic method in which a magnetic tape is placed over the area to be inspected. The tape is magnetized by the strong surface field in the vicinity of the flaw, and is then removed and inspected for magnetic anomalies. The magnetic tape records an imprint of the flux leakage from the surfaces of components and is inspected afterward using magnetometers such as Hall probes

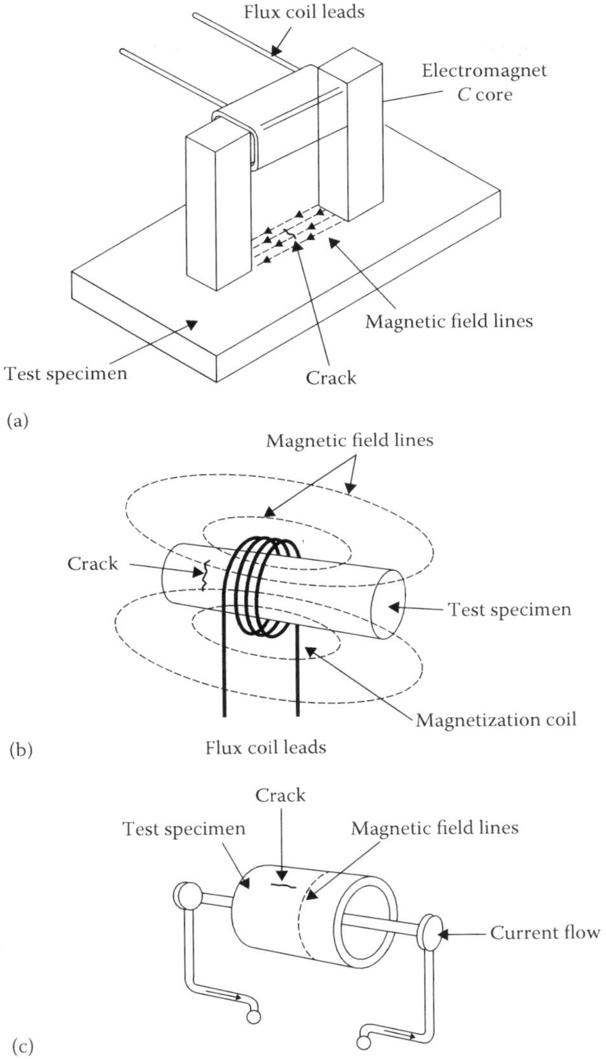

FIGURE 15.15 Magnetization methods used in magnetic particle inspection based on generation of fields in the specimen with (e.g., ac fields) or without (e.g., dc fields) generation of associated currents in the specimen: (a) the magnetic yoke method; (b) the coil magnetization method; and (c) the internal conductor method.

or fluxgates. The advantage of this method over the standard MPI technique is that the magnetograph (usually a magnetic tape) can be used in difficult locations such as underwater inspection of oil rigs. Also, the magnetic imprint on the replica can be read using a magnetometer to obtain a quantitative measure of flux leakage from regions where it would be difficult to use a magnetometer directly.

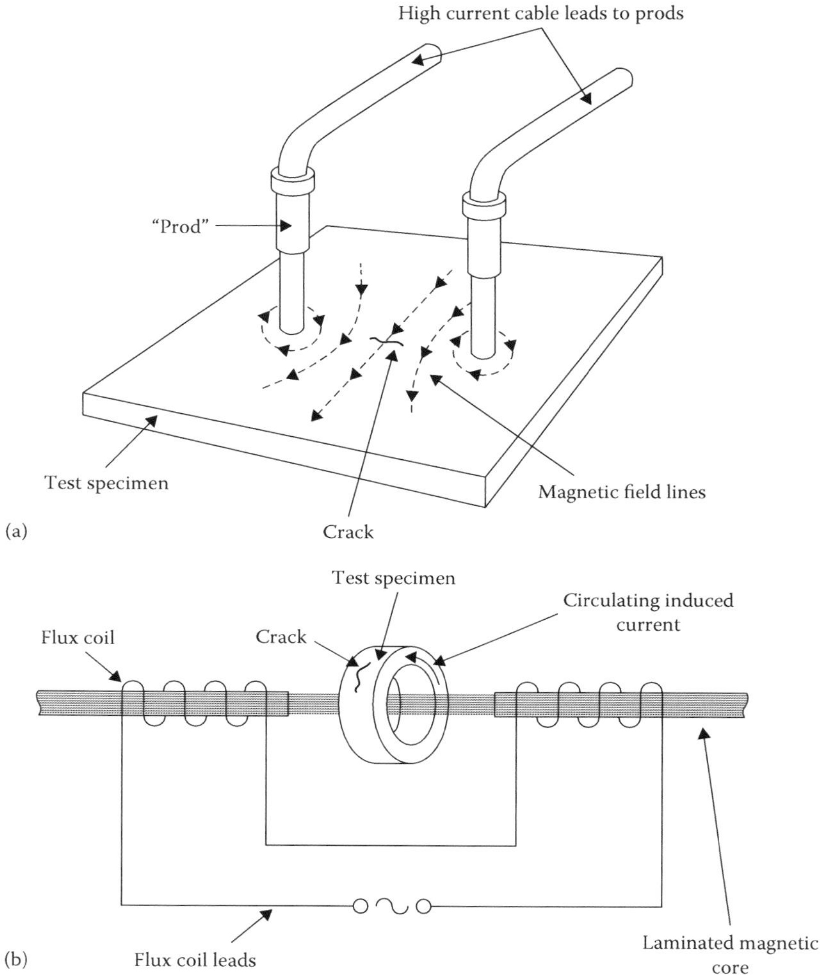

FIGURE 15.16 Magnetization methods used in magnetic particle inspection based on generation of fields in the specimen via generation of associated currents in the specimen: (a) the current flow or *prod* method and (b) the current induction method.

15.2.2 Applications of the Magnetic Particle Method

Much effort in MPI has been directed toward establishing standard procedures and conditions for applying the method. For complex geometries, the magnetic particle method is difficult to apply [47] and has sometimes failed to reveal structural faults in aerospace components because of low magnetization in some regions. More recent developments using multicircuit magnetization methods have now overcome this problem. Optimum conditions for the application of MPI to inspection of welds were described by Massa [48]. Once again the critical factor was to determine adequate levels of magnetization in order to reveal the presence of defects. No simple optimum procedure was found in that case but tables of values of critical field strength $\boldsymbol{H}$ for various geometries were given.

Other efforts in MPI have been directed toward modeling of the magnetic field throughout the whole of an object including the leakage fields in the vicinity of flaws. These calculations use Maxwell's equations and finite-element methods of computation [49–51] and are discussed in Section 15.2.5. The computer-aided design of MPI systems, especially multicircuit units for inspection of automotive parts, is now widely used.

Some standard procedures have been recommended in the United Kingdom, where a field strength of at least 2400 A m^{-1} (30 Oe) was suggested for using MPI on steels according to British Standard 6072. However, opinions differ over the necessity of the recommendation. Work on MPI applications to pressure vessels and pipelines has shown that the recommendations of British Standard 6072 do not seem to be generally applicable. Studies indicated that field strengths lower than 2400 A m^{-1} were quite adequate for satisfactory MPI indications. Edwards and Palmer [52] have investigated procedures for applying the method to tubular specimens threaded on a current-carrying conductor and for a cylindrical bar using prod magnetization. It was shown that care is needed to generate sufficient field strength for adequate magnetization of the specimen in these cases.

It is sometimes incorrectly assumed that the optimum magnetizing condition corresponds to the maximum permeability; however, this was not found to be true in the work of Oehl and Swartzendruber [53]. They found that for cylindrical defects with square cross-sections, the ratio of leakage field to applied field reached a maximum at an applied field of between 800 and 2400 A m^{-1} in steels, depending on the lift-off. This was well removed from the maximum permeability of the material, which occurred at $\boldsymbol{H} = 120$ A m^{-1}.

Although it is possible to apply the MPI technique successfully to magnetically harder materials using solely the remanent magnetization of a specimen and its accompanying field, it is not as easy to control the relative orientation of field and flaw. Therefore, it is preferable to apply a controlled magnetic field to the specimen, using one of the five methods discussed above. Enhancements of the MPI method include fully automated scanning of steel components for crack indications using optoelectronic devices, followed by computer-operated digital image processing techniques to enhance the results [54,55]. The automation of the inspection method has the advantage of eliminating subjective evaluation of results and is therefore desirable.

The *wet method* in which a magnetic colloid (a ferrofluid), or a suspension of larger magnetic particles in a carrier fluid, is used has many similarities with the Bitter pattern technique for magnetic domain observations described in Section 6.1.7. Because of the finer particles used in the wet method compared with the dry method, it gives improvement in spatial resolution over the dry powder method, and therefore, it can be successfully used for detection of *smaller* flaws. However, in the case of larger castings with relatively wide cracks, the coarse-grained dry powder method remains the appropriate technique. Other developments in the MPI technique have focused on improving measurement techniques and control procedures for current, field strength, and light intensity in order to enhance the capabilities of fully automated inspection systems. Expert systems have been developed to further improve the automation of measurements and to develop evaluation algorithms.

MPI is therefore a well-established technique and many of the problems associated with its use appear to have been solved. The remaining research efforts are directed toward optimizing conditions for its use by computer modeling and simulation, and increasing automation. The standard reference work on MPI is *Principles of Magnetic Particle Testing* by Betz [56], which includes the history of its development; the underlying principles; methods of generating fields in the materials; and descriptions of the dry powder methods, the wet, or magnetic ink, method, and the fluorescent powder method. A more recent reference work on MPI is to be found in Volume 6 of the NDT Handbook [57].

15.2.3 Magnetic Flux Leakage

The magnetic flux leakage method is derived from the MPI method. Both depend on the perturbation of magnetic flux in a ferromagnetic material caused by surface or near-surface flaws. While in the MPI method, detection relies on the accumulation of magnetic powder, or sometimes the use of a magnetic recording tape to indicate the presence of a defect, the flux leakage technique utilizes a magnetometer.

As with the MPI method, the flux leakage signal depends on the level of magnetization within the material. The advantage of the magnetic flux leakage method is that the use of a magnetometer allows a quantitative measurement of the leakage field in the vicinity of a flaw to be obtained. The disadvantage compared to the MPI method is that it takes a much longer time to scan the field around a component than it does to dust it with magnetic particles.

The flux leakage method is therefore better suited to situations where the location of flaws is known or where the location can be predicted with a reasonable chance of success. Under those circumstances, a careful magnetic inspection can be conducted over a confined area. Another situation where flux leakage magnetometry has advantages over the particle method is where the part to be tested is not easily accessible for a visual inspection, for example, the inside surface of a long tube such as a pipeline.

The field components in three directions, perpendicular and parallel to the flaw on the surface and normal to the surface, can be measured although often only components parallel to the surface are usually measured. The method only began to gain wide acceptance after the design of a practical flux leakage measuring system by Hastings [58]. This was capable of detecting surface and subsurface flaws on the inner diameter of steel tubes, a location that was unsuitable for the standard MPI method. The magnetometer probe is most often a Hall probe or induction coil, although high-resolution magnetometers such as SQUIDs can be used, as described in Section 15.3.2. The magnetometer is scanned across the surface of the component looking for anomalies in the flux density $\boldsymbol{B}$, which indicate the location of a flaw, as in Figure 15.17a, or the location of a region of different permeability, as in

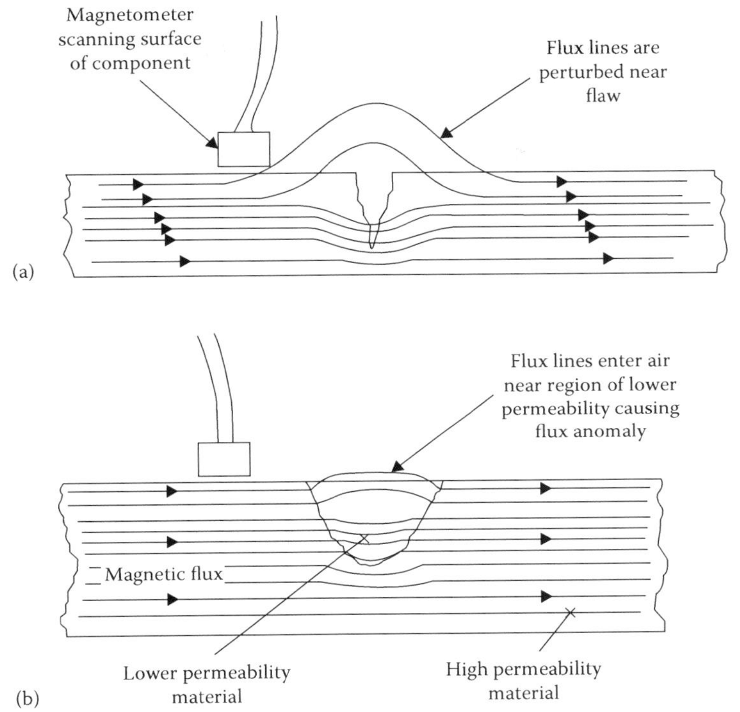

FIGURE 15.17 Scanning the surface of a specimen with a magnetometer (a) to detect flaws and (b) to detect regions of low permeability by the magnetic flux leakage method.

Figure 15.17b. The leakage flux as a function of distance across a crack is shown in Figure 15.18, together with the search coil output as it is moved across the crack.

The magnetic flux leakage magnetometry has been an important and highly successful technique, both for detection of flaws and for stresses. Work on detection of flux anomalies in pipelines indicated that these could be related to stresses within the pipe wall. The method has been particularly useful for the inspection of defects in pipelines as in the work of Atherton [59].

A review of the flux leakage method has been given by Forster [60], which discusses both the experimental and theoretical foundations of the method. The leakage fields of course give rise to demagnetizing fields in a way that is similar to that discussed in Chapter 2, Section 2.4.3. The demagnetizing fields $\boldsymbol{H}_{\mathrm{d}}$ of defects were calculated for ellipsoidal flaws inside the material, and consequently the *true* or *internal* field $\boldsymbol{H}_{\mathrm{in}}$ in the defect was calculated from the measured applied field $\boldsymbol{H}_{\mathrm{app}}$ using the equation

$$\boldsymbol{H}_{\mathrm{in}} = \boldsymbol{H}_{\mathrm{app}} - \boldsymbol{H}_{\mathrm{d}} \tag{15.8}$$

This procedure is equivalent to assuming that the defect behaves like a simple magnetic dipole. Such an assumption is only valid as a first approximation. If the

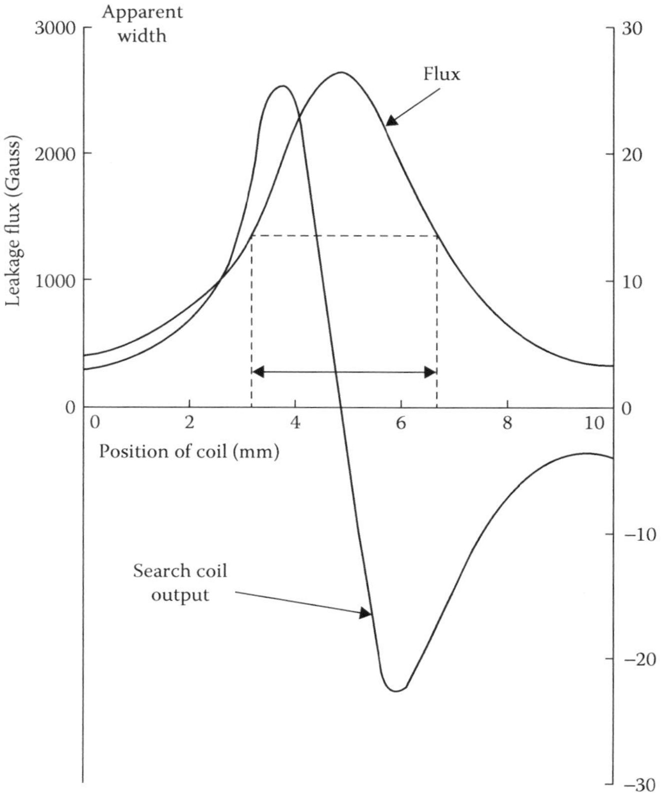

FIGURE 15.18 Variation of flux leakage with distance across a crack.

permeability of the material is μ_{iron} and the permeability of the flaw is μ_{flaw} and the demagnetizing factor, which depends on the shape of the flaw, is N_d, then the internal field is [60]

$$\boldsymbol{H}_{in} = \boldsymbol{H}_{app}\left\{\frac{\mu_{iron}}{\mu_{iron} + N_d\left(\mu_{flaw} - \mu_{iron}\right)}\right\} \tag{15.9}$$

One of the difficulties here is that in general the magnetic induction $\boldsymbol{B}$ inside a ferromagnet is not known as a function of field $\boldsymbol{H}$, and hence μ_{iron} is not known as a function $\boldsymbol{H}$.

Experimental results on the variation of leakage flux with position above the crack and also as a function of depth within the crack are shown in Figure 15.19. These show that as the distance above the crack increases, the leakage field becomes less sensitive to crack depths, particularly at distances above 1 mm for crack depths of 2.5 mm with varying widths of 0.2 to 2 mm.

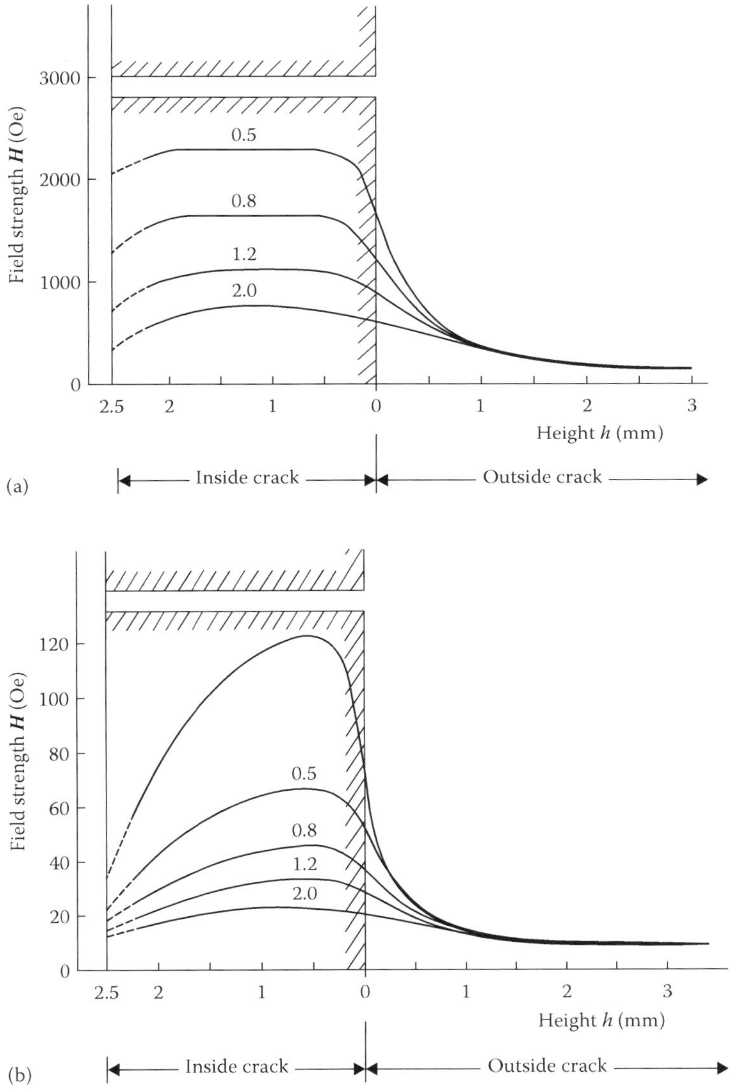

FIGURE 15.19 Variation of leakage flux with position within and above a crack for various crack widths: (a) for a carbon steel and (b) for a chromium steel. (After Forster, F., *Mater. Eval.*, 43, 1154, 1985.)

15.2.4 Applications of the Flux Leakage Method

Applications of the flux leakage method are mostly concerned with large-scale structures. Stumm [61] has described several devices based on the flux leakage technique, which can be used for testing of ferromagnetic tubes for flaws. One of the problems identified with these instruments was the difference in defect signal between an external (near side) flaw and an internal (far side flaw). The signal from an internal flaw was much smaller than the signal from a comparable external flaw, as expected because of the differences in distance between the field sensor and the flaw. The difficulty arises in deciding whether a given signal is due to a small external flaw or a much larger internal flaw. In other NDT methods such as ultrasonics, a distance amplitude correction curve is used [62] to normalize detected signals for different distances. This is much more difficult in flux leakage because unlike in ultrasound, there is no time measurement to indicate how far the defect is from the sensor.

A device called the *rotomat* enabled a tube to be passed through a rotating magnetizing yoke, which generated a circumferential magnetic field for detecting longitudinal defects in the tube wall. The leakage flux was measured using an array of Hall probes. Tube diameters of 20–450 mm could be inspected. A second device, the *tubomat*, rotated the tube while maintaining the detection system stationary. The detection and field generation systems were identical to those used in the rotomat system. The method of magnetization was, however, dependent upon the size of the pipe. For large pipes, the pipe was threaded with a central conductor, which carried the high current required to obtain the optimum field strength in the pipe wall. For smaller pipes, the magnetic yoke method was employed. A third and somewhat different system was the *discomat*, which was used primarily for weld inspection. The detection system containing five Hall probes was rotated on a disk at typically 50 rev s^{-1} and the tube was magnetized transversely to the direction of the weld using a magnetizing yoke.

Owston [63] has reported on measurement of leakage flux from fatigue cracks and artificial flaws such as saw slots in mild steel. These he attempted to interpret in terms of a simple dipole model. The magnetic leakage field parallel to the surface and perpendicular to the cracks was measured as a function of lift-off (i.e., distance from the field detector to the surface of the test material). Results indicated that the leakage field increased linearly with slot depths of up to 0.2 in, while the derivative $d^2\boldsymbol{H}/dz^2$, where $\boldsymbol{H}$ is the magnetic field and z is the distance from the center of the slot measured in the surface of the specimen, was found to be proportional to $1/l^4$, where l was the lift-off distance. Barton, Lankford, and Hampton [64] investigated the flux leakage method for testing of bearings, which revealed that magnetic signatures associated with pits, voids, and inclusions could be identified in the leakage fields of bearing races.

Beissner, Matzkanin, and Teller have provided a review of the applications of magnetic flux leakage as an NDE tool in the monograph *NDE Applications of Magnetic Leakage Field Methods* [65].

15.2.5 Leakage Field Calculations

Once quantitative measurements of leakage fields became a routine procedure, the next step was to interpret the signals in terms of flaw size and shape. Therefore, leakage field calculations for specific flaw shapes were made so that the theoretical leakage field profiles expected on the basis of particular flaw characteristics such as flaw length, width, depth, and orientation could be compared with experimental observations with the objective of characterizing the flaws on the basis of their leakage fields. The early calculations were based on analytic model equations. The advantages of an analytical expression are that the fields due to defects can be rapidly calculated, and in addition, the equations can be differentiated to find the forces on magnetic particles.

Two of the landmark works in the development of leakage field calculations were by Schcherbinin and Zatsepin [66,67]. These calculations were based on approximating surface defects by magnetic dipoles and calculating the associated dipole magnetic fields. In this way, expressions were obtained for the leakage fields associated with flaws including the field component normal to the surface of the part $\boldsymbol{H}_y$ and the field component parallel to the surface of the part but perpendicular to the flaw $\boldsymbol{H}_x$. They realized that exact calculation of leakage fields arising from real defects with complicated geometries presented an extremely difficult problem, which was intractable given the numerical methods and the state of computer technology at the time. They therefore looked for a simplification whereby the leakage field of flaws could be approximated as a point, line, or strip dipole. They then calculated the magnetic fields using analytical dipole expressions. The results were compared with observations of leakage fields due to real defects as shown in Figure 15.20. Notice that the

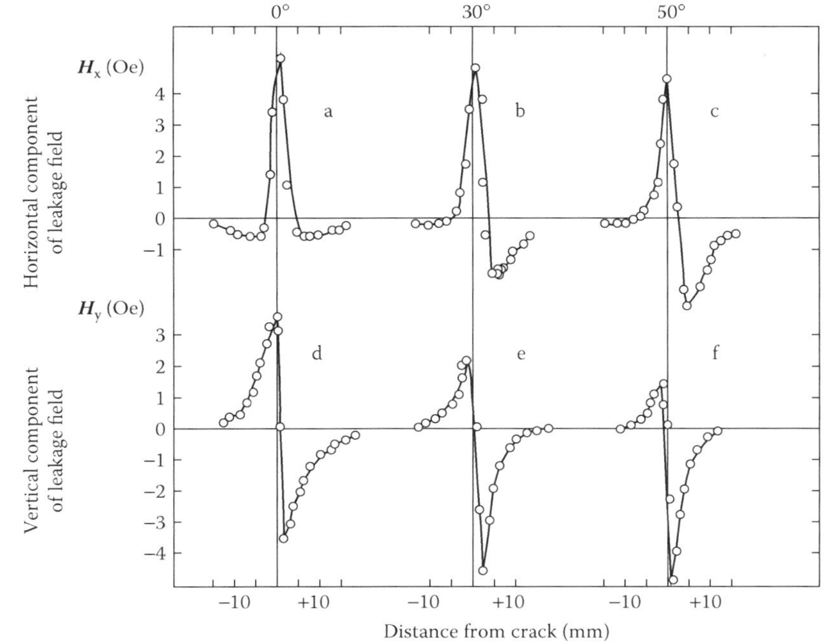

FIGURE 15.20 Calculated leakage field due to flat-bottomed flaw using the dipole model: (a), (b), and (c) are horizontal field components and (d), (e), and (f) are vertical field components. The spot is of length 4 mm, width 0.4 mm, and depth 2 mm. (After Zatsepin, N. N., and Schcherbinin, V. E., *Sov. J. NDT*, 2, 50, 1966; Schcherbinin, V. E., and Zatsepin, N. N., *Sov. J. NDT*, 2, 59, 1966.)

horizontal component of field across the flaw has a unipolar profile, while the normal component has a bipolar profile.

Subsequent works on leakage field calculations were based on the same model. Experimental measurements were made on specimens with artificial cracks machined in the surfaces. These cracks were typically 0.2 to 3.0 mm deep, 1 mm wide, and 1 to 30 mm long. The magnetic fields were generated using the magnetic yoke method, with a maximum magnetic field of 12 kA m^{-1}. Magnetic field components tangential to the surface of the specimen and perpendicular to the flaw, $\boldsymbol{H}_x$, and magnetic field components normal to the surface of the specimen, $\boldsymbol{H}_y$, were measured.

It was found that for a given crack size the relationship between the field component $\boldsymbol{H}_y$ normal to the surface and the field $\mathbf{H}_0$ inside the crack remained almost linear. However, as the crack size increased, the ratio $\boldsymbol{H}_y/\boldsymbol{H}_0$ increased and appeared to reach a saturation level for very large crack lengths. The tangential field $\boldsymbol{H}_x$ perpendicular to the crack was also measured. For a given crack size, it was found to decay with displacement from the center of the flaw. $\boldsymbol{H}_x$ was also found to increase with crack length, although this variation was dependent upon the magnetizing field $\boldsymbol{H}_0$.

Another analytical solution for the leakage field of surface-breaking cracks was shown by Edwards and Palmer [68], in which the crack was approximated as a semi-elliptic slot in the material. From these results, Edwards and Palmer calculated the magnetic field strength $\boldsymbol{H}$ needed to detect 1 and 10 μm wide artificial cracks for a range of crack depths and different permeabilities of the material. They concluded that fields in the range of 10^2–10^3 A m^{-1} are required for crack detection. This is in agreement with those used in practice, for example, BS 6072, which recommends a magnetic induction of 0.72 T, corresponding to fields of 5700 A m^{-1} in a steel of relative permeability 100.

Important progress in the calculation of leakage fields using numerical methods, as distinct from analytical methods, was made by Hwang and Lord [69] using finite-element methods. This was the first attempt to use numerical methods to calculate the magnetic fields around defects. It represented a breakthrough in the development of the subject because it enabled the leakage fields to be calculated from the existing field and permeability in the bulk of the material. Some results of their calculations are shown in Figure 15.21. The leakage field profiles obtained for the case of a simple rectangular slot were in excellent agreement with observation. This work was therefore instrumental in demonstrating how the finite-element technique could be used for modeling magnetic fields of defects.

It was clear from their work that the finite-element method was sufficiently flexible, so that its successful application in the case of a simple slot defect indicated its likely successful application in the case of complex defect geometries, an extension to which the analytical model Schcherbinin and Zatsepin was not capable. Lord and Hwang [70] extended the application of the finite-element method to a variety of more complicated flaw shapes. It was deduced from this that the finite-element method provided the possibility of defect characterization from the leakage field profile. It was found, for example, that the peak-to-peak value of the leakage flux density $\boldsymbol{B}_N$ increased with increasing flaw depth, while the separation between the peaks in $\boldsymbol{B}_N$ depended on flaw width. These results were in agreement

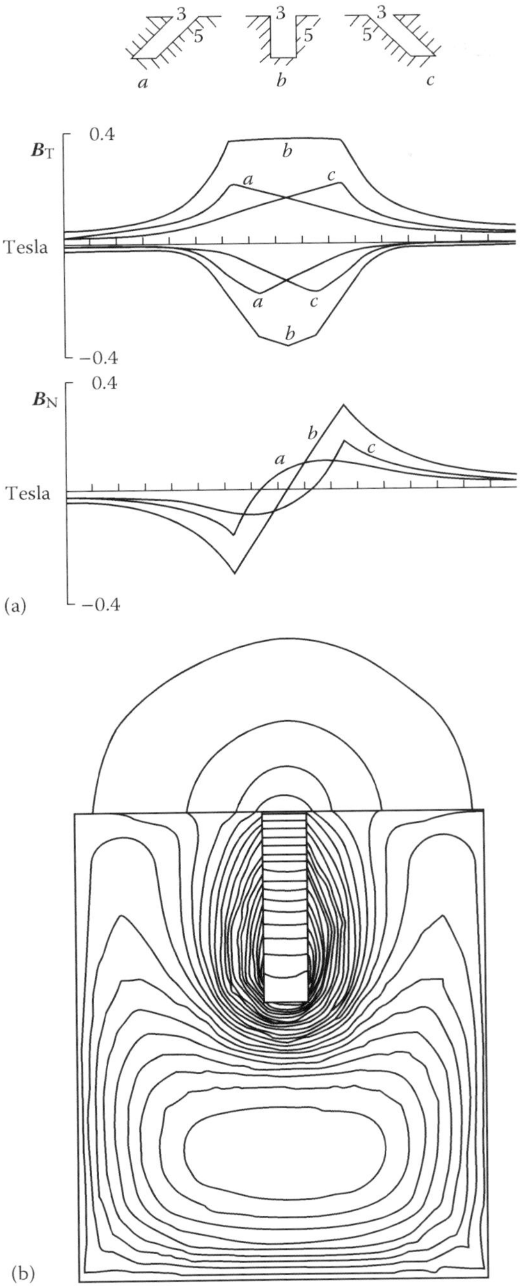

FIGURE 15.21 Calculated leakage field due to a flaw using finite-element techniques. (After Hwang, J. H., and Lord, W., *J. Test. Eval.*, 3, 21, 1975.)

with experimental observations and indicated the usefulness of finite-element techniques for interpreting leakage field measurements with the objective of characterizing the defects.

As shown by Lord et al. [71], the key to the application of magnetic flux leakage techniques to NDT is the development of an adequate mathematical model for magnetic field/defect interactions. However, there is a complication because the modeling of the $\boldsymbol{B}$, $\boldsymbol{H}$, or hysteresis characteristics in ferromagnetic materials adds another dimension to this problem, since even if the internal magnetic field $\boldsymbol{H}_0$ is known there is still a range of possible values of the flux density $\boldsymbol{B}$ inside the material, which arises from field exposure and hysteresis. These different possible values of $\boldsymbol{B}$ will certainly give different values of leakage flux density even for the same type of defect with the same magnetizing field $\boldsymbol{H}_0$ inside the material. They concluded that the complex defect geometries, together with the nonlinear magnetization characteristics of ferromagnetic steels, made closed form or analytical solutions of magnetic field defect interactions virtually impossible. However, they were able to show a number of examples of successful application of numerical finite-element modeling and to conclude optimistically about the use of finite element methods for defect characterization in NDT.

Lord [72] gave a review of the application of the finite-element numerical method to calculation of leakage fields arising from magnetic field-defect interactions. In this, he showed that progress has been made in theoretical modeling of leakage field signatures using numerical methods, but that there are many complex problems still to be solved. One of the most significant of which is taking into account the hysteresis in $\boldsymbol{B}$ and $\boldsymbol{H}$ characteristics of ferromagnetic materials before applying the finite-element, or other numerical calculation of the leakage field. However, the work of Lord et al. firmly established the use of finite-element calculations, as a technique for characterization of defects from leakage flux measurements.

Reviews of the subject of magnetic leakage field calculations and the interpretation of experimental measurements have been given by Holler and Dobmann [73]. Dobmann [74] has discussed the problems of both detection and sizing of defects and has attempted to relate these to theoretical model predictions. Theoretical work on flux leakage has lagged considerably behind experimental development.

15.3 MAGNETIC IMAGING METHODS

In order to evaluate the variation of materials properties from one location to the next, imaging techniques can be used. For example, the variation of a particular magnetic property, or the variation of magnetic field strength across the surface of a material can be used to produce a two-dimensional image, which can identify unusual regions of a material, which may have defects or be otherwise damaged; or simply to investigate the spatial variation of magnetization, which is of interest particularly in magnetic recording media. Many different magnetic measurements could be used for this. For example, imaging based on Barkhausen effect emissions has been reported [75]. However, we shall look here at only two magnetic NDE imaging methods—specifically magnetic force microscopy and scanning SQUID microscopy.

15.3.1 Magnetic Force Microscopy

The magnetic force microscope (MFM), which was first reported by Martin and Wickramasinghe [76], was developed from the earlier atomic force microscope of Binnig, Quayte, and Gerber [77] in which the alternating voltage on the probe tip is replaced with an alternating magnetic field. The device samples the magnetic field gradients close to the surface of the specimen. The method is therefore basically a scanning force probe technique based on magnetic forces between the magnetized tip of the microscope and a specimen, as shown in Figure 15.22. The MFM can be used to detect both static and dynamic magnetic fields on the surface of a specimen. In the original device of Martin and Wickramasinghe, a spatial resolution of 100 nm was demonstrated. Today, spatial resolutions of 40 nm are routine, and claims of resolutions down to 10 nm have been made in some cases.

There are two main modes of operation: (1) constant height and (2) variable height. In the first technique, the magnetic field is varied at a frequency ω, resulting in vibration of the tip. The tip is then scanned across the surface of the specimen. The changes in vibration amplitude, phase, and even frequency of the probe tip are detected by a laser interferometer. The change in phase angle of the vibration of the tip can be related to the magnetic force gradient at the surface, as shown below. The typical scanning height in this mode of operation is 100–200 nm.

The second technique relies on moving the tip closer to the surface by controlling the tip-sample spacing. This can be used to image static magnetic fields. The magnetization in the tip is modulated and the force on the tip is detected in the same way as before, using the laser interferometer to measure vibration amplitude, phase, and frequency of the tip. A block diagram of the typical experimental arrangement for the MFM is shown in Figure 15.23.

The image formed as a result of magnetic force microscopy is obtained by measuring the force or the force gradient, and these images can be used to study magnetic structures such as domain walls, closure domains, Bloch lines and, in the case of magnetic recording media, intentionally magnetized regions of a material. The MFM is widely used in evaluating magnetic recording media.

There are a variety of analytical techniques that can be used to extract information about the magnetic field at the surface of the specimen from the MFM response.

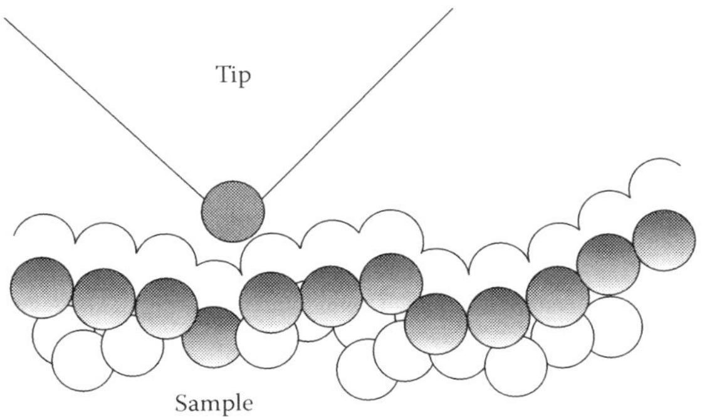

FIGURE 15.22 Magnetic probe scanning over the surface of a material.

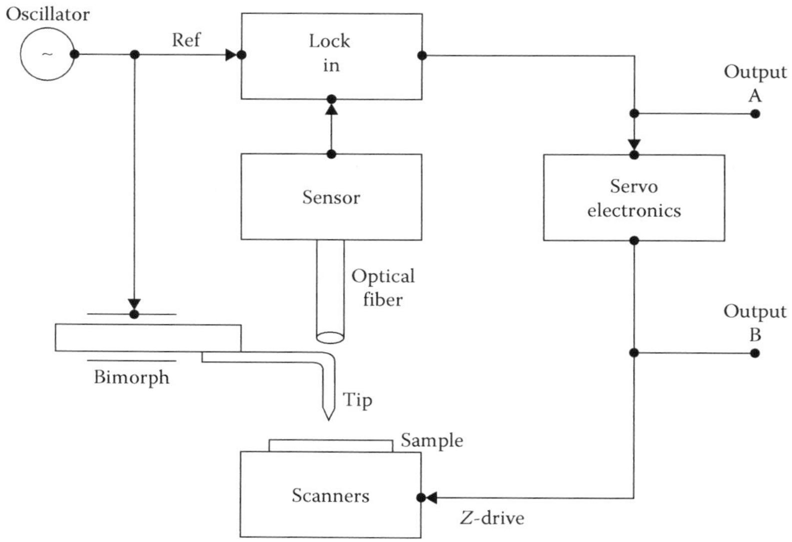

FIGURE 15.23 A schematic of the magnetic force microscope electronics.

These have been described by Rugar et al. [78]. For example, the force gradient $\partial F/\partial x$ at the surface of a specimen will lead to changes in the frequency of oscillation of the tip, $\Delta\omega$, given by

$$\frac{\Delta\omega}{\omega_0} = \frac{1}{2k}\left(\frac{\partial F}{\partial x}\right) \tag{15.10}$$

where:

ω_0 is the natural resonance frequency

k is the restoring force coefficient or *spring constant* of the cantilever

Therefore, under ac conditions, the force gradient can be determined from measurement of the frequency of oscillation. Alternatively, the change in phase $\Delta\phi$ can be measured, since for a system vibrating close to its resonant frequency with quality factor Q, we have $\Delta\phi = -2Q\,\Delta\omega/\omega_0$. Therefore,

$$\Delta\phi = -\frac{Q}{k}\left(\frac{\partial F}{\partial x}\right) \tag{15.11}$$

The form of the cantilever tip is critical to the operation of an MFM. Ideally, it should be a small single-domain particle with constant magnetization. Problems can arise if the magnetization in the tip changes as a result of exposure to an external field, or alternatively if the magnetic field generated by the tip itself changes the magnetization in the specimen. Therefore, difficulties are encountered in the imaging of either extremely hard or extremely soft magnetic materials, and these types of applications require careful consideration in order to achieve meaningful results.

Hartmann [79] developed a type of MFM probe that used spherical single-domain particles of diameter 10–100 nm, instead of the usual sharp-tipped probes. The reduced stray field from the tip enabled him to image domain walls in extremely soft iron single crystal whiskers without adversely affecting the magnetization conditions in the test specimen. The domain walls were then identified as regions with localized high-field gradients. According to Hartmann, force sensitivity down to 10^{-10} N can be achieved in this way even under ambient atmosphere conditions in the dc or *force sensing* mode of operation. Force gradients are of course detected in the ac mode of operation as discussed above.

Modeling of the MFM response is also important in order to interpret the signals obtained. This is achieved using classical magnetostatics by considering the tip as a dipole in an external magnetic field, as shown in Figure 15.24, and then calculating the forces on the tip. Proksch and Dahlberg [80] have modeled the response to bit transitions in magnetic recording media. The magnetic interactions between tip and specimen, which can change the magnetization of both, make it difficult to model the response of the MFM. If either the field from the tip is greater than the coercivity in the specimen, or the stray field at the surface of the specimen is greater than the

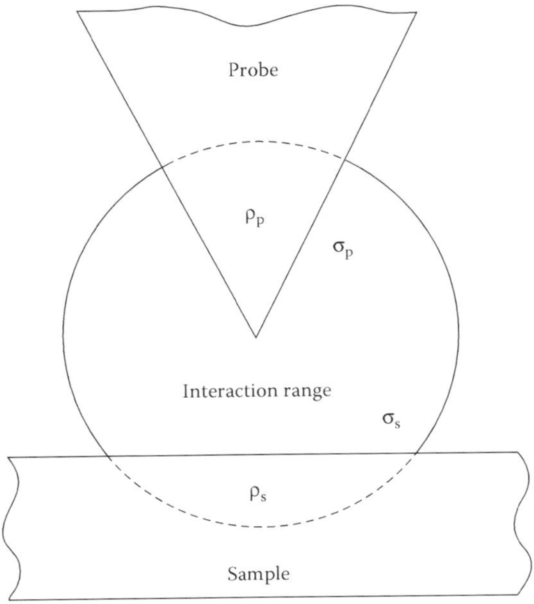

FIGURE 15.24 Interactions between the probe tip and the specimen can lead to unanticipated changes in the *MFM* image: (a) *Field tip* $\boldsymbol{H}_t$ affects magnetization in sample, (b) stray field from sample $\boldsymbol{H}_s$ affects magnetization of tip, and (c) interactions are nonlinear and therefore difficult to model.

coercivity of the tip, then these interactions will be nonlinear, and determining the significance of the response of the tip becomes more difficult.

Gomez et al. [81] reported a variation of the MFM called the *magnetic force scanning tunneling microscope*, in which a flexible magnetic probe is placed close to the surface of the specimen of interest, and a bias potential is applied leading to a tunneling current between probe and specimen. The distance between the probe and specimen is altered by magnetic interaction forces. These changes in distance lead to changes in the tunneling current, which can therefore be used to provide a map of the magnetic forces across the surface of the specimen, and hence an image of magnetic structure.

Gomez et al. [82] also investigated the performance of the MFM in the presence of external fields. At low-field strengths, of course, the response of the tip is caused by the component of the local surface field that is along the direction of magnetization of the single domain in the probe tip, as discussed in Section 3.2.2,

$$F_x = \mu_0 \left(\boldsymbol{m} \frac{d\boldsymbol{H}}{dx} \right) \hat{x} \tag{15.12}$$

and therefore there is no translational force in directions orthogonal to $\boldsymbol{m}$ even if there is a field gradient in these directions. However, reorientation of the magnetization in the probe tip by a controlled applied field allows selection of the component of the surface field of the specimen for imaging.

Magnetic force microscopy became relevant even on the industrial scale rather soon after its initial development according to Hartmann [83] because of several advantages it has over other scanning techniques and because of the need to provide evaluation of materials on the submicrometer scale particularly in the magnetic recording industry. The major advantages can be summarized as follows:

1. The long-range magnetic interactions are not sensitive to surface contamination; therefore, special surface preparation that is often required in other methods is not needed here.
2. It can be performed in various environments such as ambient atmosphere, ultra high vacuum, under liquids, and at different temperatures, including room temperature.
3. It is completely nondestructive with respect to crystal structure of the specimen.
4. The specimen does not need to be an electrical conductor.
5. Thin nonmagnetic layers on the surface do not invalidate the result.
6. It can be readily combined with other techniques such as scanning tunneling microscopy.

In addition, there are at least two drawbacks of the method, which need to be kept in view. First, the image obtained is strongly dependent on the type of probe used because of the tip/specimen interaction. Second, the magnetic fields of the tip and specimen can change each other's magnetization locally, leading to nonlinear interactions. For these two reasons, interpretation of the MFM image can be difficult in many cases. As an example, we can consider two extreme cases of probe tip: the sharp probe tip and the spherical tip. Each has some advantages and disadvantages. The sharp tip has high spatial resolution and high-field gradients emanating from the tip. This makes the sharp tip somewhat less useful for scanning very soft materials because of the increased likelihood that the high-field gradient will cause a change in the domain structure of the specimen under examination. The spherical tip, however, has lower stray fields and is better adapted for scanning soft materials but has lower spatial resolution.

Babcock et al. [84] have studied the performance of MFM probe tips under different applied fields. In order to do this without affecting the magnetization in a specimen, the sample fields were generated by a small-scale current-carrying conductor. This ensured that the tip field and sample fields were in this case noninteracting. It was found that while in principle the application of vertical and/or horizontal magnetic fields changes the direction of magnetization in the probe tip, nevertheless in practice, it is difficult to know the final orientation of the magnetization in the tip when the field is removed. Therefore, it remained difficult to entirely separate the measurement of horizontal and vertical components of the surface field from the magnetization of the sample.

In conclusion, magnetic force microscopy is one of a class of techniques known as scanning probe techniques. The instruments are modified versions of the atomic force microscope in which the cantilever is equipped with a ferromagnetic tip. These devices can be used to probe the magnetic structure at the surfaces of specimens down to spatial resolutions of 40 nm, by generating a two-dimensional image of the local magnetic field gradients at the surface. The magnetic interactions between the tip and the specimen that are used to produce the image can be nonlinear in some cases, leading to difficulties in the interpretation of the image if either the magnetization of the tip or the domain configuration in the specimen is changed. However, in all other respects the method can be considered as nondestructive, and it provides a quantitative, high-resolution alternative to the Bitter colloid technique. The ability to resolve features at length scales down to 40 nm, with little or no surface preparation required and under ambient atmosphere conditions, make this technique ideal for practical high-resolution magnetic evaluation of material surfaces.

15.3.2 Scanning SQUID Microscopy

Some of the drawbacks of the MFM mentioned above can be overcome using the scanning SQUID microscope (SSM). This does not suffer, for example, from magnetic interactions between the detector and the specimen. In addition, the SSM has unequalled field sensitivity and can be used in scanning mode to produce a magnetic field image of the surface of a specimen. On the other hand, it does have some drawbacks, including a relatively poor spatial resolution compared with MFM, and of course the need to provide cryogenic cooling for the superconducting components of the system. A comparison of the performance characteristics of the MFM and SSM is shown in Table 15.1.

The SSM is in essence a device that scans a specimen past a SQUID detector in order to map the normal component of the magnetic field at the surface of the

TABLE 15.1
Comparison of the Performance Characteristics of the MFM and SSM

	Field Sensitivity		Spatial Resolution (μm)
	(A m^{-1})	(T)	
MFM	10	10^{-5}	0.050
SSM	10^{-3}	10^{-9}	10

specimen as a function of position and produces an image based on the spatial variation of field strengths. The device was first developed by Vu and van Harlingen [85], in which a flux transformer coupled to the SQUID was scanned over the surface of the specimen of interest. The operation of the SQUID itself has been described in Section 3.4, its main advantage over other detection techniques being the extremely high resolution, which in this case was 10^5 flux quanta per $\sqrt{\text{Hz}}$. Vu and van Harlingen used a stepper motor manipulator with positional resolution of 2 μm, although because of the coil sizes, the spatial resolution of the field was only 20 μm. The device was first used to image vortex structures from a type II superconductor, but its applications were soon extended to include biomagnetism and NDE.

Comprehensive summaries of SQUID applications in both biomagnetism and NDE have been given by Wikswo [86] and by Kirtley [87], in which the spatial resolution was reported to be 4 μm. The disadvantage of the cryogenic housing remains a major problem for widespread applications. A schematic of the SSM configuration is shown in Figure 15.25.

15.4 SENSITIVITY TO MICROSTRUCTURE AND MATERIAL TREATMENT

A variety of magnetic methods for characterization and evaluation of materials is available, and several of these have been discussed here. In order to be useful for materials, evaluation of these methods must be sensitive to changes in structure of materials, including the emergence of defects, and to external condition such as the application of stress.

The first group of techniques is designed to determine intrinsic properties of materials and these include hysteresis measurements for determination of bulk properties, the Barkhausen effect for surface characterization, and magnetoacoustic emission. These methods all offer wide scope for future development because of their sensitivity to material treatment, particularly the presence of different material phases, chemical composition, and material deformation in its various forms: elastic and plastic strain, fatigue, dislocation density, and creep stress.

Another major group of magnetic techniques has been developed to detect and characterize defects and flaws in materials. These include MPI and quantitative magnetic flux leakage determination. The relative advantages of these methods in different situations have been discussed, and a summary of the literature on these has been given. In view of the diverse range of applications, only a few guiding principles have

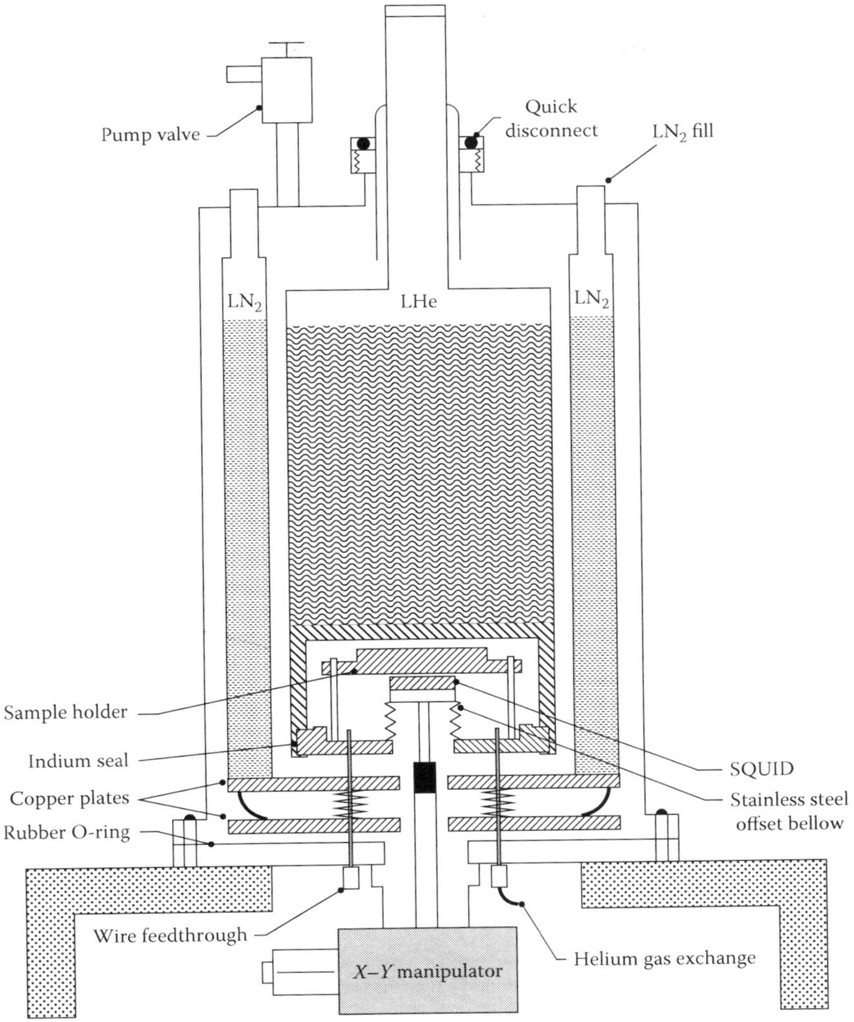

FIGURE 15.25 Schematic of a scanning SQUID microscope system. (After Wikswo, J. P., *IEEE Trans. Appl. Supercond.*, 5, 74, 1995.)

been given here, and in particular instances, the necessary additional information should be obtained from more specialized references.

Finally, two imaging methods have been discussed: MFM and scanning SQUID microscopy. These two techniques can be used to produce an image of the magnetic field at the surface of a test specimen. This is especially useful for characterizing magnetic recording media. The two methods have relative advantages and disadvantages depending on the spatial resolution on the field sensitivity that is required, and these have been compared and discussed in the text.

REFERENCES

1. Jiles, D. C., and Melikhov, Y. Modelling of non-linear behavior and hysteresis in magnetic materials, in *Handbook of Magnetism and Advanced Magnetic Materials, Vol. 2: Micromagnetism* (eds. H. Kronmuller and S.S. Parkin), John Wiley & Sons, Chichester, West Sussex, 2007, pp. 1059–1079.
2. Jiles, D. C. Dynamics of domain magnetization and the Barkhausen effect, *Czech. J. Phys.*, 50, 893, 2000.
3. Mikheev, M. N., Morozova, V. M., Morozov, A. P. et al. *Sov. J. NDT*, 14, 9, 1978.
4. Mikheev, M. N. *Sov. J. NDT*, 19, 1, 1983.
5. Kuznetsov, I. A., Somova, V. M., and Bashkirov, Y. P. *Sov. J. NDT*, 8, 506, 1982.
6. Zatsepin, N. N., Ashtashenko, P. P., Potapova, N. A. et al. *Sov. J. NDT*, 19, 158, 1983.
7. Rodigin, N. M., and Syrochkin, V. P. *Sov. J. NDT*, 9, 453, 1983.
8. Vekser, N.A., Smimov, A. S., Yu Fadeev, A. et al., *Sov. J. NDT*, 11, 183, 1975.
9. Langman, R. *IEEE Trans. Magn.*, 21, 1314, 1985.
10. Novikov, V. F., and Fateev, I. G. *Sov. J. NDT*, 18, 489, 1982.
11. Atherton, D. L., and Jiles, D. C. *IEEE Trans. Magn.*, 19, 2021, 1983.
12. Burkhardt, G. L., and Kwun, H. *J. Appl. Phys.*, 61, 1576, 1987.
13. Jiles, D. C., and Atherton, D. L. *J. Magn. Magn. Mater.*, 61, 48, 1986.
14. Szpunar, B., and Szpunar, J. A. *IEEE Trans. Magn.*, 20, 1882, 1984.
15. Sablik, M. J., Kwun, H., Burkhardt, G. L., and Jiles, D. C. *J. Appl. Phys.*, 63, 3930, 1988.
16. Garikepati, P., Chang, T. T., and Jiles, D. C., Theory of ferromagnetic hysteresis: Evaluation of stress from hysteresis curves, *IEEE Trans. Magn.*, 24, 2922, 1988.
17. Kaminski, D. A., Jiles, D. C., and Sablik, M. J. *JMMM*, 104, 382, 1992.
18. Jiles, D. C. *J. Phys. D: Appl. Phys.*, 28, 1537, 1995.
19. Shah, M. B., and Bose, M. S. C. *Phys. Stat. Solidi.*, 86, 275, 1984.
20. Barkhausen, H. *Phys. Z.*, 29, 401, 1919.
21. Raghunathan, A., Melikhov, Y., Snyder, J. E., and Jiles, D. C. *J. Magn. Magn. Mater.*, 324, 20, 2012.
22. Tiitto, S. *Acta Polytech. Scand.*, 119, 1, 1977.
23. Sundstrom, O., and Torronen, K. *Mater. Eval.*, 37, 51, 1979.
24. Karjalainen, L. P., Moilanen, M., and Rautioaho, R. *Mater. Eval.*, 37, 45, 1979.
25. Lomaev, G. V., Malyshev, V. S., and Degterev, A. P. *Sov. J. NDT*, 20, 189, 1984.
26. Mayos, M., Segalini, S., and Putignani, M., in *Review of Progress in Quantitative NDE* (eds. D. O. Thompson and D. E. Chimenti), Plenum Press, New York, 1987.
27. Theiner, W. A., and Altpeter, I., in *New Procedures in NDT* (ed. P. Holler), Springer-Verlag, Berlin, Germany, 1983.
28. Schneider, E., Theiner, W. A., and Altpeter, I., in *Nondestructive Methods for Materials Property Determination* (eds. C. O. Ruud and R. E. Green), Plenum Press, New York, 1984.
29. Jiles, D. C., Sipahi, L. B., and Williams, G. Modeling of micromagnetic Barkhausen activity using a stochastic process extension to the theory of hysteresis, *J. Appl. Phys.*, 73, 5830, 1993.
30. Jiles, D. C. Dynamics of domain magnetization and the Barkhausen effect, *Czech. J. Phys.*, 50, 893, 2000.
31. Jiles, D. C. Theory of the magnetomechanical effect, *J. Phys. D: Appl. Phys.*, 28, 1537, 1995.
32. Mierczak, L., Jiles, D. C., and Fantoni, G. A new method for evaluation of mechanical stress using the reciprocal amplitude of magnetic Barkhausen noise, *IEEE Trans. Magn.*, 47, 459, 2011.
33. Matzkanin, G. A., Beissner, R. E., and Teller, C. M. *The Barkhausen Effect and Its Applications*. SWRI Report No. NTIAC-79-2, 1979.

34. Lord, A. E., in *Physical Acoustics XI* (eds. W. P. Mason and R. N. Thurston), Academic Press, New York, 1975.
35. Kusanagi, H., Kimura, H., and Sasaki, H. *J. Appl. Phys.*, 50, 1989, 1979.
36. Ono, K., and Shibata, M. *Mater. Eval.*, 38, 55, 1980.
37. Burkhardt, G. L., Beissner, R. E., Matzkanin, G. A., and King, J. D. *Mater. Eval.*, 40, 669, 1981.
38. Theiner, W. A., and Willems, H. H., in *Nondestructive Methods for Materials Property Determination* (eds. C. O. Ruud and R. E. Green), Plenum Press, New York, 1984.
39. Edwards, C., and Palmer, S. B. *J. Acoust. Soc. Am.*, 82, 534, 1987.
40. Ranjan, R., Jiles, D. C., and Rastogi, P. K. *IEEE Trans. Magn.*, 23, 1869, 1987.
41. Higgins, F. P., and Carpenter, S. H. *Acta Metall.*, 26, 133, 1978.
42. Jiles, D. C., and Atherton, D. L. *J. Phys. D: Appl. Phys.*, 17, 1265, 1984.
43. Konovalov, O. S., Golovko, A. S., and Roitman, V. I. *Sov. J. NDT*, 18, 554, 1982.
44. Suzuki, M., Komura, I., and Takahashi, H. *Int. J. Pres. Ves. Pip.*, 6, 255, 1981.
45. Langman, R. *NDT Int.*, 14, 255, 1981.
46. Mikheev, M. N., Bida, G. V., Tsarkova, T. P., and Kastin, V. N. *Sov. J. NDT*, 18, 641, 1982.
47. Gregory, C. A., Holmes, V. L., and Roehrs, R. J. *Mater. Eval.*, 30, 219, 1972.
48. Massa, G. M. *NDT Int.*, 9, 16, 1976.
49. Goebbels, K., and Simkin, J. *Review of Progress in Quantitative NDE* (eds. D. O. Thompson and D. E. Chimenti), Plenum Press, New York, 1990.
50. Goebbels, K., in *Proceedings of the 12th World Conference on NDT* (eds. J. Boogaard and G. M. VanDijk), Elsevier, Amsterdam, the Netherlands, April 1989, p. 719.
51. Goebbels, K. *Materialprufung*, 30, 327, 1988.
52. Edwards, C., and Palmer, S. B. *NDT Int.*, 14, 177, 1981.
53. Oehl, C. L., and Swartzendruber, L. J. *J. NDE*, 3, 125, 1982.
54. Chen, Y. F. *Mater. Eval.*, 42, 1506, 1984.
55. Goebbels, K., and Ferrano, G. *Proceedings of the 4th European Conference on NDT* (eds. J. M. Farley and R. W. Nichols), Pergamon Press, Oxford, September 1987, p. 2762.
56. Betz, C. E. *Principles of Magnetic Particle Testing*, Magnaflux, Chicago, 1966.
57. Schmidt, J. T., and Skeie, K. *NDT Handbook, Vol. 6: Magnetic Particle Testing* (ed. P. Mclntire), American Society for Nondestructive Testing, New York, 1989.
58. Hastings, C. H. *ASTM Proc.*, 47, 651, 1947.
59. Atherton, D. L. *NDT Int.*, 16, 145, 1983.
60. Forster, F. *Mater. Eval.*, 43, 1154, 1985.
61. Stumm, W. *Non-Destr. Test.*, 7, 251, 1974.
62. Jiles, D.C. *Introduction to the Principles of Materials Evaluation*, Taylor & Francis, Boca Raton, Florida, 2008.
63. Owston, C. N. *Brit. J. NDT*, 16, 162, 1974.
64. Barton, J. R., Lankford, J., and Hampton, P. L. *Trans. Soc. Auto. Eng.*, 81, 681, 1972.
65. Beissner, R. E., Matzkanin, G. A., and Teller, C. M. SWRI Report Number NTIAC-80-1, 1980.
66. Zatsepin, N. N., and Schcherbinin, V. E. *Sov. J. NDT*, 2, 50, 1966.
67. Schcherbinin, V. E., and Zatsepin, N. N. *Sov. J. NDT*, 2, 59, 1966.
68. Edwards, C., and Palmer, S. B. *J. Phys. D: Appl. Phys.*, 19, 657, 1986.
69. Hwang, J. H., and Lord, W. *J. Test. Eval.*, 3, 21, 1975.
70. Lord, W., and Hwang, J. H. *Brit. J. NDT*, 19, 14, 1977.
71. Lord, W., Bridges, J. M., Yen, W., and Palanisamy, S. *Mater. Eval.*, 36, 47, 1978.
72. Lord, W. *IEEE Trans. Magn.*, 19, 2437, 1983.
73. Holler, P., and Dobmann, G., in *Research Techniques in NDT*, Vol. IV (ed. R. S. Sharpe), Academic Press, London, 1980.

74. Dobmann, G., in *Electromagnetic Methods of NDT* (ed. W. Lord), Gordon and Breach, New York, 1985.
75. Negley, M., and Jiles, D. C. *IEEE Trans. Magn.*, 30, 4509, 1994.
76. Martin, Y., and Wickramasinghe, H. K. *Appl. Phys. Letts.*, 50, 1455, 1987.
77. Binnig, G., Quayte, C., and Gerber, C. *Phys. Rev. Letts.*, 56, 930, 1986.
78. Rugar, D., Mamin, H. J., Guenther, P. et al. *Appl. Phys.*, 68, 1169, 1990.
79. Hartmann, U., Goddenhenrich, T., Lemke, H., and Heiden, C. *IEEE Trans. Magn.*, 26, 1512, 1990.
80. Proksch, R., and Dahlberg, E. D. *JMMM*, 104, 2123, 1992.
81. Gomez, R. D., Adly, A. A., Mayergoyz, I. D., and Burke, E. R. *IEEE Trans. Magn.*, 29, 2494, 1993.
82. Gomez, R. D., Burke, E. R., and Mayergoyz, I. D. *J. Appl. Phys.*, 79, 6441, 1996.
83. Hartmann, U. *JMMM*, 157, 545, 1996.
84. Babcock, K. L., Elings, V. B., Shi, J. et al. *Appl. Phys. Letts.*, 69, 705, 1996.
85. Vu, L. N., and van Harlingen, D. J. *IEEE Trans. Appl. Supercond.*, 3, 1918, 1993.
86. Wikswo, J. P. *IEEE Trans. Appl. Supercond.*, 5, 74, 1995.
87. Kirtley, J. *IEEE Spectrum*, 33 (12), 40, 1996.

FURTHER READING

Betz, C. E. *Principles of Magnetic Particle Testing*, Magnaflux Corporation, Chicago, IL, 1966.

Halmshaw, R. *Non Destructive Testing*, Edward Arnold, London, 1987, Ch. 5.

Jiles, D. C. Magnetic methods for non destructive evaluation, *NDT Int.*, 21, 311, 1988.

Jiles, D. C. Magnetic methods for non-destructive evaluation (Part 2), *NDT Int.*, 23, 83, 1990.

Jiles, D. C. Dynamics of domain magnetization and the Barkhausen effect, *Czech. J. Phys.*, 50, 893, 2000.

Jiles, D. C. Magnetic methods in nondestructive testing, in *Encyclopedia of Materials Science and Technology* (eds. K. H. J. Buschow et al.), Elsevier Press, Oxford, September 2001, p. 6021.

Jiles, D. C. *Introduction to the Principles of Materials Evaluation*, Taylor & Francis, Boca Raton, Florida, 2008.

Jiles, D. C., and Lo, C. C. H. The role of new materials in the development of magnetic sensors and actuators, *Sensors and Actuat. A: Phys.*, 106, 3, 2003.

Jiles, D. C., and Melikhov, Y. Modelling of non-linear behavior and hysteresis in magnetic materials, in *Handbook of Magnetism and Advanced Magnetic Materials, Vol. 2: Micromagnetism* (eds. H. Kronmuller and S. S. Parkin), John Wiley & Sons, Chichester, West Sussex, 2007, pp. 1059–1079.

Mierczak, L., Jiles, D. C., and Fantoni G. A new method for evaluation of mechanical stress using the reciprocal amplitude of magnetic Barkhausen noise, *IEEE Trans. Magn.*, 47, 459, 2011.

Solutions to Exercises

CHAPTER 1

Exercise 1.1: Definition of Ampere

The ampere is defined as the current that, when passed along two infinite conductors lying parallel to each other 1 m apart, gives rise to a force per unit length of 2×10^{-7} N m^{-1}.

From Ampère's circuital law, the magnetic field $\boldsymbol{H}$ at a radial distance a from a long conductor carrying current i is

$$\boldsymbol{H} = \frac{i}{2\pi a} \text{ A m}^{-1}$$

Therefore, the field at a distance of 1 m from a long linear conductor carrying a current of 1 A is

$$\boldsymbol{H} = \frac{1}{2\pi} \text{ A m}^{-1}$$

The force per unit length, F/l, on a current-carrying conductor with current i caused by a perpendicular magnetic induction $\boldsymbol{B}$ is

$$\frac{F}{l} = i\boldsymbol{B}$$

and in this case $F/l = 2 \times 10^{-7}$ N m^{-1}, and $i = 1$ A. Therefore,

$$\boldsymbol{B} = 2 \times 10^{-7} \text{ T}$$

Now since by definition $\boldsymbol{B} = \mu_0 \boldsymbol{H}$, then the magnetic induction of 2×10^{-7} T corresponds to a magnetic field in free space of $1/2\pi$ A m^{-1}, and so

$$\mu_0 = \frac{\boldsymbol{B}}{\boldsymbol{H}}$$
$$= 4\pi \times 10^{-7} \text{ H}m^{-1}$$

Exercise 1.2: Difference between *H* and *B*

An essay on the difference between $\boldsymbol{B}$ and $\boldsymbol{H}$ should concentrate on the two Maxwell equations, which relate to these quantities:

$$\nabla \times H = \boldsymbol{J} + \frac{\partial \boldsymbol{D}}{\partial t}$$

$$\nabla \times \boldsymbol{E} = -\frac{\partial \boldsymbol{B}}{\partial t}$$

or equivalently the Ampère circuital law for $\boldsymbol{H}$,

$$\int \boldsymbol{H} \cdot dl = Ni$$

or its Biot-Savart law form

$$dH = \frac{1}{4\pi} \frac{i\, dl \times \hat{r}}{\boldsymbol{r}^2}$$

and the Faraday-Lenz law of induction for $\boldsymbol{B}$,

$$V = -NA \frac{d\boldsymbol{B}}{dt}$$

and the Ampère force law (Ampère's other law) in its various forms

$$\begin{aligned} \boldsymbol{F} &= i\, dl \times \boldsymbol{B} \\ &= \boldsymbol{m} \times \boldsymbol{B} \\ &= q v \times \boldsymbol{B} \end{aligned}$$

the last of these also being known as the *Lorentz force law*, although it is essentially just the Ampère force law in a slightly different form.

A field strength of 1 A m^{-1} is generated at the center of a circular coil of conductor of diameter 1 m when it carries a current of 1 A.

A magnetic induction of 1 T generates a force of 1 N m^{-1} on a conductor carrying a current of 1 A perpendicular to the direction of the induction.

The main difference between these quantities is then clear. $\boldsymbol{H}$ is the *magnetic field*, which is generated any time that there is an electric current, and its magnitude and direction can be determined from integrating the effects of all the currents. $\boldsymbol{B}$ is the *magnetic induction* and this is related to the force experienced by a charge in motion (or a magnetic dipole). This quantity $\boldsymbol{B}$ can be dependent partly on the electric currents, but it also has another component determined by the magnetization in a material. Therefore, $\boldsymbol{B}$ is not simply another way of measuring $\boldsymbol{H}$.

Exercise 1.3: Units in Magnetism

Any discussion of the subject of units in magnetism ought to mention that the magnetic units are always derived units determined by the choice of the base units (either meter, kilogram, second, and ampere or centimeter, gram, and second) and as such are likely to have inconvenient values. One of the greatest difficulties is that in SI units, the permeability of free space is not 1 (as it is in CGS units) and this is often an inconvenience particularly in conversion from magnetic field to flux density or magnetic induction. Another problem concerns the somewhat inconveniently small size of the ampere per meter for practical field measurements on the everyday scale. As a result of this, commercial equipment and reported results are often in the older CGS unit system. Furthermore, conversion from CGS to SI units is extremely cumbersome, principally because the unit of $\boldsymbol{H}$, the oersted, is $1000/4\pi$ A m^{-1}, although the conversion between gauss and tesla is somewhat easier, since 1 T = 10,000 G.

Finally, the philosophies on which the two magnetic units systems have been developed are totally different, with CGS being developed from magnetostatic considerations (magnetic poles), while SI has been developed from electro-dynamic considerations (current loops), and this is reflected in the values of the units chosen for each system. While the CGS units could be developed on an electrodynamic basis, the units would not arise naturally, that is to say, arbitrary coefficients would need to be chosen for the currents in CGS and for the pole strengths in SI, to obtain the correct values of the units. Nevertheless, the fact that this can be done is evident, since exact conversions between CGS and SI units are possible.

The main difficulty in the failure to resolve this is that results are going to continue to be reported in different units with the attendant inconvenience of unit conversion in order to make comparisons.

Exercise 1.4: Magnetic Field at the Center of a Long Solenoid

Using the result for the field on the axis of a single turn obtained in Exercise 1.1, consider the field due to an elemental length of solenoid dx with n turns per unit length, each carrying a current i

$$d\boldsymbol{H} = \frac{ni}{2a}\sin^3\alpha\, dx$$

We now need to express either α in terms of x and integrate over the range $x = -\infty$ to $x = \infty$, or to express x in terms of α and integrate from $\alpha = -\pi$ to $\alpha = \pi$.

$$\tan\alpha = \frac{a}{x}$$

$$x = a\cot\alpha$$

Therefore,

$$dx = -a\,\text{cosec}^2\,\alpha\, d\alpha$$

$$d\boldsymbol{H} = -\left(\frac{ni}{2}\right)\sin^3\alpha\,\text{cosec}^2\alpha\, d\alpha$$

$$= -\left(\frac{ni}{2}\right)\sin\alpha\, d\alpha$$

Now integrating from $\alpha = -\pi$ to $\alpha = +\pi$

$$H = -\left(\frac{ni}{2}\right)\int_{\pi}^{-\pi} \sin\alpha \, d\alpha$$

$$H = ni \text{ A m}^{-1}$$

Exercise 1.5: Force on a Current-Carrying Conductor

a. The magnetic field at a distance a from a long conductor carrying a current i is

$$\boldsymbol{H} = \frac{i}{2\pi a}$$

The force per meter exerted on a current carrying conductor by a field $\boldsymbol{H}$ is

$$\boldsymbol{F} = \mu_0 i dl \times \boldsymbol{H}$$

In this case $\boldsymbol{H}$ and dl are perpendicular so that the force per unit length is

$$\boldsymbol{F} = \frac{\mu_0 i^2}{2\pi a}$$

When the conductors both carry 1 A and are 1 m apart

$$\boldsymbol{F} = \frac{\left(4\pi\times10^{-7}\right)(1)(1)}{2\pi}$$

$$= 2\times10^{-7} \text{ N m}^{-1}$$

This is a surprisingly small force given that the currents seem so large.

b. The force exerted, as given by the above equation, is

$$\boldsymbol{F} = \mu_0 i dl \times \boldsymbol{H}$$

and in this case the current is perpendicular to the field, so

$$\boldsymbol{F} = \mu_0 i l \boldsymbol{H}$$

$$= \left(4\pi\times10^{-7}\right)(5)(0.35)\left(160\times10^{3}\right)$$

$$\boldsymbol{F} = 0.0352 \text{ N}$$

Exercise 1.6: Torque on a Current-Loop Dipole

$$\tau = \mu_0 \boldsymbol{m} \times \boldsymbol{H}$$

$$= \boldsymbol{m} \times \boldsymbol{B}$$

and if the $\boldsymbol{m}$ and $\boldsymbol{B}$ vectors are perpendicular and since $\boldsymbol{m} = ANi$

$$\tau = ANi\mathbf{B}$$

$$= \left(4\times10^{-4}\right)\left(100\right)\left(1\times10^{-3}\right)\left(0.2\right)$$

$$= 8\times10^{-6}\ \mathrm{N\,m}$$

Exercise 1.7: Force between Two Flat Coaxial Coils

Assuming that the coils are flat (large diameter but with small separation distance along the common axis), the interaction between them can be approximated to that of two infinitely long linear conductors a small distance apart.

The flux density B at a radial distance d from a long conductor is

$$\boldsymbol{B} = \frac{\mu_0 Ni}{2\pi d}$$

$$\frac{d\boldsymbol{F}}{dl} = i\times\boldsymbol{B}$$

and if $\boldsymbol{i}$ is perpendicular to $\boldsymbol{B}$

$$\frac{d\boldsymbol{F}}{dl} = \frac{\mu_{0Ni^2}}{2\pi d}$$

Therefore, integrating around the circumference

$$\boldsymbol{F} = \int_0^{2\pi a} \frac{\mu_0 Ni^2}{2\pi d}\, dl$$

$$\boldsymbol{F} = \frac{\mu_0 Ni^2}{2\pi d} 2\pi a$$

And since there are N turns the force on a coil due to the field $\boldsymbol{B}$ is

$$\boldsymbol{F} = \mu_0 N^2 i^2 \frac{a}{d}$$

$$\boldsymbol{F} = \frac{\left(4\pi\times10^{-7}\right)\left(100\right)^2\left(0.5\right)^2\left(0.05\right)}{0.005}$$

$$\boldsymbol{F} = 0.0314\ \mathrm{N}$$

or

$$\boldsymbol{F} = \frac{\pi}{100} N$$

Exercise 1.8: Forces on a Dipole

The on-axis field is

$$\boldsymbol{H} = \frac{ia^2}{2\left(a^2 + x^2\right)^{3/2}}$$

$$\boldsymbol{B} = \mu_0 \boldsymbol{H}$$

and

$$\tau = \boldsymbol{m} \times \boldsymbol{B}$$

$$\tau = \boldsymbol{m} \times \mu_0 \frac{ia^2}{2\left(a^2 + x^2\right)^{3/2}}$$

(1) If $\boldsymbol{m}$ is oriented along the axis, then the cross product of $\boldsymbol{m}$ and $\boldsymbol{B}$ is zero

$$\tau = 0$$

$$\boldsymbol{F} = 0$$

(2) If $\boldsymbol{m}$ is oriented perpendicular to the axis, then

$$\tau = \frac{\boldsymbol{m}\mu_0 ia^2}{2\left(a^2 + x^2\right)^{3/2}}$$

Substituting in values,

$$\tau = \frac{\left(5 \times 10^{-4}\right)\left(4\pi \times 10^{-7}\right)(0.5)(0.05)^2}{2\left[(0.05)^2 + x^2\right]^{3/2}}$$

$$\tau = \frac{3.925 \times 10^{-13}}{\left(0.0025 + x^2\right)^{3/2}}$$

So, for example, at $x = 0$, the torque reaches a maximum value

$$\tau_{max} = \frac{3.925 \times 10^{-13}}{(0.0025)^{3/2}}$$

$$\tau_{max} = 3.142 \times 10^{-9} \text{ N m}$$

$$\tau_{max} = \pi \times 10^{-9} \text{ N m}$$

For an arbitrary position along the axis of the coil, choose $x = 0.1$ m

$$\tau = \frac{3.925 \times 10^{-13}}{\left[0.0025 + (0.1)^2\right]^{3/2}}$$

$$= \frac{3.925 \times 10^{-13}}{(0.0125)^{3/2}}$$

$$\tau = 2.808 \times 10^{-10}\ \mathrm{N\,m}$$

Finally, the translational force on a dipole is

$$\boldsymbol{F} = \mu_0 \boldsymbol{m} \cdot \frac{d\boldsymbol{H}}{dx}$$

and there is a field gradient caused by the coil

$$\boldsymbol{F} = \left(4\pi \times 10^{-7}\right)\left(5 \times 10^{-4}\right)\frac{d\boldsymbol{H}}{dx}$$

$$\boldsymbol{F} = 6.28 \times 10^{-10}\frac{d\boldsymbol{H}}{dx}$$

The gradient $d\boldsymbol{H}/dx$ at the plane of the coil $(x = 0)$ is zero. But a slight deviation from the position in the plane of the coil will give a gradient. Calculate field at $x = 0$ and $x = \boldsymbol{l}$

At $x = 0$,

$$\boldsymbol{H} = \frac{i}{2a}$$

$$\boldsymbol{H} = 5\ \mathrm{A\,m^{-1}}$$

Assuming a dipole length of 0.01 m

at $x = 0.01$

$$\boldsymbol{H} = \frac{(0.5)(0.05)^2}{2\left[(0.05)^2 + (0.01)^2\right]^{3/2}}$$

$$= 4.7\ \mathrm{A\,m^{-1}}$$

the gradient is therefore

$$\frac{d\boldsymbol{H}}{dx} = \frac{4.71 - 5.00}{0.01}$$

$$\frac{d\boldsymbol{H}}{dx} = 29 \text{ A m}^{-2}$$

so there will be a translational force $\boldsymbol{F}$ of

$$\boldsymbol{F} = \mu_0 m \cdot \frac{d\boldsymbol{H}}{dx}$$

$$= (4\pi \times 10^{-7})(5 \times 10^{-4})(29)$$

$$\boldsymbol{F} = 18.2 \times 10^{-9} \text{ N}$$

At the exact center, the forces on the two poles will be equal and opposite so that $\boldsymbol{F}_{\text{net}} = 0$.

But if one pole is at the center and the other pole is at a distance l, then the forces will not balance and $\boldsymbol{F}_{\text{net}}$ will have a finite value

$$\boldsymbol{F} = \mu_0 \boldsymbol{m} \frac{d\boldsymbol{H}}{dx}$$

We now calculate $d\boldsymbol{H}/dx$ at the location $x = 0$

$$\boldsymbol{H} = \frac{i}{2a}$$

$$= 5 \text{ A m}$$

At the location $x = \boldsymbol{l} = 0.025$ m say

$$\boldsymbol{H} = \frac{(0.5)(0.05)^2}{2\left[(0.05)^2 + (0.025)^2\right]^{3/2}}$$

$$\boldsymbol{H} = 3.578 \text{ A m}^{-1}$$

So the field gradient is

$$\frac{d\boldsymbol{H}}{dx} = \frac{5.00 - 3.578}{0.025}$$

$$= 57 \text{ A m}^{-2}$$

$$F = \mu_0 m \frac{dH}{dx}$$

$$= \left(4\pi\times10^{-7}\right)\left(5\times10^{-4}\right)(57)\ \text{N}$$

$$F = 3.6\times10^{-8}\ \text{N}$$

Exercise 1.9: Field Generated by a Square Coil

(1) By the Biot-Savart law

$$\delta \boldsymbol{H} = \frac{1}{4\pi r^2} i\, dl \times \hat{\boldsymbol{r}}$$

and

$$dl \times r = dl \sin(90-\alpha)$$

$$\delta \boldsymbol{H} = \frac{i}{4\pi r^2} dl \sin(90-\alpha)$$

$$\delta \boldsymbol{H} = \frac{i}{4\pi r^2} dl \cos\alpha$$

To get the entire field at the center due to this one side, we need to integrate from $\alpha = -\pi/4$ to $\alpha = +\pi/4$. (This is similar to the case of the infinite conductor when we have to integrate from $\alpha = -\pi/2$ to $\alpha = +\pi/2$.)
But the value of r changes with the angle α

$$r\cos\alpha = \frac{a}{2}$$

Therefore,

$$\boldsymbol{H} = \int_{-\pi/4}^{+\pi/4} \frac{i}{4\pi r^2} \cos\alpha\, dl$$

$$\boldsymbol{H} = \frac{i}{4\pi} \int_{-\pi/4}^{+\pi/4} \frac{\cos\alpha}{(a/2\cos\alpha)^2} dl$$

$$\boldsymbol{H} = \frac{i}{4\pi} \int_{-\pi/4}^{+\pi/4} 4\frac{\cos^3\alpha}{a^2} dl$$

Now we need to convert dl into a function of α

$$dl\cos\alpha = r\,d\alpha$$

$$\boldsymbol{H} = \frac{i}{\pi}\int_{-\pi/4}^{+\pi/4}\frac{\cos^2\alpha}{a^2}\,r\,d\alpha$$

$$\boldsymbol{H} = \frac{i}{\pi}\int_{-\pi/4}^{+\pi/4}\frac{\cos^2\alpha}{a^2}\cdot\frac{a}{2\cos\alpha}\,d\alpha$$

$$\boldsymbol{H} = \frac{i}{2\pi}\int_{-\pi/4}^{+\pi/4}\frac{\cos\alpha}{a}\,d\alpha$$

$$\boldsymbol{H} = \frac{i}{2\pi a}\Big[\sin\alpha\Big]_{-\pi/4}^{+\pi/4}$$

$$\boldsymbol{H} = \frac{i}{2\pi a}\left[\sin\frac{\pi}{4} - \sin\left(-\frac{\pi}{4}\right)\right]$$

Note here that if the integration went from $-\pi/2$ to $+\pi/2$, the answer would be $\boldsymbol{H} = i/\pi a = i/2\pi(a/2)$, which is the expected result for the infinite current-carrying conductor.

and

$$\sin\frac{\pi}{4} = \frac{1}{\sqrt{2}}$$

so

$$\boldsymbol{H} = \frac{i}{2\pi a}\left[\frac{1}{\sqrt{2}} - \left(-\frac{1}{\sqrt{2}}\right)\right]$$

$$\boldsymbol{H} = \frac{i}{2\pi a}\cdot\frac{2}{\sqrt{2}}$$

$$\boldsymbol{H} = \frac{1}{\sqrt{2}}\cdot\frac{i}{\pi a}$$

Compared with the infinite conductor, the field from one side of the square is $1/\sqrt{2}$ times the size or 0.707 times the size.

Now because there are four sides each contributing equally, the total field at the center of the square is

$$H = \frac{4}{\sqrt{2}} \cdot \frac{i}{\pi a}$$

$$H = 2\sqrt{2} \cdot \frac{i}{\pi a}$$

$$i = 1\ \text{A}$$

$$a = 1\ \text{m} \qquad H = \frac{2.82}{3.142} = 0.90\ \text{A m}^{-1}$$

$$H = \frac{4ia^2}{\pi\left(a^2 + 4x^2\right)} \cdot \frac{1}{\sqrt{2a^2 + 4x^2}}$$

(2)

$$H = \frac{4ia^2}{\pi}\left[\frac{1}{a^2 + 4x^2} \cdot \frac{1}{\sqrt{2a^2 + 4x^2}}\right]$$

$$H = \frac{ia^2}{2\pi}\left[\frac{1}{\left(a^2/4\right) + x^2} \cdot \frac{1}{\sqrt{\left(a^2/2\right) + x^2}}\right]$$

(3) For a circular conductor, the field at the center is given by

$$H = \int_{\alpha=0}^{2\pi} \delta H \, d\alpha$$

and from the Biot-Savart law

$$\delta H = \frac{1}{4\pi r^2} i \, dl \times \hat{r}$$

In this case r is a constant value of a, and $dl = a\,d\alpha$

$$\delta H = \frac{1}{4\pi a^2} i \, a \, d\alpha$$

$$H = \frac{1}{4\pi a^2} \int_0^{2\pi} a \, d\alpha$$

$$H = \frac{i}{4\pi a}(2\pi - 0)$$

$$H = \frac{i}{2a}$$

$$H = 0.5\ \text{A m}^{-1}$$

(4) Numerical values for cases (1)–(3)

(1)
$$H = 2\sqrt{2}\frac{i}{\pi a}$$

$$i = 1, a = 1$$

$$H = 0.900\ \text{A m}^{-1}$$

$$H = \frac{4ia^2}{\pi}\left[\frac{1}{a^2+4x^2}\cdot\frac{1}{\sqrt{2a^2+4x^2}}\right]$$

(2)
$$H = \frac{4}{\pi}\cdot\frac{1}{5}\cdot\frac{1}{\sqrt{6}}$$

$$H = 0.104\ \text{A m}^{-1}$$

(3)
$$H = \frac{i}{2a}$$

$$H = 0.500\ \text{A m}^{-1}$$

Consistency check look at limiting case when $x = 0$

(2)
$$H = \frac{4ia^2}{\pi}\left[\frac{1}{a^2+4x^2}\cdot\frac{1}{\sqrt{2a^2+4x^2}}\right]$$

So if $x = 0$,

$$H = \frac{4ia^2}{\pi}\left[\frac{1}{a^2}\cdot\frac{1}{a\sqrt{2}}\right]$$

$$H = 2\sqrt{2}\cdot\frac{i}{\pi a}$$

Therefore, the component of the equation that represents the variation of the field along the axis is the second term on the right-hand side of

$$H = \frac{2i}{\pi a}\left[\frac{2a^3}{a^2+4x^2}\cdot\frac{1}{\sqrt{2a^2+4x^2}}\right]$$

$$\frac{2a^3}{a^2+4x^2}\cdot\frac{1}{\sqrt{2a^2+4x^2}}$$

when $x = 0$, this becomes $\sqrt{2}$

Exercise 1.10: Force on a Rod Carrying a Current

$$\boldsymbol{F} = il \times \boldsymbol{B}$$
$$\boldsymbol{F} = il\boldsymbol{B}$$

Equilibrium is reached when

$$mlg\ \cos(90 - \theta) = il\boldsymbol{B}$$

$$\sin\theta = \frac{i\boldsymbol{B}}{mg}$$

$$\theta = \text{Arcsin}\left(\frac{iB}{mg}\right)$$

$$\theta = \text{Arcsin}\left[\frac{(1)(0.1 \times 10^{-3})}{0.2 \times 9.8}\right]$$

$$\theta = \text{Arcsin}\left(\frac{0.1 \times 10^{-3}}{1.96}\right)$$

$$\theta = \text{Arcsin}(0.05 \times 10^{-3})$$

$$\theta = \sin\theta$$

$$\theta \cong 0.00005\,\text{rads}$$

CHAPTER 2

Exercise 2.1: Currents and Poles

The *equivalent current* and the *magnetic pole* models are just two different approaches for calculating the magnetic properties of materials and the magnetic fields generated by magnetic materials. The equivalent current model uses the Biot-Savart law

$$\boldsymbol{H} = \frac{1}{4\pi}\int \frac{i\,dl \times \hat{\boldsymbol{r}}}{r^2}$$

and determines a current distribution that is responsible for a particular field distribution. The pole model uses the magnetic equivalent of Coulomb's law to determine the pole distribution that is responsible for the field distribution

$$\boldsymbol{H} = \frac{1}{4\pi}\int \frac{dp}{r^2}\,\hat{\boldsymbol{r}}$$

It can be seen that these expressions are essentially equivalent, although this equivalence is only exact when the distance r from the source of the field, whether dipole or current, is large. In the *near field* region, they give different results. Therefore, these models are only appropriate on the macroscopic or classical scale.

Dipole Field

$$\boldsymbol{H}(\boldsymbol{r}) = \frac{1}{4\pi}\sum_{i=1,2}\frac{pi}{r_i^2}\hat{\boldsymbol{r}}_i$$

and summing the contributions at any point distant r from the dipole ($r \gg \ell$) gives

$$\boldsymbol{H}_r = \frac{2\boldsymbol{m}\cos\theta}{4\pi r^3}$$

$$\boldsymbol{H}_\theta = \frac{\boldsymbol{m}\sin\theta}{4\pi r^3}$$

$$\boldsymbol{H}_\phi = 0$$

Field Due to a Current Loop

In this case, an analytic solution can only be obtained along the axis of the loop

$$\boldsymbol{H} = \frac{ia^2}{2\left(a^2 + r^2\right)^{3/2}}$$

$\pi a^2 = A$ the cross-sectional area of the loop. For $r \gg a$

$$\boldsymbol{H} = \frac{iA}{2\pi r^3}$$

If we have a cylindrical magnet whose length is long compared with the cross-sectional area, we can obtain the field due to a circulating current. In this case, the field on the axis of the *solenoid* is

$$\boldsymbol{H} = \frac{Ni}{2L}\left(\cos\theta_1 - \cos\theta_2\right)$$

and at the end face, $\theta_1 = \pi/2$ and $\theta_2 = 0$, giving

$$\boldsymbol{H} = \frac{Ni}{2L}$$

So the field strength at the end face is half the value in the center of an infinite solenoid. If we take the magnetic pole approach, with magnetization $\boldsymbol{M}$ and magnetic moment $\boldsymbol{M}Al$ (where l could be less than L), then

$$p = \frac{\phi}{\mu_0} = \boldsymbol{M}A$$

and the field at the surface is given by

$$\boldsymbol{H} = \frac{1}{4\pi}\sum_i \frac{pi}{r_i^2}\,\hat{\boldsymbol{r}}_i$$

If the poles are located symmetrically a distance d inside the magnet from the end faces,

$$\boldsymbol{H} = \frac{1}{4\pi}\left[\frac{p}{d^2} - \frac{p}{(L-d)^2}\right]$$

$$= \frac{\boldsymbol{M}A}{4\pi}\left[\frac{1}{d^2} - \frac{1}{(L-d)^2}\right]$$

$$= \frac{\boldsymbol{M}AL}{4\pi}\left[\frac{L-2d}{d^2(L-d)^2}\right]$$

Exercise 2.2: Demagnetizing Field

When magnetized to saturation, the demagnetizing field $\boldsymbol{H}_\mathrm{d}$ is

$$\boldsymbol{H}_\mathrm{d} = N_\mathrm{d}\boldsymbol{M}_\mathrm{s}$$

For $\mu_\mathrm{r} = 5$, $l/d = 5{:}1$ and $N_\mathrm{d} = 0.025$; therefore

$$\boldsymbol{H}_\mathrm{d} = 20\times 10^3\ \mathrm{A\,m^{-1}}$$

For $\mu_\mathrm{r} = 5$, $l/d = 2{:}1$ and $N_\mathrm{d} = 0.11$; therefore

$$\boldsymbol{H}_\mathrm{d} = 88\times 10^3\ \mathrm{A\,m^{-1}}$$

Now the total field needed to saturate the material is determined by the permeability. Clearly the total field needs to be

$$\boldsymbol{H}_\mathrm{tot} = \frac{\boldsymbol{M}_\mathrm{s}}{(\mu_r - 1)} = \frac{\boldsymbol{M}_\mathrm{s}}{4}$$

and

$$\boldsymbol{H}_\mathrm{tot} = \boldsymbol{H}_\mathrm{app} - N_\mathrm{d}\boldsymbol{M}_\mathrm{s}$$

$$\boldsymbol{H}_\mathrm{app} = \frac{\boldsymbol{M}_\mathrm{s}}{4} + N_\mathrm{d}\boldsymbol{M}_\mathrm{s}$$

$$= \left(\frac{1}{4} + N_\mathrm{d}\right)\boldsymbol{M}_\mathrm{s}$$

In the case of a 5:1 aspect ratio,

$$\boldsymbol{H}_{\mathrm{app}} = 220 \times 10^3 \ \mathrm{A\,m^{-1}}$$

and in the case of a 2:1 aspect ratio,

$$\boldsymbol{H}_{\mathrm{app}} = 288 \times 10^3 \ \mathrm{A\,m^{-1}}$$

Exercise 2.3: Demagnetizing Field Calculation

$$\boldsymbol{H}_{\mathrm{in}} = \boldsymbol{H}_{\mathrm{app}} - N_{\mathrm{d}}\boldsymbol{M}$$

and for a sphere $N_{\mathrm{d}} = 1/3$. Since $M_{\mathrm{s}} = 1.69 \times 10^6$ A m^{-1}

$$\boldsymbol{H}_{\mathrm{app}} = \boldsymbol{H}_{\mathrm{in}} + \left(\frac{1}{3}\right)\left(1.69 \times 10^6\right) \mathrm{A\,m^{-1}}$$

and since we are assuming $N_{\mathrm{d}}\boldsymbol{M} \gg \boldsymbol{H}_{\mathrm{in}}$, this gives

$$\boldsymbol{H}_{\mathrm{app}} = 5.62 \times 10^5 \ \mathrm{A\,m^{-1}}$$

Exercise 2.4: Demagnetizing Effects at Different Field Strengths

$$\boldsymbol{M} = \frac{\boldsymbol{B}}{\mu_0} - \boldsymbol{H}$$

At $\boldsymbol{H} = 80$ kA m^{-1}

$$\boldsymbol{M} = 7.16 \times 10^5 - 0.8 \times 10^5 \ \mathrm{A\,m^{-1}}$$

$$= 6.36 \times 10^5 \ \mathrm{A\,m^{-1}}$$

The internal field is given by

$$\boldsymbol{H}_{\mathrm{in}} = \boldsymbol{H}_{\mathrm{app}} - N_{\mathrm{d}}\boldsymbol{M}$$

$$= 8 \times 10^4 - (0.02)\left(6.36 \times 10^5\right)$$

$$= 6.73 \times 10^4 \mathrm{A\,m^{-1}}$$

At 160 kA m^{-1}

$$\boldsymbol{M} = 8.75 \times 10^5 - 1.6 \times 10^5 \ \mathrm{A\,m^{-1}}$$

$$= 7.15 \ \times 10^5 \ \mathrm{A\,m^{-1}}$$

and

$$\boldsymbol{H}_{\text{in}} = 16.0 \times 10^4 - (0.02)(7.15 \times 10^5)$$

$$= 14.57 \times 10^4 \text{ A m}^{-1}$$

Fractional errors obtained if fields are not corrected for demagnetizing effects are at 80 kA m^{-1}

$$\frac{(\boldsymbol{H}_{\text{app}} - \boldsymbol{H}_{\text{in}})}{\boldsymbol{H}_{\text{app}}} = 0.159$$

at 160 kA m^{-1}

$$\frac{(\boldsymbol{H}_{\text{app}} - \boldsymbol{H}_{\text{in}})}{\boldsymbol{H}_{\text{app}}} = 0.089$$

Therefore, demagnetizing effects become proportionally less of a problem at higher field strengths.

Exercise 2.5: Flux Density in an Iron Ring with and without an Air Gap

In the continuous ring, the flux density is given by

$$\Phi = \frac{Ni}{(l/\mu A)}$$

$$= 6.0 \times 10^{-4} \text{ Wb}$$

When there is a saw cut in the ring, the total reluctance is the sum of reluctance of the iron and the air, so the relation becomes

$$Ni = \Phi\left(\frac{l_a}{\mu_a A_a} + \frac{l_i}{\mu^i A_i}\right)$$

where:

l_i = 0.4995 m
l_a = 0.0005 m
A_i = 2 × 10^{-4} m^2
μ_i = 0.001885 H m^{-1}
μ_a = 12.57 × 10^{-7} H m^{-1}

Under these conditions, the flux will be

$$\Phi = \frac{(800)(2 \times 10^{-4})}{(397.77 + 264.98)}$$

$$= 2.4 \times 10^{-4} \text{ Wb}$$

The additional current required to restore the flux to its original value is

$$N\delta i = \frac{\Phi l_a}{\mu_a A_a}$$

$$\delta i = \frac{(6.0\times 10^{-4})(0.0005)}{(0.0002\times 4\pi\times 10^{-7})}$$

$$= 1.5\ A$$

The total current needed to restore the flux density is therefore $i + \delta i = 2.5$ A.

Exercise 2.6: Force between Two Permanent Magnets

H is the field
Φ is the flux
$\boldsymbol{B}$ is the induction
A is the surface area
μ is the permeability
$\Phi = \boldsymbol{B}A = \mu_0 \boldsymbol{H}A$ is the total flux
$p = \Phi/\mu_0 = \boldsymbol{M}A$ is the pole strength
d is the separation between faces

Therefore, using the pole model and assuming the poles are located at the end faces,

$$\boldsymbol{F} = \frac{\mu_0}{4\pi}\frac{p^2}{d^2}$$

Since there are two ends to each magnet, each attracting the other magnet with the same force, the total force will be doubled:

$$\boldsymbol{F}_{tot} = \frac{\mu_0}{2\pi}\frac{p^2}{d^2}$$

$$= \frac{\mu_0}{2\pi}\frac{\Phi^2}{\mu_0^2 d^2}$$

or finally

$$\boldsymbol{F}_{tot} = \frac{\mu_0}{2\pi}\frac{A^2\boldsymbol{M}^2}{d^2}$$

Exercise 2.7: Magnetic Field and Magnetic Flux Density

N = 500 turns, d = 0.1 m, circumference = πd = 0.3142 m, $A = 1.0 \times 10^{-5}$ m^2, i = 1 A, $\Delta\phi = 0.2 \times 10^{-4}$ Wb

By Ampère's circuital law,

$$Ni = 500 \text{ A turns}$$

$$\boldsymbol{H} = \frac{Ni}{\pi d} = 1591\,\text{A m}^{-1}$$

$$\Delta \boldsymbol{B} = \frac{\Delta\phi}{A} = 2.0 \text{ T}$$

$$\boldsymbol{M} = \frac{\boldsymbol{B}}{\mu_0} - \boldsymbol{H} = 1.59 \times 10^6\,\text{A m}^{-1}$$

$$\mu_0\mu_r = \frac{\boldsymbol{B}}{\boldsymbol{H}} = 0.00126\,\text{H m}^{-1}$$

$$\mu_r = \frac{\boldsymbol{B}}{\mu_0 \boldsymbol{H}} = 1000$$

Exercise 2.8: Change in *H* and *B* Caused by Demagnetizing Field for Different Shapes

(1) $r_a : r_b = 1:1$; (2) $r_a : r_b = 5:1$; (3) $r_a : r_b = 100:1$

$$\text{For } 1:1\ N_d = 0.333$$

$$\text{For } 5:1\ N_d = 0.04$$

$$\text{For } 100:1\ N_d = 0.0003$$

Relative permeability $\mu_r = 1000$ in all cases

$$\boldsymbol{H}_{in} = \boldsymbol{H}_{app} - N_d \boldsymbol{M}$$

$$= \boldsymbol{H}_{app} - N_d\left(\frac{\boldsymbol{B}}{\mu_0} - \boldsymbol{H}_{in}\right)$$

$$\boldsymbol{H}_{in} = \frac{\boldsymbol{H}_{app}}{1 + N_d(\mu_r - 1)}$$

$$\text{For } 1:1\ \frac{\boldsymbol{H}_{in}}{\boldsymbol{H}_{app}} = 0.003$$

$$\text{For } 5:1\ \frac{\boldsymbol{H}_{in}}{\boldsymbol{H}_{app}} = 0.0244$$

$$\text{For } 100:1\ \frac{\boldsymbol{H}_{in}}{\boldsymbol{H}_{app}} = 0.769$$

$$\frac{\boldsymbol{B}_{\text{in}}}{\boldsymbol{B}_{\text{out}}} = \frac{\mu_{\text{r}}\boldsymbol{H}_{\text{in}}}{\boldsymbol{H}_{\text{app}}}$$

$$\text{For } 1:1 \; \frac{\boldsymbol{B}_{\text{in}}}{\boldsymbol{B}_{\text{out}}} = (1000)(0.003) = 3$$

$$\text{For } 5:1 \; \frac{\boldsymbol{B}_{\text{in}}}{\boldsymbol{B}_{\text{out}}} = (1000)(0.0244) = 24.4$$

$$\text{For } 100:1 \; \frac{\boldsymbol{B}_{\text{in}}}{\boldsymbol{B}_{\text{out}}} = (1000)(0.769) = 769$$

Exercise 2.9: Divergence of *B*

$$\boldsymbol{B} = a \times i + ayj + 0k$$

A plot actually shows that the **B** vector is diverging. It can similarly be shown by algebra

$$\nabla \cdot \mathbf{B} = \left(\frac{\partial}{\partial x}\boldsymbol{i} + \frac{\partial}{2y}\boldsymbol{j} + \frac{\partial}{\partial z}\boldsymbol{k} \right) \cdot \left(ax\boldsymbol{i} + ay\,\boldsymbol{j} + 0\boldsymbol{k} \right)$$

so,

$$\nabla \cdot \mathbf{B} = \frac{\partial ax}{\partial x} + \frac{\partial ay}{\partial y} + 0$$

$$\nabla \cdot \mathbf{B} = a + a$$

So clearly, $\nabla \cdot \mathbf{B} \neq 0$ except in the special case $a = 0$, which is a trivial case of zero field.

$$\nabla \cdot \mathbf{B} = 2a$$

Monopole

This would have

$$\boldsymbol{B} = \frac{\boldsymbol{B}_0}{r^2}x\boldsymbol{i} + \frac{\boldsymbol{B}_0}{\boldsymbol{r}^2}\boldsymbol{y}\boldsymbol{j} + \frac{\boldsymbol{B}_0}{r^2}z\boldsymbol{k}$$

where:

$$r = \sqrt{x^2 + y^2 + z^2}$$

$$\boldsymbol{B} = \frac{\boldsymbol{B}_0}{\sqrt{x^2 + y^2 + z^2}}\left(x\boldsymbol{i} + y\boldsymbol{j} + z\boldsymbol{k} \right)$$

This is the form that satisfies Gauss's law for a monopole.

Exercise 2.10: Demagnetizing Field and the Directions of *H* and *B*

The demagnetizing field is given by

$$\boldsymbol{H}_{\text{demag}} = -N_{\text{d}}\boldsymbol{M}$$

The circumstances under which there will be no demagnetizing field are either

$$M = 0 \text{ (demagnetized)}$$

or

$$N_{\text{d}} = 0 \text{ (closed loop or } \textit{infinite} \text{ sample)}$$

Demagnetizing Field

If the material is uniformly magnetized with a magnetization $\boldsymbol{M}$ within, and no external $\boldsymbol{H}$ field, then inside the magnetized material

$$\boldsymbol{B} = \mu_0(\boldsymbol{H} + \boldsymbol{M})$$

The demagnetizing field acts in the opposite direction to $\boldsymbol{M}$, so that

$$\boldsymbol{B} = \mu_0(\boldsymbol{M} - N_{\text{d}}\boldsymbol{M})$$

$$\boldsymbol{B} = \mu_0(1 - N_{\text{d}})\boldsymbol{M}$$

and as N_{d} is always less than 1, $\boldsymbol{B}$ and $\boldsymbol{M}$ point in the same direction, which is opposite to $\boldsymbol{H}$.

$$\text{For iron, } \boldsymbol{M}_{\text{s}} = 1.7 \times 10^6 \text{ A m}^{-1}$$

If the length to diameter ratio is 5:1, then the demagnetizing factor N_{d} is

$$N_{\text{d}} = 0.04$$

Therefore, inside the magnetic material, we have

$$\boldsymbol{H} = -N_{\text{d}}\boldsymbol{M}_{\text{s}}$$

$$\boldsymbol{M} = \boldsymbol{M}_{\text{s}}$$

$$\boldsymbol{M}_{\text{s}} = 1.76 \times 10^6 \text{ A m}^{-1}$$

$$\boldsymbol{B} = \mu_0(1 - N_{\text{d}})\boldsymbol{M}_{\text{s}}$$

Therefore,

$$\boldsymbol{H} = -0.068 \times 10^6 \text{ A m}^{-1}$$

$$\boldsymbol{M} = 1.7 \times 10^6 \text{ A m}^{-1}$$

$$\boldsymbol{B} = 2.05 \text{ T}$$

CHAPTER 3

Exercise 3.1: Search Coil Methods

Number of turns $N = 50$, resistance $R = 250\ \Omega$, and change in flux $\Delta\Phi = 3$ m Wb. Using Faraday's law,

$$V = -\frac{Nd\Phi}{dt}$$

and since $V = iR$

$$iR\,dt = -Nd\Phi$$

$$q = i\,dt = -\left(\frac{N}{R}\right)d\Phi$$

$$= 6 \times 10^{-4} \text{ C}$$

Exercise 3.2: Dipole Oscillations

The force on the moment caused by an applied field H in free space is

$$\tau = \mu_0 \boldsymbol{m} \times \boldsymbol{H}$$

$$= \mu_0 \boldsymbol{mH} \sin\theta$$

and for small θ, sin θ ≅ θ. Therefore,

$$\tau = \mu_0 \boldsymbol{mH}\,\theta$$

If I is the moment of inertia of the dipole, then

$$I\frac{d^2\theta}{dt^2} = -\mu_0 \boldsymbol{mH}\theta$$

$$\frac{d^2\theta}{dt^2} = -\frac{\mu_0 \boldsymbol{mH}\,\theta}{I}$$

where:

- $\mu_0 mH/I = \omega^2$
- ω is the frequency of oscillation

Earth's Field Alone

If $\boldsymbol{H}_e$ is the Earth's field, let ω_1 be the frequency of oscillation,

$$\omega_1^2 = \frac{\mu_0 \boldsymbol{mH}_e}{I}$$

Earth's Field and Solenoid Field

If the solenoid field is $\boldsymbol{H}_s$, then let ω_2 be the frequency of oscillation

$$\omega_2^2 = \mu_0 \frac{\boldsymbol{m}}{I}\left(\boldsymbol{H}_s \pm \boldsymbol{H}_e\right)$$

Therefore, the ratio of the total field to the Earth's field will be

$$\frac{\boldsymbol{H}_s \pm \boldsymbol{H}_e}{\boldsymbol{H}_e} = \frac{\omega_2^2}{\omega_1^2} = 11.11$$

$$\boldsymbol{H}_s = \left(11.11 \pm 1.0\right)\boldsymbol{H}_e$$

$$= 141.5 \text{ or } 169.5 \text{ A m}^{-1}$$

and since

$$\boldsymbol{H}_s = ni$$

$$i = \frac{\boldsymbol{H}_s}{n}$$

$$= 94.3 \text{ or } 113 \text{ mA}$$

Exercise 3.3: Torque Magnetometer

$$\tau = \mu_0 \boldsymbol{m} \times \boldsymbol{H} = \mu_0 \boldsymbol{mH} \sin\theta$$

At 90°

$$\tau = \left(4\pi \times 10^{-7}\right)\left(0.318\right)\left(14\right) \text{ N m}$$

$$= 5.6 \times 10^{-6} \text{ N m}$$

At 30°

$$\tau = \left(5.6 \times 10^{-6}\right)\sin 30°$$

$$= 2.8 \times 10^{-6} \text{ N m}$$

Exercise 3.4: Magnetic Resonance

A paramagnet with $S = 1/2$ is a system with one unpaired electron. Since this is a dilute paramagnet, then we may consider the electrons on neighboring ionic sites as noninteracting. The magnetic moment of a single electron is $\boldsymbol{m} = 9.27 \times 10^{-24}$ A m^2; therefore, since the gyromagnetic ratio γ is

$$\gamma = -\mu_B g\left(\frac{2\pi}{h}\right)$$

for an electron $g = 2$,

$$\gamma = \frac{(9.27\times10^{-24})(2)(6.284)}{(6.63\times10^{-34})}$$

$$= 1.76\times10^{11}\ \text{rad s}^{-1}\,\text{T}^{-1}$$

Therefore, using the expression for the resonant frequency ω_0

$$\begin{aligned}\omega_0 &= \gamma B\\ &= (1.76\times10^{11})(0.1)\ \text{rad s}^{-1}\\ &= 1.76\times10^{10}\ \text{rad s}^{-1}\\ &= 17.6\times10^{9}\ \text{rad s}^{-1}\end{aligned}$$

Exercise 3.5: Induction Coil Method

The applied field in amperes per meter can be found simply by multiplying the current in the windings by 400.

Reversing the current in the solenoid causes the flux density to reverse completely, so if the prevailing value is $\boldsymbol{B}_0$ T the change in flux density caused by reversing the field is $2\boldsymbol{B}_0$.

The flux density change is found by multiplying the deflection d (in mm) 0.17×10^{-4} Wb mm^{-1}, and then dividing the result by the product of the number of turns, 40, and the cross-sectional area 0.196×10^{-4} m^2.

$$H = 400i\ \text{A m}^{-1}$$

$$\Delta B = 2B = 0.0217d\ \text{T}$$

i_r (A)	H_a (A m^{-1})	d (mm)	B (T)
1.5	600	24.0	0.26
3.1	1240	49.2	0.53
4.9	1960	77.6	0.84
8.5	3400	103.7	1.12
11.0	4400	107.5	1.17
12.7	5080	109.1	1.18

From this table a plot of $\boldsymbol{B}$ against $\boldsymbol{H}_a$ gives the apparent magnetization curve.

From charts of demagnetizing factors given in Chapter 2, we find that $N_d = 0.0015$ for a cylinder with length to diameter ratio 40:1 and relative permeability greater than 150.

The demagnetizing field $\boldsymbol{H}_d$ is given by

$$\boldsymbol{H}_d = N_d\left(\frac{\boldsymbol{B}}{\mu_0} - \boldsymbol{H}_a\right)$$

and hence the true internal field $\boldsymbol{H}_i$ is given by

$$\begin{aligned}\boldsymbol{H}_i &= \boldsymbol{H}_a - \boldsymbol{H}_d \\ &= \boldsymbol{H}_a - N_d\left(\frac{\boldsymbol{H}}{\mu_0} - \boldsymbol{H}_a\right)\end{aligned}$$

B (T)	H_a (A m^{-1})	H_i (A m^{-1})
0.26	600	291
0.53	1240	609
0.84	1960	960
1.12	3400	2068
1.17	4400	3009
1.18	5080	3678

At $\boldsymbol{B} = 1$ T, by linear interpolation, the permeability is 6.276×10^{-4} H m^{-1}. This corresponds to a relative permeability of 499.7.

Exercise 3.6: Hall Effect

$$\begin{aligned}R_H &= \frac{1}{ne} \\ &= 1.6 \times 10^{-10}\ \text{m}^3\ \text{C}^{-1}\end{aligned}$$

and, since we know that $e = 1.602 \times 10^{-19}$ C

$$\begin{aligned}n &= \frac{1}{R_H e} \\ &= 3.90 \times 10^{28}\ \text{m}^{-3}\end{aligned}$$

This then is the total number of electrons per unit volume in the conduction band of indium. The number of atoms per unit volume is

$$n_A = \frac{7280 \times 6.025 \times 10^{26}}{115} = 3.81 \times 10^{28}\ \text{m}^{-3}$$

Therefore, the number of charge carriers per atom is

$$\frac{n}{n_A} = 1.02$$

and, since R_H is positive, these are holes.

Exercise 3.7: Hall Effect Sensor

Cube of semiconductor 1 ***mm*** × 1 ***mm*** × 1 ***mm***

$$l = 1\ \text{mm}$$

$$\text{Required sensitivity} = 1\ \text{V}\,\text{T}^{-1}$$

The Hall field E_{Hall} is given by

$$E_{\text{Hall}} = \boldsymbol{V} \times \boldsymbol{B}$$

$$E_{\text{Hall}} = \frac{\boldsymbol{J}}{ne} \times \boldsymbol{B}$$

So, the Hall voltage is V_{Hall}

$$V_{\text{Hall}} = E_{\text{Hall}} l$$

$$V_{\text{Hall}} = l\boldsymbol{V} \times \boldsymbol{B}$$

$$V_{\text{Hall}} = \frac{l\boldsymbol{J} \times \boldsymbol{B}}{ne}$$

$$\frac{V_{\text{Hall}}}{\boldsymbol{B}} = \frac{lJ}{ne}$$

and since it is a cube,

$$J = \frac{i}{l^2}$$

$$\frac{V_{\text{Hall}}}{\boldsymbol{B}} = \frac{i}{nel}$$

To get the required sensitivity,

$$n = \frac{i}{el\left(V_{\text{Hall}}/\boldsymbol{B}\right)}$$

$$i = 10\ \text{mA}, \left(\frac{V_{\text{Hall}}}{\boldsymbol{B}}\right) = 1\ \text{V}\,\text{T}^{-1}, e = 1.6 \times 10^{-19}\ \text{C}, \text{ and } l = 1\ \text{mm}$$

$$n = \frac{10 \times 10^{-3}}{\left(1.6 \times 10^{-19}\right)\left(1 \times 10^{-3}\right)(1)}$$

$$n = 6.25 \times 10^{19}\ \text{m}^{-3}$$

Hall Effect Sensor—Alternative Solution Using Mobility

$$\mu_e = \frac{\text{Velocity}}{\text{Electric field}} = \frac{v}{E_H}$$

and

$$J = nev$$

therefore,

$$J = ne\mu_e E_H$$

Assuming that the required voltage is 1 V mm^{-1}

$$E_H = 1000\ \text{V m}^{-1}$$

Therefore,

$$n = \frac{J}{e\mu_e E_H}$$

and if i = 10 mA and the cross-sectional area is 1 mm^2

$$J = \frac{10\times10^{-3}}{10^{-6}} = 1\times10^{4}\ \text{A m}^{-2}$$

$$n = \frac{1\times10^{4}}{(1.6\times10^{-19})(0.5)(1000)}$$

$$n = 1.25\times10^{20}\ \text{m}^{-3}$$

Exercise 3.8: Susceptibility Balance

The translational force on a dipole arises only when there is a field gradient $d\boldsymbol{H}/dx$ present

$$\boldsymbol{F}_{p+} = \mu_0 p\boldsymbol{H}_{p+}$$

$$\boldsymbol{F}_{p-} = -\mu_0 p\boldsymbol{H}_{p-}$$

and if there is a uniform field gradient

$$\boldsymbol{H}_{p+} = \boldsymbol{H}_{p-} + l\frac{d\boldsymbol{H}}{dx}$$

Therefore,

$$\boldsymbol{F}_{\text{net}} = \mu_0 pl\frac{d\boldsymbol{H}}{dx}$$

This equation can then be recast in terms of susceptibility, volume, field, and field gradient

$$F_{\mathrm{net}_{\mathrm{mag}}} = \mu_0 \chi V H \frac{dH}{dx}$$

The force due to gravity on the specimen of paramagnetic material is

$$F_g = mg \qquad \text{Note: This } m \text{ is mass}$$

$$F_g = \rho V g \qquad \text{where } \rho \text{ is the density}$$

Therefore, the error in the measurement of the mass is

$$\frac{F_{\mathrm{meas}}}{F_{\mathrm{true}}} = \frac{F_g - F_{\mathrm{net}_{\mathrm{mag}}}}{F_g}$$

$$= \frac{\rho V g - \mu_0 \chi V H (dH/dx)}{\rho V g}$$

The volume V cancels out so that

$$\frac{F_{\mathrm{meas}}}{F_{\mathrm{true}}} = 1 - \frac{\mu_0 \chi H}{\rho g} \frac{dH}{dx}$$

The error in the weight is therefore

$$\frac{\Delta w}{w} = \frac{\mu_0 \chi H}{\rho g} \frac{dH}{dx}$$

where:

$$\mu_0 = 4\pi \times 10^{-7} \ \mathrm{H\,m^{-1}}$$

$$\chi = 2 \times 10^{-4}$$

$$H = 2 \times 10^{6} \ \mathrm{A\,m^{-1}}$$

$$\frac{dH}{dx} = 1 \times 10^{7} \ \mathrm{A\,m^{-2}}$$

and a reasonable density is $\rho = 8000\,\mathrm{kg\,m^{-3}}$, so

$$\frac{\Delta w}{w} = \frac{(4\pi \times 10^{-7})(2 \times 10^{-4})(2 \times 10^{6})(1 \times 10^{7})}{(8000)(9.81)}$$

$$\frac{\Delta w}{w} = 0.064$$

or

$$\frac{\Delta w}{w} = 6.4\ \%$$

The actual error in $\boldsymbol{F}$ is given by

$$\Delta \boldsymbol{F} = \boldsymbol{F}_{\text{net}_{\text{mag}}}$$

$$\Delta \boldsymbol{F} = \mu_0 \chi V \boldsymbol{H} \frac{d\boldsymbol{H}}{dx}$$

$$V = 1 \times 10^{-6}\ \text{m}^3$$

$$\Delta \boldsymbol{F} = \left(4\pi \times 10^{-7}\right)\left(2 \times 10^{-4}\right)\left(1 \times 10^{-6}\right)\left(2 \times 10^{6}\right)\left(1 \times 10^{7}\right)$$

$$\Delta \boldsymbol{F} = 0.0050\ \text{N}$$

$$\Delta w = \frac{0.005}{9.81}\ \text{kg}$$

$$\Delta w = 0.00051\ \text{kg}$$

Exercise 3.9: Lorentz Force and Cyclotron Resonance

The force on a moving charge e is

$$\boldsymbol{F} = ev \times \boldsymbol{B}$$

Tangential velocity is v. Acceleration in circular motion is a

$$\boldsymbol{F} = ma$$

$$\boldsymbol{F} = mw^2 r = \frac{mv^2}{r}$$

where $\boldsymbol{r}$ is the radius of circular motion.
Therefore,

$$ev\boldsymbol{B} = \frac{mv^2}{r}$$

Therefore,

$$e\boldsymbol{B} = \frac{mv}{r}$$

So the radius of curvature is simply

$$r = \frac{mv}{e\boldsymbol{B}}$$

The time taken to complete a single circuit is

$$\tau = \frac{2\pi r}{v}$$

Therefore, the effective current is

$$i = \frac{e}{\tau}$$

$$i = \frac{ev}{2\pi r}$$

So the magnetic moment is

$$\boldsymbol{m}_{\text{mag}} = iA = \frac{ev}{2\pi r} \cdot \pi r^2 = \frac{evr}{2}$$

$$\boldsymbol{m}_{\text{mag}} = \frac{ev}{2} \cdot \frac{\boldsymbol{m}v}{e\boldsymbol{B}} = \frac{\boldsymbol{m}v^2}{2\boldsymbol{B}}$$

and given the values

$$\boldsymbol{m} = 9.1 \times 10^{-31}\ \text{kg}$$

$$v = 10^4\ \text{ms}^{-1}$$

$$\boldsymbol{B} = 10\ \text{T}$$

the result is

$$\boldsymbol{m}_{\text{mag}} = \frac{\boldsymbol{m}v^2}{2\boldsymbol{B}} = \frac{\left(9.1 \times 10^{-31}\right)\left(10^8\right)}{2 \times 10} = 4.55 \times 10^{-24}\ \text{A m}^2$$

Orbital angular momentum

$$l = mvr$$

and

$$r = \frac{\boldsymbol{m}v}{eB}$$

so

$$l = \frac{\boldsymbol{m}^2 v^2}{eB}$$

$$= \frac{\left(9.1 \times 10^{-31}\right)^2 \left(10^8\right)}{\left(1.6 \times 10^{-19}\right)10}$$

$$l = 5.18 \times 10^{-35}\ \text{kg m}^2\,\text{s}^{-1}$$

Note that

$$\frac{el}{2m} = \boldsymbol{m}_{\text{mag}}$$

Exercise 3.10: Flux Measurement Using Coil and Integrating Voltmeter

$$\boldsymbol{M}_s = 0.875 \times 10^6\ \text{A m}$$

$$\boldsymbol{H} = 1000\ \text{A m}^{-1}$$

and

$$B = \mu_0 (H + M)$$

so

$$B = (4\pi \times 10^{-7})(876 \times 10^3)$$

$$B = 1.10\ \mathrm{T}$$

and

$$\phi = BA,\ A = 10^4\ \mathrm{m}^2$$

$$\phi = 0.11 \times 10^{-3}\ \mathrm{Wb}$$

Since the flux is reversed in one second, the rate of change is

$$\frac{\Delta\phi}{\Delta t} = \frac{0.22 \times 10^{-3}}{1}\ \mathrm{Wb\,s^{-1}}$$

$$V = N\frac{d\phi}{dt}\ \mathrm{V}$$

$$N = \frac{V}{d\phi/dt} = \frac{0.5\ \mathrm{V}}{0.22 \times 10^{-3}}$$

$$N = 2.3 \times 10^3\ \text{turns}$$

CHAPTER 4

Exercise 4.1: Properties of Ferromagnets

Ferromagnets exhibit remanent magnetization, paramagnets do not. Paramagnets have relative permeabilities close to 1, ferromagnets have permeability much greater than 1.

Saturation magnetization is the upper limit to magnetization in a magnetic material. It is not affected even when the material is heated through its Curie temperature. However spontaneous magnetization within domains is reduced effectively to zero as the material is heated through its Curie temperature.

Coercivity is used to distinguish between hard and soft magnetic materials. For an electromagnet material high permeability and low coercivity are the most important magnetic property considerations. For a transformer high permeability, low hysteresis loss, and low eddy current loss are the most important magnetic property considerations.

Exercise 4.2: Use of Initial Magnetization Curve to Find Flux in Core

If the material is in the form of a toroid with mean circumference 40 cm and a cross-sectional area 4 cm^2, with a coil of 400 turns, then the magnetic field generated by a current i in the coils will be

$$\begin{aligned} \boldsymbol{H} &= \frac{Ni}{l} \\ &= \frac{400i}{0.4} \\ &= 1000i \text{ A m}^{-1} \end{aligned}$$

This will correspond to a certain magnetic induction at each field strength. The magnetic induction $\boldsymbol{B}$ (in Wb m^{-2} or equivalently tesla) can be read from the graph of $\boldsymbol{B}$ against $\boldsymbol{H}$ for the calculated values of $\boldsymbol{H}$.

i (A)	*H* (A m^{-1})	*B* (T)
0.1	100	0.52
0.2	200	1.10
0.3	300	1.24
0.4	400	1.33
0.5	500	1.40

and since the flux Φ is given by

$$\Phi = \boldsymbol{B}A$$

where $A = 4 \times 10^{-4}$ m^2 we arrive at the following values

i (A)	*B* (T)	Φ (10^{-3} Wb)
0.1	0.52	0.208
0.2	1.10	0.440
0.3	1.24	0.496
0.4	1.33	0.532
0.5	1.40	0.560

Exercise 4.3: Calculation of Atomic Magnetic Moment

1 m^3 of iron at saturation will have a magnetic moment of 1.7×10^6 A m^2. This 1 m^3 will have a mass of 7970 kg and therefore will contain N atoms

$$\begin{aligned} N &= \frac{7970 \times 6.025 \times 10^{26}}{56} \\ &= 8.58 \times 10^{28} \text{ atoms} \end{aligned}$$

Therefore, the magnetic moment per atom is $\boldsymbol{m}_{\mathrm{A}}$ given by

$$\begin{aligned} \boldsymbol{m}_{\mathrm{A}} &= \frac{\left(1.7 \times 10^{6}\right)}{\left(8.58 \times 10^{28}\right)} \text{A m}^2 \\ &= 1.98 \times 10^{-23} \text{A m}^2 \\ &= 2.49 \times 10^{-29} \text{J A m}^{-1} \\ &= 2.14 \mu_{\mathrm{B}} \end{aligned}$$

Exercise 4.4: Force Due to a Field Gradient

The force on a dipole under the action of a field gradient $d\boldsymbol{H}/dx$ is

$$\boldsymbol{F}_{\text{net}} = \boldsymbol{F}_1 - \boldsymbol{F}_2$$

$$= \mu_0 p_1 \boldsymbol{H}_1 - \mu_0 p_2 \boldsymbol{H}_2$$

and assuming the pole strengths are equal

$$\boldsymbol{F}_{\text{net}} = \mu_0 p(\boldsymbol{H}_1 - \boldsymbol{H}_2)$$

Now consider the field at the center of the dipole $\boldsymbol{H}_0$,

$$\boldsymbol{H}_1 = \boldsymbol{H}_0 + \frac{d\boldsymbol{H}}{dx}\frac{l}{2}$$

$$\boldsymbol{H}_2 = \boldsymbol{H}_0 - \frac{d\boldsymbol{H}}{dx}\frac{l}{2}$$

$$\boldsymbol{H}_1 - \boldsymbol{H}_2 = l\frac{d\boldsymbol{H}}{dx}$$

$$\boldsymbol{F}_{\text{net}} = \mu_0 pl\frac{d\boldsymbol{H}}{dx}$$

Now $pl = \boldsymbol{m}$ the magnetic moment, which is the magnetization multiplied by the volume:

$$\boldsymbol{m} = pl = \boldsymbol{M}_s V$$

$$\boldsymbol{F}_{\text{net}} = \mu_0 \boldsymbol{M}_s V\frac{d\boldsymbol{H}}{dx}$$

$$= \mu_0 \boldsymbol{M}_s \frac{4}{3}\pi r^3 \frac{d\boldsymbol{H}}{dx}$$

and when the net force is 1 N

$$\frac{d\boldsymbol{H}}{dx} = \frac{1.0}{\mu_0 \boldsymbol{M}_s (4/3)\pi r^3}$$

$$= \frac{1.0}{(12.56\times10^{-7})(1.7\times10^{6})(5.24\times10^{-10})}$$

$$= 8.9\times10^{8}\ \text{A m}^{-2}$$

Exercise 4.5: Elementary Dipole Moments

The macroscopic magnetic moment of a magnetic material is ultimately the vector sum of the magnetic moments of all the atoms in the material. The alignment of their atomic magnetic moments can be altered by the application of a magnetic field.

In the case of diamagnets, there is no net permanent magnetic moment on each atom, and therefore, the change in magnetization with field, the susceptibility, results from the magnetic moments on the atoms induced by the applied field. Since this results from Lenz's law the susceptibilities of diamagnets are negative.

Both paramagnets and ferromagnets have permanent magnetic moments per atom, and these are large compared with the moments induced on diamagnetic atoms by an applied field. Therefore, whenever paramagnetism or ferromagnetism occurs, their effects dominate any effects due to diamagnetism. Paramagnets consist of an array of approximately noninteracting magnetic moments, which therefore respond independently to the field. They therefore have small positive susceptibilities because the permanent magnetic moments on the atoms rotate into the field direction as the field is increased. Ferromagnets consist of arrays of interacting magnetic moments on the atoms, which exhibit cooperative behavior by rotating together into the field direction as the field is increased. As a result of this, ferromagnets have large, positive, susceptibilities.

Exercise 4.6: Curie's Law

The Langevin equation for magnetization of a paramagnet is

$$\boldsymbol{M} = \boldsymbol{M}_{\mathrm{s}}\left[\coth\left(\frac{\mu_0 \boldsymbol{mH}}{k_{\mathrm{B}}T}\right) - \frac{k_{\mathrm{B}}T}{\mu_0 \boldsymbol{mH}}\right]$$

If the probability of the magnetic moment $\boldsymbol{m}$ on an atom occupying an energy state E is given by

$$P = P_0 \exp\left(\frac{-E}{k_{\mathrm{B}}T}\right)$$

at a given temperature T, then by integrating this function over all possible directions and over all moments, we can obtain the net magnetization. The energy of an individual moment $\boldsymbol{m}$ in a field $\boldsymbol{H}$ is

$$E = -\mu_0 \boldsymbol{m}\cdot\boldsymbol{H} = -\mu_0 \boldsymbol{mH}\cos\theta$$

$$\boldsymbol{M} = \frac{N\boldsymbol{m}\int_0^{\pi} \exp\left(\mu_0 \boldsymbol{mH}\cos\theta / k_{\mathrm{B}}T\right)\cos\theta\sin\theta\, d\theta}{\int_0^{\pi} \exp\left(\mu_0 \boldsymbol{mH}\cos\theta / k_B T\right)\sin\theta\, d\theta}$$

$N\boldsymbol{m} = \boldsymbol{M}_{\mathrm{s}}$, and integrating

$$\boldsymbol{M} = \boldsymbol{M}_{\mathrm{s}}\left[\coth\left(\frac{\mu_0 \boldsymbol{mH}}{k_{\mathrm{B}}T}\right) - \frac{k_{\mathrm{B}}T}{\mu_0 \boldsymbol{mH}}\right]$$

The series expansion for cot h (x) is

$$\cot hx = \frac{1}{x} + \frac{x}{3} + \frac{x^3}{45} \cdots$$

For high temperatures T and hence small values of $\mu_0 \boldsymbol{mH}/k_B T$, this series can be truncated as

$$\boldsymbol{M} \cong \boldsymbol{M}_s \frac{\mu_0 \boldsymbol{mH}}{3k_B T}$$

so that

$$\chi = \frac{d\boldsymbol{M}}{d\boldsymbol{H}}$$

$$= \frac{\mu_0 N\boldsymbol{m}^2}{3k_B T}$$

and putting $C = \mu_0 Nm^2/3k_B$, this gives Curie's law

$$\chi = \frac{C}{T}$$

Exercise 4.7: Differences between Diamagnets, Paramagnets, and Ferromagnets

Diamagnets have a small negative susceptibility. Paramagnets have a small positive susceptibility. Ferromagnets have a large positive susceptibility. Diamagnets have no net permanent magnetic moment per atom. Paramagnets have a permanent magnetic moment per atom but these are randomly aligned. Ferromagnets have a net magnetic moment per atom and these are aligned parallel.

For diamagnets,

$$\chi = -\frac{\mu_0 Ze^2 n\left(r^2\right)}{6\boldsymbol{m}_e}$$

For paramagnets,

$$\chi = \frac{\mu_0 n\boldsymbol{m}^2}{3k_B T}$$

Exercise 4.8: Susceptibility of a Paramagnetic Gas

The number of particles in a volume of 22.4 L is

$$N = \frac{6.02 \times 10^{23}}{22.4 \times 10^{-3}} \text{ m}^{-3}$$

$$N = 2.69 \times 10^{25} \text{ m}^{-3}$$

According to Curie's Law for paramagnets, consisting of noninteracting magnetic moments

$$\chi = \frac{\mu_0 N \boldsymbol{m}^2}{3k_B T}$$

If $\boldsymbol{m}$ = 1, and the Bohr magneton $= 9.27 \times 10^{-24}$ A m^2 and $k_B = 1.38 \times 10^{-23}$ J K^{-1}

$$\chi = \frac{\left(4\pi \times 10^{-7}\right)\left(2.69 \times 10^{25}\right)\left(9.27 \times 10^{-24}\right)}{(3)\left(1.38 \times 10^{-23}\right)(273)}$$

$$\chi = 2.57 \times 10^{-7}$$

Exercise 4.9: Spontaneous Magnetization

Magnetic moment per atom $\boldsymbol{m} = 2\mu_B$

$$\boldsymbol{m} = 18.54 \times 10^{-24} \text{ A m}^2$$

At absolute zero of temperature, all magnetic moments will be perfectly aligned parallel

$$\boldsymbol{M}_s(0\text{K}) = \text{N m}$$

where the number of atoms per unit volume N is given by

$$N = \frac{N_A \rho}{w_A}$$

$$N = \frac{\left(6.02 \times 10^{26}\right)(8000)}{56}$$

$$N = 8.6 \times 10^{28} \text{ m}^{-3}$$

$$\boldsymbol{M}_s(0\text{ K}) = \left(8.6 \times 10^{28}\right)\left(18.54 \times 10^{-24}\right)$$

$$\boldsymbol{M}_s(0\text{ K}) = 1.59 \times 10^6 \text{ A m}^{-1}$$

This assumes all magnetic moments are aligned parallel. The value of magnetization $\boldsymbol{M}$ is therefore the value of magnetization for a single domain at absolute zero of temperature.

In a large enough sample, the material will subdivide into a number of magnetic domains to minimize magnetostatic energy. In those cases, there will be a deviation from the magnetization within a single domain (spontaneous magnetization), and then it is not possible to use $\boldsymbol{M} = N\boldsymbol{m}$ to calculate magnetization.

As the temperature increases then

1. The magnetic moment per atom remains the same.
2. The spontaneous magnetization within a single domain $\boldsymbol{M}_s$ will decrease from $\boldsymbol{M}_s = N\boldsymbol{m}$ because thermal energy disrupts the alignment of magnetic moments due to precession.
3. The plot of $\boldsymbol{M}_s/\boldsymbol{M}_0$ as a function of temperature will be as shown in Figure 9.5

   where:

$$\boldsymbol{M}_0 = \boldsymbol{M}_s(0\ \text{K})$$

The number of Bohr magnetons per atom does not change with temperature.

A piece of iron does not typically exhibit high magnetization at room temperature because it divides into a number of domains for which the vector sum of moments is less than saturation magnetization as it tries to minimize magnetostatic energy. Only when it has been magnetized does it exhibit signification bulk magnetization.

The result is that the bulk magnetization $\boldsymbol{M}$ may be close to zero even though the spontaneous magnetization within a domain is close to saturation.

Exercise 4.10: Relationship between Curie Temperature and Internal Exchange Coupling

$$E_{\text{mag}} = 0.27\,\text{eV} = 0.432\times10^{-19}\ \text{J}$$

and

$$3k_\text{B}T_\text{c} = E_{\text{mag}}$$

so

$$T_\text{c} = \frac{E_{\text{mag}}}{3k_\text{B}}$$

$$T_\text{c} = \frac{0.432\times10^{-19}}{4.14\times10^{-23}}$$

$$= 0.1043\times10^{4}$$

$$T_\text{c} = 1043\ \text{K} = 770\ °\text{C}$$

CHAPTER 5

Exercise 5.1: Fröhlich-Kennelly Equation

$$\boldsymbol{M} = \frac{\alpha\boldsymbol{H}}{1+\beta\boldsymbol{H}}$$

Therefore,

$$1+\beta\boldsymbol{H} = \frac{\alpha\boldsymbol{H}}{\boldsymbol{M}}$$

and

$$\frac{\boldsymbol{M}}{\boldsymbol{H}} = \chi = (\mu_r - 1)$$

Therefore,

$$1 + \beta \boldsymbol{H} = \frac{\alpha}{(\mu_r - 1)} = \frac{\alpha \mu_0}{(\mu - \mu_0)}$$

$$\frac{1 + \beta \boldsymbol{H}}{\alpha} = \frac{\mu_0}{\mu - \mu_0}$$

If a = $1/\mu_0\alpha$ and $b = \beta/\mu_0\alpha$, then

$$a + b\boldsymbol{H} = \frac{1}{\mu - \mu_0}$$

which is the Kennelly equation.

The Fröhlich equation

$$M = \frac{\alpha \boldsymbol{H}}{(1 + \beta \boldsymbol{H})}$$

can be expanded as a binomial series

$$\alpha \boldsymbol{H}\left(\frac{1}{\beta \boldsymbol{H}} - \frac{1}{(\beta \boldsymbol{H})^2} + \frac{1}{(\beta \boldsymbol{H})^3} \cdots\right)$$

and, if $\alpha/\beta = \boldsymbol{M}_s$,

$$\boldsymbol{M} = \boldsymbol{M}_s\left(1 - \frac{1}{\beta \boldsymbol{H}} + \frac{1}{(\beta \boldsymbol{H})^2} \cdots\right)$$

Expanding the Kennelly equation,

$$\frac{1}{\mu - \mu_0} = a + b\boldsymbol{H}$$

$$\boldsymbol{M} = \frac{1}{\mu_0}\left(\frac{\boldsymbol{H}}{a + b\boldsymbol{H}}\right)$$

Taking the limit as $\boldsymbol{H} \to \infty$

$$\frac{1}{\mu_0 b} = \boldsymbol{M}_s$$

$$\boldsymbol{M} = \boldsymbol{M}_s\left(1 - \frac{a}{b\boldsymbol{H}} + \left(\frac{a}{b\boldsymbol{H}}\right)^2 \cdots\right)$$

and the law of approach is

$$M = M_s\left(1 - \frac{A}{H} - \frac{B}{H^2} \cdots\right) + kH$$

Therefore, all three equations have similar form at high fields, neglecting the small kH term.

Exercise 5.2: Magnetization of Paramagnets

Aluminum has a permeability of 1.00002 and hence a susceptibility of $\chi = 2 \times 10^{-5}$. The magnetization and field are related by

$$M = \chi H = 0.5 \times 10^6 \text{ A m}^{-1}$$

The field required to give aluminum the magnetization as nickel is therefore

$$H = \frac{M}{\chi} = 2.5 \times 10^{10} \text{ A m}^{-1}$$

In the case of copper, $\chi = -1 \times 10^{-5}$, and so the required field is

$$H = \frac{M}{\chi} = 5.0 \times 10^{10} \text{ A m}^{-1}$$

These field levels are far above anything presently attainable.

Exercise 5.3: Determination of Rayleigh Coefficients at Low Fields

Use the Rayleigh equation to determine the coefficients from the first two data points

$$B = \mu H + \nu H^2$$

At $H = 5$ A m^{-1} and $B = 0.0019$ T

$$0.0019 = 5\mu + 25\nu$$

At $H = 10$ A m^{-1} and $B = 0.0042$ T

$$0.0042 = 10\mu + 100\nu$$

From these two equations, we find $\mu = 3.4 \times 10^{-4}$ H m^{-1} and $\nu = 8 \times 10^{-6}$ H A^{-1}. Substituting these into the equation and using the data points from the table, it is apparent that the Rayleigh region does not extend to 40 A m^{-1}, but that the Rayleigh approximation is valid at 20 A m^{-1}.

The remanence B_R in the Rayleigh region is given by

$$B_R = \frac{1}{2}\nu H_m^2$$

therefore at $\boldsymbol{H}_{\rm m} = 10$ A m^{-1}

$$\boldsymbol{B}_{\rm R} = 0.0004\ {\rm T}$$

and at $\boldsymbol{H}_{\rm m} = 20$ A m^{-1}

$$\boldsymbol{B}_{\rm R} = 0.0016\ {\rm T}$$

The hysteresis loss in the Rayleigh region is given by

$$W_{\rm H} = \frac{4}{3} v \boldsymbol{H}_{\rm m}^2$$

So at $\boldsymbol{H}_{\rm m} = 20$ A m^{-1}

$$W_{\rm H} = 85.3 \times 10^{-3}\ {\rm J\,m^{-3}}$$

Exercise 5.4: Magnetostriction of a Material without Stress

Since the specimen of terfenol does not have a preferred orientation and is not subject to stress, we can use the equation for the averaged magnetostriction of a randomly oriented polycrystal.

$$\lambda_{\rm s} = \left(\frac{2}{5}\right)\lambda_{100} + \left(\frac{3}{5}\right)\lambda_{111}$$

Consequently, for the values given for this particular piece of terfenol

$$\lambda_{\rm s} = 36 \times 10^{-6} + 960 \times 10^{-6}$$
$$= 996 \times 10^{-6}$$

Now considering the observed magnetostriction when the field is changed from perpendicular to parallel to a given direction,

$$\lambda_{\rm s\parallel} - \lambda_{\rm s\perp} = \lambda_{\rm s} + \frac{1}{2}\lambda_{\rm s} = \frac{3}{2}\lambda_{\rm s}$$
$$= 1494 \times 10^{-6}$$

Exercise 5.5: Magnetostriction of a Material under Compressive Load

In this case, the uniaxial compressive stress causes all domain magnetic moments to align perpendicular to the stress axis and hence to the field axis.

The magnetization will take place entirely by rotation and hence

$$\boldsymbol{M} = \boldsymbol{M}_{\rm s} \cos\theta$$

and therefore

$$\lambda = \frac{3}{2}\lambda_s \cos^2\theta$$

$$\lambda \boldsymbol{M} = \frac{3}{2}\lambda_s \left(\frac{\boldsymbol{M}}{\boldsymbol{M}_s}\right)^2$$

$$= \left(1494 \times 10^{-6}\right)\left(\frac{\boldsymbol{M}}{\boldsymbol{M}_s}\right)^2$$

Consequently, when $\boldsymbol{M} = \boldsymbol{M}_s/2$

$$\lambda = 373.5 \times 10^{-6}$$

and when $\boldsymbol{M} = \boldsymbol{M}_s$

$$\lambda = 1494 \times 10^{-6}$$

Exercise 5.6: Hysteresis and Power Losses

The hysteresis loss is $\int \boldsymbol{H}\, d\boldsymbol{B}$, or the enclosed area within the $\boldsymbol{B}$-$\boldsymbol{H}$ hysteresis loop. Approximate hysteresis loss W_H is

$$W_H = 4\boldsymbol{H}_c \boldsymbol{B}_R$$

$$= 200 \text{ J m}^{-3} \text{ cycle}^{-1}$$

which is comparable with grain-oriented Fe-Si. If the volume of material is V, then the total hysteresis loss is $W = W_H V$ J cycle^{-1}. $V = 1.2 \times 10^{-4}$ m^3, so

$$W = 0.024 \text{ J cycle}^{-1}$$

At a frequency of v = 60 Hz, this will therefore correspond to the power loss $P_H = Wv$ (watts)

$$P_H = 1.44 \text{ W}$$

Eddy current power loss. The expression for classical eddy current loss is

$$P_{EC} = \frac{\pi^2 d^2 v^2 B_{max}^2}{\rho\beta} \text{ W m}^{-3}$$

$$= 106.4 \times 10^{-3} \text{ W m}^3$$

and the volume is 1.2×10^{-4} m^3; therefore, the total classical eddy current loss in the transformer is

$$P_{EC} = 12.8 \text{ W}$$

Exercise 5.7: Mathematical Models for Magnetization Processes in Ferromagnets

a. Low field model—Rayleigh law

$$B = \mu(0)H + vH^2$$

or equivalently

$$M = \chi(0)H + \frac{vH^2}{\mu_0}$$

b. High field model—law of approach to saturation

$$M = M_s\left(1 - \frac{a}{H} - \frac{b}{H^2} - \dots\right) + kH$$

c. Anhysteretic magnetization model—Fröhlich-Kennelly relation

$$M = \frac{\alpha H}{1 + \beta H}$$

or the Langevin-Weiss equation

$$M = M_s\left\{\cot e\left[\frac{\mu_0 m(H + \alpha M)}{k_B T}\right] - \frac{k_B T}{\mu_0 m(H + \alpha M)}\right\}$$

Consider the variation of magnetization with field in the presence of anisotropy. At sufficiently high field, the magnetization process is by coherent rotation against anisotropy.

Let θ be the angle between the spontaneous magnetization in a domain and the field direction.

The component of M_s along the field direction is

$$M = M_s \cos\theta$$

If the magnetization is close to saturation, then θ will be small so that

$$\cos\theta \cong 1 - \frac{\theta^2}{2}$$

Therefore,

$$M = M_s\left(1 - \frac{\theta^2}{2}\right)$$

The torque exerted by the magnetic field H on this domain will be

$$\tau_H = \mu_0 M_s H \sin\theta$$

and the torque exerted by the anisotropy energy E_a by the anisotropy will be

$$\tau_a = -\frac{\partial E_a}{\partial \theta}$$

So equating these and making the approximation ***sin*** $\theta \cong \theta$ for small angles

$$\mu_0 \boldsymbol{M}_s \boldsymbol{H} \theta = -\frac{\partial E_a}{\partial \theta}$$

Therefore,

$$\theta = \frac{1}{\mu_0 \boldsymbol{M}_s \boldsymbol{H}} \cdot \left(-\frac{\partial E_a}{\partial \theta}\right)$$

Therefore,

$$\boldsymbol{M} = \boldsymbol{M}_s \left[1 - \frac{1}{2}\left(\frac{\partial E_a/\partial \theta}{\mu_0 \boldsymbol{M}_s \boldsymbol{H}}\right)^2 \cdots\right]$$

and

$$b = \frac{1}{2}\frac{(\partial E_a/\partial \theta)^2}{(\mu_0 \boldsymbol{M}_s)^2}\frac{1}{\boldsymbol{H}^2}$$

The torque due to anisotropy can by expressed in the simplest case by axial anisotropy

$$E_a = K \sin^2 \theta$$

$$\frac{dE_a}{d\theta} = 2K \sin\theta \cos\theta$$

$$\frac{dE_a}{d\theta} = K \sin 2\theta$$

and for small angles

$$\frac{dE_a}{d\theta} = 2K\theta$$

and therefore

$$b = 2\left(\frac{K\theta}{\mu_0 \boldsymbol{M}_s}\right)^2$$

and so the b coefficient, the second-order term in the law of approach, is determined by the magnetocrystalline anisotropy.

The assumptions were that θ is small so that approximation can be used for the value of sin θ and cos θ in the equation for torque and magnetization.

Exercise 5.8: Magnetostriction along Different Directions Relative to the Magnetization

Magnetostriction is the fractional charge in length caused by an applied magnetic field (here field included magnetostriction) or by the spontaneous magnetization within a domain (it arises as a result of the strain dependence of the anisotropy energy).

The equation for anisotropic magnetostriction for a cubic material is

$$\lambda_s(\theta) = \frac{3}{2}\lambda_{100}\left(\alpha_1^2\beta_1^2 + \alpha_2^2\beta_2^2 + \alpha_3^2\beta_3^2 - \frac{1}{3}\right) + 3\lambda_{111}\left(\alpha_1\alpha_2\beta_1\beta_2 + \alpha_2\alpha_3\beta_2\beta_3 + \alpha_3\alpha_1\beta_3\beta_1\right)$$

where:

the direction cosines α_1, α_2, and α_3 describe/define the direction of the magnetization vector

the direction cosines $\beta_{1,}$ $\beta_{2,}$ and β_3 describe/define the direction alone, where the magnetostriction is measured (which is also usually the direction along which the field is applied)

Note that in anisotropic materials, the magnetization vectors do not necessarily point along the direction of the magnetic field.

So with strain along the $\langle 110 \rangle$ direction,

$$\beta_1 = \frac{1}{\sqrt{2}},\ \beta_2 = \frac{1}{\sqrt{2}},\ \beta_3 = 0$$

and when the magnetization is at some arbitrary angle γ to the $\langle 100 \rangle$ direction, but remains in the (001) plane.

$$\alpha_1 = \cos\gamma,\ \alpha_2 = \cos(90 - \gamma),\ \alpha_3 = 0$$

$$= \sin\gamma$$

Substituting these values into the saturation magnetostriction equation gives

$$\lambda_{\langle 110 \rangle}(\gamma) = \frac{3}{2}\lambda_{100}\left(\frac{1}{2}\cos^2\gamma + \frac{1}{2}\sin^2\gamma + 0 - \frac{1}{3}\right) + 3\lambda_{111}\left(\frac{1}{2}\cos\gamma\sin\gamma + 0 + 0\right)$$

$$\lambda_{\langle 110 \rangle}(\gamma) = \frac{1}{4}\lambda_{100} + \frac{3}{4}\lambda_{111}\sin 2\gamma$$

When the magnetization is along the $\langle 100 \rangle$ direction, then $\gamma_{\langle 110 \rangle}(\gamma) = (1/4)\gamma_{100}$ because $\sin 2\gamma = 0$

So the change in strain along the $\langle 110 \rangle$ direction, when the moments rotate from $\langle 100 \rangle$ to the $\langle 110 \rangle$ direction, $\gamma = 45°$

$$\lambda_{110}(0) = \frac{1}{4}\lambda_{100}$$

$$\lambda_{110}(45) = \frac{1}{4}\lambda_{100} + \frac{3}{4}\lambda_{111}$$

Exercise 5.9: Differences in Magnetization Curves for the Same Material in a Toroid and in a Sphere

In the case of a toroid, the applied field $\boldsymbol{H}_{\text{app}}$ is the same as the true field.

In the case of a sphere the true field, $\boldsymbol{H}$ is smaller than the applied field $\boldsymbol{H}_{\text{app}}$ according to the following equation:

$$\boldsymbol{H} = \boldsymbol{H}_{\text{app}} - N_{\text{d}}\boldsymbol{M}$$

So, at the origin $\boldsymbol{H}_{\text{app}} \cong 0$ and $\boldsymbol{M} \cong 0$ so that $\boldsymbol{H} = 0$ and

$$\boldsymbol{H}_{\text{app}} \cong N_{\text{d}}\boldsymbol{M}$$

$$\frac{\boldsymbol{H}_{\text{app}}}{N_{\text{d}}} = \boldsymbol{M}$$

and therefore the measured differential susceptibility is

$$\frac{d\boldsymbol{M}}{d\boldsymbol{H}_{\text{app}}} \cong \frac{1}{N_{\text{d}}}$$

This arises because the term $N_{\text{d}}\boldsymbol{M}$ will be much larger than $\boldsymbol{H}$ and so to overcome the tendency of the material to remain demagnetized, we must supply a magnetic field $\boldsymbol{H}_{\text{app}}$ that is at least as great as $N_{\text{d}}\boldsymbol{M}$,

$$\boldsymbol{H}_{\text{app}} \cong N_{\text{d}}\boldsymbol{M}$$

in order to cause magnetization to change.

$$\frac{d\boldsymbol{M}}{d\boldsymbol{H}_{\text{app}}} = \frac{1}{N_{\text{d}}} = 3$$

Assuming that $d\boldsymbol{M}/d\boldsymbol{H}$ is very large compared with $d\boldsymbol{M}/d\boldsymbol{H}_{\text{app}}$, we can write

$$N_{\text{d}}\boldsymbol{M} = \boldsymbol{H}_{\text{app}} - \boldsymbol{H}$$

$$\boldsymbol{M} = \frac{1}{N_{\text{d}}}\left(\boldsymbol{H}_{\text{app}} - \boldsymbol{H}\right)$$

$$\frac{d\boldsymbol{M}}{d\boldsymbol{H}_{\text{app}}} = \frac{1}{N_{\text{d}}} - \frac{1}{N_{\text{d}}}\frac{d\boldsymbol{H}}{d\boldsymbol{H}_{\text{app}}}$$

and in these cases, $\boldsymbol{H}_{\text{app}} \gg \boldsymbol{H}$, so that $d\boldsymbol{H}/d\boldsymbol{H}_{\text{app}}$ is negligible.

Assuming that dM/dH is very large and when N_d is small, this is the same as dM/dH_{app}.

$$x' = \left(\frac{dM}{dH}\right)_{\substack{M=0\\H=0}} \gg 1$$

and

$$H = H_{app} - N_d M$$

$$1 = \frac{dH_{app}}{dH} - N_d \frac{dM}{dH}$$

and if dM/dH is large, then so is dH_{app}/dH.

$$\frac{dH}{dH_{app}} = 1 - N_d \frac{dM}{dH_{app}}$$

$$N_d \frac{dM}{dH_{app}} \cong 1$$

$$\frac{dM}{dH_{app}} \cong \frac{1}{N_d}$$

Exercise 5.10: Classical Dipole Magnetic Field of an Array of Atoms

The equations describing the magnetic field H surrounding a magnetic dipole are, at larger values of r,

$$H_r(r,\theta) = \frac{2m}{4\pi r^3}\cos\theta$$

$$H_\theta(r,\theta) = \frac{m\sin\theta}{4\pi r^3}$$

$$H_\phi(r,\theta,\phi) = 0$$

So if there are atoms with small magnetic moments on a simple cubic lattice,

$$\begin{matrix} \uparrow_{11} & \uparrow_{12} & \uparrow_{13} \\ \uparrow_{21} & \uparrow_{22} & \uparrow_{23} \\ \uparrow_{31} & \uparrow_{32} & \uparrow_{33} \end{matrix}$$

Consider the field due to the dipole m_{22} at the locations m_{12} and m_{23} nearest neighbors.

(1) Field at $\boldsymbol{m}_{12}$ due to moment at $\boldsymbol{m}_{22}$

$$\boldsymbol{H}_{\mathrm{r}}(r,\theta)=\frac{2\boldsymbol{m}}{4\pi r^{3}}f$$

and $\boldsymbol{m}$ = 1 Bohr magneton $\left(=9.27\times10^{-24}\ \mathrm{A\,m^{2}}\right)$ while $r = 0.3\ \boldsymbol{nm}$

Therefore,

$$\boldsymbol{H}_{\mathrm{r}}(r,\theta)=\frac{2\left(9.27\times10^{-24}\right)}{4\times3.142\times\left(0.3\times10^{-9}\right)^{3}}$$

$$\boldsymbol{H}_{\mathrm{r}}(r,\theta)=\frac{18.54}{0.339}\times\frac{10^{-24}}{10^{-27}}$$

$$\boldsymbol{H}_{\mathrm{r}}(r,\theta)=54\times10^{3}\ \mathrm{A\,m^{-1}}$$

The other components of the field will be zero for the location along the axis of the dipole.

(2) Field at $\boldsymbol{m}_{23}$ due to the dipole at $\boldsymbol{m}_{22}$ is

$$\boldsymbol{H}_{\theta}(r,\theta)=\frac{\boldsymbol{m}}{4\pi r^{3}}$$

$$\boldsymbol{H}_{\mathrm{r}}(r,\theta)=27\times10^{3}\ \mathrm{A\,m^{-1}}$$

The other components of the field will be zero for the location perpendicular to the axis of the dipole.

These values of the magnetic field at the two nearest neighbor locations are the classically expected magnetostatic field levels due to the dipoles associated with the atoms.

The strength of these fields are far too small to explain magnetic order in ferromagnets, which requires field strength of the order of 10^{9} A m^{-1}, so the classically expected dipole field strengths are about five orders of magnitude (100,000 times) too small to explain magnetic order and the observed Curie temperatures.

CHAPTER 6

Exercise 6.1 Magnetic Domains

The essay should contain discussion of internal exchange coupling (or Weiss interaction field) to explain magnetic ordering, and needs to consider the magnetostatic energy of a material magnetized throughout its volume and how this ensures the presence of domains despite exchange coupling and domain-wall surface energy. The Curie-Weiss law gave the first evidence for the internal coupling. Later the observation of domains, which are a highly ordered state and hence have a low entropy, can be explained most reasonably on the basis of an internal coupling field. The various methods for domain observation include the Kerr effect, the Barkhausen effect, the Bitter pattern method, neutron diffraction, and Lorentz microscopy using electron diffraction. The observation of domains on the surface tells us virtually nothing about the magnetization processes taking place in the bulk of the material. Even the domain-wall motion on the surface is atypical of the bulk material.

Exercise 6.2: Magnetocrystalline Anisotropy

Magnetocrystalline anisotropy simply refers to the differences in energy of magnetic moments when oriented along certain crystallographic directions. It is usually represented as an energy density, in joules per cubic meter. In the case of cubic magnetocrystalline anisotropy, the energy can be represented as

$$E_{\text{an}} = K_0 + \frac{K_1}{2} \sum_{\substack{i \neq j \\ i=1,2,3}} \cos^2\theta_i \cos^2\theta_j$$

However, the constant term K_0 has no influence on the orientation of the magnetic moments, and is therefore ignored, since it is only the difference in energy that is important:

$$\Delta E_{\text{an}} = \frac{K_1}{2} \sum_{\substack{i \neq j \\ i=1,2,3}} \cos^2\theta_i \cos^2\theta_j$$

The direction of the magnetization, θ_1, θ_2, and θ_3 is determined by minimization of the anisotropy energy in the absence of other effects. Therefore, if $K_1 > 0$, the anisotropy energy reaches a minimum value of $E_{\text{an}} = 0$ when either $\cos\theta_1 = 0$, $\cos\theta_2 = 0$, or $\cos\theta_3 = 0$. In other words, $K_1 > 0$ favors alignment along the $\langle 100 \rangle$ family of directions. When $K_1 < 0$, the anisotropy energy reaches a minimum of $E_{\text{an}} = -K_1/3$ when the magnetic moments are aligned along the $\langle 111 \rangle$ family of directions.

Exercise 6.3: Effects of Anisotropy on Rotation of Magnetization

If the anisotropy is such that the $\langle 100 \rangle$ directions are the magnetically easy axes, then the anisotropy constant K will be positive. The anisotropy energy E_a can then be written

$$E_{\text{an}} = \frac{1}{4} K_1 \sin^2 \phi$$

where is the angle of the magnetization away from the nearest $\langle 100 \rangle$ direction. The field energy is E_{H}

$$\begin{aligned} E_{\text{H}} &= \mu_0 \boldsymbol{M}_0 \boldsymbol{H} \cos\theta \\ &= \mu_0 \boldsymbol{M}_s \boldsymbol{H} \cos\theta \end{aligned}$$

where θ is the direction of the magnetization in any given domain away from the $\langle 110\rangle$ direction of the field.

Clearly, as θ becomes larger the anisotropy energy increases while the field energy decreases.

After all domain-wall processes have been completed, the magnetization in any domain will lie along the $\langle 010\rangle$ and $\langle 100\rangle$ directions closest to the field direction $\langle 110\rangle$. Under these conditions, we must have

$$\theta = \frac{\pi}{4} - \phi$$

and

$$\begin{aligned} E_{\text{tot}} &= E_{\text{a}} + E_{\text{H}} \\ &= \left(\frac{K_1}{4}\right)\sin^2 2\phi - \mu_0 \boldsymbol{M}_{\text{s}}\boldsymbol{H}\cos\left(\frac{\pi}{4} - \phi\right) \end{aligned}$$

The magnetization will take up a direction such that $dE_{\text{tot}}/d\phi = 0$

$$\frac{dE_{\text{tot}}}{d\phi} = K_1 \sin 2\phi \cos 2\phi - \mu_0 \boldsymbol{M}_{\text{s}}\boldsymbol{H}\sin\left(\frac{\pi}{4} - \phi\right)$$

The component of magnetization in the field direction is of course

$$\boldsymbol{M} = \boldsymbol{M}_{\text{s}}\cos\left(\frac{\pi}{4} - \phi\right)$$

Consequently, we arrive at the following relation between magnetization $\boldsymbol{M}$ and field $\boldsymbol{H}$

$$\begin{aligned} \boldsymbol{H} &= \left(\frac{K_1}{\mu_0 \boldsymbol{M}_{\text{s}}}\right)\left[\frac{\sin 2\phi \cos 2\phi}{\sin\left[(\pi/4) - \phi\right]}\right] \\ &= \left(\frac{K_1}{\mu_0 \boldsymbol{M}_{\text{s}}}\right)\frac{\boldsymbol{M}}{\boldsymbol{M}_{\text{s}}}\sin 2\phi\left[\frac{2\cos 2\phi}{\sin\left[(\pi/2) - 2\phi\right]}\right] \end{aligned}$$

and since $\cos 2\phi = \sin(\pi/2 - 2\phi)$

$$\boldsymbol{H} = \left(\frac{K_1}{\mu_0 \boldsymbol{M}_{\text{s}}}\right)\left(\frac{\boldsymbol{M}}{\boldsymbol{M}_{\text{s}}}\right)2\sin 2\phi$$

$$\sin 2\phi = 2\left(\frac{\boldsymbol{M}}{\boldsymbol{M}_{\text{s}}}\right)^2 - 1 = 2\cos^2\left(\frac{\pi}{4} - \phi\right) - 1$$

$$\boldsymbol{H} = \left(\frac{4K_1}{\mu_0 \boldsymbol{M}_{\text{s}}}\right)\left(\frac{\boldsymbol{M}}{\boldsymbol{M}_{\text{s}}}\right)\left[\left(\frac{\boldsymbol{M}}{\boldsymbol{M}_{\text{s}}}\right)^2 - \frac{1}{2}\right]$$

The field needed to saturate the magnetization under these conditions is therefore

$$\boldsymbol{H} = \left(\frac{2K_1}{\mu_0 \boldsymbol{M}_s}\right)$$

Exercise 6.4: Critical Fields as Determined by Anisotropy

The anisotropy energy in the case of uniaxial materials such as cobalt is given by

$$E_a = K_{ul}\sin^2\theta$$

while the field energy is

$$E_H = \quad \mu_0 \boldsymbol{M}_s \boldsymbol{H}\cos\phi$$

where ϕ is the direction of moments from the field direction. Given the situation we have chosen $\theta + \phi = \pi/2$ and therefore

$$E_{tot} = K_{ul}\sin^2\theta \quad \mu_0 \boldsymbol{M}_s \boldsymbol{H}\cos\varphi$$

$$= K_{ul}\sin^2\theta \quad \mu_0 \boldsymbol{M}_s \boldsymbol{H}\sin\theta$$

The moments will reach equilibrium when $dE_{tot}/d\theta = 0$ so that

$$\frac{dE_{tot}}{d\theta} = 2K_{ul}\sin\theta\cos\theta - \mu_0 \boldsymbol{M}_s \boldsymbol{H}\cos\theta = 0$$

$$0 = \left(2K_{ul}\sin\theta - \mu_0 \boldsymbol{M}_s \boldsymbol{H}\right)\cos\theta$$

so either $\cos\theta = 0$, meaning $\theta = \pi/2$, or

$$2K_{ul}\sin\theta \ = \mu_0 \boldsymbol{M}_s \boldsymbol{H}$$

and hence

$$\boldsymbol{H} = \left(\frac{2K_{ul}}{\mu_0 \boldsymbol{M}_s}\right)\sin\theta$$

So when $\theta = \pi/2$, meaning the magnetization is deflected into the base plane

$$\boldsymbol{H} = \frac{2K_{ul}}{\mu_0 \boldsymbol{M}_s} = 4.58\times 10^5 \text{ A m}^{-1}$$

and when $\theta = \pi/4$, meaning the moment are at 45° to the unique axis, we have

$$\boldsymbol{H} = \frac{K_{ul}\sqrt{2}}{\mu_0 \boldsymbol{M}_s} = 3.24\times 10^5 \text{ A m}^{-1}$$

Exercise 6.5: Stress-Induced Anisotropy

The stress anisotropy energy $E\sigma$ generated by a stress σ will be

$$E_\sigma = \frac{3}{2}\lambda_s \sigma \cos^2 \phi$$

where ϕ is the angle between the magnetization vector and the stress direction.

The anisotropy energy is E_a given by

$$E_a = K_1\left(\cos^2\theta_1\cos^2\theta_2 + \cos^2\theta_2\cos^2\theta_3 + \cos^2\theta_3\cos^2\theta_1\right)$$

where is the θ_i are the angles that the magnetization vector makes relative to the crystal axes. We note that the anisotropy energies along certain axes are

$$E_{a(111)} = \frac{1}{3}K_1$$

$$E_{a(110)} = \frac{1}{4}K_1$$

$$E_{a(100)} = 0$$

and remember, K_1 is negative.

In order to rotate all the magnetization vectors into the plane perpendicular to the stress axis, consider the case which requires the most energy. This will be when the magnetization begins along an easy axis aligned coaxially with the stress.

To rotate this magnetization vector, we must simply find the path of rotation, between the easy axis along the stress direction and an easy axis perpendicular to this axis, which has the smallest maximum anisotropy energy.

Clearly, this energy maximum cannot be any larger then $K_1/3$, which is the anisotropy energy along the hard axis. In fact, if the rotation of magnetization is restricted to one of the (100) planes, the maximum anisotropy energy experienced will be (1/4) K_1

$$E_{a\max} = \frac{1}{4}K_1$$

and we can write the anisotropy in this plane as a function of the angle θ from the $\langle 100\rangle$axes or the angle ϕ from the stress axis, which is coincident with the $\langle 111\rangle$ axis in this case

$$E_a = \frac{1}{4}K_1\left(\sin^2\theta - 1\right)$$

$$= \frac{1}{4}K_1\left[\sin^2\left(\frac{\pi}{2} - \phi\right) - 1\right]$$

$$= \frac{1}{4}K_1\left(\cos^2\phi - 1\right)$$

where the additional constant term $K_1/4$ has been included for convenience only to ensure that the anisotropy along the $\langle 111 \rangle$ direction is zero. The moments must be rotated to overcome the maximum anisotropy in this plane of rotation. Therefore the stress anisotropy at the angle ϕ corresponding to $E_{\text{a max}}$ must equal $K_1/4$

$$E_\sigma(\phi) = E_{\text{a max}}$$

$$-\frac{3}{2}\lambda_s \sigma \cos^2\left(\frac{\pi}{4}\right) \geq -\frac{1}{4}K_1$$

or

$$\sigma \leq \frac{K_1}{3\lambda_s}$$

remembering that this is a compressive stress, and hence σ is negative.

Now substituting in the values of the saturation magnetostriction and the anisotropy constant

$$\sigma = \frac{-(2\times 10^4)}{(3)(1067\times 10^{-6})}$$

$$= -6.25\times 10^6 \text{ Pa}$$

Therefore, σ must be compressive with magnitude greater than 6.25 MPa, which is equivalent to 899 p.s.i.

Exercise 6.6: Simple Ising Lattice

Consider the array of nine magnetic moments, each of identical magnitude, which can point in any direction

$$\begin{array}{cccccc} \boldsymbol{m}_{11} & \uparrow & \boldsymbol{m}_{12} & \uparrow & \boldsymbol{m}_{13} & \downarrow \\ \boldsymbol{m}_{21} & \downarrow & \boldsymbol{m}_{22} & \downarrow & \boldsymbol{m}_{23} & \uparrow \\ \boldsymbol{m}_{31} & \uparrow & \boldsymbol{m}_{32} & \downarrow & \boldsymbol{m}_{33} & \uparrow \end{array}$$

If only nearest-neighbor interactions are significant, the *coupling* energy or *interaction* energy is

$$E = -\mu_0 \sum_{\text{nearest neighbors}} \alpha_{ij,kl} \boldsymbol{m}_{ij} \cdot \boldsymbol{m}_{kl}$$

If the coupling coefficient $\alpha_{ij,kl}$ is identical, say α, for all pairs

$$E = -\mu_0 \alpha \sum_{\text{nearest neighbors}} \boldsymbol{m}_{ij} \cdot \boldsymbol{m}_{kl}$$

In this case, the nearest-neighbor interaction leaves us with two sublattices:

Sublattice **A**: $\boldsymbol{m}_{ij}$ for which $i + j = \text{odd}$
Sublattice **B**: $\boldsymbol{m}_{ij}$ for which $i + j = \text{even}$

Further, all corner lattice sites are equivalent and all edge lattice sites are equivalent. Therefore,

$$\begin{aligned} E = & -4\alpha\mu_0 m_{11}\left(m_{21} + m_{12}\right) \\ & -4\alpha\mu_0 m_{12}\left(m_{11} + m_{22} + m_{13}\right) \\ & -\alpha\mu_0 m_{22}\left(m_{12} + m_{21} + m_{23} + m_{32}\right) \end{aligned}$$

Case I $\alpha < 0$

In this case, we obtain

$$\begin{aligned} E = & \, 4|\alpha|\mu_0 \boldsymbol{m}_{11}\left(\boldsymbol{m}_{21} + \boldsymbol{m}_{12}\right) \\ & + 4|\alpha|\mu_0 \boldsymbol{m}_{12}\left(\boldsymbol{m}_{11} + \boldsymbol{m}_{22} + \boldsymbol{m}_{13}\right) \\ & + |\alpha|\mu_0 \boldsymbol{m}_{22}\left(\boldsymbol{m}_{12} + \boldsymbol{m}_{21} + \boldsymbol{m}_{23} + \boldsymbol{m}_{32}\right) \end{aligned}$$

and this energy is minimized when $\boldsymbol{m}_{21}$ and $\boldsymbol{m}_{12}$ are antiparallel to $\boldsymbol{m}_{11}$; $\boldsymbol{m}_{22}$ is antiparallel to $\boldsymbol{m}_{12}$, $\boldsymbol{m}_{21}$, $\boldsymbol{m}_{23}$, and $\boldsymbol{m}_{32}$; and $\boldsymbol{m}_{13}$ is antiparallel to $\boldsymbol{m}_{12}$ and $\boldsymbol{m}_{23}$; and so on. In other words, the energy is minimized with antiferromagnetic ordering.

Case II $\alpha > 0$

In this case, we have the following expression for the energy:

$$\begin{aligned} E = & -4|\alpha|\mu_0 \boldsymbol{m}_{11}\left(\boldsymbol{m}_{21} + \boldsymbol{m}_{12}\right) \\ & -4|\alpha|\mu_0 \boldsymbol{m}_{12}\left(\boldsymbol{m}_{11} + \boldsymbol{m}_{22} + \boldsymbol{m}_{13}\right) \\ & -|\alpha|\mu_0 \boldsymbol{m}_{22}\left(\boldsymbol{m}_{12} + \boldsymbol{m}_{21} + \boldsymbol{m}_{23} + \boldsymbol{m}_{32}\right) \end{aligned}$$

and this energy is minimized when $\boldsymbol{m}_{11}$ is parallel to $\boldsymbol{m}_{21}$ and $\boldsymbol{m}_{12}$; $\boldsymbol{m}_{22}$ is parallel to $\boldsymbol{m}_{12}$, $\boldsymbol{m}_{21}$, $\boldsymbol{m}_{23}$, and $\boldsymbol{m}_{32}$; and $\boldsymbol{m}_{13}$ is parallel to $\boldsymbol{m}_{12}$ and $\boldsymbol{m}_{23}$; and so on. In other words, the energy is minimized with ferromagnetic ordering.

Relating the ordering temperature to the coupling coefficient, we equate the internal coupling energy to the thermal energy at the Curie (or Néel) temperature.

a. *Mean field coupling*
   If each moment in the array interacts equally with all others, the coupling energy is

$$E = -\mu_0 \alpha N(N-1)\boldsymbol{m}^2 = -72\mu_0\alpha\boldsymbol{m}^2$$

and the thermal energy, allowing three degrees of freedom, is

$$E = 3Nk_BT$$

Equating these at the Curie temperature T_c gives the well-known result

$$T_c = \frac{\mu_0\alpha(N-1)m^2}{3k_B} = \frac{8\mu_0\alpha m^2}{3k_B}$$

b. *Nearest-neighbor coupling*
In this case, the procedure is identical, except that the internal coupling energy is

$$E = -24\alpha\mu_0\boldsymbol{m}^2$$

while the thermal energy is again

$$E = 3Nk_BT$$

and consequently, if we equate these at the Curie temperature

$$T_c = \frac{8\mu_0 m^2}{9k_B}$$

Exercise 6.7: Exchange Energy and Domains

The Weiss (mean) interaction field is given by

$$\boldsymbol{H}_e = \alpha\boldsymbol{M}_s = \alpha\,\mathrm{N}\,\mathrm{m}$$

$$= \frac{3k_BT_c}{\mu_0\boldsymbol{m}}$$

Cobalt:
Using $T_c = 1140°\mathrm{C} = 1413\,\mathrm{K}$, and with a magnetic moment per atom of

$$1.72\mu_B = 1.59\times10^{-23}\,\mathrm{A\,m^2}$$

$$\text{Cobalt: } \boldsymbol{H}_e = 2.9\times10^{9}\,\mathrm{A\,m^{-1}}$$

Iron:
Using $T_c = 770\,°\mathrm{C} = 1043\,\mathrm{K}$, and with a magnetic moment per atom of $2.2\mu_B = 2.04\times10^{-23}\,\mathrm{A\,m^2}$,

$$\text{Iron: } \boldsymbol{H}_e = 1.69\times10^{9}\,\mathrm{A\,m^{-1}}$$

Using nearest-neighbor coupling, there are 12 nearest neighbors for cobalt and 8 nearest neighbors for iron. Therefore, we can say that

$$\boldsymbol{H}_{\mathrm{e}} = \frac{Z}{V} \cdot \alpha_{ij}\, \boldsymbol{m}_i$$

where:
  Z is the number of nearest neighbors
  V is the volume of unit cell

and the energy of the exchange interaction is

$$E = \mu_0 \boldsymbol{m}\boldsymbol{H}_{\mathrm{e}}$$

Cobalt:

$$E = \left(4\pi \times 10^{-7}\right)\left(1.59 \times 10^{-23}\right)\left(2.9 \times 10^{9}\right)$$

$$E = 5.79 \times 10^{-20}\ \mathrm{J}$$

and the energy of their interaction per nearest-neighbor is $E_{\mathrm{nn}} = E/12$

$$E_{\mathrm{nn}} = 4.8 \times 10^{-21}\ \mathrm{J}$$

Iron:

$$E = \left(4\pi \times 10^{-7}\right)\left(2.04 \times 10^{-23}\right)\left(1.69 \times 10^{9}\right)$$

$$E = 4.33 \times 10^{-20}\ \mathrm{J}$$

and the energy of this interaction per nearest-neighbor pair is $E_{\mathrm{nn}} = E/8$

$$E_{\mathrm{nn}} = 5.41 \times 10^{-21}\ \mathrm{J}$$

The interaction between magnetic moments ϑ can be defined by

$$\vartheta = \frac{3k_{\mathrm{B}}T_{\mathrm{c}}}{2\mu_0 n m^2}$$

Cobalt:

$$\vartheta = \frac{3 \times 1.38 \times 10^{-23} \times 1412}{2 \times 4\pi \times 10^{-7} \times 12 \times \left(1.59 \times 10^{-23}\right)^2}$$

$$\vartheta = 0.76 \times 10^{31}\ \mathrm{m}^{-3}$$

Iron:

$$\vartheta = \frac{3 \times 1.38 \times 10^{-23} \times 1043}{2 \times 4\pi \times 10^{-7} \times 8 \times \left(2.04 \times 10^{-23}\right)^2}$$

$$\vartheta = 0.52 \times 10^{31}\ \mathrm{m}^{-3}$$

and therefore the nearest-neighbor coupling coefficient α_{ij} is

$$\alpha_{ij} = \boldsymbol{H}_e \cdot \frac{V}{Z\boldsymbol{m}}$$

The volume of the unit cells in these two cases are as follows:

$$\text{Hexagonal: } V = a^2 c \cdot 3\sqrt{3}$$

$$\text{Cubic: } V = a^3$$

Cobalt:

$$a = 0.25 \times 10^{-9} \text{ m}$$

$$c = 0.41 \times 10^{-9} \text{ m}$$

and therefore

$$\alpha_{ij} = \frac{a^2 c \cdot 3\sqrt{3}}{Z} \cdot \frac{\boldsymbol{H}_e}{\boldsymbol{m}}$$

$$\alpha_{ij} = 1.11 \times 10^{-29} \cdot 2.6 \times 10^{31}$$

$$\alpha_{ij} = 288$$

Iron:

$$a = 0.29 \times 10^{-9} \text{ m}$$

and therefore

$$\alpha_{ij} = \frac{a^2}{Z} \cdot \frac{\boldsymbol{H}_e}{\boldsymbol{m}}$$

$$\alpha_{ij} = 3.05 \times 10^{-30} \cdot 8.2 \times 10^{31}$$

$$\alpha_{ij} = 253$$

Exercise 6.8: Magnetocrystalline Anisotropy

The equation for the anisotropy energy E_{an} of a cubic material is given on page 149 of the book, Equation 6.29.

$$E_{an} = K_1\left(\cos^2\theta_1 \cos^2\theta_2 + \cos^2\theta_2 \cos^2\theta_3 + \cos^2\theta_3 \cos^2\theta_1\right)$$

and the first-order anisotropy coefficient, K_1, has units of joules per cubic meter.

Since K_1 is negative for nickel, the minimum of the anisotropy energy occurs for values of θ_1, θ_2, and θ_3, which maximize the expression

$$\cos^2\theta_1 \cos^2\theta_2 + \cos^2\theta_2 \cos^2\theta_3 + \cos^2\theta_3 \cos^2\theta_1$$

This can be demonstrated by differentiation, remembering that θ_1, θ_2, and θ_3 are not independent because the sums of the square of the direction cosines are equal to 1

$$\cos^2\theta_1 + \cos^2\theta_2 + \cos^2\theta_3 = 1$$

This means that there are only two degrees of freedom in deciding the orientation in three dimensions. For example, if θ_1 and θ_2 are already decided, then θ_3 is also decided because of the mathematical constraint above concerning the sum of the squares of the direction cosines.

The maximum of the expression occurs when $\cos\theta_1 = \cos\theta_2 = \cos\theta_3 = 1/\sqrt{3}$. Then

$$E_{\text{an}} = K_1\left(\frac{1}{3}\cdot\frac{1}{3} + \frac{1}{3}\cdot\frac{1}{3} + \frac{1}{3}\cdot\frac{1}{3}\right)$$

$$E_{\text{an}} = \frac{K_1}{3}$$

$$E_{\text{an}} = -1.6667 \times 10^3\ \text{J m}^{-3}$$

To determine the *equivalent field* that would exert the same force as the anisotropy, we need to consider the torque due to a magnetic field on a magnetic moment $\boldsymbol{m}$.

$$E_{\text{H}} = -\mu_0 \boldsymbol{M}_s \boldsymbol{H} \cos\theta$$

where θ is the angle between the field and the magnetic moment. The torque is then the derivative of this energy with respect to angle, $\tau_{\text{H}} = \partial E_{\text{H}}/\partial\theta$.

$$\tau_{\text{H}} = \mu_0 \boldsymbol{M}_s \boldsymbol{H} \sin\theta$$

and for small angle deviations from the easy axis $\sin\theta \cong \theta$, so

$$\tau_{\text{H}} = \mu_0 \boldsymbol{M}_s \boldsymbol{H} \theta$$

The anisotropy energy is given by

$$E_{\text{an}} = K\left(\cos^2\theta_1\cos^2\theta_2 + \cos^2\theta_2\cos^2\theta_3 + \cos^2\theta_3\cos^2\theta_1\right)$$

and similarly the torque due to the anisotropy τ_{an} is the derivative of this energy with respect to angle

$$\tau_{\text{an}} = \frac{\partial E_{\text{an}}}{\partial\theta}$$

We know that the easy axis is the $\langle 111\rangle$ direction in nickel so let $\cos\theta_2 = 1/\sqrt{3}$ and $\cos\theta_3 = 1/\sqrt{3}$. Then to a first approximation,

$$E_{\text{an}} = K_1\left(\frac{2}{3}\cos\theta_1 + \frac{1}{9}\right)$$

for small angular deviations from the ⟨111⟩ direction. Therefore,

$$\tau_{an} = \frac{4K_1}{3}\cos\theta\sin\theta$$

and recognizing that 2 cos θ sin θ = sin 2θ

$$\tau_{an} = \frac{2K_1}{3}\sin 2\theta$$

and since sin 2 θ ≈ 2θ

$$\tau_{an} = \frac{4}{3}K_1\theta$$

Equating this with the torque under a magnetic field $\boldsymbol{H}$, if

$$\tau_{an} = \tau_H$$

$$\frac{4}{3}K\theta = \mu_0 \boldsymbol{M}_s \boldsymbol{H}\theta$$

$$\boldsymbol{H} = \frac{4K}{3\mu_0 \boldsymbol{M}_s}$$

which shows the equivalent field that would produce the same torque as the anisotropy in terms of the anisotropy coefficient and the saturation magnetization. For nickel, $K = -0.05\times10^5\ \mathrm{J\,m^{-3}}$ and $\boldsymbol{M}_s = 0.48\times10^6\ \mathrm{Am^{-1}}$, so the equivalent field is

$$\boldsymbol{H}_{eq} = -\frac{(4)(0.05\times10^5)}{(3)(12.56\times10^{-7})(0.48\times10^6)}$$

$$\boldsymbol{H}_{eq} = 11.05\times10^3\ \mathrm{A\,m^{-1}}$$

Exercise 6.9: Magnetocrystalline Anisotropy, Domain-Wall Energy, and Domain-Wall Thickness

$$K = 330\times10^3\ \mathrm{J\,m^{-3}}$$

$$\boldsymbol{M}_s = 0.4\times10^6\ \mathrm{A\,m^{-1}}$$

$$T_c = 470\,^\circ\mathrm{C}(743\,\mathrm{K})$$

$$a_c = 3\ \mathrm{nm}$$

$$a_a = 0.6\ \mathrm{nm}$$

Crystal structure is hexagonal so number of nearest neighbors = 12

$$H_a = \frac{2K}{\mu_0 M_s}$$

$$H_a = \frac{(2)(330\times10^3)}{(12.56\times10^{-7})(0.4\times10^6)}$$

$$H_a = 1.31\times10^6 \text{ A m}^{-1}$$

$$\delta = \pi\sqrt{\frac{A}{K}}$$

$$\gamma = 2\pi\sqrt{AK}$$

Therefore, we need to calculate A according to

$$A = \frac{3k_B T_c}{az}$$

where z is the number of nearest neighbors.

$$A = \frac{(3)(1.38\times10^{-23})(743)}{(12)(3\times10^{-9})} = 85\times10^{-14}$$

$$A = 0.85\times10^{-12} \text{ J m}^{-1}$$

$$\delta = (3.142)\sqrt{\frac{(0.85\times10^{-12})}{(330\times10^3)}} = 3.142\sqrt{2.575\times10^{-18}}$$

$$\delta = 5.04\times10^{-9} \text{ m}$$

$$\gamma = 2\pi\sqrt{(0.85\times10^{-12})(330\times10^3)} = 6.284\sqrt{28.05\times10^{-8}}$$

$$\gamma = 3.3\times10^{-3} \text{ J m}^{-2}$$

Some have argued that because of the large c-axis lattice spacing, the number of nearest neighbors is actually 6 not 12. In that case,

$$A = 1.7\times10^{-12} \text{ J m}^{-1}$$

and

$$\delta = 7.13\times10^{-9} \text{ m}$$

and

$$\gamma = 4.66\times10^{-3} \text{ J m}^{-2}$$

Finally, if the lattice parameter used to calculate $\boldsymbol{a}$ is 0.6×10^{-9} m, then

$$A=\frac{3k_BT_c}{az}$$

$$A=\frac{(3)(1.38\times10^{-23})(743)}{(0.6\times10^{-9})(12)}$$

$$A=4.25\times10^{-12}\ \mathrm{J\,m^{-1}}$$

$$\delta=11.27\times10^{-9}\ \mathrm{m}$$

$$\gamma=7.38\times10^{-3}\ \mathrm{J\,m^{-2}}$$

Exercise 6.10: Critical Radius for Single-Domain Particles

The demagnetizing field is given by

$$\boldsymbol{H}_\mathrm{d}=-N_d\boldsymbol{M}_\mathrm{s}$$

and the magnetostatic energy per unit volume is

$$\frac{\Delta E_\mathrm{ms}}{V}=-\int\mu_0\boldsymbol{H}_\mathrm{d}\,d\boldsymbol{M}_\mathrm{s}$$

The magnetostatic energy due to the demagnetizing field is

$$\frac{\Delta E_\mathrm{ms}}{V}=+\mu_0N_\mathrm{d}\int\boldsymbol{M}_\mathrm{s}\,d\boldsymbol{M}_\mathrm{s}=\frac{\mu_0N_\mathrm{d}\boldsymbol{M}_\mathrm{s}^{\,2}}{2}$$

So for a sphere of radius r, the magnetostatic energy is

$$E_\mathrm{ms}=\frac{4\pi r^2}{3}\cdot\frac{\mu_0N_\mathrm{d}\boldsymbol{M}_\mathrm{s}^{\,2}}{2}$$

The domain-wall energy that is associated with a domain wall across the diameter of the sphere is

$$E_\mathrm{wall}=\pi r^2\gamma$$

Therefore, at the critical radius r_c, we have

$$E_\mathrm{ms}=E_\mathrm{wall}$$

$$\frac{4\pi r^2}{3}\cdot\frac{\mu_0N_\mathrm{d}\boldsymbol{M}_\mathrm{s}^{\,2}}{2}=\pi r_\mathrm{c}^{\,2}\gamma$$

Therefore,

$$\frac{2}{9}r_c^{\,3}\mu_0 \boldsymbol{M}_s^{\,2} = r_c^{\,2}\gamma$$

$$r_c = \frac{9\gamma}{2\mu_0 \boldsymbol{M}_s^{\,2}}$$

and

$$\gamma = 2\pi\sqrt{AK}$$

and

$$A = \frac{k_B T_c}{az}$$

where $z = 12$ for a hexagonal lattice and we shall use $\boldsymbol{a} = 0.3$ nm for the lattice parameter

$$A = \frac{(1.38\times10^{-23})(1404)}{(0.3\times10^{-9})(12)}$$

$$A = 5.38\times10^{-12}\ \mathrm{J\,m^{-1}}$$

$$\gamma = 2\pi\sqrt{AK} = (6.284)\sqrt{(5.38\times10^{-12})(0.5\times10^{6})}$$

$$\gamma = 10.3\times10^{-3}\ \mathrm{J\,m^{2}}$$

$$r_c = \frac{9\gamma}{2\mu_0 \boldsymbol{M}_s^{\,2}}$$

$$r_c = \frac{(9)(10.3\times10^{-3})}{(2)(12.56\times10^{-7})(1.4\times10^{6})^2} = 18.8\times10^{-9}\ \mathrm{m}$$

$$r_c = 19\ \ \mathrm{nm}$$

CHAPTER 7

Exercise 7.1: Magnetostatics

If we consider the additional energy caused by the presence of the domain walls, then

$$E_{\text{wall}} = \gamma A$$

where A is the total area of walls inside the cube. It is clear that with a wall spacing of d, the total wall area must be $A = l^3/d$ if all the walls are parallel to a face.

The magnetostatic energy is given by the equation

$$E_{ms} = \frac{1}{2}\mu_0 N \boldsymbol{M}_s^2 \frac{d}{l} l^3$$

and if we add these to get the total energy and divide by the volume

$$E_{tot} = \frac{\gamma}{d} + \frac{1}{2}\mu_0 N \boldsymbol{M}_s^2 \frac{2}{s} \text{ J m}^{-3}$$

The minimum energy will occur when $dE_{Tot}/dd = 0$.

$$\frac{dE_{tot}}{dd} = \frac{\gamma}{d^2} + \frac{1}{2}\mu_0 N \boldsymbol{M}_s^2 \frac{1}{l} = 0$$

$$\gamma = \frac{1}{2}\mu_0 N \boldsymbol{M}_s^2 \frac{d^2}{l}$$

and, substituting in the values,

$$\gamma = 0.785 \times \text{J m}^{-2}$$

Exercise 7.2: Magnetostatics

From the previous value of γ and knowing the anisotropy and lattice parameter, we can calculate the exchange energy E_{ex} for a lattice with $a = 0.3$ nm,

$$\gamma = 2\pi\sqrt{\frac{E_{ex}K}{a}}$$

$$E_{ex} = \frac{\gamma^2 a}{4\pi^2 K}$$

$$= 4.74 \times 10^{-23} \text{ J}$$

Exercise 7.3: Critical Dimensions of Single-Domain Particles in Nickel

The magnetostatic energy per unit volume is E_{ms}/V

$$\frac{E_{ms}}{V} = \int \mu_0 \boldsymbol{H}_d dM$$

and for a sphere magnetized to saturation

$$\boldsymbol{H}_d = \frac{1}{3}\boldsymbol{M}_s$$

Therefore,

$$\frac{E_{ms}}{V} = \frac{\mu_0 \boldsymbol{M}_s^2}{6}$$

and $V = 4\pi r^3/3$, therefore

$$E_{ms} = \frac{2\pi\mu_0 \boldsymbol{M}_s^2 r^3}{9}$$

The reduction of magnetostatic energy caused by dividing the particle into two domains is

$$\Delta E_{ms} = \frac{\pi\mu_0 \boldsymbol{M}_s^2 r^3}{9}$$

A Bloch wall through the center of the particle will have an area of πr^2, so the wall energy will be,

$$E_{wall} = \gamma\pi r^2$$

where γ is the wall energy per unit area. The energy reduction caused by the splitting of such a particle into two domains will be

$$\Delta E = \gamma\pi r^2 - \frac{\pi\mu_0 \boldsymbol{M}_s^2 r^3}{9}$$

We therefore need to find the critical condition under which we just fail to save energy by dividing into two domains. Let r_c be the critical radius under this condition

$$\Delta E = \gamma\pi r_c^2 - \frac{\pi\mu_0 \boldsymbol{M}_s^2 r_c^3}{9} = 0$$

$$r_c = \frac{9\gamma}{\mu_0 \boldsymbol{M}_s^2}$$

$$= 1.92 \times 10^{-8} \text{ m}$$

Exercise 7.4: Calculation of Wall Energy and Thickness from Anisotropy Energy, Saturation Magnetization, and Exchange Energy

The material is barium ferrite. Following the equations given in the text, we know that

$$\gamma = \frac{\mu_0 j m^2 \pi^2}{al} + Kl$$

$$= \frac{JS^2\pi^2}{al} + Kl$$

At equilibrium, we must have $d\gamma/dl = 0$, and consequently,

$$l = \left(\frac{JS^2\pi^2}{Ka}\right)^{1/2}$$

$$= 86.5 \times 10^{-10} \text{ m}$$

The wall energy is

$$\gamma = \frac{JS^2\pi^2}{al} + Kl$$

$$= 5.69 \times 10^{-3}\ \mathrm{J\ m^{-2}}$$

The critical radius for a spherical single-domain particle is then

$$r_c = \frac{9}{2\mu_0 \boldsymbol{M}_s^2}$$

$$= 14.1 \times 10^{-8}\ \mathrm{m}$$

Exercise 7.5: Estimation of Domain Spacing in Cobalt

The magnetostatic energy per unit area of the surface is given as

$$E_{ms} = 0.135\,\mu_0 \boldsymbol{M}_s^2 d$$

Suppose the other dimensions of the sample are x and y meters, respectively, the total magnetostatic energy will be

$$E_{ms} = 0.135\mu_0 \boldsymbol{M}_s^2 xyd$$

The total number of walls N in a cuboid of side x, y, and d with domains perpendicular to the x-y plane and lying in the l-y plane is given by

$$N = \frac{x}{d}$$

Total wall energy is then

$$E_{wall} = Nly\gamma$$

$$= \frac{l\gamma xy}{d}$$

The total energy is the sum of wall energy and magnetostatic energy

$$E_{tot} = \left(0.135\,\mu_0 \boldsymbol{M}_s^2 d + \frac{l\gamma}{d}\right) xy$$

$$\frac{dE_{tot}}{dd} = \left(0.135\,\mu_0 \boldsymbol{M}_s^2 d - \frac{l\gamma}{d}\right) xy$$

At equilibrium,

$$\frac{dE_{tot}}{dd} = 0$$

and so

$$0.135\mu_0 \boldsymbol{M}_s^2 - \frac{l\gamma}{d^2} = 0$$

The equilibrium spacing of the domains therefore does not depend on x and y, and is given by

$$d = \left(\frac{l\gamma}{0.135\mu_0 \boldsymbol{M}_s^2} \right)$$

$$= 1.49 \times 10^{-4} (l)^{1/2}$$

and if $l = 1$ cm, (0.01 m), then

$$d = 1.49 \times 10^{-5}\,\text{m}$$

Exercise 7.6: Exchange Energy

In both cases, the materials have axial anisotropy, so they will contain only 180° domain walls. Therefore, we can begin the analysis by considering the equations for width δ and energy per unit area γ of 180° domain walls.

First, calculate the exchange energy per spin, which corresponds to the given Curie temperature:

$$\sum_{nn} E_{ex} \cong 3k_B T_c = \mu_0 jzm^2$$

and for cobalt $T_c = 1400$ K, and for neodymium-iron-boron $T_c = 585$ K, therefore

$$\text{Co:} \quad \sum_{nn} E_{ex} = 5.8 \times 10^{-20}\ \text{J}$$

$$\text{NdFeB:} \sum_{nn} E_{ex} = 2.4 \times 10^{-20}\ \text{J}$$

and remembering that hexagonal close packed (hep) cobalt has $Z = 12$ nearest neighbors, while tetragonal neodymium-iron-boron has four nearest neighbors, the pair interaction energies are, $E_{ex} = \mu_0 \boldsymbol{J} \boldsymbol{m}^2$

$$\text{Co:} \quad E_{ex} = 4.8 \times 10^{-21}\ \text{J}$$

$$\text{NdFeB:}\ E_{ex} = 6.0 \times 10^{-21}\ \text{J}$$

Now, with these values of the exchange coupling it is possible to calculate A, the exchange stiffness:

$$A = \frac{E_{ex}}{a}$$

$$\text{Co:} \quad A = 1.6 \times 10^{-11}\ \text{J m}^{-1}$$

$$\text{NdFeB:}\ A = 0.6 \times 10^{-11}\ \text{J m}^{-1}$$

and, from these values of A and the anisotropy coefficient K, we can calculate the domain-wall width δ and domain-wall energy γ.

$$\delta = \pi\sqrt{\frac{A}{K}}$$

$$\gamma = 2\pi\sqrt{AK}$$

So, for cobalt, $K = 4.5 \times 10^5$ J m^{-3}.

$$\delta = 1.87 \times 10^{-8} \text{ m}$$

$$\gamma = 0.017 \text{ J m}^{-2}$$

and for NdFeB, $K = 4 \times 10^6$ J m^{-3}

$$\delta = 3.8 \times 10^{-9} \text{ m}$$

$$\gamma = 0.031 \text{ J m}^{-2}$$

Exercise 7.7: Existence of Domain Walls

Domain boundaries form the transition regions between domains that have their spontaneous magnetization oriented in different directions.

The domain boundaries have a finite thickness as a result of energy minimization. The exchange energy is minimized when the boundaries are infinitely thick, while the anisotropy energy is minimized when the boundaries are infinitesimally thin.

As a result of the competition between the anisotropy and exchange energies, a minimum energy is achieved at a finite thickness.

$$\delta = \pi\sqrt{\frac{A}{K}}$$

where:

δ is the domain-boundary thickness
A is the exchange stiffness
K is the anisotropy energy

Therefore, if A increases, the domain-wall thickness increases, while if K increases the domain-wall thickness decreases.

The transition region between two domains results in the magnetic moments in the domain boundaries lying along directions that are neither optimal for exchange energy minimization nor for anisotropy energy minimization. Both of these energies contribute to the domain boundary surface energy of which is given by the equation.

$$\gamma = 2\pi\sqrt{AK}$$

So exchange stiffness and anisotropy are the main factors that determine domain boundary energy. However, other factors can play a role, the most significant of these other factors being defects or inhomogeneities, which can cause localized minima in domain boundary surface energy, leading to pinning of the domain boundaries on the defects.

Exercise 7.8: Exchange Energy

Domain walls exist to provide a transition between domains with moments oriented in different directs. The domains exist in order to minimize magnetostatic energy by reducing the magnetization. Domain walls have a finite thickness in order to minimize the sum of the anisotropy and exchange energies. This minimum energy is the domain-wall surface energy, which although minimized is not zero.

The exchange energy per atoms magnetic moment E_{ex} is determined by the Curie temperature T_c

$$E_{ex} = 3k_B T_c$$

$$E_{ex} = (3)(1.38\times10^{-23})(700+273)$$

$$E_{ex} = 4.03\times10^{-20}\ \text{J}$$

$$E_{ex} = 0.25\ \text{eV}$$

The exchange energy between two nearest-neighbor spins (magnetic moments) is the above energy divided by the number of nearest neighbors z. For a simple cubic material, z = 6

$$E_{nn} = \frac{E_{ex}}{Z}$$

$$E_{nn} = 6.72\times10^{-21}\ \text{J}$$

$$E_{nn} = 0.042\ \text{eV}$$

The domain-wall thickness δ is given by

$$\delta = \pi\sqrt{\frac{A}{K}}$$

and

$$A = \frac{E_{nn}}{a}$$

$$A = \frac{6.72\times10^{-21}}{0.25\times10^{-9}}$$

$$A = 2.69\times10^{-11}\ \text{J}\cdot\text{m}^{-1}$$

and therefore,

$$\delta = (3.142)\sqrt{\frac{2.69\times10^{-11}}{4\times10^{4}}}$$

$$\delta = 81\ \text{nm}$$

The domain surface energy γ is given by

$$\gamma = 2\pi\sqrt{AK}$$

$$\gamma = (6.284)\sqrt{(2.69\times10^{-11})(4\times10^{4})}$$

$$\gamma = 6.5\times10^{-3}\ \mathrm{J\cdot m^{-2}}$$

Exercise 7.9: Exchange Energy, Anisotropy, and Domain-Wall Properties

For a simple cubic lattice, the number of nearest neighbors is z = 6.

a. The exchange energy per spin is

$$E_{\mathrm{ex}} = 3k_{\mathrm{B}}T_{\mathrm{c}}$$

$$E_{\mathrm{ex}} = (3)(1.38\times10^{-23})(1023)$$

$$E_{\mathrm{ex}} = 4.24\times10^{-20}\ \mathrm{J}$$

b. The exchange stiffness is given by

$$A = \frac{E_{\mathrm{ex}}}{az}$$

$$A = \frac{4.24\times10^{-20}}{(0.3\times10^{-9})(6)}$$

$$A = 2.35\times10^{-11}\ \mathrm{J\,m^{-1}}$$

c. The exchange energy between any two nearest-neighbor pairs is

$$\frac{E_{\mathrm{ex}}}{z} = \frac{4.24\times10^{-20}}{6}\ \mathrm{J}$$

$$\frac{E_{\mathrm{ex}}}{z} = 7.06\times10^{-21}\ \mathrm{J}$$

d. The domain-wall thickness is given by

$$\delta = \pi\sqrt{\frac{A}{K}}$$

$$\delta = 3.142\sqrt{\frac{2.35\times10^{-11}}{5\times10^{4}}} = 68\times10^{-9}\ \mathrm{m}$$

$$\delta = 68\ \mathrm{nm}$$

e. The domain-wall surface energy is given by

$$\gamma = 2\pi\sqrt{AK}$$

$$\gamma = 6.284\sqrt{(2.35\times10^{-11})(5\times10^{4})} = (6.284)(1.08\times10^{-3})$$

$$\gamma = 6.8\times10^{-3}\ \ \mathrm{J\,m^{-2}}$$

Exercise 7.10: Stress and Magnetoelastic Energy

Anisotropy and exchange energies determine the domain-wall surface energy. Anisotropy can be changed by the application of stress, so the domain-wall surface energy is determined by a combination of magnetocrystalline anisotropy and stress-induced anisotropy.

Tensile stress can generate a contribution to anisotropy, as discussed in Section 8.1.2, Equation 8.2. The stress-induced anisotropy energy will be K_σ in joules per cubic meter as given by

$$K_\sigma = \frac{3}{2}\lambda\sigma$$

Since stress causes an axial anisotropy the angular dependence of the energy will be

$$E(\phi) = K_\sigma \sin^2\phi$$

where ϕ is the angle to the unique axis.

For a 180° domain wall, the surface energy is given by

$$\gamma = 2\pi\sqrt{AK}$$

and in this case, the magnetocrystalline anisotropy is given as negligible, so that $K \cong K_\sigma$

The exchange stiffness A is the exchange energy per nearest-neighbor pair divided by the lattice parameter a.

$$A = \frac{E_{\mathrm{nn}}}{a}$$

$$A = \frac{3k_\mathrm{B}T_\mathrm{c}}{az}$$

where for nickel it is known that

$$a = 0.35\times10^{-9}\ \ \mathrm{m}$$

assuming the lattice parameter doesn't change with the application of stress,

$$T_\mathrm{c} = 631\ \mathrm{K}$$

$$z = 12$$

Therefore,

$$A = \frac{(3)(1.38\times10^{-23})(631)}{(0.35\times10^{-9})(12)}$$

$$A = 6.22\times10^{-12}\ \text{J m}^{-1}$$

and

$$\gamma = 2\pi\sqrt{AK}$$

and

$$K = \frac{3}{2}\lambda\sigma$$

We are given that σ = 100 MPa. The value of λ for nickel inside a domain is $\lambda = \lambda_s = 35\times10^{-6}$

$$K = \frac{(3)(35\times10^{-6})(100\times10^{6})}{2}$$

$$K = 5250\ \text{J m}^{-3}$$

$$\gamma = (6.284)\sqrt{(6.22\times10^{-12})(5.25\times10^{3})}$$

$$\gamma = 1.1\times10^{-3}\ \text{J m}^{-2}$$

But taking into account that nickel has a finite magnetocrystalline anisotropy energy of $K_1 = -5\times10^{3}\ \text{J m}^{-3}$, the anisotropy of nickel under stress will be the sum of these

$$K = K_1 + K_\sigma$$

$$K = (-5\times10^{3}) + (5.25\times10^{3})$$

$$K = 0.25\times10^{3}\ \text{J m}^{-3}$$

which means that the application of stress has almost eliminated the anisotropy

$$\gamma = 2\pi\sqrt{AK}$$

$$\gamma = (6.284)\sqrt{(6.22\times10^{-12})(0.25\times10^{3})}$$

$$\gamma = 0.25\times10^{-3}\ \text{J m}^{2}$$

CHAPTER 8

Exercise 8.1: Magnetization Mechanisms

An essay on magnetization mechanisms must include descriptions of domain-wall motion and domain rotation in both their reversible and irreversible modes. Reversible domain-wall motion occurs for small changes in applied field and amounts to domain-wall bending for strongly pinned walls, or translation of the wall through a region with no pinning sites. When the pinning sites are strong, the wall motion occurs principally by bending; when the pinning sites are weak, the wall motion occurs by planar translation. Irreversible domain-wall motion occurs when the domain wall breaks away from pinning sites. Reversible domain rotation occurs in the presence of anisotropy when the domain magnetization is deflected by the action of the field into directions away from the original easy axis, but sufficiently close to the original direction, which when applied, field is removed; they relax back to the original direction. Irreversible rotation occurs when the domain magnetizations are deflected far enough that they overcome the maximum restraining force provided by anisotropy and suddenly deflect into a different easy axis closer to the field direction. Reversible rotation can also occur at higher field strengths when the increase in field deflects the moments from the easy axis closest to the applied field, but when the field is removed the moments relax back to this easy axis.

Exercise 8.2: Magnetostatic Energy Associated with a Void

The magnetostatic energy per unit volume is given by

$$E_{\text{ms}} \equiv -\mu_0 \int \boldsymbol{H}_{\text{d}} \boldsymbol{M}$$

In this case, we have the demagnetizing field for a sphere, $\boldsymbol{H}_{\text{d}} = -(1/3)\boldsymbol{M}$.

$$E_{\text{ms}} = -\mu_0 \int_0^{\boldsymbol{M}_{\text{s}}} \frac{1}{3}\boldsymbol{M}\, d\boldsymbol{M}$$

$$= \frac{\mu_0 \boldsymbol{M}_{\text{s}}^2}{6}$$

The volume of the sphere is $(4/3)\pi r^3$, so the magnetostatic energy is

$$E_{\text{ms}} = -\left(\frac{2\pi}{9}\right)\mu_0 \boldsymbol{M}_{\text{s}}^2 r^3$$

We assume that the reduction in magnetostatic energy caused when a domain wall intersects the void in $\Delta E_{\text{ms}} = (\pi/9)\mu_0 \boldsymbol{M}_{\text{s}}^2 r^3$.

For iron with a spherical void of $r = 5 \times 10^{-8}$ m,

$$\Delta E_{\text{ms}} = \left(\frac{\pi}{9}\right)\mu_0 \boldsymbol{M}_s^2 r^3$$

$$= 15.87 \times 10^{-17} \text{ J}$$

The reduction in wall area is πr^2

$$\Delta A = 7.8 \times 10^{-15} \text{m}^2$$

and hence the reduction in wall energy is

$$\Delta E_{\text{wall}} = \left(7.8 \times 10^{-15}\right)\gamma$$

where γ is the wall energy per unit area.

$$\Delta E_{\text{wall}} = 1.57 \times 10^{-17} \text{J}$$

For a spherical void of radius 10^{-6} m,

$$\Delta E_{\text{ms}} = \left(\frac{\pi}{9}\right)\mu_0 \boldsymbol{M}_s^2 r^3$$

$$= 1.27 \times 10^{-17} \text{ J}$$

The reduced in wall area is

$$\Delta A = \pi r^2$$

$$= 3.14 \times 10^{-12} \text{m}^2$$

The reduction in wall energy is

$$\Delta E_{\text{wall}} = \left(2 \times 10^{-3}\right)\left(3.12 \times 10^{-12}\right) \text{ m}^2$$

$$= 6.28 \times 10^{-15} \text{ J}$$

In both cases, the reduction is magnetostatic energy is more significant.

Exercise 8.3: Reduction of Domain-Wall Energy by Voids

Let ΔE_{ms} be the reduction in magnetostatic energy when a domain wall intersects a void of radius *r*.

$$\Delta E_{\text{ms}} = \left(\frac{\pi}{9}\right)\mu_0 \boldsymbol{M}_s^2 r^3$$

If there are *N* such voids and the separation between the planar walls is *d*, then each domain wall will intersect n_w voids

$$n_w = N^{2/3}$$

and since the wall separation is d, there will be $1/d$ walls per unit volume. Therefore, the total number of voids intersected by domain walls throughout the volume will be n

$$n = \frac{N^{2/3}}{d}$$

The total decrease in magnetostatic energy is therefore

$$\Delta E_{\text{ms}} = \left(\frac{\pi}{9}\right)\mu_0 \boldsymbol{M}_s^2 r^3 \left(\frac{N^{2/3}}{d}\right)$$

The total energy associated with domain walls in this unit volume is

$$\Delta E_{\text{wall}} = \frac{\gamma}{d}$$

If the total reduction is magnetostatic energy due to the voids is equal to the increase in wall energy, we must have

$$\Delta E_{\text{ms}} = \Delta E_{wall}$$

$$\left(\frac{\pi}{9}\right)\mu_0 \boldsymbol{M}_s^2 r^3 \frac{N^{2/3}}{d} = \frac{\gamma}{d}$$

$$N^{2/3} = \frac{9\gamma}{\left(\pi\mu_0 \boldsymbol{M}_s^2 r^3\right)}$$

which gives

$$N = 1.98 \times 10^9 \,\text{m}^{-3}$$

Exercise 8.4: Effect of Stress on Anhysteretic Susceptibility

If a 180° wall is moved through a distance dx, the change in field energy will be

$$dE_{\text{H}} = 2\mu_0 \boldsymbol{M}_s \boldsymbol{H} dx$$

The change in magnetoelastic energy will be

$$dE_{\text{me}} = \frac{3}{2} l_s ds = 3 l_s a x dx$$

At equilibrium, we must have $dE_{\text{tot}}/dx = 0$

$$\frac{dE_{\text{tot}}}{dx} = \frac{dE_{\text{H}}}{dx} + \frac{dE_{\text{me}}}{dx} = 0$$

$$= -2\mu_0 \boldsymbol{M}_s \boldsymbol{H} + 3\lambda_s a x$$

Consequently,

$$x = \frac{2\mu_0 \boldsymbol{M}_s \boldsymbol{H}}{3\lambda_s a}$$

The magnetization will be

$$\boldsymbol{M} = 2A\boldsymbol{M}_s x$$

where A is the wall area.

$$\boldsymbol{M} = \frac{4A\mu_0 \boldsymbol{M}_s^2 \boldsymbol{H}}{3\lambda_s a}$$

and the initial susceptibility will therefore be

$$\frac{d\boldsymbol{M}}{d\boldsymbol{H}} = \frac{4A\mu_0 \boldsymbol{M}_s^2}{3\lambda_s a}$$

In determining the anhysteretic, we will use the Fröhlich-Kennelly equation modified for the inclusion of the effects of stress. The original unstressed anhysteretic magnetization can be modeled using the following equation:

$$\boldsymbol{M}_{an} = \frac{\alpha \boldsymbol{H}}{1 + \beta \boldsymbol{H}}$$

Since we know that $\boldsymbol{M}_s = 0.9 \times 10^6 = \alpha/\beta$, and that the limit as $\boldsymbol{H}$ tends to zero of $d\boldsymbol{M}_{an}/d\boldsymbol{H} = 1000$ is equal to α, we can determine the two coefficients α, β

$$\alpha = 1000$$

$$\beta = 1.1 \times 10^{-3}$$

Now turning to the anhysteretic magnetization under stress, we replace $\boldsymbol{H}$ by the effective field, giving

$$\boldsymbol{M}_{an} = \frac{\alpha\left(\boldsymbol{H} + \frac{3}{2}\frac{\sigma}{\mu_0}\frac{d\lambda}{dM}\right)}{1 + \beta\left(\boldsymbol{H} + \frac{3}{2}\frac{\sigma}{\mu_0}\frac{d\lambda}{dM}\right)}$$

If we make the approximation $\lambda = b\boldsymbol{M}^2$, then

$$\frac{d\lambda}{d\boldsymbol{M}} = 2b\boldsymbol{M}$$

$$\boldsymbol{M}_{an} = \frac{\alpha\left[\boldsymbol{H} + (3b\sigma/\mu_0)\boldsymbol{M}_{an}\right]}{\left\{1 + \beta\left[\boldsymbol{H} + (3b\sigma/\mu_0)\boldsymbol{M}_{an}\right]\right\}}$$

$$\frac{d\boldsymbol{M}_{an}}{d\boldsymbol{H}} = \frac{\alpha - \beta \boldsymbol{M}_{an}}{1 + \beta \boldsymbol{H} - 3(\alpha b\sigma/\mu_0) + 6(\beta b\sigma/\mu_0)\boldsymbol{M}_{an}}$$

so that in the limit as $\boldsymbol{H}$ approaches zero and $\boldsymbol{M}_{\text{an}}$ approaches zero

$$\frac{d\boldsymbol{M}_{\text{an}}}{d\boldsymbol{H}} = \frac{\alpha}{1-3\alpha\left(b\sigma/\mu_0\right)}$$

When $\sigma = 0$,

$$\frac{d\boldsymbol{M}_{\text{an}}}{d\boldsymbol{H}} = \chi_{\text{an}}\left(0\right) = 1000$$

When $\sigma = -20$ MPa,

$$\frac{d\boldsymbol{M}_{\text{an}}}{d\boldsymbol{H}} = \frac{1000}{1+0.11}$$

$$\chi_{\text{an}}\left(\sigma\right) = 901$$

Therefore, the compressive stress reduces the susceptibility along the direction of the stress of a ferromagnet with positive magnetostriction coefficient provided $d\lambda/d\boldsymbol{M}$ is positive.

Exercise 8.5: Susceptibility and Domain-Wall Motion

Assuming that the domain-wall surface energy is relatively low and the pinning strength high, so that domain-wall bending occurs, a solution can be obtained for cylindrical deformation of the domain walls for small displacements.

For a 180° domain wall, the force per unit area on the domain wall will be

$$\frac{\boldsymbol{F}}{A} = 2\mu_0\boldsymbol{M}_s \cdot \boldsymbol{H}$$

and if the deformation is cylindrical with a domain-wall surface energy γ, then the excess pressure across the wall is

$$\frac{\boldsymbol{F}}{A} = \frac{\gamma}{r}$$

where r is the radius of curvature. Now consider the volume swept out by the deformation of the wall

$$\Delta V = Al$$

and using the approximate expression $A = \left(2/3\right)l^2x$, for the area of a segment of a circle of height x

$$\Delta V = \frac{2}{3}l^2x$$

Assuming small displacement x of the wall, the approximation $x = l^2/8r$ can be used, giving

$$\Delta V = \frac{1}{12}\frac{l^4}{r}$$

and consequently for one wall, the change in magnetic moment due to bending will be $\Delta \boldsymbol{m} = 2\boldsymbol{M}_s \Delta V$:

$$\Delta m = \frac{\boldsymbol{M}_s l^4}{6r}$$

The equation for the forces on the wall due to wall bending and applied field can be used to obtain an expression for r

$$r = \frac{\gamma}{2\mu_0 \boldsymbol{M}_s \boldsymbol{H}}$$

so that

$$\Delta m = \frac{\mu_0 \boldsymbol{M}_s^2 l^4 \boldsymbol{H}}{3\gamma}$$

In order to convert from change in magnetic moment $\boldsymbol{m}$ to change in magnetization, we need to assume a certain volume density of these domain walls. Supposing that this density is $1/l^3$ per unit volume,

$$\boldsymbol{M} = \frac{\Delta m}{l^3} = \frac{\mu_0 \boldsymbol{M}_s^2 l \boldsymbol{H}}{3\gamma}$$

and the susceptibility is therefore

$$\chi_{\text{in}} = \frac{d\boldsymbol{M}}{d\boldsymbol{H}}$$

$$= \frac{\mu_0 \boldsymbol{M}_s^2 l}{3\gamma}$$

assuming that for the purposes of calculating χ_{in}, the displacement of domain walls is small enough to allow the approximations made.

Exercise 8.6: Rotation against Anisotropy

In order to obtain the field strength along the ⟨010⟩ direction needed to rotate the moments from the ⟨100⟩ easy axis in the presence of a two-dimensional anisotropy, it is necessary to find the angle at which the maximum torque opposing rotation of moments due to anisotropy occurs, and then find the field strength that is just great enough to overcome this.

Since the situation is defined as two dimensional, the anisotropy in the x-y plane is

$$E_a = K \cos^2\theta_1 \cos^2\theta_2$$

and $\theta_2 = \pi/2 - \theta_1$,

$$E_a = K\cos^2\theta_1 \sin^2\theta_1$$

where now we can simply use θ as the angle between the moments and the $\langle 100\rangle$ axis, and the subscripts can be dropped. If a magnetic field $\boldsymbol{H}$ is applied along the $\langle 010\rangle$ axis, the energy per unit volume of the magnetization $\boldsymbol{M}_s$ is given by $E_H = -\mu_0 \boldsymbol{M}_s \cdot \boldsymbol{H}$

and in this case the angle between $\boldsymbol{H}$ and $\boldsymbol{M}_s$ is $\pi/2 - \theta$, so

$$E_H(\theta) = -\mu_0 \boldsymbol{M}_s \boldsymbol{H} \sin\theta$$

Therefore, the free energy term of interest is

$$E(\theta) = \frac{K}{4}\sin^2 2\theta - \mu_0 \boldsymbol{M}_s \boldsymbol{H} \sin\theta$$

and equilibrium occurs when the derivative of this with respect to angle is zero

$$\frac{dE(\theta)}{d\theta} = \left[2K\sin\theta\left(1-\sin^2\theta\right) - \mu_0 \boldsymbol{M}_s \boldsymbol{H}\right]\cos\theta$$

$$= 0$$

so that either $\cos\theta = 0$, meaning $\theta = \pi/2$, which is the trivial solution along the field direction, or

$$2K\sin\theta\left(1-2\sin^2\theta\right) - \mu_0 \boldsymbol{M}_s \boldsymbol{H} = 0$$

and therefore,

$$\boldsymbol{H}_c = \frac{2K}{\mu_0 \boldsymbol{M}_s}\sin\theta\,\cos 2\theta$$

Now we need to find the location at which the opposing torque due to anisotropy is greatest

$$\tau_a = -\frac{dE_a}{d\theta}$$

$$= 2K\cos\theta\sin\theta\left(\cos^2\theta - \sin^2\theta\right)$$

$$= \frac{K}{2}\sin 4\theta$$

This torque has a maximum when $\theta = 22.5°$. Therefore, we need only find the field strength sufficient to reach $\theta = 22.5°$. Putting this value of θ into the equation for $\boldsymbol{H}_c$ gives

$$\boldsymbol{H}_c = 12.68\times 10^3\,\mathrm{A\,m^{-1}}$$

Exercise 8.7: Effects of Stress on Domain-Wall Motion and Domain Rotation

Long-range stress alters the anisotropy through the magnetostatic coupling. This results in changes in permeability along different directions in the material and along different directions relative to the stress axis.

Short-range stress can cause localized changes in the energy of domain walls, and therefore alters the *potential landscape* that the domain wall sees as it moves. This leads to changes in pinning of the domain walls, and hence when averaged over a macroscopic volume to changes in coercivity.

Exercise 8.8: Saturation Magnetostriction along Different Crystallographic Directions

For a cubic material such as nickel, the saturation magnetostriction λ_s varies along different directions according to the Equation 5.17 on page 123.

$$\lambda_s = \frac{3}{2}\lambda_{100}\left(\alpha_1^2\beta_1^2 + \alpha_2^2\beta_2^2 + \alpha_3^2\beta_3^2 - \frac{1}{3}\right) + 3\lambda_{111}\left(\alpha_1\alpha_2\beta_1\beta_2 + \alpha_2\alpha_3\beta_2\beta_3 + \alpha_3\alpha_1\beta_3\beta_1\right)$$

where:

the α_i represents direction cosines of the direction along which the magnetic moments lie

the β_i represent the direction in which the magnetostriction is measured (which is usually the direction along which the magnetic field is applied)

If the magnetic moments are aligned along the ⟨100⟩ directions, then

$$\alpha_1 = 1$$

$$\alpha_2 = 0$$

$$\alpha_3 = 0$$

and if the magnetostriction is measured along the ⟨110⟩ directions, then

$$\beta_1 = \frac{1}{\sqrt{2}}$$

$$\beta_2 = \frac{1}{\sqrt{2}}$$

$$\beta_3 = 0$$

with the result

$$\lambda_{s110} = \frac{3}{2}\lambda_{100}\left(\frac{1}{2} - \frac{1}{3}\right) + 3\lambda_{111}(0)$$

$$\lambda_{s110} = \frac{\lambda_{100}}{4}$$

and $\lambda_{100} = -50 \times 10^{-6}$, so

$$\lambda_{s110} = -12.5 \times 10^{-6}$$

If the magnetic moments are along the $\langle 111 \rangle$ directions, then

$$\alpha_1 = \frac{1}{\sqrt{3}}$$

$$\alpha_2 = \frac{1}{\sqrt{3}}$$

$$\alpha_3 = \frac{1}{\sqrt{3}}$$

and in that case

$$\lambda_{s110} = \frac{3}{2}\lambda_{100}\left(\frac{1}{6} + \frac{1}{6} + 0 - \frac{1}{3}\right) + 3\lambda_{111}\left(\frac{1}{6} + 0 + 0\right)$$

$$\lambda_{s110} = \frac{1}{2}\lambda_{111}$$

and

$$\lambda_{111} = -25 \times 10^{-6}$$

$$\lambda_{s110} = -12.5 \times 10^{-6}$$

If the magnetic moments are aligned along the $\langle 110 \rangle$ directions, then

$$\alpha_1 = \frac{1}{\sqrt{2}}$$

$$\alpha_2 = \frac{1}{\sqrt{2}}$$

$$\alpha_3 = 0$$

and in that case

$$\lambda_{s110} = \frac{3}{2}\lambda_{100}\left(\frac{1}{2}-\frac{1}{3}\right)+3\lambda_{111}\left(\frac{1}{4}\right)$$

$$\lambda_{s110} = \frac{\lambda_{100}}{4}+\frac{3}{4}\lambda_{111}$$

and

$$\lambda_{100} = -50\times 10^{-6}$$

$$\lambda_{111} = -25\times 10^{-6}$$

so

$$\lambda_{s110} = -31.25\times 10^{-6}$$

In general, high magnetostriction is disadvantageous in soft magnetic materials because it can cause increased anisotropy in the presence of stress (which can reduce permeability and increase coercivity) or it can cause increased domain-wall pinning, which also results in lower permeability and higher coercivity.

Exercise 8.9: Magnetocrystalline Anisotropy and Domain Magnetization Rotation

$$K_u = 5\times 10^5 \text{ J m}^{-3} \text{ (uniaxial)}$$

$$\boldsymbol{M}_s = 1.44\times 10^6 \text{ A m}^{-1}$$

a. Antiparallel field

The torque opposing the change in magnetization is due to the anisotropy

$$\tau_{an} = -\frac{dE_{an}}{d\theta}$$

$$\tau_{an} = -\frac{d}{d\theta}\left(K\sin^2\theta\right)$$

$$\tau_{an} = -2K\sin\theta\cos\theta$$

This reaches a maximum when $\theta = \pi/4$ so $\cos\theta = \sin\theta = 1/\sqrt{2}$

The torque due to the field is

$$\tau_H = \mu_0 \boldsymbol{M}_s \times \boldsymbol{H}$$

$$\tau_H = \mu_0 M_s \boldsymbol{H}\sin(180-\theta)$$

The switching of magnetization occurs when the magnetization reaches an angle of $45°(\pi/4)$ and at equilibrium

$$\tau_H = \tau_{an}$$

$$\mu_0 \boldsymbol{M}_s \boldsymbol{H} \sin(180 - \theta) = -2K \sin\theta \cos\theta$$

$$\sin(180 - \theta) = \sin\theta$$

$$\mu_0 \boldsymbol{M}_s \boldsymbol{H} \sin\theta = 2K \sin\theta \cos\theta$$

Therefore,

$$\boldsymbol{H}_c = \frac{2K}{\mu_0 \boldsymbol{M}_s} \cos\theta$$

At $\theta = \pi/4$, $\cos\theta = 1/\sqrt{2}$
Therefore,

$$\boldsymbol{H}_c = \frac{K\sqrt{2}}{\mu_0 \boldsymbol{M}_s}$$

$$\boldsymbol{H}_c = \frac{(5 \times 10^5)(1.414)}{(12.56 \times 10^{-7})(1.44 \times 10^6)}$$

$$\boldsymbol{H}_c = 0.391 \times 10^6 \ \text{A m}^{-1}$$

b. On application of a field $\boldsymbol{H}$ perpendicular to the axis of magnetization
Torque due to field τ_H

$$\tau_H = \mu_0 \boldsymbol{M}_s \times \boldsymbol{H}$$

$$\tau_H = \mu_0 \boldsymbol{M}_s \boldsymbol{H} \sin\phi$$

$$\tau_H = \mu_0 \boldsymbol{M}_s \boldsymbol{H} \cos\theta$$

Torque due to anisotropy

$$\tau_{an} = -\frac{d E_{an}}{d\theta}$$

$$\tau_{an} = -\frac{d}{d\theta}\left(K \sin^2\theta\right)$$

$$\tau_{an} = -2K \sin\theta \cos\theta$$

and at equilibrium, the torques cancel

$$\mu_0 \boldsymbol{M}_s \boldsymbol{H} \cos\theta = 2K \sin\theta \cos\theta$$

Therefore,

$$\mu_0 \boldsymbol{M}_s \boldsymbol{H} = 2K \sin\theta$$

To calculate the susceptibility χ, we need the component of magnetization along the field direction

$$\boldsymbol{M} = \boldsymbol{M}_s \cos\phi$$

$$\boldsymbol{M} = \boldsymbol{M}_s \sin\theta$$

Therefore,

$$\mu_0 \boldsymbol{M}_s \boldsymbol{H} = \frac{2K\boldsymbol{M}}{\boldsymbol{M}_s}$$

and

$$\chi = \frac{\boldsymbol{M}}{\boldsymbol{H}} = \frac{\mu_0 \boldsymbol{M}_s^2}{2K}$$

$$\chi = \frac{\left(12.56\times10^{-7}\right)\left(1.44\times10^{6}\right)^2}{(2)\left(5\times10^{5}\right)}$$

$$\chi = 2.60$$

c. Since

$$\mu_0 \boldsymbol{M}_s \boldsymbol{H} = 2K\sin\theta$$

Then, when the magnetization reaches saturation along the field direction $\sin\theta = 1$, the field needed to achieve this is

$$\boldsymbol{H}_s = \frac{2K}{\mu_0 \boldsymbol{M}_s} = \frac{(2)\left(5\times10^{5}\right)}{\left(12.56\times10^{-7}\right)\left(1.44\times10^{6}\right)}$$

$$\boldsymbol{H}_s = 0.55\times10^{6}\ \text{A m}^{-1}$$

Exercise 8.10: Exchange Energy, Anisotropy Energy, Thermal Energy, and Superparamagnetic Particle Size

Exchange energy causes alignment of magnetic moments on the atoms inside the material. In the case of ferromagnetic materials, this is a parallel alignment of magnetic moments.

Anisotropy energy causes the aligned magnetic moments to prefer certain crystallographic axes over others for this alignment. In iron, for example, the ⟨100⟩ cube-edge directions are favored, whereas in nickel, the ⟨111⟩ cube diagonal axes are favored.

At a temperature ***T*** K, the thermal energy of each magnetic moment is

$$E_{\text{Th}} = 3k_B T\ \text{J}$$

whereas the anisotropy energy is

$$E_{an} = KV$$

where:

K is the anisotropy coefficient in $\mathrm{J\,m^{-3}}$
V is the volume of the particle in $\mathrm{m^3}$

Assuming a spherical particle, then $V = 4/3\,\pi r^3$, so that

$$E_{an} = \frac{4}{3}\pi r^3 K$$

and at the critical radius r_c, above which the particle is ferromagnetic and below which it is paramagnetic, can be found when the anisotropy energy just equals the thermal energy

$$3k_B T = \frac{4}{3}\pi r_c^{\,3} K$$

and therefore,

$$r_c = \left(\frac{9k_B T}{4\pi K}\right)^{1/3}$$

From this equation, it can be seen that the critical radius for establishing ferromagnetic order increases with temperature, so that $r_c \propto T^{1/3}$

Assuming a temperature of 300 K, the critical radius will be

$$r_c = \left(\frac{(9)(1.38\times10^{-23})(300)}{(4)(3.142)(1\times10^{4})}\right)^{1/3} = \left(296\times10^{-27}\right)^{1/3}$$

$$r_c = 6.67\times10^{-9}\ \mathrm{m}$$

$$= 6.67\ \mathrm{nm}$$

CHAPTER 9

EXERCISE 9.1: PARAMAGNETIC SUSCEPTIBILITY OF OXYGEN

Since the substance is a gas, we can assume no interactions between the electrons and neighboring atoms, and hence no exchange interaction. According to the classical Langevin theory of paramagnetism, the magnetization **M** is given by

$$\boldsymbol{M} = N\boldsymbol{m}\left[\cot h\left(\frac{\mu \boldsymbol{m}\boldsymbol{H}}{k_B T}\right) - \frac{k_B T}{\mu_0 \boldsymbol{m}\boldsymbol{H}}\right]$$

At 0°C or 273 K we know that for the realistic values of **H**

$$\mu_0 \boldsymbol{m}\boldsymbol{H} = k_B T$$

and hence that

$$\boldsymbol{M} = \frac{N\mu_0 \boldsymbol{m}^2 \boldsymbol{H}}{3k_B T}$$

Consequently,

$$\chi = \frac{N\mu_0 \boldsymbol{m}^2}{3k_B T}$$

$$\chi = \frac{(4\pi \times 10^{-7})(2.69 \times 10^{25})(2.78 \times 10^{-23})^2}{(3)(273)(1.38 \times 10^{-23})}$$

$$= 2.31 \times 10^{-6} \text{ (dimensionless)}$$

Exercise 9.2: Magnetic Mean Interaction Field for Iron

In the Curie-Weiss law, we have

$$\chi = \frac{\boldsymbol{M}}{\boldsymbol{H} + \alpha \boldsymbol{M}}$$

$$= \frac{C}{T - T_c}$$

$$T_c = \alpha C$$

and using the Langevin equation, the Curie constant is given by

$$C = \frac{N\mu_0 \boldsymbol{m}^2}{3k_B}$$

$$T_c = \frac{\alpha N\mu_0 \boldsymbol{m}^2}{3k_B}$$

So the value of α is given by

$$a = \frac{3k_B T_c}{N\boldsymbol{m}_0 \boldsymbol{m}^2}$$

$$a = \frac{(3)(1.38 \times 10^{-23})(1043)}{(8.57 \times 10^{28})(4p \times 10^{-7})(2.2)^2 (9.27 \times 10^{-23})^2}$$

$$a = 964$$

and since $\boldsymbol{H}_c = \alpha \boldsymbol{M}_s$

$$H_c = 1.64 \times 10^9 \text{ Am}^{-1}$$

Exercise 9.3: Critical Behavior of Spontaneous Magnetization

The classical Weiss-type ferromagnet for a system with two possible micro-states leads to the following expression for magnetization:

$$\boldsymbol{M} = \boldsymbol{M}_0 \tan h\left(\frac{\mu_0 m\alpha \boldsymbol{M}_s}{k_B T}\right)$$

Close to the Curie point $T \simeq T_c$, and so

$$\frac{\mu_0 m\alpha \boldsymbol{M}_s}{k_B T} \ll 1$$

The series expansion for tan$h$ (x) is

$$\tan h(x) = x - \frac{x^3}{3} + \ldots$$

Therefore, using the first few terms of this series as the value of tan$h$(x) when x is small

$$\boldsymbol{M}_s = N\boldsymbol{m}\left[\frac{\mu_0 m\alpha \boldsymbol{M}_s}{k_B T} - \frac{1}{3}\left(\frac{\mu_0 m\alpha \boldsymbol{M}_s}{k_B T}\right)^3\right]$$

For the tan$h$(x) expression, the Curie temperature is given by

$$T_c = \frac{N\mu_0 \boldsymbol{m}^2 \alpha}{k_B}$$

Substituting these values into the above equation leads to

$$\boldsymbol{M}_s = \left(\frac{T_c}{T}\right)\boldsymbol{M}_s - \left(\frac{T_c}{3T}\right)\left(\frac{\mu_0 m\alpha}{k_B T}\right)^2 \boldsymbol{M}_s^3$$

Therefore,

$$1 = \left(\frac{T_c}{T}\right) - \left(\frac{T_c}{3T}\right)\boldsymbol{M}_s^2\left(\frac{\mu_0 \boldsymbol{m}\alpha}{k_B T}\right)^2$$

$$T = T_c - \left(\frac{T_c}{3}\right)\boldsymbol{M}_s^2\left(\frac{\mu_0 \boldsymbol{m}\alpha}{k_B T}\right)^2$$

$$M_s^2 = \frac{3(T_c - T)}{T_c}\left(\frac{k_B T}{\mu_0 m\alpha}\right)^2$$

$$M_s = \left(\frac{k_B T}{\mu_0 m\alpha}\right)\frac{\sqrt{3}\cdot\sqrt{(T_c - T)}}{\sqrt{T_c}}$$

Exercise 9.4: Spontaneous Magnetization

According to the classical Weiss model, the spontaneous magnetization decays with temperature according to

$$\frac{M_s}{M_0} = \coth\left(\frac{\mu_0 \alpha m M_s}{k_B T}\right) - \frac{k_B T}{\mu_0 \alpha m M_s}$$

and this has the form shown in Figure 9.5, although the numerical values will be slightly different. Solving the equation for $M_s/M_0 = 0.9$ gives $T/T_c = 0.25$, and for $M_s/M_0 = 0.1$ gives $T/T_c = 0.99$. Therefore, the required values of temperature are 158 and 625 K, respectively.

The thermal energy equals Weiss mean-field interaction energy at T_c.

Exercise 9.5: Magnetic Heat Capacity

The volume specific heat of a material is simply given by

$$C_v = \left(\frac{dU}{dT}\right)_v$$

which is the rate of change of internal energy U with temperature T. If we apply this to a magnetic material such as nickel, then there is a magnetic component that arises because of the internal magnetic energy

$$U_{mag} = -\mu_0 \alpha M_s^2$$

where:

α is the coupling coefficient, which can be obtained from the Curie temperature
M_s is the spontaneous magnetization within the domain

Therefore, the magnetic contribution will be

$$C_{mag} = -2\mu_0 \alpha M_s^2 \left(\frac{dM_s}{dT}\right)$$

At this point, a suitable function for M_s is needed. Using a one-dimensional model of a ferromagnet, we obtain

$$M_s = M_0 \tan h\left(\frac{\mu_0 m\alpha M_s}{k_B T}\right)$$

so that

$$\frac{dM_s}{dT} = -\frac{\mu_0 N m^2 \alpha}{k_B T}\sec h^2\left(\frac{\mu_0 m\alpha M_s}{k_B T}\right)\left(\frac{M_s}{T} - \frac{dM_s}{dT}\right)$$

and $T_c = \mu_0 N m^2 \alpha / k_B$, therefore

$$\frac{dM_s}{dT} = -\left(\frac{M_s}{T}\right)\frac{\sec h^2\left(\mu_0 m\alpha M_s/k_B T\right)}{\left[\left(T/T_c\right) - \sec h^2\left(\mu_0 m\alpha M_s/k_B T\right)\right]}$$

which has a singularity when $T/T_c = \mathrm{sech}^2(T_c/T)$. This occurs as T approaches T_c. Therefore, the magnetic contribution to the specific heat has an anomaly at T_c.

Exercise 9.6: Magnetic Contribution to Heat Capacity

The magnetic contribution to the specific heat is given by

$$C_{mag} = -2\mu_0 \alpha M_s\left(\frac{dM_s}{dT}\right)$$

and

$$\alpha = \frac{k_B T_c}{\mu_0 N m^2}$$

so

$$C_{mag} = -\frac{2k_B T_c}{m}\left(\frac{dM_s}{dT}\right)$$

$$= 2.84 \times 10^6 \text{ J m}^{-3}$$

or equivalently 4.5 cal mol^{-1} K^{-1}.

Exercise 9.7: Classical Theory of Paramagnetism

The equation that should be used to determine the variation of magnetization with field and temperature is

$$M = Nm\left[\cot h\left(\frac{\mu_0 m H}{k_B T}\right) - \frac{k_B T}{\mu_0 m H}\right]$$

where we know that m = 2 Bohr magneton and therefore

$$m = 1.85 \times 10^{-23} \text{ A m}^2$$

The number of atoms per unit volume is

$$N = \frac{\rho N_A}{W_A}$$

$$N = \frac{(8000)(6.02\times10^{26})}{56}$$

$$N = 8.6\times10^{28}\ \ \text{m}^{-3}$$

and therefore complete saturation $\boldsymbol{M}_0 = N\boldsymbol{m}$ is given by

$$\boldsymbol{M}_0 = 1.59\times10^{6}\ \ \text{A m}^{-1}$$

Now calculate the variation of $\boldsymbol{M}$ with temperature and field according to the above equation

$$\boldsymbol{M} = 1.59\times10^{6}\left\{\coth\left[1.68\times10^{-6}\left(\frac{\boldsymbol{H}}{T}\right)\right] - 0.595\times10^{6}\left(\frac{T}{\boldsymbol{H}}\right)\right\}$$

$$\frac{\mu_0 \boldsymbol{m}}{k_B} = \frac{(12.56\times10^{-7})(1.85\times10^{-23})}{(1.38\times10^{-23})}$$

$$\frac{\mu_0 \boldsymbol{m}}{k_B} = 1.68\times10^{-6}$$

H/T	$x = \mu_0 mH/k_B T$	$\coth(\mu_0 mH/k_B T)$	$(M/M_0)\coth(x) - (1/x)$
0	0	0	0
297,619	0.5	2.164	0.164
595,238	1	1.314	0.314
892,857	1.5	1.105	0.438
1,190,476	2	1.037	0.537
1,785,714	3	1.005	0.672
2,380,952	4	1.0007	0.751
2,976,190	5	1.00009	0.80
5,952,380	10	1.0000	0.90

The graph of $\boldsymbol{M}/\boldsymbol{M}_0$ against $\boldsymbol{H}/T$ should be drawn.

Exercise 9.8: Magnetization Equation for a Uniaxial Magnet

The magnetic moments are constrained to align parallel or antiparallel to the field $\boldsymbol{H}$.

The energy of the moments in the field are as follows:

$$E = -\mu_0 \boldsymbol{m}\cdot\boldsymbol{H}$$

and so the energies of a moment aligned parallel (E+) or antiparallel (E−) to the field are

$$E_+ = -\mu_0 \boldsymbol{mH}$$

$$E_- = \mu_0 \boldsymbol{mH}$$

The probabilities of lying parallel or antiparallel to the field are

$$P_+ = \frac{\exp(\mu_0 \boldsymbol{mH}/k_B T)}{\exp(\mu_0 \boldsymbol{mH}/k_B T) + \exp(-\mu_0 \boldsymbol{mH}/k_B T)}$$

and

$$P_- = \frac{\exp(-\mu_0 \boldsymbol{mH}/k_B T)}{\exp(\mu_0 \boldsymbol{mH}/k_B T) + \exp(-\mu_0 \boldsymbol{mH}/k_B T)}$$

So the magnetization of an assembly of N such moments is

$$\boldsymbol{M} = N\boldsymbol{m}(P_+ - P_-)$$

$$\boldsymbol{M} = N\boldsymbol{m}\left[\frac{\exp(\mu_0 \boldsymbol{mH}/k_B T) - \exp(-\mu_0 \boldsymbol{mH}/k_B T)}{\exp(\mu_0 \boldsymbol{mH}/k_B T) + \exp(-\mu_0 \boldsymbol{mH}/k_B T)}\right]$$

$$\boldsymbol{M} = N\boldsymbol{m} \tan h(\mu_0 \boldsymbol{mH}/k_B T)$$

Exercise 9.9: Classical Theory of Ferromagnetism

The Langevin-Weiss theory uses the basic formalism of the classical theory of paramagnetism and introduces a new term, the *mean field* $\alpha\boldsymbol{M}$, which is then added to the applied field $\boldsymbol{H}$ as a *perturbation* so that $\boldsymbol{H}$ becomes $\boldsymbol{H} + \alpha\boldsymbol{M}$.

In order to calculate the spontaneous magnetization, which of course arises when $\boldsymbol{H} = 0$, it is necessary to first calculate the mean field coupling.

$$\alpha = \frac{3k_B T_c}{\mu_0 N \boldsymbol{m}^2}$$

where

$$\boldsymbol{m} = 0.927 \times 10^{-25}\ \text{Am}^2,\ N = 1 \times 10^{29}\ \text{m}^{-3},\ k_B = 1.38 \times 10^{-23}\ \text{J K}^{-1},\ \text{and } T_c = 773\,\text{K}$$

$$\alpha = \frac{(3)(1.38 \times 10^{-23})(773)}{(12.56 \times 10^{-7})(1 \times 10^{29})(0.927 \times 10^{-25})}$$

$$\alpha = \frac{3.2 \times 10^{-20}}{1.08 \times 10^{-23}}$$

$$\alpha = 2963$$

The spontaneous magnetization $\boldsymbol{M}_s$ is then given by

$$M_s = Nm\left[\cot h\left(\frac{\mu_0 \alpha m M_s}{k_B T}\right) - \left(\frac{k_B T}{\mu_0 \alpha m M_s}\right)\right]$$

where $Nm = 0.927 \times 10^6$ A m^{-1}.

The above equation is implicit in M_s and therefore needs to be solved numerically. Another way to solve it is to use Figure 9.5, on page 227. The value of T/T_c is $300/773 = 0.39$, which from the graph in Figure 9.5 suggest a value of M_s of about $0.95 Nm = 0.88 \times 10^6$ A m.

Solutions from $0.75 - 0.88 \times 10^6$ Am^{-1} are reasonable.

Exercise 9.10: Calculation of Spontaneous Magnetization within a Domain

Calculate spontaneous magnetization at 20°C

$$m = 0.6\mu_B = 5.56 \times 10^{-24} \text{ Am}^2$$

The coupling coefficient is

$$\alpha = \frac{3k_B T_c}{\mu_0 N m^2}$$

$$\alpha = \frac{(3)(1.38 \times 10^{-23})(631)}{(12.56 \times 10^{-7})(5.56 \times 10^{-24})(9 \times 10^{28})}$$

$$\alpha = 7469$$

Spontaneous magnetization is given by

$$M_s = Nm\left[\cot h\left(\frac{\mu_0 \alpha m M_s}{k_B T}\right) - \left(\frac{k_B T}{\mu_0 \alpha m M_s}\right)\right]$$

$$M_s = 0.5 \times 10^6\left[\cot h(6.45) - 0.155\right]$$

$Nm = 0.5 \times 10^6$ A m^{-1}. Solving this implicit equation in M_s gives

$$M_s \cong 0.81 Nm$$

$$M_s = 0.41 \times 10^6 \text{ A m}^{-1}$$

CHAPTER 10

Exercise 10.1: Diamagnetic Moment

The change of orbital frequency with applied magnetic field is as follows:

$$m = iA$$

Therefore,

$$\Delta \boldsymbol{m} = \Delta(iA)$$

and the current is $i = e\nu$ where ν is the frequency of rotation, $\nu = \omega/2\pi$

$$\Delta \boldsymbol{m} = \frac{e}{2\pi}\Delta(\omega A)$$

suppose that only the frequency changes. Then the area A remains $A = \pi r^2$, where r is the radius of orbit. Then

$$\Delta \boldsymbol{m} = \frac{er^2}{2}\Delta\omega$$

and from the Langevin theory of diamagnetism in Section 9.1.2

$$\Delta \boldsymbol{m} = \frac{\mu_0 e^2 r^2 \boldsymbol{H}}{4\boldsymbol{m}} = \frac{e^2 r^2 \boldsymbol{B}}{4\boldsymbol{m}}$$

Therefore,

$$\frac{e^2 r^2 \boldsymbol{B}}{4\boldsymbol{m}} = \frac{er^2}{2}\Delta\omega$$

$$\Delta\omega = \frac{e\boldsymbol{B}}{2\boldsymbol{m}}$$

Exercise 10.2: Relation between Magnetic Moment and Angular Momentum

The magnetic moment of a current i passing around a closed loop of material with cross-sectional area A is

$$\boldsymbol{m} = iA$$

Now consider an electron moving along an orbit. The angular momentum will be

$$\boldsymbol{p}_0 = m_e r^2 \frac{d\phi}{dt}$$

where:
- m_e is the electron mass
- r is the instantaneous radius
- $d\phi/dt$ is the rate of change of angle

If we integrate this equation,

$$\int_0^\tau \boldsymbol{p}_0 dt = m_e \int_0^{2\pi} r^2 d\phi$$

$$P_0 \tau = 2m_e A$$

and therefore

$$A = \frac{\boldsymbol{p}_0 \tau}{2m_e}$$

so that the magnetic moment then becomes

$$\boldsymbol{m}_0 = \boldsymbol{p}_0 \frac{i\tau}{2m_e}$$

and $i\tau = -e$, the electronic charge, so that

$$\boldsymbol{m}_0 = -\left(\frac{e}{2m_e}\right)\boldsymbol{p}_0$$

which is the required relationship between the orbital angular momentum of the electron and its orbital magnetic moment.

In the case of the spin angular momentum, we can consider first a spherical approximation to the electron. In this case, the spin angular momentum is given by

$$\boldsymbol{p}_s = \boldsymbol{I}\omega$$

where the moment of inertia $I = (2/5)m_e a^2$ for a sphere with radius a and the mass evenly distributed over the volume. In this case, a is the *classical electron radius*,

$$\boldsymbol{p}_s = \frac{2}{5} m_e a^2 \omega$$

If we consider all of the charge to be distributed over a classical spherical surface,

$$\text{Radius of elemental band } R = a \sin\phi$$

$$\text{Surface area of elemental band } dA = 2\pi R a d\phi = 2\pi a^2 \sin\phi d\phi$$

$$\text{Electric charge on surface of elemental band } dq = \frac{-e dA}{4\pi a^2} = \frac{-e \sin\phi d\phi}{2}$$

The contribution to the magnetic moment of an elemental band at an angle ϕ is

$$d\boldsymbol{m} = \pi R^2 \frac{dq}{\tau}$$

$$= -\pi a^2 \sin^3\phi \frac{e}{2\tau} d\phi$$

then the integral of current and cross-sectional area over the entire volume becomes

$$\int d\boldsymbol{m} = -\frac{\pi a 2 e}{2\tau} \int_0^\pi \sin^3\phi d\phi$$

and $\tau = 2\pi/\omega$

$$\boldsymbol{m} = -\frac{\omega e a^2}{4}\int_0^{\pi} \sin^3\phi d\phi = -\frac{\omega e a^2}{3}$$

and since $\boldsymbol{p}_s = I\omega$

$$\boldsymbol{m} = -\frac{5}{6}\left(\frac{e}{m_e}\right)\boldsymbol{p}_s$$

which is clearly at variance with experimental observations. If the mass of the *classical spherical electron* is not evenly distributed over its volume, for example, if the mass is all concentrated at the classical electron radius

$$I = \frac{2}{3} m_e a^2$$

and therefore,

$$\boldsymbol{m} = \frac{1}{2}\frac{e}{m_e}\boldsymbol{p}_s$$

which also does not agree with observations. However, if we consider a disk with charge confined to the circumference, we can obtain the expected result

$$\begin{aligned}\boldsymbol{m}_s &= iA \\ &= -\frac{e}{\tau}\pi a^2 \\ &= -\frac{e\omega}{2}a^2\end{aligned}$$

and the angular momentum of such a disk is

$$\begin{aligned}\boldsymbol{p}_s &= I\omega \\ &= \frac{m_e a^2}{2}\omega\end{aligned}$$

so that

$$\boldsymbol{m}_s = -\left(\frac{e}{m_e}\right)\boldsymbol{p}_s$$

The only way to get the correct result classically is to assume either (1) that the electronic mass is not evenly distributed over the volume, (2) that the charge is not evenly distributed, or (3) that the electron behaves more like a charged spinning disk.

Exercise 10.3: Tangential Speed of Electron—Classical Model

Consider the magnetic moment of a sphere with uniformly distributed charge over the surface. This is a poor approximation to an electron, but the calculation serves to illustrate the difficulty with our classical Ampèrian current model of magnetization. We will calculate the magnetic moment of an elemental ring of the sphere and integrate. It is assumed that charge will be on the surface only

$$\text{Area of ring section} = 2\pi a r d\theta$$

$$dA = 2\pi r^2 \sin\theta d\theta$$

If the charge density on the surface is *p*, then the total charge on this elemental ring is *dq*

$$dq = \rho dA$$

$$= 2\pi\rho r^2 \sin\theta d\theta$$

The *current di* passing as this elemental ring rotates about a diameter is

$$di = v dq$$

$$= \frac{\omega}{2\pi} dq = \omega\rho r^2 \sin\theta d\theta$$

and the magnetic moment of the ring section is

$$d\boldsymbol{m} = A di$$

$$= \pi a^2 \frac{\omega}{2\pi} dq$$

$$= \pi\omega\rho r^4 \sin^3\theta d\theta$$

This can now be integrated over the entire sphere, from $\theta = 0$ to $\theta = \pi$ to get the magnetic moment of the sphere

$$\boldsymbol{m} = \pi\omega\rho r^4 \int_0^{\pi} \sin^3\theta d\theta$$

$$= \frac{4}{3}\pi r^3 \omega\rho r$$

Applying the above result to the magnetic moment of an electron

$$\boldsymbol{m} = 9.27 \times 10^{-24} \text{ A m}^{-2}$$

the tangential velocity of this classical electron is then

$$v = \omega r = \frac{3\boldsymbol{m}}{4\pi r^3 \rho}$$

and $\rho^4\pi r^2 = e$ the charge on an electron

$$v = \frac{3\boldsymbol{m}}{er}$$

$$= 6.2 \times 10^{10} \text{ m s}^{-1}$$

clearly an impossible velocity. These last two problems show that the classical description of the magnetic moment of the electron is deficient.

Exercise 10.4: Orbital and Spin Angular Momentum of an Electron

a. The principal quantum number n determines the shell of the electron, and hence its energy.

Its allowed values are $n = 1, 2, 3, \ldots$.

The orbital angular momentum l defines the orbital angular momentum $\boldsymbol{p}_0$ of the electron when multiplied by $(h/2\pi)$. It can have values of $l = 0, 1, 2, 3, \ldots, (n-1)$, where n is the principal quantum number.

The magnetic quantum number $\boldsymbol{m}_l$ gives the component of the orbital angular momentum l along the z-axis when a magnetic field is applied along that axis. Its values are restricted to $\boldsymbol{m}_l = -1, \ldots, -2, -1, 0, 1, 2, \ldots, +l$.

The spin quantum number $\boldsymbol{s}$ defines the spin angular momentum of an electron. The value of the spin quantum number is always 1/2 for an electron. The angular momentum due to spin $\boldsymbol{s}h/2\pi$ is therefore always an integer multiple of $h/4\pi$. The component or resultant of s in a magnetic field is represented by $\boldsymbol{m}_s$ and is restricted to $\boldsymbol{m}_s = \pm 1/2$.

The various angular momenta of a electron can be calculated from the two quantum numbers $\boldsymbol{l}$ and $\boldsymbol{s}$

$$\boldsymbol{p}_0 = \boldsymbol{l}\left(\frac{h}{2\pi}\right)$$

$$\boldsymbol{p}_s = \boldsymbol{s}\left(\frac{h}{2\pi}\right)$$

The total angular momentum $\boldsymbol{j}$ can then be calculated from the vector sum of $\boldsymbol{l}$ and $\boldsymbol{s}$, remembering that for a single electron the magnitude of j must be a half integer.

$$\boldsymbol{p}_j = \mathbf{j}\left(\frac{h}{2\pi}\right)$$

The values of $\boldsymbol{l}$, $\boldsymbol{s}$, and $\boldsymbol{j}$ differ from the classically expected values in practice and this can be explained by quantum mechanics. The solution of the Schroedinger equation in the simple case of a single electron orbiting a nucleus only permits solution (states) for which the angular momentum is $\langle \bar{i} \rangle = \sqrt{[l(l+1)]}$. A similar argument holds for $\boldsymbol{s}$ and hence j.

b. The allowed solutions for the resultant orbital angular momentum $\boldsymbol{L}$ of the vector sum of two orbital angular momentum vectors of length 2 and 3 are

$$\boldsymbol{L} = 5,\ 4,\ 3,\ 2,\ 1$$

c. The Co^{2+} ion in a $3d^7$ state has its electrons in the following state, $l = 2$ and 3d electrons

	$\boldsymbol{m}_l$	2	1	0	−1	−2	2	1	0	1	2
	$\boldsymbol{m}_s$	$\frac{1}{2}$	$\frac{1}{2}$	$\frac{1}{2}$	$\frac{1}{2}$	$\frac{1}{2}$	$-\frac{1}{2}$	$-\frac{1}{2}$			
Occupancy	s	↑	↑	↑	↑	↑	↓	↓			

So by summing the spins $S = \sum \boldsymbol{m}_l = 3/2$ and summing the components of orbital angular momentum $L = \sum \boldsymbol{m}_l = 3$. The shell is more than half full so $\boldsymbol{J} = |\boldsymbol{L} + \boldsymbol{S}| = 9/2$.

Exercise 10.5: Zeeman Effect

The cadmium singlet observed at $\lambda = 643.8$ nm is a result of an electronic transition from $6^1D_2(L = 2, S = 0)$ to $5^1P_1(L = 1, S = 0)$. On application of a magnetic field this will exhibit the normal Zeeman effect since the net spin of the state is zero.

The shift in energy upon application of a magnetic field $\boldsymbol{H}$ is

$$\begin{aligned}\Delta E &= \left(\frac{eh}{4\pi m_e}\right)\mu_0 \boldsymbol{H} \\ &= \mu_B \mu_0 \boldsymbol{H} \\ &= 5.8 \times 10^{-5}\ \text{eV T}^{-1} \\ &= 0.9273 \times 10^{-23}\ \text{J T}^{-1} \\ &= 1.165 \times 10^{-29}\ \text{J A}^{-1}\ \text{m}^{-1}\end{aligned}$$

The equation for the energy shift with field is

$$\begin{aligned}\Delta E &= 1.165 \times 10^{-29} \boldsymbol{H}\ \text{J} \\ &= 0.9273 \times 10^{-23} \mu_0 \boldsymbol{H}\ \text{J}\end{aligned}$$

When $\mu_0\boldsymbol{H} = 0.5$ T, this gives

$$\Delta E = 0.464 \times 10^{-23}\ \text{J}$$

When $\mu_0\boldsymbol{H} = 1.0$ T

$$\Delta E = 0.927 \times 10^{-23}\ \text{J}$$

When $\mu_0 \boldsymbol{H} = 2.0$ T

$$\Delta E = 1.85 \times 10^{-23} \text{ J}$$

The shift in frequency can be calculated from the relation

$$\Delta E = h(v - v_0)$$

$$\Delta v = \frac{\Delta E}{h}$$

$$= \frac{(0.9273 \times 10^{-23})}{(6.626 \times 10^{-34})} \text{s}^{-1} \text{ T}^{-1}$$

$$= 1.399 \times 10^{10} \text{s}^{-1} \text{ T}^{-1}$$

and using the relation

$$\lambda v = c$$

where c is the speed of light.

$$\frac{\Delta v}{\Delta \lambda} = -\frac{c}{\lambda^2}$$

$$\Delta \lambda = -\frac{\lambda^2}{c} \Delta v$$

Substituting in the values for the shift in frequency at an induction of 1 T

$$\Delta \lambda = -\frac{(643.8 \times 10^{-9})^2 (1.399 \times 10^{10})}{2.98 \times 10^8}$$

$$= -1.946 \times 10^{-11} \text{ m}$$

At 1 T,

$$\Delta \lambda = -0.01946 \text{ nm}$$

At 0.5 T,

$$\Delta \lambda = -0.009730 \text{ nm}$$

At 2.0 T,

$$\Delta \lambda = -0.03892 \text{ nm}$$

Exercise 10.6: Determination of Atomic Angular Momentum

The values of $\boldsymbol{J}$ can be obtained by vector addition of $\boldsymbol{L}$ and $\boldsymbol{S}$, remembering that if $\boldsymbol{S}$ is an integer, then $\boldsymbol{J}$ is an integer and if $\boldsymbol{S}$ is a half integer, then $\boldsymbol{J}$ is a half integer.

a. $\boldsymbol{L} = 2$ and $\boldsymbol{S} = 3$. Possible values of $\boldsymbol{J}$ are 5, 4, 3, 2, and 1.
b. $\boldsymbol{L} = 3$ and $\boldsymbol{S} = 2$. Possible values of $\boldsymbol{J}$ are 5, 4, 3, 2, and 1.
c. $\boldsymbol{L} = 3$ and $\boldsymbol{S} = 5/2$. Possible values of $\boldsymbol{J}$ span the range from $|\boldsymbol{L} + \boldsymbol{S}|$ to $|\boldsymbol{L} - \boldsymbol{S}|$, that is, 11/2, 9/2, 7/2, 5/2, 3/2, and 1/2.
d. $L = 2$, $S = 5/2$. Possible values of J are 9/2, 7/2, 5/2, 3/2, 1/2.

The ground state of the carbon atom can be determined from Hund's rules. Carbon has four electrons in the $n = 2$ shell. The s subshell is full with two electrons, the p subshell, which has $l = 1$ has two electrons. The occupancy of available states must be such as to maximize $\boldsymbol{S}$ by Hund's rules.

	$\boldsymbol{m}_l$	1	0	−1	1	0	1
	$\boldsymbol{m}_s$	$\frac{1}{2}$	$\frac{1}{2}$	$\frac{1}{2}$	$-\frac{1}{2}$	$-\frac{1}{2}$	$-\frac{1}{2}$
Occupancy	s	↑	↑				

Therefore, $\boldsymbol{S} = \Sigma \boldsymbol{m}_s = 1$ and $\boldsymbol{L} = \Sigma \boldsymbol{m}_l = 1$ since the subshell is less than half full $\boldsymbol{J} = |\boldsymbol{L} - \boldsymbol{S}|$

$$\boldsymbol{J} = 0$$

Exercise 10.7: Relationship between Angular Momentum and Magnetic Moment

An electron has a magnetic moment as a result of having charge $e = 1.6 \times 10^{-19}$ C and angular momentum. The angular momentum has two components: orbital and spin.

The relationship between orbital angular momentum $\boldsymbol{P}_0$ and the orbital magnetic moment $\boldsymbol{m}_0$ is

$$\boldsymbol{m}_0 = -\left(\frac{e}{2m_e}\right)\boldsymbol{P}_0$$

where $\boldsymbol{m}_e$ is the mass of the electron.

The relationship between spin angular momentum $\boldsymbol{P}_s$ and the spin magnetic moment $\boldsymbol{m}_s$ is

$$\boldsymbol{m}_s = -\left(\frac{e}{m_e}\right)\boldsymbol{P}_s$$

These two components can then be added vectorially to give the total magnetic moment $\boldsymbol{m}_{\text{tot}}$

$$\boldsymbol{m}_{tot} = \boldsymbol{m}_0 + \boldsymbol{m}_s$$

The orbital and spin angular momenta are related to the orbital and spin angular momentum quantum numbers l and s by the following relations:

$$\boldsymbol{P}_0 = l\left(\frac{h}{2\pi}\right)$$

and

$$\boldsymbol{P}_s = s\left(\frac{h}{2\pi}\right)$$

so that the orbital and spin contributions to the magnetic moment can be expressed directly in terms of the orbital and spin quantum numbers.

$$\boldsymbol{m}_0 = -\left(\frac{eh}{4\pi m_e}\right)l$$

and

$$\boldsymbol{m}_s = -2\left(\frac{eh}{4\pi m_e}\right)s$$

In these equations, the term $\left(eh/4\pi m_e\right)$ corresponds to 1 Bohr magneton.

Exercise 10.8: Magnetic Moment of a Classical Rotating Sphere

Show that provided $\boldsymbol{M} = \sigma\omega r$ identical magnetic fields are produced by a sphere of radius r with uniform magnetization $\boldsymbol{M}$ and a nonmagnetic charged rotating sphere also of radius r, carrying a uniform charge density σ and rotating about a diameter with angular velocity ω.

The problem reduces itself to finding conditions under which the magnetized sphere and the rotation sphere have the same magnetic moment. Both have the same radius and so two identical magnetic moments $\boldsymbol{m}$ will produce the same magnetic field $\boldsymbol{B}$ with components

$$\boldsymbol{B}_r(r,\theta) = \mu_0 \frac{2\boldsymbol{m}\cos\theta}{4\pi r^3}$$

$$B_\theta(r,\theta) = \mu_0 \frac{\boldsymbol{m}\sin\theta}{4\pi r^3}$$

$$B_\phi = 0$$

and

$$\boldsymbol{m} = \boldsymbol{M}\frac{4}{3}\pi r^3$$

for the uniformly magnetized sphere.

For the rotating charged sphere, there will be contributions to the magnetic moment from every elemental current loop on the surface of the sphere.

The contribution due to the element shown is

$$d\boldsymbol{m} = A di$$

$$d\boldsymbol{m} = \pi a^2 di$$

$$a = r\sin\theta$$

and

$$di = \frac{w}{2\pi}\, dq$$

$$d\boldsymbol{m} = \pi r^2 \sin^2\theta \cdot \frac{w}{2\pi}\, dq$$

and the charge on an elemental ring at an angle θ is given by

$$dq = \sigma 2\pi a r d\theta$$

Therefore,

$$d\boldsymbol{m} = \pi w \sigma r^4 \sin^3\theta d\theta$$

This must now be interpreted over the range θ = 0 to π to get the magnetic moment of the whole sphere

$$\boldsymbol{m} = \pi w \sigma r^4 \int_0^\pi \sin^3\theta d\theta$$

and

$$\int_0^\pi \sin^2\theta \, d\theta = \left[-\frac{1}{3}\cos\theta\left(\sin^2\theta + 2\right)\right]_0^\pi$$

$$\int_0^\pi \sin^2\theta \, d\theta = \frac{4}{3}$$

Therefore,

$$\boldsymbol{m} = \frac{4}{3}\pi w \sigma r^4$$

which equals the magnetic moment of the magnetized sphere when $\boldsymbol{M} = \sigma\omega r$.

Exercise 10.9: Failure of the Classical Model of Electron Magnetic Moment Based on Spinning Charge

If we suppose that the electron is a classical charged particle with charge $e = 1.6 \times 10^{-19}$ C and a classical radius $r = 2.8 \times 10^{-15}$ m, then if all the charge is on the surface of the sphere, the contribution ***dm*** to magnetic moment given by an elemental ring slice of the electron at angle θ is

$$d\boldsymbol{m} = A\,di$$

$$d\boldsymbol{m} = \pi\left(r\sin\theta\right)^2 v\,dq$$

And $v = w/2\pi$, and ***q*** is the amount of charge residing on the elemental ring slice

$$dq = \left(\frac{e}{4\pi r^2}\right) \cdot 2\pi r \sin\theta \cdot r d\theta$$

$$d\boldsymbol{m} = \pi\left(r\sin\theta\right)^2 \cdot \frac{w}{2\pi}\left(\frac{e}{4\pi r^3}\right) \cdot 2\pi r^2 \sin\theta\, d\theta$$

Next integrate over all elemental ring slices from $\theta = 0$ to $\theta = \pi$

$$\boldsymbol{m} = \int_{\theta=0}^{\theta=\pi} d\boldsymbol{m}$$

$$\boldsymbol{m} = \pi w r^4 \left(\frac{e}{4\pi r^2}\right) \int_0^{\pi} \sin^3\theta\, d\theta$$

$$\boldsymbol{m} = \frac{wer^2}{4} \cdot \frac{4}{3}$$

$$\boldsymbol{m} = \frac{1}{3} erv$$

Therefore, the required velocity to achieve a magnetic moment of 1 Bohr magneton under these conditions is

$$v = \frac{3\boldsymbol{m}}{er}$$

$$v = \frac{3 \times 0.927 \times 10^{-23}}{\left(1.6 \times 10^{-19}\right)\left(2.8 \times 10^{-15}\right)}$$

$$v = 62 \times 10^9 \text{ m s}^{-1}$$

or about 200 times the speed of light. Since the actual radius of the electron is less than the given classical radius of an electron, the velocity would have to be even higher than this. The classical model of an electron as a rotating sphere of charge therefore has to be abandoned.

Exercise 10.10: Magnetic Moment of an Isolated Atom in Terms of Orbital and Spin Quantum Numbers

The electron configuration for an isolated Mn^{2+} ion is $3d^5$, so

$$Mn^{2+} : 1s^2; 2s^2 2p^6; 3s^2 3p^6 3d^5$$

So in terms of the orbit and spin contributions to magnetic moment, $\boldsymbol{m}_l$ and $\boldsymbol{m}_s$

$\boldsymbol{m}_l$	2	1	0	−1	−2	2	1	0	−1	−2
$\boldsymbol{m}_s$	$+\frac{1}{2}$	$+\frac{1}{2}$	$+\frac{1}{2}$	$+\frac{1}{2}$	$+\frac{1}{2}$	$-\frac{1}{2}$	$-\frac{1}{2}$	$-\frac{1}{2}$	$-\frac{1}{2}$	$-\frac{1}{2}$
	↑	↑	↑	↑	↑					

Therefore, Mn^{2+} has $L = 0$ and $S = 5/2$

In this case, since the orbital moment is zero, it does not matter if the orbital moment is quenched

$$\boldsymbol{m} = g\sqrt{s(s+1)}$$

$$\boldsymbol{m} = 2\sqrt{8.75}$$

$$\boldsymbol{m} = 5.92 \text{ Bohr magnetons}$$

$$\boldsymbol{m} = 5.48 \times 10^{-23} \text{ A m}^2$$

CHAPTER 11

Exercise 11.1: The Exchange Interaction

The exchange interaction J arises from the determination of the total energy of a two-electron system in which the electrons can never have the same set of quantum numbers by virtue of the Pauli exclusion principle. (This means that the two-electron wave function must be antisymmetric.)

Under these circumstances, there arises an energy term, which results from the electrons *exchanging* places. Classically, this would not alter the energy of the system (since the particles could occupy the same states in classical physics) but in quantum mechanics, the situation is different and the electrons therefore have an additional energy, which acts in many respects like a strong magnetic field.

Exercise 11.2: Magnetic Moment of Dysprosium Ions

The dysprosium ion Dy^3+ has nine electrons in its 4f shell. For this shell, $n = 4$ and $l = 3$, so the electron distribution is

m_l	3	2	1	0	−1	−2	−3	3	2	1	0	−1	−2	−3
s	↑	↑	↑	↑	↑	↑	↑	↓	↓					
m_s	$\frac{1}{2}$	$\frac{1}{2}$	$\frac{1}{2}$	$\frac{1}{2}$	$\frac{1}{2}$	$\frac{1}{2}$	$\frac{1}{2}$	$-\frac{1}{2}$	$-\frac{1}{2}$					

Summing these leads to $S = \Sigma \boldsymbol{m}_s = 5/2$ and $L = \Sigma \boldsymbol{m}_l = 5$. Since the shell is more than half full $J = |L + S| = 15/2$. To calculate the susceptibility of a salt containing 1 g mole of Dy^3+ at 4 K first calculate the magnetic moment per ion

$$\boldsymbol{m} = -g\mu_B J$$

$$= -g\mu_B\sqrt{\left[J(J+1)\right]}$$

$$g = 1 + \frac{J(J+1)+S(S+1)-L(L+1)}{2J(J+1)}$$

and in this case, when the values of L, S, and J are substituted in, we obtain $9 = 4/3$.

$$\boldsymbol{m} = -\frac{4}{3}\left(9.27\times10^{-24}\right)\sqrt{\frac{255}{4}}$$

$$= -9.87\times10^{-23} \text{ A m}^2$$

and

$$\chi = \frac{\boldsymbol{M}}{\boldsymbol{H}}$$

$$= \frac{N\mu_0\boldsymbol{m}^2}{3k_BT}$$

$$= \frac{\left(6.02\times10^{23}\right)\left(4\pi\times10^{-7}\right)\left(9.87\times10^{-23}\right)^2}{3\left(1.38\times10^{-23}\right)4}$$

$$= 4.45\times10^{-5}$$

Exercise 11.3: Paramagnetism of $S = 1$ System

The magnetization can be expressed as a function of temperature and magnetic field using the Brillouin function $B_j(x)$ where the argument of the function x is $\mu_0\boldsymbol{mH}/k_BT$

$$\boldsymbol{M} = N\boldsymbol{m}B_J\left(\frac{\mu_0\boldsymbol{mH}}{k_BT}\right)$$

where:

N is the number of atoms per unit volume
$\boldsymbol{m}$ is the magnetic moment per atom

For a system with spin only, we have $g = 2$ and $J = S$. In this case therefore, $J = 1$. Substituting these values into the Brillouin function,

$$B_J(x) = \frac{\left[(2J+1)/2J\right]\cot h\left[x(2J+1)/2J\right]-(1/2J)}{\cot h(x/2J)}$$

$$= \frac{3}{2}\cot h\left(\frac{3x}{2}\right) - \frac{1}{2}\cot h\left(\frac{x}{2}\right)$$

and $\coth(x) = 1/x + x/3$ for small x. If we make this approximation then, in this case,

$$B_J(x) = \frac{2x}{3}$$

So if $\mu_0 \boldsymbol{mH}/k_B T = 1$, we can make this approximation and consequently

$$B_J\left(\frac{\mu_0 \boldsymbol{mH}}{k_B T}\right) = \frac{2\mu_0 \boldsymbol{mH}}{3k_B T}$$

and the magnetization $\boldsymbol{M}$ is given by

$$\boldsymbol{M} = N\boldsymbol{m}B_J\left(\frac{\mu_0 \boldsymbol{mH}}{k_B T}\right)$$

$$= \frac{N\boldsymbol{m}2\mu_0 \boldsymbol{mH}}{3k_B T}$$

$$= \frac{2N\mu_0 \boldsymbol{m}^2 \boldsymbol{H}}{3k_B T}$$

Exercise 11.4: Magnetic Moments of Various Ions

These ions may be conveniently separated into two groups according to their electron configuration:

Ions	Outer Electron Configuration
d series ions	
Nb^+	$4d^4$
Fe^{3+}	$3d^5$
Fe^{2+}	$3d^6$
Pt^{2+}	$5d^8$
f series ions	
Gd^{3+}	$4f^7.5s^2.p^6$

The magnetic moment of an atom arises only from the unfilled shell. In the first group of ions, this is the d shell, in the second group, this is the f shell. The d shell has $\lambda = 2$, while the f shell has $\lambda = 3$. So, for the d series elements,

No. d Electrons		=	**1**	**2**	**3**	**4**	**5**	**6**	**7**	**8**	**9**	**10**
$l = 2$	m_e	=	2	1	0	−1	−2	2	1	0	−1	−2
	m_s		1/2	1/2	1/2	1/2	1/2	−1/2	−1/2	−1/2	−1/2	−1/2
	L	=	2	3	3	2	0	2	3	3	2	0
	S	=	1/2	1	3/2	2	5/2	2	3/2	1	1/2	0
	J	=	3/2	2	3/2	0	5/2	4	9/2	4	5/2	0
	G	=				0	2	3/2		1.25		
						Nb^+	Fe^{3+}	Fe^{2+}		Pt^{2+}		

These d electrons have $\lambda = 2$ irrespective of whether in the third, fourth, or fifth shell.

For the f series elements

No. d Electrons		=	**1**	**2**	**3**	**4**	**5**	**6**	**7**	**8**	**9**	**10**	**11**	**12**	**13**	**14**
$l = 3$	m_e	=	3	2	1	0	−1	−2	−3	3	2	1	0	−1	−2	−3
	m_s	=	1/2	1/2	1/2	1/2	1/2	1/2	1/2	−1/2	−1/2	−1/2	−1/2	−1/2	−1/2	−1/2
	L	=	3	5	6	6	5	3	0	3	5	6	6	5	3	0
	S	=	1/2	1	3/2	2	5/2	3	7/2	3	5/2	2	3/2	1	1/2	0
	J	=	5/2	4	9/2	4	5/2	0	7/2	6	15/2	8	15/2	6	7/2	0
	G	=							2							
									Gd^{3+}							

Remembering that

$$g = 1 + \frac{J(J+1) + S(S+1) - L(L+1)}{2J(J+1)}$$

So we have the following results:

For the d series elements the net magnetic moments are $m = g\sqrt{S(S+1)}$:

Ion	$g\sqrt{J(J+1)}$	$g\sqrt{S(S+1)}$
Nb^+	0	4.90
Fe^{3+}	5.92	5.92
Fe^{2+}	3.67	4.90
$Pt2^+$	1.80	2.83

For the 4f series elements the net magnetic moments are $\boldsymbol{m} = g\sqrt{J(J=1)}$:

Ion	$g\sqrt{J(J+1)}$
Gd^{3+}	7.94

Exercise 11.5: Saturation Magnetization of Ferrite

Magnetite, Fe_3O_4 crystallizes in the inverse spinel structure. The metal ions are distributed among the A sites (tetrahedral) and the B sites (octahedral) and the number of Fe^{3+} ions on each type of site is equal. These align with magnetic moments antiparallel leaving only the Fe^{2+} ions as the uncompensated contribution to the magnetic moment of the unit cell. Schematically, therefore, in each unit cell, we have the following magnetic configuration:

$$Fe^{3+}(\uparrow)Fe^{3+}(\downarrow)Fe^{2+}(\uparrow)$$

Therefore, only the one Fe^{2+} magnetic moment in each cell accounts for the net magnetic properties. Fe^{2+} ions have a magnetic moment of 5.4 Bohr magnetons or 5.0×10^{-23}A m^2. The formula weight of Fe_3O_4 is 232, and the density is 5.18×10^3 kg m^{-3}. Therefore,

$$\text{lm}^3 \text{ contains } N = \frac{5.18\times10^3}{232}\times6.02\times10^{26} = 1.34\times10^{28} \text{ unit cells}$$

and the volume V of a unit cell is therefore,

$$V = 7.5 \times 10^{-29}\text{m}^3$$

The saturation magnetization is

$$\boldsymbol{M}_\text{s} = \frac{\boldsymbol{M}}{V}$$

$$= 0.67\times10^6 \text{ A m}^{-1}$$

If Fe^{2+} is replaced by Mn^{2+} without changing the size of the unit cell the only difference is that Mn^{2+} has a magnetic moment of 5.9 Bohr magneton or 5.5×10^{-23} A m^2. Therefore,

$$\boldsymbol{M}_\text{s} = \frac{\boldsymbol{m}}{V}$$

$$= 0.73\times10^6 \text{ A m}^{-1}$$

Exercise 11.6: Atomic Moments from Crystal Data

For iron the volume V of the unit cell is

$$V = a^3 = 2.44\times10^{-29}\text{m}^3$$

and iron has body-centered cubic (bcc) lattice structure, so the number of atoms per unit cell is $N = 2$, and the saturation magnetization is

$$\boldsymbol{M}_\text{s} = \frac{N\boldsymbol{m}}{V} = 1.7\times10^6 \text{ A m}^{-1}$$

The magnetic moment per atom $\boldsymbol{m}$ is therefore

$$\boldsymbol{m} = \frac{V}{N} \times 1.7 \times 10^{6} \text{ A m}^{2}$$

$$= 2.60 \times 10^{-23} \text{ A m}^{2}$$

$$1\mu_{B} = 9.27 \times 10^{-24} \text{ A m}^{2}$$

$$\boldsymbol{m} = 2.23 \text{ Bohr magneton per atom}$$

For cobalt, with a hexagonal lattice

$$V = a^{2}c\frac{\sqrt{3}}{2} = 2.40 \times 10^{-29} \text{ m}^{3}$$

and for a hexagonal close-packed (HCP) the number of atoms per unit cell (of this volume) is 2, so we arrive at

$$\boldsymbol{m} = \frac{V}{N} \times 1.4 \times 10^{6} \text{ A m}^{-1}$$

$$= 1.68 \times 10^{-23} \text{ A m}^{2}$$

$$= 1.81 \text{ Bohr magneton per atom}$$

For nickel, which is face-centered cubic (FCC), the number of atoms per unit cell is $N = 4$ and the volume is $V = a^{3} = 4.29 \times 10^{-29}$ m^{3}, therefore

$$\boldsymbol{m} = 0.5 \times 10^{6} \frac{V}{N} \text{ A m}^{2}$$

$$= 5.63 \times 10^{-24} \text{ A m}^{2}$$

$$= 0.58 \text{ Bohr magneton per atom}$$

There are several ways to solve the second part of the problem, including the use of the quantum mechanical Brillouin function. The easiest solution, however, is simply to use the classical Langevin function. The internal field can be determined from the high temperature approximation of the Langevin function by equating exchange and thermal energies at T_{c}

Internal coupling field $\boldsymbol{H}_{e}$ (A m^{-1}) is given by

$$\boldsymbol{H}_{e} = a\boldsymbol{M}_{s} = \frac{3k_{B}T_{c}}{m_{0}m}$$

Fe: $\boldsymbol{H}_{e} = 1.66 \times 10^{9}$ A m^{-1}

Co: $\boldsymbol{H}_{e} = 2.75 \times 10^{9}$ A m^{-1}

Ni: $\boldsymbol{H}_{e} = 3.87 \times 10^{9}$ A m^{-1}

Exchange energy per unit volume $E_{ex,voi}$(J m^{-3}):

$$E_{ex,vol} = \mu_0 \boldsymbol{M}_s \cdot \boldsymbol{H}_e$$

$$\text{Fe}: E_{ex,vol} = 3.54 \times 10^9 \text{ J m}^{-3}$$

$$\text{Co}: E_{ex,vol} = 4.83 \times 10^9 \text{J m}^{-3}$$

$$\text{Ni}: E_{ex,vol} = 2.43 \times 10^9 \text{J m}^{-3}$$

The exchange energy per atom E_{ex} in joules is the exchange energy per unit volume obtained from the coupling field divided by the number of atoms per unit volume:

$$E_{ex} = 3k_B T = \frac{E_{ex,vol}}{N} \text{ joules per atom}$$

$$\text{Fe:} \quad E_{ex} = 4.30 \times 10^{-20} \text{ joules per atom}$$

$$\text{Co:} \quad E_{ex} = 5.86 \times 10^{-20} \text{ joules per atom}$$

$$\text{Ni:} \quad E_{ex} = 2.61 \times 10^{-20} \text{ joules per atom}$$

The exchange energy $E_{ex,nn}$ between nearest-neighbor atoms is

$$E_{ex,nn} = \frac{E_{ex}}{Z}$$

and the number of nearest neighbors z for iron is 8; for cobalt is 12; and for nickel is 12. Therefore,

$$\text{Fe: } E_{ex} = 5.38 \times 10^{-21} \text{ J}$$

$$\text{Co: } E_{ex} = 4.83 \times 10^{-21} \text{ J}$$

$$\text{Ni: } E_{ex} = 2.18 \times 10^{-21} \text{ J}$$

Appendix A

MAGNETIC FIELD AS A RELATIVISTIC CORRECTION TO THE ELECTRIC FIELD

Consider two charges, each of q coulombs moving at constant velocity v along the x-direction in an arbitrary laboratory reference frame S, with the charges separated by a distance r along the y-axis as shown in Figure A.1.1. These two charges are at rest, and also a fixed distance r apart along the y-axis in a reference frame S', which itself is moving with velocity v along the x-axis relative to the laboratory reference frame S. The forces between the two charges will be different in the two reference frames. This difference, which is velocity dependent, is a relativistic correction to the electrostatic force and is more commonly known as the *magnetic force*.

Let the components of the forces between the charges be F_x, F_y, and F_z in the laboratory reference frame, and F'_x, F'_y, and F'_z in the moving reference frame. Since the charges are stationary in reference frame S', the forces are

$$F'_x = 0$$

$$F'_y = \frac{1}{4\pi\epsilon_0}\frac{q^2}{r^2} \tag{A.1.1}$$

$$F'_z = 0$$

So the only term of any significance is the coulomb electrostatic term along the y-axis. In the laboratory reference frame S, the forces can be obtained using the relativistic Lorentz transformation on the forces in the moving reference frame S'. This is given by

$$F_x = F'_x = 0$$

$$F_y = \frac{1}{\gamma}F'_y \tag{A.1.2}$$

$$F_z = \frac{1}{\gamma}F'_z = 0$$

where $\gamma = 1\big/\sqrt{1-\left(v^2/c^2\right)}$ and c is the velocity of light. The motion along the x-direction causes the forces along the orthogonal directions to be different from the classically expected electrostatic force. The only term of significance here is F_y

$$F_y = \sqrt{\left(1-\frac{v^2}{c^2}\right)}\frac{q^2}{4\pi\epsilon_0 r^2} \tag{A.1.3}$$

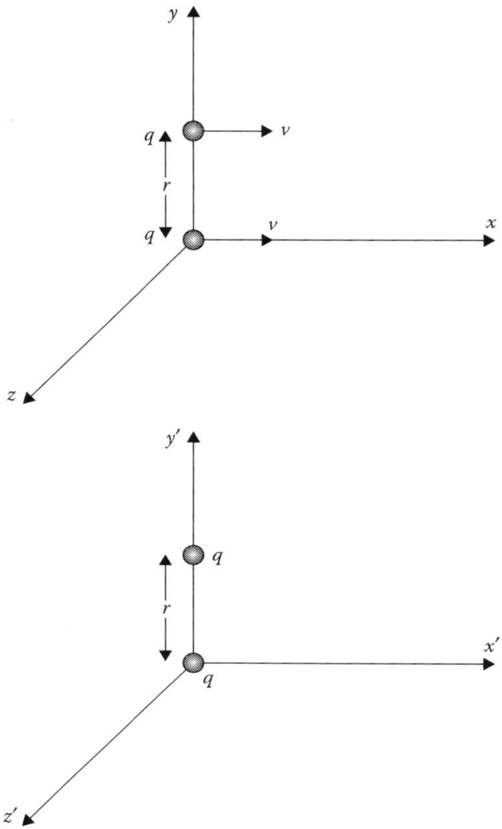

FIGURE A.1.1 Two isolated charges moving with velocity v in the laboratory reference frame S generate an additional force, which is not observed in the reference frame S' in which the charges are stationary.

The displacement r in the S frame of reference and r' and the S' frame of reference are identical since the motion is defined here as along the x-axis only. Hence

$$F_y = \left(1 - \frac{v^2}{c^2}\right)\gamma \frac{q^2}{4\pi\epsilon_0 r^2} \tag{A.1.4}$$

$$F_y = \gamma q \left(\frac{q}{4\pi\epsilon_0 r^2} - \frac{qv^2}{4\pi\epsilon_0 c^2 r^2}\right) \tag{A.1.5}$$

and $\mu_0 = 1/\epsilon_0 c^2$, so

$$F_y = \gamma q \left(\frac{q}{4\pi\epsilon_0 r^2} - \frac{\mu_0}{4\pi}\frac{qv^2}{r^2}\right) \tag{A.1.6}$$

For low velocities v, the value of $\gamma \approx 1$, and therefore, provided the velocity of the charges is not comparable to the velocity of light c; the γ term can be ignored here. In fact, typically for electrons moving in a wire, $v \approx 10^{-4}$ m s^{-1}, so that $\gamma = 1$ to better than one apart in 10^{24}. Therefore,

$$F = \frac{q^2}{4\pi\epsilon_0 r^2} - q\frac{\mu_0}{4\pi}\frac{qv^2}{r^2} \tag{A.1.7}$$

$$F_{\text{net}} = F_{\text{electric}} + F_{\text{magnetic}}$$

Since we determined that the force on the moving charge was due to a combination of external factors (i.e., a field) and internal factors (i.e., those specific to the moving charge that is subjected to the force), we can separate out the only factors relating to the moving charge: its charge q and velocity v. All other factors can be collected together into field terms,

$$F_{\text{net}} = q\left(\frac{q}{4\pi\epsilon_0 r^2}\right) + qv\left(\frac{-\mu_0}{4\pi}\frac{qv}{r^2}\right) \tag{A.1.8}$$

The second q and the second v inside the brackets belong to the other charged particle and are therefore *external*. The terms in brackets are therefore the field terms. The electric field is $q/4\pi\epsilon_0 r^2$ and the magnetic field $(-\mu_0/4\pi)(qv/r^2)$. Converting this to a vector form,

$$\mathbf{F} = q\left(\frac{q}{4\pi\epsilon_0 r^2}\right)\mathbf{j} + qv\left(\frac{-\mu_0}{4\pi}\frac{qv}{r^2}\right)\mathbf{j} \tag{A.1.9}$$

From the cross-products of the unit vectors, $\mathbf{j} = -\mathbf{i} \times \mathbf{k}$. Therefore,

$$\mathbf{F}_y = q\left[\mathbf{E} - \frac{\mu_0}{4\pi}\frac{qv^2}{r^2}(-\mathbf{i}\times\mathbf{k})\right] \tag{A.1.10}$$

and since $v = v\mathbf{i}$,

$$= q\left(\mathbf{E} + \mathbf{v}\times\frac{\mu_0}{4\pi}\frac{qv}{r^2}\mathbf{k}\right) \tag{A.1.11}$$

The additional term $(\mu_0/4\pi)(qv/r^2)\mathbf{k}$ is a correction to the electric field. So the net force on the moving charge is due to a combination of the classically expected electrostatic coulomb force and an extra force, which is the result of the charge and the vector product of the velocity of the charge with the magnetic field. This equation can also be expressed in terms of the current i due to the moving charge, since $qv = i\delta l$, where δl is a current element swept out by a charge q in unit time when moving with velocity v. If r is a unit vector along the radial direction, from the current element to the field point,

$$\mathbf{F}_y = q\left(\mathbf{E} + v \times \frac{\mu_0}{4\pi}\frac{i\delta\mathbf{l}\times\mathbf{r}}{r^2}\right) \qquad \text{(A.1.12)}$$

This expression for the magnetic field is known as the *Biot-Savart law* and contains terms that describe the distribution of the current $(1/4\pi)(i\delta\mathbf{l}\times\mathbf{r}/r^2)$ and a term μ_0, which relates to the permeability of the medium, in this case free space.

FORCE ON A CHARGE MOVING IN AN EXTERNAL MAGNETIC FIELD OF KNOWN STRENGTH

More generally, it is observed experimentally that an electric charge q moving in an applied magnetic field in free space is subjected to a magnetic force $\boldsymbol{F}$, which is the vector product of the velocity v and the magnetic flux density $\boldsymbol{B}$ times the magnitude of the charge q. The force can also be expressed in terms of the magnetic field $\boldsymbol{H}$.

$$\boldsymbol{F} = qv\times\boldsymbol{B} = \mu_0 qv\times\boldsymbol{H} \qquad \text{(A.1.13)}$$

Equation A.1.13 for the magnetic force is the same equation that emerges from relativity. From this, the magnetic force can be equated with the additional field term in Equation A1.12. The field in free space can therefore be expressed either in terms of the geometrical distribution of the current producing it $\left[\boldsymbol{H} = (1/4\pi)(i\delta\mathbf{l}\times\mathbf{r}/r^2)\right]$ or, by including the permeability of free space μ_0, in terms of the resulting flux density $\left[\boldsymbol{B} = (\mu_0/4\pi)(i\delta\mathbf{l}\times\mathbf{r}/r^2)\right]$. In free space, of course, these two terms are synonymous, but in a magnetic medium, there arises a real distinction when $\mu_r \neq 1$.

Appendix B

DERIVATION OF MAXWELL'S EQUATIONS FROM THE RELATIVISTIC LORENTZ TRANSFORMATION

There are four equations of the electromagnetic field known as *Maxwell's equations*, which give relations between the magnetic field $\boldsymbol{H}$, the electric field $\boldsymbol{E}$, the magnetic flux density $\boldsymbol{B}$, and the electric flux density $\boldsymbol{D}$.

Faraday-Lenz law of induction:

$$\nabla \times \boldsymbol{E} = -\frac{\partial \boldsymbol{B}}{\partial t} \qquad \text{(B.2.1)}$$

Coulomb-Gauss law for magnetic flux:

$$\nabla \cdot \boldsymbol{B} = 0 \qquad \text{(B.2.2)}$$

Ampére's circuital law:

$$\nabla \times \boldsymbol{H} = \boldsymbol{J} + \frac{\partial \boldsymbol{D}}{\partial t} \qquad \text{(B.2.3)}$$

Coulomb-Gauss law for electric flux:

$$\nabla \cdot \boldsymbol{D} = \rho \qquad \text{(B.2.4)}$$

In these equations, $\boldsymbol{J}$ is the area current density in amperes per square meter that generates the field and ρ is the charge density in coulombs per cubic meter that generates the electric flux.

There is one equation for each of the quantities $\boldsymbol{H}$, $\boldsymbol{E}$, $\boldsymbol{D}$, and $\boldsymbol{B}$. Here, $\boldsymbol{H}$ is measured in amperes per meter, $\boldsymbol{E}$ is measured in volts per meter, $\boldsymbol{B}$ is measured in webers per square meter (kg A^{-1} s^{-2}), and $\boldsymbol{D}$ is measured in coulomb per square meter (A s m^{-2}). Both $\boldsymbol{D}$ and $\boldsymbol{E}$ are treated as flux densities, although in truth, nothing is actually flowing. From the above equations, there is some equivalence between $\boldsymbol{H}$ and $\boldsymbol{E}$ (the magnetic and electric fields) and between $\boldsymbol{B}$ and $\boldsymbol{D}$ (the magnetic and electric flux densities). This is further emphasized by the fields in magnetizable and polarizable matter as shown later.

The vectors **H** and **B** are clearly distinct in several ways. One distinction is that the normal component of $\boldsymbol{B}$ is continuous across an interface, and the tangential component of $\boldsymbol{H}$ is continuous across an interface (unless there is a surface current). However, the tangential component of $\boldsymbol{B}$ and the normal component of $\boldsymbol{H}$ are not necessarily continuous across an interface. Similarly, the tangential component of $\boldsymbol{E}$ is continuous across an interface, while the normal component of $\boldsymbol{D}$ is continuous across an interface.

CURL OF *E*

The Maxwell equation for the curl of $\boldsymbol{E}$ shows that it is determined by the rate of change of magnetic induction $\boldsymbol{B}$. In general, the expression for the electric and magnetic fields at a point $x\boldsymbol{i} + y\boldsymbol{j} + z\boldsymbol{k}$ generated by a charge moving with velocity v along the x-axis of a coordinate system are

$$\boldsymbol{E}(x, y, z, t) = \frac{\gamma q}{4\pi\varepsilon_0}\frac{(x - vt)\boldsymbol{i} + y\boldsymbol{j} + z\boldsymbol{k}}{r^3} \tag{B.2.5}$$

and

$$\boldsymbol{B}(x, y, z, t) = \frac{\gamma\mu_0 q v}{4\pi}\frac{(-z\boldsymbol{j} + y\boldsymbol{k})}{r^3} \tag{B.2.6}$$

where $\gamma = 1/\sqrt{(1 - v^2/c^2)}$ and so

$$r = \left[\gamma^2 (x - vt)^2 + y^2 + z^2\right]^{1/2} \tag{B.2.7}$$

Taking the curl of the electric field $\sigma \times \boldsymbol{E}$,

$$\nabla \times \boldsymbol{E} = \left(\frac{\partial E_z}{\partial y} - \frac{\partial E_y}{\partial z}\right)\boldsymbol{i} + \left(\frac{\partial E_x}{\partial z} - \frac{\partial E_z}{\partial x}\right)\boldsymbol{j} + \left(\frac{\partial E_y}{\partial x} - \frac{\partial E_x}{\partial y}\right)\boldsymbol{k} \tag{B.2.8}$$

$$\nabla \times \boldsymbol{E} = \frac{3\gamma q}{4\pi\epsilon_0}\left[-\frac{(x - vt)(1 - \gamma^2)z}{r^5}\boldsymbol{j} + \frac{(x - vt)(1 - \gamma^2)y}{r^5}\boldsymbol{k}\right] \tag{B.2.9}$$

and

$$\frac{\partial \boldsymbol{B}}{\partial t} = \frac{\gamma\mu_0 q v}{4\pi}\left[\frac{3\gamma^2 v(x - vt)}{r^5}(-z\boldsymbol{j} + y\boldsymbol{k})\right] \tag{B.2.10}$$

$$= \frac{3\gamma q\mu_0}{4\pi}\left[\frac{\gamma^2 v^2 (x - vt)}{r^5}(-z\boldsymbol{j} + y\boldsymbol{k})\right] \tag{B.2.11}$$

and since $\mu_0 = 1/\epsilon_0 c^2$

$$\frac{\partial \boldsymbol{B}}{\partial t} = \frac{3\gamma q}{4\pi\epsilon_0}\left[\frac{\gamma^2 v^2}{c^2}\frac{(x - vt)}{r^5}(-z\boldsymbol{j} + y\boldsymbol{k})\right] \tag{B.2.12}$$

and $-\left(v^2/c^2\right)\gamma^2 = 1-\gamma^2$ so that from B.2.9 and B.2.12

$$\nabla \times \boldsymbol{E} = -\frac{\partial \boldsymbol{B}}{\partial t} \tag{B.2.13}$$

DIVERGENCE OF *B*

The Maxwell equation for the divergence of $\boldsymbol{B}$, σ … $\boldsymbol{B} = 0$, can be shown from the equation for the magnetic induction at a general point $\boldsymbol{B}(x, y, z, t)$ caused by a charge q moving with velocity v along the x-axis of an arbitrary coordinate system.

$$\boldsymbol{B}(x,y,z,t) = \frac{\mu_0}{4\pi}\gamma q v\left[\frac{-z\boldsymbol{j} + y\boldsymbol{k}}{r^3}\right] \tag{B.2.14}$$

and therefore,

$$\nabla \cdot \boldsymbol{B} = \frac{\mu_0 \gamma q v}{4\pi}\left[\frac{3yz}{r^5} - \frac{3yz}{r^5}\right] = 0 \tag{B.2.15}$$

CURL OF *H*

The Maxwell equation for the curl of $\boldsymbol{H}$, $\nabla \times \boldsymbol{H} = \boldsymbol{J} + \left(\partial \boldsymbol{D}/\partial t\right)$, or Ampére's law, can also be determined from the equation for the magnetic field generated by a charge q moving with velocity v along the x-axis. In the laboratory frame, the field at x, y, z, t is

$$\boldsymbol{H}(x,y,z,t) = \frac{1}{4\pi}\gamma q v\left[\frac{-z\boldsymbol{j} + y\boldsymbol{k}}{r^3}\right] \tag{B.2.16}$$

so that

$$\nabla \times \boldsymbol{H} = \begin{vmatrix} \boldsymbol{i} & \boldsymbol{j} & \boldsymbol{k} \\ \dfrac{\partial}{\partial x} & \dfrac{\partial}{\partial y} & \dfrac{\partial}{\partial z} \\ \boldsymbol{H}_x & \boldsymbol{H}_y & \boldsymbol{H}_z \end{vmatrix} \tag{B.2.17}$$

$$= \left(\frac{\partial \boldsymbol{H}_z}{\partial y} - \frac{\partial \boldsymbol{H}_y}{\partial z}\right)\boldsymbol{i} + \left(\frac{\partial \boldsymbol{H}_x}{\partial z} - \frac{\partial \boldsymbol{H}_z}{\partial x}\right)\boldsymbol{j} + \left(\frac{\partial \boldsymbol{H}_y}{\partial x} - \frac{\partial \boldsymbol{H}_x}{\partial y}\right)\boldsymbol{k}$$

and,

$$\frac{\partial \boldsymbol{H}_x}{\partial y} = 0 \tag{B.2.18}$$

$$\frac{\partial \boldsymbol{H}_x}{\partial z} = 0 \tag{B.2.19}$$

$$\frac{\partial \boldsymbol{H}_y}{\partial z} = \frac{\gamma q v}{4\pi}\left(\frac{3z^2 - r^2}{r^5}\right) \tag{B.2.20}$$

$$\frac{\partial \boldsymbol{H}_y}{\partial x} = \frac{\gamma q v}{4\pi}\left(\frac{3\gamma^2 z(x - vt)}{r^5}\right) \tag{B.2.21}$$

$$\frac{\partial \boldsymbol{H}_z}{\partial y} = \frac{\gamma q v}{4\pi}\left(\frac{r^2 - 3y^2}{r^5}\right) \tag{B.2.22}$$

$$\frac{\partial \boldsymbol{H}_z}{\partial x} = \frac{\gamma q v}{4\pi}\left[\frac{-3\gamma^2 y(x - vt)}{r^5}\right] \tag{B.2.23}$$

$$\nabla \times \boldsymbol{H} = \frac{\gamma q v}{4\pi r^5}\left[\left(2r^2 - 3y^2 - 3z^2\right)\boldsymbol{i} + 3\gamma^2 y(x - vt)\boldsymbol{j} + 3\gamma^2 z(x - vt)\boldsymbol{k}\right] \tag{B.2.24}$$

and the electric flux density $\boldsymbol{D}$ at an arbitrary field point (x, y, z, t) due to a charge q moving with velocity v along the x-axis is

$$\boldsymbol{D}(x, y, z, t) = \frac{\gamma q}{4\pi}\left[\frac{(x - vt)\boldsymbol{i} + y\boldsymbol{j} + z\boldsymbol{k}}{r^3}\right] \tag{B.2.25}$$

so,

$$\frac{\partial \boldsymbol{D}}{\partial t} = \frac{\gamma q v}{4\pi r^5}\left\{\left[3\gamma^2 (x - vt)^2 - r^2\right]\boldsymbol{i} + 3\gamma^2 y(x - vt)\boldsymbol{j} + 3\gamma^2 z(x - vt)\boldsymbol{k}\right\} \tag{B.2.26}$$

where $r = \left[\gamma^2 (x - vt)^2 + y^2 + z^2\right]^{1/2}$ and consequently from Equations B.2.24 and B.2.26,

$$\nabla \times \boldsymbol{H} = \frac{\partial \boldsymbol{D}}{\partial t} \tag{B.2.27}$$

which means that for a single point charge moving with fixed velocity, the curl of the magnetic field at any point is equal to the time derivative of the electric flux density.

A more general expression for the curl of $\boldsymbol{H}$ is obtained by considering current flow in addition to the motion of single isolated charges, the currents give rise to a current density term $\boldsymbol{J}$, while the isolated charges give rise only to a displacement current term $\partial D/\partial t$. (It is important that the contribution to $\sigma \times \boldsymbol{H}$ from the individual moving charges to expressed in terms of $\partial D/\partial t$ and not $\epsilon_0 \partial \boldsymbol{D}/\partial t$ as some authors show. In the latter case, one must also at some point add the polarization current density $\partial P/\partial t$ to get the correct expression, whereas $\partial D/\partial t$ includes both contributions.) Both contribute to the curl of $\boldsymbol{H}$. In the above analysis, there was no charge distribution from which to determine a current density. If we consider a distribution of charges, then

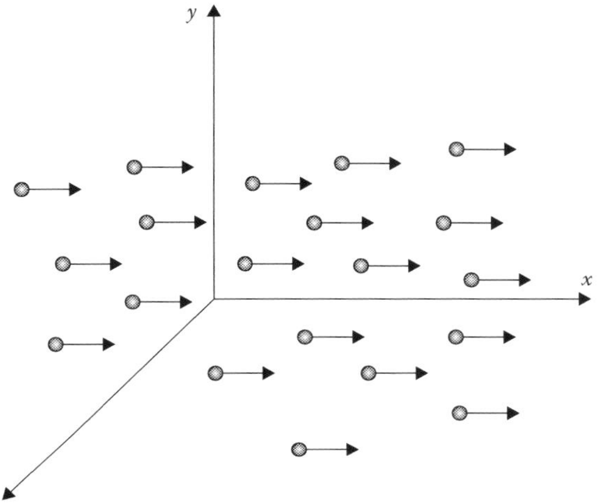

FIGURE B.2.1 A uniform distribution of charges, with charge density ρ all moving with constant velocity v along the x-axis in the laboratory reference frame S, while the charges are stationary in the S' reference frame.

we obtain the macroscopic Ampére's law equation. Consider, therefore, a uniform distribution of charges with charge density ρ, moving with constant velocity v along the x-direction of two reference frames S and S' as shown in Figure B.2.1.

A derivation can be obtained in several ways, but one of the most economical is to invoke the requirement that, since all physical laws must remain valid under a relativistic Lorentz transformation, the Gauss's law must be valid in both the laboratory reference frame S and the moving frame S'.

In the reference frame S, there will be a current density $\boldsymbol{J} = \rho v \boldsymbol{i}$, while in the S' reference frame the current density will be $\boldsymbol{J} = 0$. It also follows from the Lorentz transformation that the charge density in the S' reference frame is $\rho' = \rho/\gamma$.

This is due to the total charge remaining constant under the Lorentz transformation but the length along the x-axis contracting. Therefore, Gauss's law in the two frames of reference are

$$\nabla \cdot \boldsymbol{D} = \rho \tag{B.2.28}$$

$$\nabla \cdot \boldsymbol{D}' = -\frac{\rho}{\gamma} \tag{B.2.29}$$

The relationship between the components of $\boldsymbol{D}$ and $\boldsymbol{D}'$ observed in the two reference frames are

$$\begin{aligned} \boldsymbol{D} &= \boldsymbol{D}_x \boldsymbol{i} + \boldsymbol{D}_y \boldsymbol{j} + \boldsymbol{D}_z \boldsymbol{k} \\ \boldsymbol{D}' &= \boldsymbol{D}_x \boldsymbol{i} + \gamma\left(\boldsymbol{D}_y - \frac{v}{c^2}\boldsymbol{H}_z\right)\boldsymbol{j} + \gamma\left(\boldsymbol{D}_z - \frac{v}{c^2}\boldsymbol{H}_y\right)\boldsymbol{k} \end{aligned} \tag{B.2.30}$$

and taking the divergence gives

$$\nabla \cdot \boldsymbol{D} = \gamma\left(\frac{\partial}{\partial x} + \frac{v}{c^2}\frac{\partial}{\partial t}\right)\boldsymbol{D}_{x} + \gamma\frac{\partial}{\partial y}\left(\boldsymbol{D}_{y} - \frac{v}{c^2}\boldsymbol{H}_{z}\right) + \gamma\frac{\partial}{\partial z}\left(\boldsymbol{D}_{z} + \frac{v}{c^2}\boldsymbol{H}_{y}\right) \quad \text{(B.2.31)}$$

and the Lorentz transformation gives $\partial/\partial x' = \gamma\left[\partial/\partial x + \left(v/c^2\right)\left(\partial/\partial t\right)\right]$, $\partial/\partial y' = \partial/\partial y$ and $\partial/\partial z' = \partial/\partial z$

$$\gamma\nabla \cdot \boldsymbol{D} + \frac{\gamma v}{c^2}\frac{\partial D_{x}}{\partial t} - \frac{\gamma v}{c^2}\left(\frac{\partial \boldsymbol{H}_{z}}{\partial y} - \frac{\partial \boldsymbol{H}_{y}}{\partial z}\right) = \frac{\rho}{\gamma} \quad \text{(B.2.32)}$$

and consequently,

$$\frac{\partial \boldsymbol{D}_{x}}{\partial t} - \left(\frac{\partial \boldsymbol{H}_{z}}{\partial y} - \frac{\partial \boldsymbol{H}_{y}}{\partial z}\right) = \frac{c^2}{v}\left(\frac{1}{\gamma^2} - 1\right)\rho = -\rho v \quad \text{(B.2.33)}$$

$\rho v = J$ the current density, so

$$\frac{\partial \boldsymbol{H}_{z}}{\partial y} - \frac{\partial \boldsymbol{H}_{y}}{\partial z} = J + \frac{\partial \boldsymbol{D}_{x}}{\partial t} \quad \text{(B.2.34)}$$

Equation B.2.34 has been derived under the conditions of a steady current constrained to move along the x-axis. Under more general coordinates, we obtain

$$\nabla \times \boldsymbol{H} = \boldsymbol{J} + \frac{\partial \boldsymbol{D}}{\partial t} \quad \text{(B.2.35)}$$

Note that the displacement current $\partial D/\partial t$ has nothing to do with the fictitious *bound* or *ampérian* currents, which are sometimes invoked in order to derive this equation in terms of $\sigma \times \boldsymbol{B}$.

DIVERGENCE OF *D*

The final Maxwell equation relates only to the electric flux density $\boldsymbol{D}$, and so is not of immediate interest in the discussion of magnetic fields. It can be derived from Coulomb's law and Gauss's theorem.

Index

Note: Locators followed by '*f*' and '*t*' refer to figures and tables, respectively.

E

Made in the USA
Coppell, TX
03 December 2021

66999133R00343